Modern Developments in Flood Modelling

Modern Developments in Flood Modelling

Editors

Aristoteles Tegos
Alexandros Ziogas
Vasilis Bellos

MDPI • Basel • Beijing • Wuhan • Barcelona • Belgrade • Manchester • Tokyo • Cluj • Tianjin

Editors
Aristoteles Tegos
School of Civil Engineering
National Technical University
Athens
Greece

Alexandros Ziogas
School of Civil Engineering
University of Patras
Patra
Greece

Vasilis Bellos
Department of
Environmental Engineering
Democritus University of
Thrace
Xanthi
Greece

Editorial Office
MDPI
St. Alban-Anlage 66
4052 Basel, Switzerland

This is a reprint of articles from the Special Issue published online in the open access journal *Hydrology* (ISSN 2306-5338) (available at: www.mdpi.com/journal/hydrology/special_issues/flood_modelling).

For citation purposes, cite each article independently as indicated on the article page online and as indicated below:

LastName, A.A.; LastName, B.B.; LastName, C.C. Article Title. *Journal Name* **Year**, *Volume Number*, Page Range.

ISBN 978-3-0365-7809-5 (Hbk)
ISBN 978-3-0365-7808-8 (PDF)

Cover image courtesy of Aristoteles Tegos

Contents

About the Editors

Aristoteles Tegos

Dr. Aristoteles Tegos is a senior engineer at Ryan Hanley Ltd., Ireland, and a researcher at the School of Civil Engineering, National Technical University of Athens. He has a PhD in the field of hydrology, and his primary interests are related to evapotranspiration, flooding, and environmental flow modelling. His research activity focuses on the definition of data-driven methods with applications in multiple fields, including fluvial flood risk assessment, hydrological models, and river monitoring.

Alexandros Ziogas

Alexandros Ziogas is a civil engineer (PhD, University of Patras, Greece) that completed his postgraduate studies and conducts research on water resources and the environment. He has more than 17 years of experience as a researcher and professional engineer in the fields of surface and groundwater resource modelling, hydrology, and climate change. He is head of the Water Resources and Climate Change Unit at EMVIS S.A., Consultant Engineers-Environmental Services Research Information Technology & Services.

Vasilis Bellos

Dr. Vasilis Bellos is an assistant professor in the Department of Environmental Engineering of Democritus University of Thrace. His position is related to integrated water resource management, as well as the design and environmental management of hydraulic works. His main research interest is the numerical modelling of water-related problems using mechanistic simulators or data-driven approaches. He is an expert in hydrology, hydraulic engineering, river hydraulics, computational hydraulics, hydroinformatics, and flood modelling.

hydrology

MDPI

Editorial

Modern Developments in Flood Modelling

Aristoteles Tegos [1,2,*], Alexandros Ziogas [3] and Vasilis Bellos [4]

1 Laboratory of Hydrology and Water Resources Development, School of Civil Engineering, National Technical University of Athens, Heroon Polytechneiou 9, 15780 Zographou, Greece
2 Ryan Hanley Ltd. Ireland, 170/173 Ivy Exchange, Granby Pl, Parnell Square W, D01 N938 Dublin, Ireland
3 EMVIS S.A., Consultant Engineers-Environmental Services Research Information Technology & Services, 21, Paparrigopoulou Str., 15343 Agia Paraskevi, Greece; alzio@emvis.gr
4 Laboratory of Ecological Engineering and Technology, Department of Environmental Engineering, School of Engineering, Democritus University of Thrace, Vas. Sofias 12, 67100 Xanthi, Greece; vbellos@env.duth.gr
* Correspondence: tegosaris@yahoo.gr

1. Introduction

Flood modelling is among the most challenging scientific task because it covers a wide area of complex physical phenomena associated with highly uncertain and non-linear processes where the development of physically interpretive solutions usually suffers from the lack of recorded data.

The objective of the Special Issue, titled "Modern Developments in Flood Modelling", is to define and discuss several related topics, aiming to provide new insights within the geoscientific domain on the use of new remote sensing datasets in the service of flood modelling, on new methodologies addressing complex problems such as joint probability theory and rainfall maximum modelling at different temporal scales, and on strategies for reproducing catastrophic events in data-scarce areas and modelling flood risk with new tools in coastal areas.

This Special Issue comprises thirteen contributions tackling the above-mentioned goals. Our issue received a high number of diverse submissions, with an 82% acceptance rate.

2. Contributed Papers

The articles in this Special Issue address a wide variety of topics reflecting the challenges mentioned above. Their details are briefly presented below.

The paper "Regional Ombrian Curves: Design Rainfall Estimation for a Spatially Diverse Rainfall Regime" [1] by Theano Iliopoulou, Nikolaos Malamos and Demetris Koutsoyiannis demonstrates new insight in modelling regional ombrian curves (I.D.F curves) by providing a new parsimonious model of the extreme rainfall properties at any point in a given area. The curves were constructed following a newly revisited mathematical formulation of single-site curves coupled with a new regionalization approach. The results showed that the model efficiently captures the spatial variability of extreme rainfall in the area, covering scales from 5 min to 48 h.

The paper "Forensic Hydrology: A Complete Reconstruction of an Extreme Flood Event in Data-Scarce Area" [2] by Aristoteles Tegos, Alexandros Ziogas, Vasilis Bellos and Apostolos Tzimas presents a state-of-the-art approach to reconstructing catastrophic flooding events in data-scarce areas. The study focused on the recent catastrophic flooding event, namely medicane Ianos, which substantially affected the town of Karditsa, Greece. A rainfall–runoff CN-unit hydrograph model was combined with a hydrodynamic model based on a 2D shallow water equations model. Having used numerous remote sensing rainfall datasets along with satellite flooding footage and videos posted to social media sites such as Facebook, the catastrophic event was reconstructed efficiently in a high-complexity area associated with low-lying flooding fluvial and pluvial water paths.

Citation: Tegos, A.; Ziogas, A.; Bellos, V. Modern Developments in Flood Modelling. *Hydrology* **2023**, *10*, 112. https://doi.org/10.3390/hydrology10050112

Received: 2 May 2023
Accepted: 10 May 2023
Published: 15 May 2023

The paper "Predicting Urban Flooding Due to Extreme Precipitation Using a Long Short-Term Memory Neural Network" [3] by Raphaël A. H. Kilsdonk, Anouk Bomers and Kathelijne Wijnberg presents a long short-term memory (LSTM) neural network model to predict flood time series at 230 manhole locations present in the sewer system of the city of Amersfoort. According to the authors, it is the first time that an LSTM was applied to such a large sewer system in addition to a wide variety of synthetic precipitation events in terms of precipitation intensity. It was concluded that the LSTM could accurately predict the timing and volume of flooding for the large number of manholes for historic precipitation events and that the LSTM was able to reduce forecasting times, demonstrating the applicability of using this methodology as an early flood-warning system in urban areas.

The paper "Flood Exposure of Residential Areas and Infrastructure in Greece" [4] by Stefanos Stefanidis, Vasileios Alexandridis and Theodora Theodoridou exhibits the first nationwide spatial assessment of flood exposure in residential areas and infrastructures in Greece. Spatial analysis and open access data were used to illustrate the variations in flood exposure. The ratio of the urban fabric, transportation and social, industrial and commercial infrastructures in 100-year flood zones was evaluated, as well as the spatial pattern of the exposure. Based on the authors' view, the proposed methodology could serve as a roadmap for integrated flood risk assessment, as the results can be easily overlaid with other spatial data for further analysis.

The paper "Identifying Modelling Issues through the Use of an Open Real-World Flood Dataset" [5] by Vasilis Bellos, Ioannis Kourtis, Eirini Raptaki, Spyros Handrinos, John Kalogiros, Ioannis Sibetheros and Vassilios Tsihrintzis deals with the reconstruction of the flood wave that hit the town of Mandra (Athens, Greece) on 15 November 2017. The flash flood event was caused by a huge storm which was part of the Medicane Numa-Zeno. The works used in the reconstruction were associated with (a) the post-event collection of 44 maximum water depths and (b) hydrodynamic simulation employing the HEC-RAS and MIKE FLOOD software. Calibration strategies in computationally demanding cases were considered, and whether the calibrated parameters can be blindly transferred to another simulator (informed modeling) was tested.

The paper "Differentiated Spatial-Temporal Flood Vulnerability and Risk Assessment in Lowland Plains in Eastern Uganda" [6] by Godwin Erima, Isa Kabenge, Antony Gidudu, Yazidhi Bamutaze and Anthony Egeru was developed to map flood inundation areas along the Manafwa River, Eastern Uganda, using HEC-RAS integrated with SWAT models. The aim was to evaluate the predictive capacity of SWAT by comparisons with streamflow observations and to derive, using HECRAS, the flood inundation maps. The overall outcome demonstrated the benefits of combined modeling systems in predicting the extent of flood inundation.

The paper "Numerical and Physical Modeling of Ponte Liscione (Guardialfiera, Molise) Dam Spillways and Stilling Basin" [7] by Monica Moroni, Myrta Castellino and Paolo De Girolamo provides new insights into dam-related studies by combining computational fluid dynamics and physical models. The work deals with the 1:60 Froude-scaled numerical model of the Liscione (Guardialfiera, Molise, Italy) dam spillway and the downstream stilling basin. The model was scaled according to the Froude number, and fully developed turbulent flow conditions were reproduced at the model scale. From the analysis of the results of both the physical and the numerical models, it was found that the stilling basin is undersized with a significant impact on the erodible downstream river bottom in terms of scour depths.

The paper "Wetland Vulnerability Metrics as a Rapid Indicator in Identifying Nature-Based Solutions to Mitigate Coastal Flooding" [8] by Narcisa Gabriela Pricope and Greer Shivers presents a rapid method to quantify changes in ecosystem dynamics with the use of wetland vulnerability assessments to prioritize potential locations for NBS implementation. Exposure risk using 100- and 500-year special flood hazard areas, 1–10 ft of sea level rise scenarios and high-tide flooding and sensitivity using time series analyses of Landsat 8-derived multispectral indices were quantified. The work underlines the critical importance of conserving or restoring brackish and freshwater marshes and swamp forests, even though they represent a minority of the wetland types present in the highly populated Atlantic Coastal Plain region.

The paper "Trivariate Joint Distribution Modelling of Compound Events Using the Nonparametric D-Vine Copula Developed Based on a Bernstein and Beta Kernel Copula Density Framework" [9] by Shahid Latif and Slobodan Simonovic demonstrates the use of a D-vine copula in the nonparametric fitting procedure to model trivariate joint probability analyses of storm surges, river discharge and rainfall in compound flood risk assessments. A trivariate distribution can demonstrate the risk of compound phenomena more realistically, such as storm surges, rainfall and river discharge, rather than considering each contributing factor independently or in pairwise dependency relations. This work introduced the vine copula approach in a nonparametric setting by introducing Bernstein and Beta kernel copula density in establishing trivariate flood dependence.

The paper "Assessing the Impact of the Urban Landscape on Extreme Rainfall Characteristics Triggering Flood Hazards" [10] by Yakob Umer, Victor Jetten, Janneke Ettema and Gert-Jan Steeneveld presents a configuration of the WRF model developed for the city of Kampala, Uganda. The use of the WRF model to study the deep convection over Kampala required a special configuration, which requires the proper position and extent of the city in order to better consider the spatial contrast between the city and Lake Victoria. The study provides an explicit and alternative satellite-derived urban fraction in the WRF model. The study contributes to the emerging understanding of the usability of high-resolution urban fractions from remote sensing images to properly account for the impact of urban heterogeneity on extreme rainfall events.

The paper "Water Level Forecasting in Tidal Rivers during Typhoon Periods through Ensemble Empirical Mode Decomposition" [11] by Yen-Chang Chen, Hui-Chung Yeh, Su-Pai Kao, Chiang Wei and Pei-Yi Su demonstrates a parsimonious model that performs ensemble empirical mode decomposition (EEMD) and stepwise regression to forecast the water level of a tidal river. The proposed model is conceptually simple and highly accurate, providing reliable forecasts for a given location 1 h ahead using the observed ocean components at the down-stream gauging stations and the corresponding stream component the water stages at the upstream gauging stations.

The paper "Evaluation of Various Resolution DEMs in Flood Risk Assessment and Practical Rules for Flood Mapping in Data-Scarce Geospatial Areas: A Case Study in Thessaly, Greece" [12] by Nikolaos Xafoulis, Yiannis Kontos, Evangelia Farsirotou, Spyridon Kotsopoulos, Konstantinos Perifanos, Nikolaos Alamanis, Dimitrios Dedousis and Konstantinos Katsifarakis investigated flood modelling sensitivity against geospatial data accuracy using the following DTM resolutions in a mountainous river sub-basin of Thessaly's Water District (Greece): (a) open 5 m and (b) 2 m data from Hellenic Cadastre (HC) and (c) 0.05 m data from a topographical mission using an unmanned aerial vehicle (UAV). RAS-Mapper and HEC-RAS were used for 1D (steady state) hydraulic simulation regarding a 1000-year return period. The flood modelling results were analyzed via a statistical analysis based on the correlation matrix presenting linear relationships between input data variables (i.e., elevation, slope, sinuosity ratio) and cross section-specific results, including flow characteristics (i.e., Froude number, hydraulic radius), flood extents and flow depths. The correlation results indicated strong linearities, namely riverbed elevations vs. cross-section ID numbers, and weaker linearities (e.g., riverbed elevations and hydraulic radii and Froude number vs. flood extents).

The paper "CoastFLOOD: A High-Resolution Model for the Simulation of Coastal Inundation Due to Storm Surges" [13] by Christos Makris,, Zisis Mallios, Yannis Androulidakis and Yannis Krestenitis demonstrates a new numerical code (CoastFLOOD) with high-resolution (5 m × 5 m) raster-based, storage-cell modelling of coastal inundation via Manning-type equations in a decoupled 2D formulation at local-scale (20 km × 20 km) lowland littoral floodplains. The new model is based on the well-established LISFLOOD model and uses outputs of either regional-scale storm surge simulations or satellite altimetry data for sea level anomalies. The presented case studies demonstrated model applications at 10 selected coastal sites of the Ionian Sea (east-central Mediterranean Sea) and confirm the capability of the new model to reproduce past flooding events.

3. Conclusions

Since we have been conducting research in the field of flooding for more than a decade, and considering the remaining challenges within flooding assessment research, this Special Issue was a great opportunity to discover ideas and promote new techniques across the geosciences community.

As Guest Editors, we are enthusiastic about the successful completion of the SI, as it presents highly diverse and valuable works. We trust that the selected research papers will be a valuable contribution to the domain of geosciences in the years to come.

Author Contributions: Writing—original draft preparation, A.T.; writing—review and editing, A.T, A.Z. and V.B. All authors have read and agreed to the published version of the manuscript.

Funding: The creation of this Special Issue did not receive external funding.

Institutional Review Board Statement: Not applicable.

Informed Consent Statement: Informed consent was obtained from all subjects involved in the study.

Data Availability Statement: Not applicable.

Acknowledgments: We would like to acknowledge the efforts of all authors that contributed to the Special Issue. A special thank goes to the *Hydrology* Editors for their dedication to this project and their valuable collaboration in the setup, promotion and management of the Special Issue.

Conflicts of Interest: The authors declare no conflict of interest.

References

1. Iliopoulou, T.; Malamos, N.; Koutsoyiannis, D. Regional Ombrian Curves: Design Rainfall Estimation for a Spatially Diverse Rainfall Regime. *Hydrology* **2022**, *9*, 67. [CrossRef]
2. Tegos, A.; Ziogas, A.; Bellos, V.; Tzimas, A. Forensic Hydrology: A Complete Reconstruction of an Extreme Flood Event in Data-Scarce Area. *Hydrology* **2022**, *9*, 93. [CrossRef]
3. Kilsdonk, R.A.H.; Bomers, A.; Wijnberg, K.M. Predicting Urban Flooding Due to Extreme Precipitation Using a Long Short-Term Memory Neural Network. *Hydrology* **2022**, *9*, 105. [CrossRef]
4. Stefanidis, S.; Alexandridis, V.; Theodoridou, T. Flood Exposure of Residential Areas and Infrastructure in Greece. *Hydrology* **2022**, *9*, 145. [CrossRef]
5. Bellos, V.; Kourtis, I.; Raptaki, E.; Handrinos, S.; Kalogiros, J.; Sibetheros, I.A.; Tsihrintzis, V.A. Identifying Modelling Issues through the Use of an Open Real-World Flood Dataset. *Hydrology* **2022**, *9*, 194. [CrossRef]
6. Erima, G.; Kabenge, I.; Gidudu, A.; Bamutaze, Y.; Egeru, A. Differentiated Spatial-Temporal Flood Vulnerability and Risk Assessment in Lowland Plains in Eastern Uganda. *Hydrology* **2022**, *9*, 201. [CrossRef]
7. Moroni, M.; Castellino, M.; De Girolamo, P. Numerical and Physical Modeling of Ponte Liscione (Guardialfiera, Molise) Dam Spillways and Stilling Basin. *Hydrology* **2022**, *9*, 214. [CrossRef]
8. Pricope, N.G.; Shivers, G. Wetland Vulnerability Metrics as a Rapid Indicator in Identifying Nature-Based Solutions to Mitigate Coastal Flooding. *Hydrology* **2022**, *9*, 218. [CrossRef]
9. Latif, S.; Simonovic, S.P. Trivariate Joint Distribution Modelling of Compound Events Using the Nonparametric D-Vine Copula Developed Based on a Bernstein and Beta Kernel Copula Density Framework. *Hydrology* **2022**, *9*, 221. [CrossRef]
10. Umer, Y.; Jetten, V.; Ettema, J.; Steeneveld, G.-J. Assessing the Impact of the Urban Landscape on Extreme Rainfall Characteristics Triggering Flood Hazards. *Hydrology* **2023**, *10*, 15. [CrossRef]
11. Chen, Y.-C.; Yeh, H.-C.; Kao, S.-P.; Wei, C.; Su, P.-Y. Water Level Forecasting in Tidal Rivers during Typhoon Periods through Ensemble Empirical Mode Decomposition. *Hydrology* **2023**, *10*, 47. [CrossRef]

12. Xafoulis, N.; Kontos, Y.; Farsirotou, E.; Kotsopoulos, S.; Perifanos, K.; Alamanis, N.; Dedousis, D.; Katsifarakis, K. Evaluation of Various Resolution DEMs in Flood Risk Assessment and Practical Rules for Flood Mapping in Data-Scarce Geospatial Areas: A Case Study in Thessaly, Greece. *Hydrology* **2023**, *10*, 91. [CrossRef]
13. Makris, C.; Mallios, Z.; Androulidakis, Y.; Krestenitis, Y. CoastFLOOD: A High-Resolution Model for the Simulation of Coastal Inundation Due to Storm Surges. *Hydrology* **2023**, *10*, 103. [CrossRef]

hydrology

Article

Regional Ombrian Curves: Design Rainfall Estimation for a Spatially Diverse Rainfall Regime

Theano Iliopoulou [1,*], Nikolaos Malamos [2] and Demetris Koutsoyiannis [1]

1 Department of Water Resources, Faculty of Civil Engineering, National Technical University of Athens, Heroon Polytechneiou 5, GR-157 80 Zografou, Greece; dk@itia.ntua.gr
2 Department of Agriculture, University of Patras, Theodoropoulou Terma, GR-272 00 Amaliada, Greece; nmalamos@upatras.gr
* Correspondence: tiliopoulou@hydro.ntua.gr

Abstract: Ombrian curves, i.e., curves linking rainfall intensity to return period and time scale, are well-established engineering tools crucial to the design against stormwaters and floods. Though the at-site construction of such curves is considered a standard hydrological task, it is a rather challenging one when large regions are of interest. Regional modeling of ombrian curves is particularly complex due to the need to account for spatial dependence together with the increased variability of rainfall extremes in space. We develop a framework for the parsimonious modeling of the extreme rainfall properties at any point in a given area. This is achieved by assuming a common ombrian model structure, except for a spatially varying scale parameter which is itself modeled by a spatial smoothing model for the 24 h average annual rainfall maxima that employs elevation as an additional explanatory variable. The fitting is performed on the pooled all-stations data using an advanced estimation procedure (K-moments) that allows both for reliable high-order moment estimation and simultaneous handling of space-dependence bias. The methodology is applied in the Thessaly region, a 13,700 km^2 water district of Greece characterized by varying topography and hydrometeorological properties.

Keywords: ombrian curves; intensity–duration–frequency curves; rainfall extremes; regionalization; regional frequency analysis; spatial rainfall; design rainfall

Citation: Iliopoulou, T.; Malamos, N.; Koutsoyiannis, D. Regional Ombrian Curves: Design Rainfall Estimation for a Spatially Diverse Rainfall Regime. *Hydrology* **2022**, *9*, 67. https://doi.org/10.3390/hydrology9050067

Academic Editor: Abdullah Gokhan Yilmaz

Received: 27 March 2022
Accepted: 17 April 2022
Published: 23 April 2022

1. Introduction

Ombrian (from the Greek word 'όμβρος' meaning rainfall) curves are a standard engineering tool in the form of a mathematical relationship linking rainfall intensity to timescale and return period, usually known as 'intensity-duration-frequency' curves. This term, albeit widely used, appears to be a misnomer, considering that 'duration' does not refer to the actual duration of a rainfall event but rather to the (arbitrary) time scale of averaging the rainfall intensity, while 'frequency' is not meant to be frequency but its reciprocal, i.e., return period. To oppose this common confusion (and having in mind the Aristotelian principle that science presupposes clarity—or saphenia [1]), the term 'ombrian curves' has been used as an alternative name in the past [2–5] and has been adopted here as well.

Ombrian curves have been used in hydrology since the works of Sherman [6] and Bernard [7] and own their popularity to their practical benefits when design problems affected by rainfall extremes are of interest. Although the typical curves have been constructed mostly following an empirical fashion, over the past decades, there have been several attempts to provide a theoretical basis for their modeling, e.g., [8–10], with the most recent advance being their full upgrade to multi-scale models of rainfall intensity [2]. At this point, it could be argued that the issue of deriving the curves for single sites has been efficiently tackled both at the practical level, with diverse methodologies providing satisfactory results for small scales, of the order of minutes to a few days (see the review by

Svensson and Jones [11]), and at the theoretical level as well, achieving their full modeling validity over any scale of interest [2,3].

At the same time, there have been various efforts to produce regional ombrian curves, a considerably more demanding task that is essential when hydrological analyses for large areas are to be performed, and design rainfall estimates are required for multiple sites. This is the case when regional flooding is studied, e.g., when large-scale flood protection works and urban stormwater networks are to be constructed. Even more, the construction of regional curves is critical when rainfall data for at-site analysis are missing, as is often the case. As such, the construction of ombrian curves at regional scales and even at the national level has been a priority for various countries, with a notable early example being the construction of regional curves in the US [12].

Broadly, the construction of regional ombrian curves can follow two different approaches:

(a) the at-site, independent fitting approach, which consists of separately fitting the curves to multiple gauged sites and using spatial interpolation methods to map the parameters over the whole region.
(b) the regional, simultaneous fitting approach, which consists of appropriately pooling the data together and obtaining a single model valid over the entire area, which is, in essence, the inverse approach to (a).

Frameworks based on the first approach are older yet still in use, e.g., [12–14], as this approach is easier to apply. Indeed, the task of constructing the curves for a single site is more straightforward, and the same is true for the mapping of the parameters in space, given the ample availability of geostatistical software. However, an important limitation of this approach is that results are very sensitive to data from single stations, which are often short and fragmented and can be impacted by large sampling uncertainty. A further issue with automated spatial interpolation methods is that their structure is obscure and cannot be easily modified, making their results not always interpretable. For instance, it is known that both the number and spatial distribution of available stations impact the reliability of the geostatistical method, but the former is not easy to assess. In this respect, Malamos and Koutsoyiannis [15] note that kriging requires a large number of available data points (at least 100, according to Oliver and Webster, [16], or 50–100 according to other studies [17]), in order to produce a reliable estimation of the variogram. The same authors propose a bilinear surface smoothing (BSS) model [18,19] that is shown to have superior performance to Universal kriging in terms of bias and heteroscedastic behaviors of the data, as well as a more interpretable theoretical structure.

On the other hand, having a single spatial model is a theoretically more powerful approach since it allows simultaneously using all observations to limit uncertainty by 'substituting space for time'; i.e., the principle behind the well-known 'regional frequency analysis' [20]. Nonetheless, the simultaneous use of all stations in the estimation process is challenging, for it requires a rigorous estimation framework to handle the underlying assumptions that control the information content of the pooled records. These are related to the different record lengths of the stations and the presence or not of spatial dependence among them. The most well-known framework of this type is the regional frequency estimation based on the L-moments [20,21]. Yet the latter is founded on the inter-site independence assumption and has been shown to decrease in accuracy as the level of dependence increases [22]. To explicitly address the effect of the spatial dependence in the data, a new regional frequency estimation framework has been proposed that allows high-order properties estimation from spatially correlated data by means of 'knowable' moments' (K-moments of high order [2,23]). Separate space dependence models have been employed in other approaches (e.g., [24]). Evidently, the issue of regional frequency analysis is still an evolving research subject.

Aside from the choice of the theoretical framework for the regional estimation, the formulation of a regionalization approach for the extreme rainfall properties is an additional demanding task that has to be performed in approach (b). The literature on the different types of regionalization techniques for frequency analysis is vast and extensively covered by previous works [11,25]. The simpler approach considered is the estimation of a common frequency distribution by directly pooling together all stations of a homogenous region [26]. This is a reasonable approach for small areas or areas with limited spatial variability but is less efficient for complex regimes. In such cases, an extension of the same concept is to delineate the region by identifying homogenous sub-regions and pool the data within them, allowing for the possible variation only of a site-specific scale parameter, known as the 'index-rainfall'. This is a widely used approach [11,20,25,27,28], inspired by a similar method in flood frequency analysis, the so-called 'index-flood' method [29]. Considerable limitations of the approach relate to the subjectivity of detecting clusters in complex regimes and the arising of spatial discontinuities at the clusters' boundaries. Remediation to these issues is the 'region of influence approach', which, instead of using fixed boundaries regions, employs individual regions of varying boundaries centered at the site of interest [30]. Irrespective of the arrangements of each method, the need to model the spatial variability of the site-specific parameter(s) still emerges and is often treated through standard interpolation and geostatistical methods as in approach (a).

In this sense, several frameworks resort to a combination of the two approaches (e.g., [28,31,32]). Namely, the data are pooled together for the distribution fitting after being proportionally adjusted by a site-specific parameter, which itself is obtained for a given grid by interpolation methods. This approach bypasses the difficulty of identifying sub-regions based on hydrological similarity and is characteristically called 'regionless' or 'boundaryless' [25,32].

The scope of this work is to construct regional ombrian curves by exploiting and combining recent advances in the different methodological components of the analysis and integrating them into a new framework for regional frequency analysis of rainfall extremes. The latter is based on:

1. An advanced estimation method of extreme values based on high-order moment estimation while respecting space dependence [2].
2. A couple of flexible spatial smoothing models [18,19] to describe the spatial variability of extreme rainfall without resorting to uncontrolled interpolation.
3. A formulation of ombrian curves, which is revisited through recent theoretical developments in the field [2].

This is the first time that the K-moments framework, by now used in frequency estimation for several hydrometeorological variables [23,33,34], has been put into practice in regional frequency estimation of rainfall extremes. This is also the first time that the BSS model framework has been employed for the regionalization of extreme rainfall. As a proof-of-concept, the framework is applied in the region of Thessaly (Greece), utilizing data from 55 stations over a 13,700 km^2 basin area. The large spatial scale, together with the hydrological complexity of the case study, form a challenging test for the methodology, from which new insights into regional rainfall frequency analysis are gained.

2. Methodology

2.1. Mathematical Form of the Ombrian Relationship

Koutsoyiannis ([2]; Chapter 8) recently developed a framework advancing typical ombrian curves to stochastic models of rainfall intensity, valid over any scale supported by the data. This type of ombrian model arises directly from the stochastic properties (dependence structure and marginal distribution) of rainfall intensity and their (non-simple) scaling behavior, and, as such, it can also be applied for simulation of the rainfall process [35]. The framework in [2] can be applied at any time scale, arbitrarily large, to produce the ombrian relationship linking rainfall intensity x to any timescale k and return period T. We note, though, that for large time scales the mathematics becomes somewhat involved. Here we apply the framework only for small time scales, for which a Pareto distribution for the non-zero rainfall intensity is justified. (For larger scales, this should be replaced by a Pareto–Burr–Feller distribution.) In this case, the Pareto distribution quantile is given as [2]:

$$x = \lambda(k)\frac{\left(P_1^{(k)}T/k\right)^{\xi} - 1}{\xi}, \quad \xi > 0 \tag{1}$$

where $P_1^{(k)}$ is the probability wet, $\lambda(k)$ is a scale parameter and ξ is the tail-index of the Pareto distribution. Both $P_1^{(k)}$ and $\lambda(k)$ are functions of the timescale obtained as [2]:

$$P_1^{(k)} = \frac{1-\xi}{1/2-\xi}\frac{\mu^2}{\gamma(k)+\mu^2} \tag{2}$$

$$\lambda(k) = \frac{\mu(1-\xi)}{P_1^{(k)}} = \frac{(1/2-\xi)\left(\gamma(k)+\mu^2\right)}{\mu} \tag{3}$$

where μ is the mean intensity (constant at all time scales) and $\gamma(k)$ is the climacogram of the process, i.e., the function of the variance across timescale, which can follow different models [2]. A simplification of Equations (1)–(3) is possible based on the following assumptions stemming from the fine-scale behavior of rainfall. For small time scales, of the order of minutes to a few days:

- $P_1^{(k)} \propto k$, and hence we can set the quantity $\beta(k) := k/P_1^{(k)} = \beta =$ constant in Equation (1).
- $\gamma(k) \gg \mu^2$, and thus we can neglect the latter term in their sum.
- we select the generalized Cauchy-type model for the climacogram:

$$\gamma(k) = \lambda_1^2\left(1+\left(\frac{k}{\alpha}\right)^{2M}\right)^{\frac{H-1}{M}} \tag{4}$$

where α and λ_1 are scale parameters, with dimensions of time $[t]$ and $[x]$, respectively, and H, M are dimensionless parameters in the interval (0, 1), controlling the long-range (Hurst-Kolmogorov dynamics) and local scaling of the process (fractal behavior) of the process, respectively. For M we take the neutral value $M = 1/2$ as default. We note though that if the focus is on even smaller temporal scales, this value ($M = 1/2$) can be inappropriate.

These simplifying assumptions result in some violations of a full stochastic consistency, as detailed in [2]. However, at small scales, the violations are negligible [2,3]. By virtue of these simplifications, the ombrian relationship is given as:

$$x = \lambda_1^2\frac{(1/2-\xi)}{\xi\mu}\left(1+\frac{k}{\alpha}\right)^{2H-2}\left(\left(\frac{T}{\beta}\right)^{\xi} - 1\right) \tag{5}$$

Setting $\lambda = (1/2-\xi)\lambda_1^2 \,/\, \xi\mu$ and $\eta = 2-2H$, Equation (5) can be rewritten as:

$$x = \lambda\frac{(T/\beta)^{\xi} - 1}{(1+k/\alpha)^{\eta}}, \quad \xi > 0 \tag{6}$$

where the following five parameters are involved, λ an intensity scale parameter in units of x (e.g., mm/h), β a timescale parameter in units of the return period (e.g., years), α a timescale parameter in units of timescale (e.g., h) with $\alpha \geq 0$, η a dimensionless parameter with $0 < \eta < 1$, and $\xi > 0$ the tail index of the process.

It is easily observed that Equation (6) can be written concisely as the quotient of two separable functions $b(T)$ and $a(k)$ of the return period and the timescale, respectively, in the form:

$$x = \frac{b(T)}{a(k)} \tag{7}$$

with

$$a(k) = \left(1 + \frac{k}{\alpha}\right)^{\eta} \tag{8}$$

being a function of time scale, in wide use as an approximation [2,36] but here resulting as a consequence of the climacogram, and $b(T)$ a function of the return period that is analytically derived from the distribution function of the rainfall intensity.

In the case that the return period of the rainfall intensity is determined based on rainfall exceedances extracted from the full series (i.e., Peaks over Threshold, POT), a Pareto distribution can be generally assumed for modeling the rainfall intensity, as implied in Equation (6). However, if the return period is determined based on series of annual maxima (AM) of rainfall intensity, then both long-term empirical evidence and theoretical arguments support the use of the Extreme Value Type 2 (EV2) distribution from the Generalized Extreme Value (GEV) distribution family [37–39]:

$$F(y) = \exp\left(-\left(1 + \xi\left(\frac{y}{\nu} - \psi\right)\right)^{-\frac{1}{\xi}}\right),\ y \geq \nu\left(\psi - \frac{1}{\xi}\right) \tag{9}$$

where ψ (dimensionless), $\nu > 0$ (units same as in y) and $\xi > 0$ (dimensionless) are location, scale, and shape parameters, respectively. It should be mentioned that the case of $\xi < 0$ is not appropriate for maximum rainfall, since it presumes the existence of an upper limit for the variable, which is inconsistent with the physical reality. Furthermore, the case of $\xi = 0$, i.e., assuming a Gumbel (Extreme Value Type 1—EV1) distribution for the maximum rainfall intensity, is also not supported by worldwide empirical evidence and is to be avoided in general [37]. Therefore, it is not developed herein, but further details for this case are given in Koutsoyiannis [2].

Equivalently, the EV2 distribution as given by Equation (9) can be re-parameterized consistently to Equation (6) as follows:

$$F(y) = \exp\left(-\frac{\Delta}{\beta}\left(\frac{y}{\lambda} + 1\right)^{-\frac{1}{\xi}}\right) \tag{10}$$

where $\beta = (1 - \xi\psi)^{1/\xi}\Delta$ and $\lambda = (1 - \xi\psi)\,\nu/\xi$ and $\xi > 0$.

The variable y represents either the rainfall intensity x or, equivalently, the product $x\,a(k)$ (Equation (7)). Solving Equation (10) in terms of y and substituting $F(y) = 1 - \Delta/T$, where $\Delta = 1$ year for annual maxima, yields:

$$y_T = \lambda\left((-(\beta/\Delta)\ln(1 - \Delta/T))^{-\xi} - 1\right) \tag{11}$$

Therefore, by substituting Equations (8) and (11) in (7), the following generalized form of ombrian curves for annual maxima is derived:

$$x = \lambda\frac{(-(\beta/\Delta)\ln(1 - \Delta/T))^{-\xi} - 1}{(1 + k/\alpha)^{\eta}},\quad \xi > 0 \tag{12}$$

It is easily shown that for small return periods, Equation (6) deriving from a Pareto distribution yields higher intensity than Equation (12), whereas for larger return periods ($T > 10$ years), the two are practically indistinguishable, given that for small Δ/T holds $\ln\left[1 - (\Delta/T)\right] = -(\Delta/T) - (\Delta/T)^2 - \cdots \approx -\Delta/T$. Therefore, even in the case that the model fitting is based on Equation (12), i.e., when annual maxima are used (which is also the case here), it is safer, from an engineering point of view, to express the final model used to obtain design rainfall in the form of Equation (6). Yet, the availability of either POT or AM series determines which of Equation (6) or (12), respectively, will be used for model fitting. Following a slightly different parameterization, Equations (6) and (12) were also proposed by Koutsoyiannis et al. [8] for small scales, which nevertheless are sufficient for most engineering applications of ombrian curves, namely those involving flood analyses.

The advantage of this simplified approach is precisely the separability of $a(k)$ and $b(T)$ functions that allows for a two-step procedure for the parameter estimation. This turns out to be convenient for typical applications and even more for regional analyses, as will be shown next. It also has attractive flexibility in using different sources of data. Namely, a reliable determination of the timescale parameters α and η requires data from fine-scale records, whereas the parameters of the function $b(T)$, including the highly uncertain tail index, are better inferred from daily raingauge data, which are usually more systematic and less prone to erroneous recordings [8].

2.2. Regionalization Method: Bilinear Surface Smoothing Models for the 24 h Average Annual Rainfall Maxima

The target of the regional model is to generalize Equation (6) or (12) in space, achieving their applicability for any set of coordinates in a given area. To efficiently describe the spatial heterogeneity in a region without resorting to a great number of parameters and uncontrolled interpolation, we should make an assumption on which parameters we consider as regionally varying. It is reasonable to begin by applying diagnostics for the spatial heterogeneity of rainfall in the study area, both in terms of the $a(k)$ and $b(T)$ functions' parameters, bearing in mind that selected parameters should be reliably estimated from single-site data. Following this rationale and supported by the available data, which in our case are series of annual maxima at different temporal scales, as shown in Section 3.2, we choose to spatially model the parameter λ of Equation (12). This corresponds to the scale parameter of the EV2 distribution, which is proportional to the mean value of the process, i.e., the average annual maxima at each location for any timescale. Since the transformation between timescales is controlled by the $a(k)$ function, it suffices to model the average maxima at a single, convenient timescale. We choose the 24 h scale due to the much greater availability of daily data in the study region. Depending on the need to apply further complexity, the location parameter of the EV2 distribution could be another option for spatial modeling. Yet it is not advisable to spatially model the highly uncertain shape parameter (i.e., the tail index) based on single-site data unless long-term empirical evidence supports a spatial variation thereof. Parameters α and η, which control the timescale transformations of the curves, are very sensitive to the existence of sub-daily data [8], which are scarce in the study region, and thus, they do not constitute good choices for spatial modeling either. For these reasons, we apply common values for the rest of the parameters over the whole area. The choice of the mean of the AM distribution as the parameter to be regionalized has been proven to be a robust choice in regional frequency analysis, dating back to the index-flood method [29] and several applications thereafter (see [11]).

Having chosen to spatially model only the λ parameter related to the mean value of the annual maxima series, we have to identify a model for the spatial variation. As already discussed in the introduction, a large region with complex topography cannot be efficiently treated as a homogenous area. In this case, instead of identifying several sub-regions, which may give rise to abrupt changes in the final results with questionable physical basis, it is better to explicitly model the process at any given point in the area of interest. Towards this 'regionless' modeling approach, we apply a framework of spatial smoothing modeling that is described next.

We apply two versions of a bilinear surface smoothing model proposed by Malamos and Koutsoyiannis [18,19], generalizing in 2D a previous numerical smoothing and interpolation method [40,41]. A brief overview of the mathematical framework is presented since it is detailed in the aforementioned publications. The general idea behind both methods is to compromise the trade-off between the objectives of minimizing the fitting error and the roughness of the fitted bilinear surface, therefore termed bilinear surface smoothing (BSS). The larger the weight of the first objective, the rougher the surface will appear, while the opposite is true for a larger weight of the second objective.

The mathematical framework of BSS suggests that fitting is meant in terms of minimizing the generalized cross-validation error (GCV; [42]) between the set of the given data points and the corresponding estimates. The general estimation function, $\hat{z}_u$, for point u on a plane, according to the BSS method, is:

$$\hat{z}_u = d_u \tag{13}$$

where d_u is the value of the fitted bilinear surface d at that point.

The BSS method can be extended by the introduction of an additional explanatory variable (bilinear surface smoothing with an explanatory variable; BSSE) at a denser dataset compared to that of the main variable, as follows. We assume that at the locations of the given data points, we also know the value of an explanatory variable t, and therefore for each point z there corresponds a value t. In this case, the general estimation function for point u is:

$$\hat{z}_u = d_u + t_u e_u \tag{14}$$

where d_u, e_u are the values of two fitted bilinear surfaces at that point, namely d and e, while t_u is the value of the explanatory variable at that point. This is not a global linear relationship but a local linear one as the quantities d_u and e_u change in space.

In the case of the BSS, there are four adjustable parameters for surface d: the numbers of intervals along the horizontal and vertical direction, respectively, i.e., mx, my, and the corresponding smoothing parameters $\tau_{\lambda x}$ and $\tau_{\lambda y}$. The incorporation of the explanatory variable for the BSSE case adds two more adjustable parameters: the smoothing parameters $\tau_{\mu x}$ and $\tau_{\mu y}$ corresponding to surface e. The values of all the smoothing parameters are restricted in the interval [0, 1) for both directions [18]. When the smoothing parameters are close to 1, the resulting bilinear surfaces exhibit greater smoothness, whereas, for small values of these parameters, interpolation among the known points is obtained.

A desirable feature of the method for regional analyses is the fact that it is proven reliable even in the case of few and scarce data, in contrast to common geostatistical methods that require a denser data network to be applied reliably (e.g., to estimate a semivariogram) [15]. It is important to note that the method is also parsimonious in terms of the number of parameters and the choices involved in the modeling. This is evident when compared to the standard kriging framework, in which one has to decide among the n different types of the method, among the 4 n standard variogram types (i.e., spherical, exponential, Gaussian and power), and also identify the values for the range, sill, and nugget, resulting in a total of 12 n choices, depending on the selected number of kriging methods. This increases the complexity of the approach and may increase the subjectivity as well, given that an objective framework for basing these decisions is lacking.

2.3. Bilinear Surface Smoothing Model Parameters Estimation

As mentioned, the parameter estimation methodology for both the BSS and BSSE methods is based on the minimization of GCV error, and therefore, there is no effect in terms of heteroscedasticity. For a given combination of the bilinear surface segments, mx and my, the minimization of GCV error results in the optimal values of $\tau_{\lambda x}$, $\tau_{\lambda y}$ and $\tau_{\mu x}$, $\tau_{\mu y}$. This can be repeated for several trial combinations of mx and my values until the global minimum of GCV is reached.

Both variants of the method are applied, one taking into account only the coordinates of the stations (BSS) and the second one exploiting as well the elevation of the stations (BSSE) as an additional explanatory variable. In the second case, a digital elevation model of the wider area (Thessaly and neighboring areas) with 90 m resolution at the equator (SRTM; [43]) is employed both for extracting the elevation at the stations' coordinates and for making estimations for any point in the given grid.

For the objective evaluation of the two methods, two statistical indices, the root mean square error (RMSE) and the mean absolute error (MAE), are compared in terms of (a) performance using all data and (b) the leave-one-out cross-validation performance, i.e., when the estimation at each coordinate set is performed by omitting the known value at that position. This analysis is presented in Section 4.1.

2.4. Timescale Parameters Estimation

The simplified version of the ombrian model also allows for a simplified fitting procedure. By utilizing the separability of functions $a(k)$ and $b(T)$ in this version, an independent, two-step fitting approach can be used, as introduced by Koutsoyiannis et al. [8]. Namely, Equation (12) can be written as:

$$\left(1+\frac{k}{\alpha}\right)^{\eta} x = \lambda\left(\left(-(\beta/\Delta)\ln(1-\Delta/T)\right)^{-\xi}-1\right) \tag{15}$$

From this expression, it is easy to see that for the different timescales k_j the stochastic variables $\underline{y}_j := a(k_j)\underline{x} = (1+k/\alpha)^{\eta}\underline{x}$ have a common distribution function, with the $\underline{y}_j$ for the different k_j being samples of it. Let then, $\underline{y}_{ji} := a(k_j)\underline{x}_{ji}$ of length $n = \sum\limits_j n_j$ denote the merged sample of all sub-samples $\underline{x}_{ji}$ of size n_j corresponding to timescale k_j. Let also $\underline{r}_{ji}$ denote the rank of each sub-sample $\underline{x}_{ji}$ in the merged sample $\underline{y}_{ji}$ so that the mean rank of each sub-sample is given as $\underline{r}_j = \sum\limits_i \underline{r}_{ji}/n_j$. Replacing all $\underline{r}_{ji}$ with the mean rank value $\underline{r}_j$ we obtain a sample of n values, with n_1 equal to r_1, n_2 equal to r_2 etc. Then the mean and variance estimators are, respectively:

$$\underline{\overline{r}} := \frac{1}{n}\sum_j n_j \underline{r}_j \tag{16}$$

$$\underline{\gamma}_r := \frac{1}{n}\sum_j n_j\left(\underline{r}_j - \underline{\overline{r}}\right)^2 \tag{17}$$

If no ties are present among the different ranks, then $\underline{\overline{r}} = (n+1)/2$.

Following the assumption that the samples are from the same distribution, given by the right-hand side of Equation (15), then each $\underline{r}_j$ should be close to the mean $\underline{\overline{r}}$ while the variance should be minimal. Therefore, we can find the parameters α and η as the values that minimize the estimate of the variance γ_r from the observations x_{ji}. The original values $\underline{y}_{ji}$ could be used as well instead of the ranks, yet the use of the ranks makes the estimation process more robust to outliers. In order to improve the fit to the higher quantile region, we could also use a part of the data of each sample belonging to the highest 1/2 or 1/3 of the data [8]. In this study, the highest 1/2 is used.

The limited availability of sub-daily stations, in combination with their short records, hinder the reliable estimation of the time scale function parameters at each station separately. It is also pointed out that parameter α is very sensitive to small scales intensities and ideally requires sub-hourly data to be reliably estimated [8]. Given that few stations have data at such scales, to limit uncertainty, we apply the methodology described above simultaneously to the sample of all fine-scale raingauges. In particular, we identify the set of the time scale parameter values as the one that minimizes the sum of all stations' variances, each of which is given by Equation (17).

2.5. Regional Estimation of Distribution Parameters through K-Moments

Having estimated the α and η parameters, it remains to specify the parameters of the $b(T)$ function through the following procedure. We form the pooled sample comprising all stations' annual rainfall maxima at the 24 h scale after first standardizing (dividing) them by their theoretical mean value, given by the spatial smoothing model. To the pooled standardized sample of annual maxima, we fit the EV2 distribution using the method of the non-central K-moments [2,23]. K-moments are newly proposed moments developed with the aims of being knowable for very large orders (depending on the sample size) and also interpretable in terms of order statistics. They are closely related to the probability-weighted moments [44], but their formulation is simpler and more intuitive, which are attractive qualities similar to the ones of the L-moments [21]. The distinctive feature of K-moments, though, is that they are tailored to perform extreme-oriented analyses, as they enable reliable estimation of very high-order moments. What is more, each high-order K-moment estimate can be directly assigned a return period, which provides a direct means to empirical estimation of probability, alternative to order statistics. Furthermore, this estimation can be appropriately modified in the case that there is bias due to dependence [2]. In particular, K-moments allow straightforward estimation of high-order moments even for spatially dependent data, which is a rare property. Namely, typical regional applications of L-moments do not go beyond the 4th moment (L-kurtosis) estimation. This advantage of K-moments is exploited in the regional frequency analysis, which is particularly sensitive to high-order moments.

Koutsoyiannis [2,23] has introduced several variants of K-moments, of which here we use the simplest non-central variant, defined as:

$$K'_p := p\mathrm{E}\left[(F(\underline{x}))^{p-1}\underline{x}\right] \tag{18}$$

for the moment order $p \geq 1$. K'_p has the important property that it equals the expected value of the maximum of p independent stochastic variables identical to $\underline{x}$, i.e.,

$$K'_p = \mathrm{E}\left[\max\left(\underline{x}_1, \underline{x}_2, \ldots, \underline{x}_p\right)\right] \tag{19}$$

The estimators of the non-central K-moments are given by the following formulae:

$$\hat{\underline{K}}'_p = \sum_{i=1}^{n} b_{inp}\, \underline{x}_{(i:n)} \tag{20}$$

$$b_{inp} = \begin{cases} 0, & i < p \\ \frac{p}{n}\frac{\Gamma(n-p+1)}{\Gamma(n)}\frac{\Gamma(i)}{\Gamma(i-p+1)}, & i \geq p \geq 0 \end{cases} \tag{21}$$

where $\underline{x}_{(i:n)}$ is the ith smallest variable in a sample $\underline{x}$, of size n, (the ith item of the sample in ascending order) and p is the order of the moment, which can be any positive number $p \leq n$. In addition, the following holds:

$$\sum_{i=1}^{n} b_{inp} = 1 \tag{22}$$

The fact that $b_{inp} = 0$ for $i < p$ means that as the moment order increases, fewer data are used in the estimation, until only one is left, the maximum, when $p = n$, and $b_{nnn} = 1$. For $p > n$, $b_{inp} = 0$ for every i, $1 \leq i \leq n$, and the estimation becomes impossible. The first order non-central K-moment is the mean value of the sample.

K-moment values can be assigned a return period as follows [2]:

$$\frac{T\left(K'_p\right)}{\Delta} = p\Lambda_p \approx \Lambda_\infty p + (\Lambda_1 - \Lambda_\infty) \tag{23}$$

where Λ_1, Λ_∞ are coefficients depending on the distribution function. For the EV2 distribution it is shown [2] that the Λ coefficients are functions of the shape parameter ξ:

$$\Lambda_1 = \frac{1}{1 - \exp\left(-(\Gamma(1-\xi))^{-\frac{1}{\xi}}\right)} \tag{24}$$

$$\Lambda_\infty = \Gamma(1-\xi)^{\frac{1}{\xi}} \tag{25}$$

For validation purposes, the following relationship of empirical return periods based on order statistics is also used, which is shown to provide an unbiased estimate of the logarithm of the return period [2]:

$$\frac{T_{(i:n)}}{\Delta} = \frac{n + e^{1-\gamma} - 1}{n - i + e^{-\gamma}} = \frac{n + 0.526}{n - i + 0.561} \tag{26}$$

The procedure outlined above could be directly applied for assigning return periods to the K-moments of any single station, and the parameters of the EV2 distribution could be obtained by minimizing an error metric (e.g., MAE or RMSE) between the theoretical quantiles and the empirical K-moments, or between the corresponding return periods. However, for the pooled sample, the resulting information gain, and thus the maximum return period that can be estimated from the data, is a function of the sample's dependence structure. It is well known that in the case of cross-correlated variables, the quantity of information for the variable corresponds to a smaller sample of length compared to the case of independence but still greater than that of an individual station.

The framework of K-moments allows for the effect of dependence to be explicitly accounted for in the estimation of the return period. This is achieved through proper modification of the order of the moments of the unified sample, p', which in turn modifies the estimation of the return period. In particular, let n_1 denote the sample length of each station, m denote the number of stations, and $n = m\, n_1$ denote the size of the merged sample, and then the following methodology is applied [2]:

- For $p \leq n_1$ we set $p' = p$.
- For $p > n_1$ the following approximation is used. We estimate the equivalent Hurst parameter H, based on the spatial correlation of the stations ρ:

$$H = \frac{1}{2} + \frac{\ln(1+\rho)}{2\ln 2} \tag{27}$$

Based on the estimated H the following coefficient is used for bias correction, Θ^{HK}:

$$\Theta^{\mathrm{HK}}(n, H) \approx \frac{2H(1-H)}{n-1} - \frac{1}{2(n-1)^{2-2H}} \tag{28}$$

Then the modified orders of the moments are obtained as:

$$p' \approx 2\Theta + (1 - 2\Theta)(p - n_1 + 1)^{((1+\Theta)^2)} + n_1 - 1 \tag{29}$$

and their corresponding return periods are adjusted accordingly based on Equation (23).

It is obvious that n_1 controls the maximum moment order, which is unaffected by dependence. In the case that the stations have different lengths, n_1 can be estimated as the average record length of all stations (here, $n_1 = 42$). An increased value of n_1 suggests that the information gain is also increasing, as fewer return periods are modified downwards, while the opposite is true when n_1 decreases. In order to bypass the uncertainty regarding the modification of the return periods based on the estimated correlation structure, a good strategy is to use for model calibration only the moment orders up to n_1, and employ the higher moments for validation. In so doing, the moments used in the calibration are still much more than the ones used in regular moment fitting procedures (typically up to 3 or 4 orders), while a second-moment set is also available for validation.

To transform the parameters of the EV2 distribution (expressed as in Equation (10)) of the standardized 24 h rainfall maxima to the final ombrian $b(T)$ parameters, we use the following procedure. Let u_T denote the 24 h annual maximum rainfall value for return period T standardized by its mean value μ. Then the rainfall intensity for any station at the 24 h timescale is $x_T^{(24\text{ h})} = \mu\, u_T/24$, where μ is the mean value used in the standardization. Likewise, the quantity $y_T := x_T^{(24\text{ h})}\,(1 + 24/\alpha)^{\eta}$, whose distribution defines the function

$b(T)$ of the ombrian relationship will be $y_T = \mu\ u_T(1+24/\alpha)^{\eta}/24$. Consequently, the variable y shares the same distribution function with the variable u with the same shape and location parameters, and scale parameter proportional to the one of u by a factor of $\mu\ (1+24/\alpha)^{\eta}/24$, i.e.,

$$\xi = \xi_u,\ \ \beta = \beta_u,\ \ \lambda = \lambda_u\ \mu\ (1+24/\alpha)^{\eta}/24 \tag{30}$$

where the subscript u denotes the standardized rainfall maxima at the 24 h scale and α is expressed in h. It is recalled that the mean value μ for each location used in the standardization is derived from the BSS/BSSE models.

In this way, parameters α and η, which are estimated simultaneously from all stations, in combination with the parameters of the distribution of the standardized maximum 24 h rainfall values and the spatially modeled mean value of the process by the BSS/BSSE models fully determine the ombrian curves, as given by Equations (6) and (12).

3. Data

3.1. Study Area

The study region is the geographic area of the Water District (WD) of Thessaly (~13,700 km^2) which is one of the 14 WD of Greece. The district extends mostly within the administrative region of Thessaly, while it also includes a small part of the region of Central Greece and a small part of the Western and Central Macedonia region. The topography of the area is characterized by the existence of four mountain ranges in its perimeters, Olympus-Kamvounia in the north, Pindus in the west, Othrys in the south, and Pelion-Ossa in the east, which surround the Thessalian plain that rests in the central area (Figure 1a). The Thessalian plain contains the largest part of the water bodies of the district and is traversed by the Pineios river and its tributaries. It is also the largest agricultural area in Greece, with lowland topography making it prone to frequent and heavy flooding [45,46]. A recent flood event of the 18–19 September 2020 triggered by a Mediterranean cyclone has caused human and livestock losses and extensive structural and agricultural damages to the area, sparking a revitalization of the decade-long initiatives for improving the area's flood protection design and strategy [47]. The climate of the Western region is continental, while the Eastern region has a typical Mediterranean climate, while the rainfall pattern exhibits strong differences between the lowlands and the mountain regions [48]. These characteristics of the study region, namely its hydrometeorological diversity, vast spatial extent, and criticality of flood risk, make it a challenging case study for the application of the methodology.

3.2. Data Processing and Quality Control

The construction of ombrian curves is based on rainfall intensity data at a range of timescales, typically starting from fine scales, i.e., 5 to 60 min, and extending to the 24 or 48 h scale for common applications. To this aim, we assemble a set of 17 rainfall records from tipping buckets and telemetric stations, recording data at the 5–30 min timescale and 61 rainfall records from daily raingauges. The data are obtained from the Public Power Corporation (PPC) of Greece, the Hellenic Ministry of Environment and Energy (HMEE), the Hellenic Ministry of Agricultural Development and Food (HMADF), the Hellenic Ministry of Agriculture (HMA), the Hellenic National Meteorological Service (HNMS) and the meteo network [49] of the National Observatory of Athens. The properties of the stations are detailed in Tables S1 and S2 of the Supplementary material.

We aggregate the original series at a range of timescales from 5 min to 48 h, with k =0.08, 0.17, 0.25, 0.5, 1, 2, 6, 12, 24, 48 h (depending on data available at the finest scale), and we extract the maximum rainfall depth at each scale $h^{(k)}$ for all hydrological years. Accordingly, we compute the corresponding rainfall intensity at the given scale as $x^{(k)} = h^{(k)}/k$, thus deriving the empirical rainfall intensities corresponding to the annual maxima of the hydrological years. The reason for using AM series instead of POT or even the full data series for the estimation of the extremes is that several historical records are available only in this form.

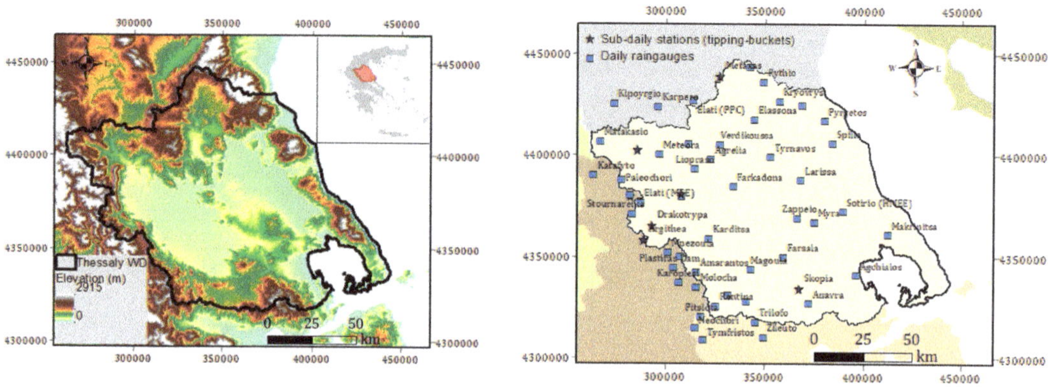

(a) (b)

Figure 1. (**a**) Spatial extent and elevation of the study region (Thessaly's Water District in Greece). (**b**) Geographic location of the 55 rainfall records, from daily raingauges and sub-daily tipping-buckets, with data at the 24 h scale (used in the BSS/BSSE models and the EV2 distribution modeling).

We note that the choice of the starting point for the aggregation is arbitrary, and a change thereof would likely result in adifferent estimate. For this reason, it was a common hydrological practice in the past to either take the maximum estimate resulting from all possible positions of the starting point or 'inflate' the given estimate by a specific factor, known as the Hershfield coefficient [50]. Although this practice aims for safer estimates from an engineering point of view, in theory, all realizations of a stochastic process are equivalent, and there is no theoretical basis to 'correct' them. In fact, by correcting the series, we distort its stochastic properties and, instead of studying $\underline{x}_\tau{}^{(k)}$, the behavior of $\underline{w}_\tau{}^{(k)} := \max_j\left(\underline{x}_{\tau+j}{}^{(k)}, j = 0, \ldots k-1\right)$ is studied, which is a different stochastic process [2]. Hence, we do not apply the Hershfield coefficient.

To ensure a good quality dataset for our analysis, we use stations that have at least 12 years of data, and we undertake quality checks based on hydrological experience in the study area. In particular, we perform spatial consistency checks excluding stations with systematically lower recordings in comparison to neighboring ones. In addition, we perform hydrological consistency checks, i.e., to ensure that single-site empirical maximum rainfall is consistent with hydrological experience worldwide, suggesting an unbounded right tail of sub-exponential type [37,38]. We note that poorly maintained raingauge records (e.g., in remote mountainous areas) sometimes exhibit maximum rainfall recordings of the same (or nearly the same) amount due to spillage effects during storm events. In this case, a bounded GEV distribution might falsely emerge.

After screening with these criteria and excluding stations with severe inconsistencies, the resulting set of stations includes 48 daily raingauge stations, 7 of which are at locations gauged by tipping buckets as well, and 14 sub-hourly tipping bucket/telemetric stations. The estimation of the extreme properties and of the average 24 h AM rainfall (Sections 2.3 and 2.5) is based on the set of daily raingauge stations due to the latter being more and of larger record lengths. Yet, maximum rainfall data at the 24 h scale are also employed from the fine-scale rainfall stations when daily raingauge data are not available at the same location. Taking the latter into account, the distribution properties of the maximum rainfall are estimated using a combined set of 55 samples of 24 h annual rainfall maxima, whose spatial distribution is depicted in Figure 1b. The set of the 14 fine-scale rainfall stations is used in the estimation of the timescale parameters (Section 2.4).

4. Results

4.1. Fitting of the Bilinear Surface Smoothing Models

Before fitting the regional model, we explore the spatial variability of the rainfall regime by identifying the variations in the mean and standard deviations of the rainfall intensity and the possible association with the elevation of the stations. In Figure 2, it can be seen that the first two moments of the rainfall intensity across scales follow a similar statistical behavior, and, therefore, it is reasonable to apply common timescale parameters (i.e., the function $a(k)$). In terms of the average annual maxima at the 24 h scale, there also appears to be a positive association with the stations' elevation, although this is not verified in all cases (Figure 3). Therefore, it seems that the elevation might serve as an explanatory variable, but it needs to be incorporated into a more general spatial model identifying additional patterns of the rainfall maxima in space.

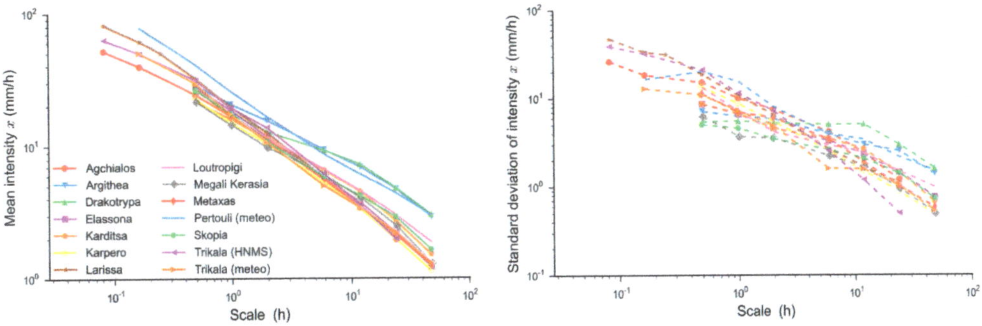

Figure 2. Mean and standard deviation of the empirical rainfall intensities for 5 min to 48 h scales for the 14 fine-scale rainfall stations.

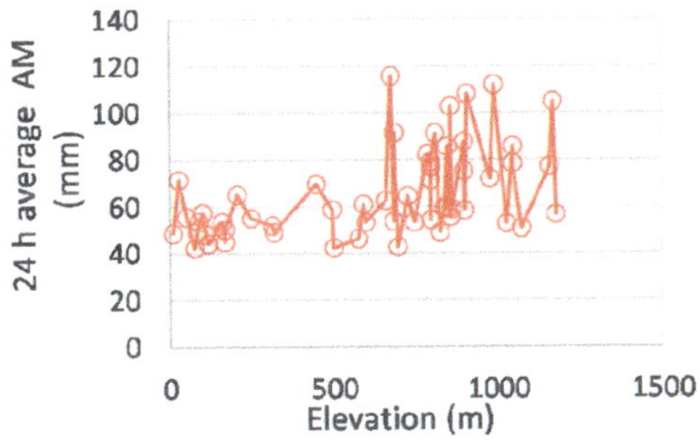

Figure 3. Average annual rainfall maxima (mm) at the 24 h scale vs. the stations' elevation.

To explore the suitability of elevation as an explanatory variable in an objective manner, we evaluate its performance within the BSS/BSSE model framework, comparing the results from both versions. In the BSSE case, the stations' altitudes are derived from a digital elevation model of the area (Thessaly and neighboring areas) with 90 m resolution at the equator (SRTM; [43]), which is also used for the estimation of the average maximum rainfall at each point in space.

The parameters deriving from the optimization are mx = 3, my = 5, $\tau_{\lambda x}$= 0.550, $\tau_{\lambda y}$= 0.005 for the BSS model and mx = 4, my = 12, $\tau_{\lambda x}$= 0.068, $\tau_{\lambda y}$= 0.001, $\tau_{\mu x}$= 0.621, $\tau_{\mu y}$= 0.451 for the BSSE model. In order to compare the model fits we compute RMSE and MAE (a) for the fit of both models using all the data and (b) for the fit applying the leave-one-out cross validation method. Results are shown in Table 1.

Table 1. Results from the BSS and BSSE models fitting.

Spatial Smoothing Model	BSS	BSSE
Parameters	mx= 3, my = 5, $\tau_{\lambda x}$= 0.550, $\tau_{\lambda y}$= 0.005	mx= 4, my= 12, $\tau_{\lambda x}$= 0.068, $\tau_{\lambda y}$= 0.001, $\tau_{\mu x}$= 0.621, $\tau_{\mu y}$= 0.451
	A. All data	
RMSE (mm)	9.0	7.0
MAE (mm)	7.3	5.3
	B. Leave-one-out cross validation	
RMSE (mm)	12.5	10.8
MAE (mm)	9.8	8.1

It is seen that the BSSE model involving elevation as an additional explanatory variable is found superior in both comparisons, which is expected by hydrological experience in the area and also supported by previous applications for annual rainfall in Central Greece [19]. Accordingly, the BSSE model is applied to estimate the 24 h average annual maxima in the center points of a 2 × 2 km grid of the study region, as shown in Figure 4. It is observed that both the Thessaly plain and the surrounding mountain ranges are strongly identified in the resulting rainfall patterns.

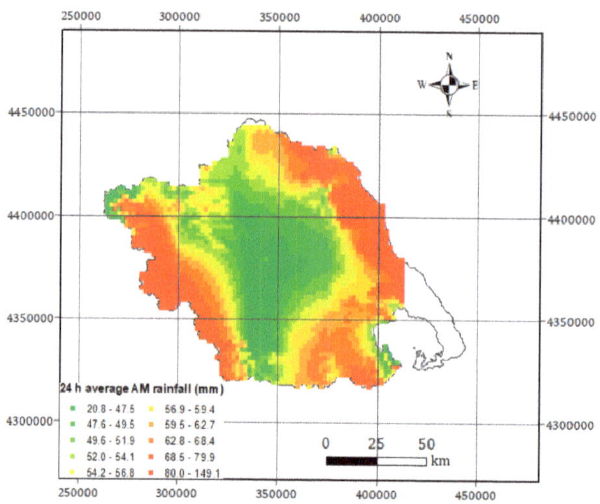

Figure 4. Spatial estimation of the 24 h average annual maximum (AM) rainfall (mm) by the BSSE model.

4.2. Construction of the Regional Ombrian Curves

To estimate a common value of the timescale parameters that would be representative of all stations, we optimize the fit for all stations by simultaneously minimizing the sum of all 14 variances as estimated from each station by Equation (17). This optimization results to parameters α = 0.03 and η = 0.64 which are considered representative for the whole region.

To estimate the distribution parameters, we first divide each annual maxima value at the 24 h scale by its modeled mean value as estimated by the BSSE model (Figure 4). To take space dependence into account, as explained in Section 2.5, we compute the spatial correlation of the 55 standardized annual maxima series at the 24 h scale. This is estimated to be $\rho = 0.17$ corresponding to $H = 0.61$ (Equation (27))—a moderate value. The standardized series are then unified into one for the estimation of the $b(T)$ parameters.

As explained above, we determine n_1 as the mean record length of the stations ($n_1 = 42$), following the rationale outlined in Section 2.5. We recall that n_1 is equal to the maximum moment order, which need not be modified for spatial dependence bias. Higher moment orders (corresponding to higher return periods) are adapted for spatial dependence. In the approach we follow herein, we choose to fit the model only up to the moment order not impacted by dependence bias. This set of 42 moments is to be used as a calibration set, while we also use the remaining higher 2305 moments as a validation set (Figure 5).

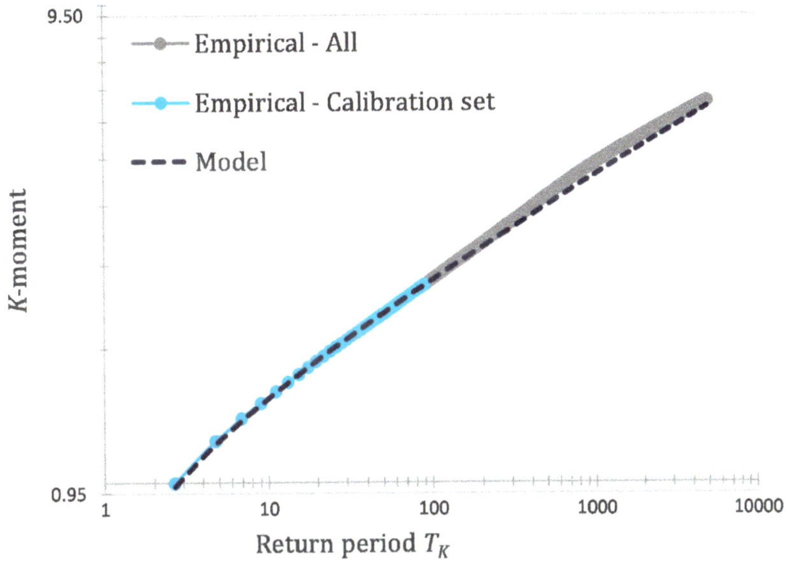

Figure 5. Empirical K-moments of the pooled standardized 24 h rainfall annual maxima sample, their theoretical values by the EV2 distribution and the corresponding return periods.

Using the method of the non-central K-moments, we fit the EV2 distribution to the unified sample of all standardized annual maxima at the 24 h scale minimizing the MAE between the empirical first 42 K-moments and the respective quantiles of the EV2 distribution. In Figure 5, it is seen that the fit is excellent for all 42 K-moments (MAE = 0.00489, RMSE = 0.00471), which constitute the calibration set, and there is also good agreement between the theoretical and empirical moments for higher orders, albeit some deviations in the area of higher return periods. Still, we have to note that very high orders are impacted by the spatial dependence structure of the data, whose estimation is in turn impacted by higher uncertainty. In any case, the use of higher return periods as a validation set proves that the fitting is robust.

Taking into account the results of the BSSE model and following the parameter transformations described in Section 2.5, the values of the four common parameters are derived as shown in Table 2, whereas the spatial distribution of the regionally varying λ parameter is shown in Figure 6. Note that the λ values are analogous to the average maxima values predicted by the BSSE model (Figure 4), as implied by Equation (30).

Table 2. Ombrian parameters α, η, ξ, β of Equations (6) and (12). Parameter λ is analytically derived at any point in space through the BSSE model.

α (h)	η (-)	ξ (-)	β (years)	λ (mm/h)
0.03	0.64	0.18	0.013	Geographic distribution shown in Figure 6 for all grid points.

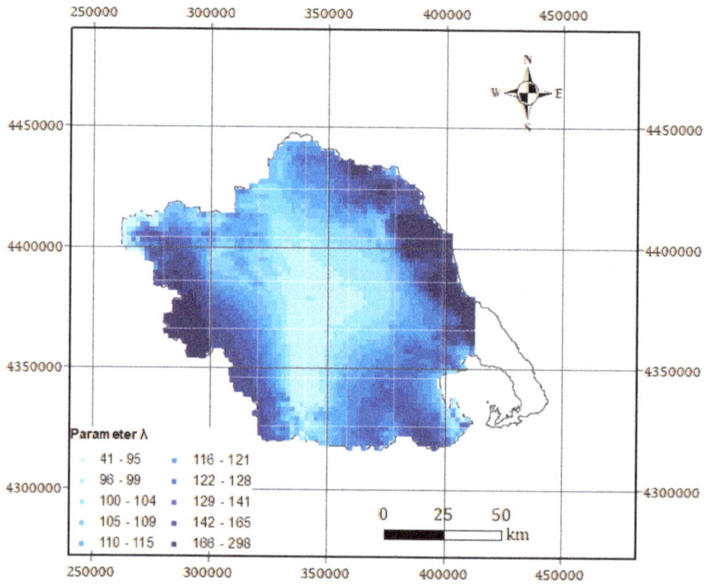

Figure 6. Spatial estimation of the scale parameter λ (mm/h).

4.3. At-Site Verification

To ensure that the spatial model is in agreement with the at-site empirical behavior, we compare the theoretical to the empirical curves as derived for various stations in characteristic locations of the WD. In Figure 7, we show the theoretical quantiles (as given by Equation (12)) of the rainfall intensities for the daily raingauges at six locations representative of the different rainfall regimes of the area, while in Figure 8, the same plots are shown for a wider range of scales that are available from the sub-daily rainfall stations. For the return period estimation of the empirical intensities, we also plot the estimates from the order statistics by Equation (26), in addition to the K-moments. Figures 7 and 8 show that the empirical distribution functions are generally in good agreement with the theoretical ones, with some notable yet non-systematic deviations in the area of higher return periods. The presence of measurement uncertainty is also evident in certain deviations, in the area of higher return periods, between the empirical intensities estimated from the daily raingauges and the sub-daily resolution gauges (Figure 8). The latter are, however, of shorter length compared to the daily gauges.

Taking into account the large sampling variability of rainfall and the spatial extent of the area, the results are deemed acceptable. Yet there are also a few cases in which the model does not capture well the single stations' behavior due to spatial differences between neighboring sites. Such an example is shown in Figure 9 for two stations in the Pertouli area, where it becomes apparent that the approach favors modeling the spatially average behavior between the two stations. It is obvious that in such cases of spatial uncertainty, a model perfectly capturing single stations' behaviors becomes less relevant; rather, the importance of robustness in the regional framework is highlighted.

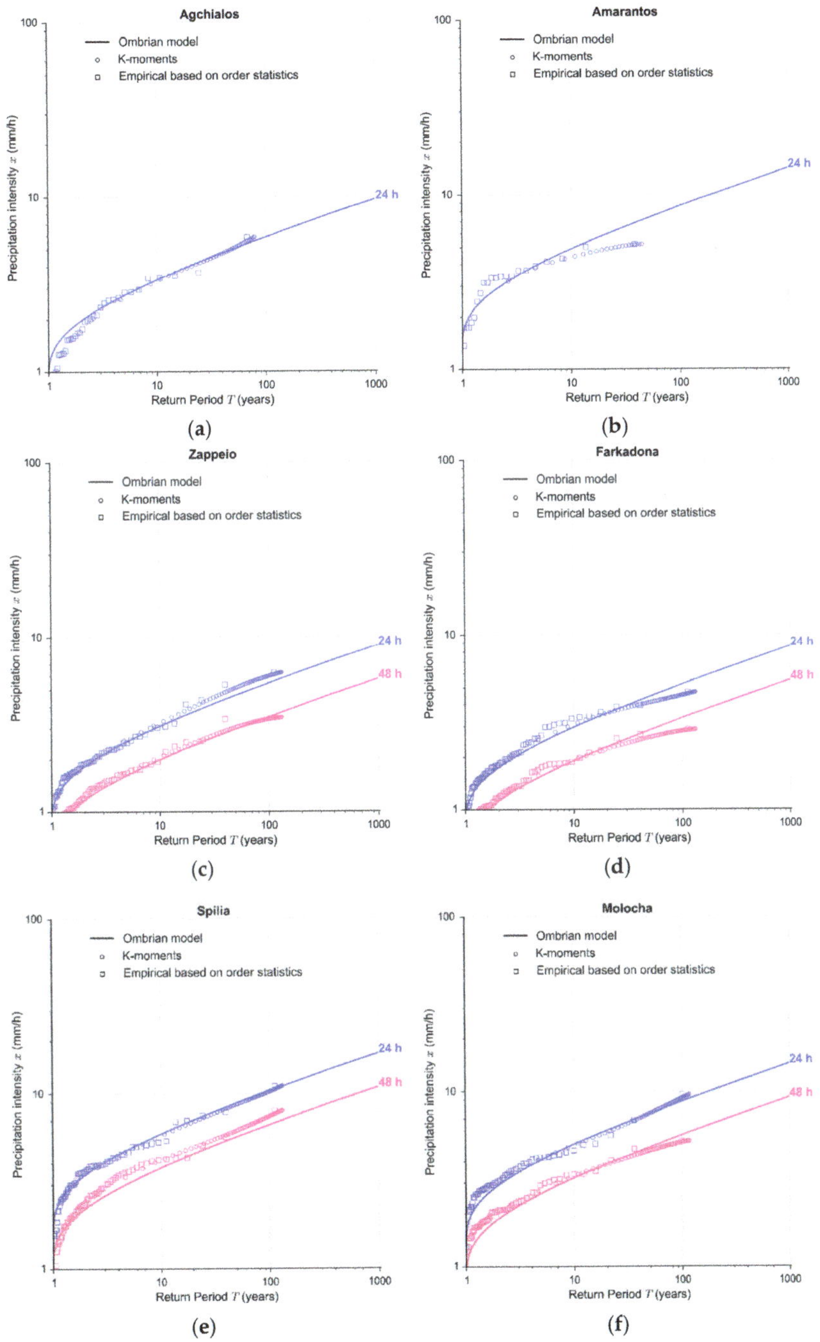

Figure 7. Theoretical and empirical distributions of 24 h and 48 h annual maximum intensities (depending on the available samples) in characteristic stations of Thessaly's WD: (**a**) Agchialos, (**b**) Amarantos, (**c**) Zappeio, (**d**) Farkadona, (**e**) Spilia and (**f**) Molocha. The empirical intensities plotted based on order statistics are also shown for validation.

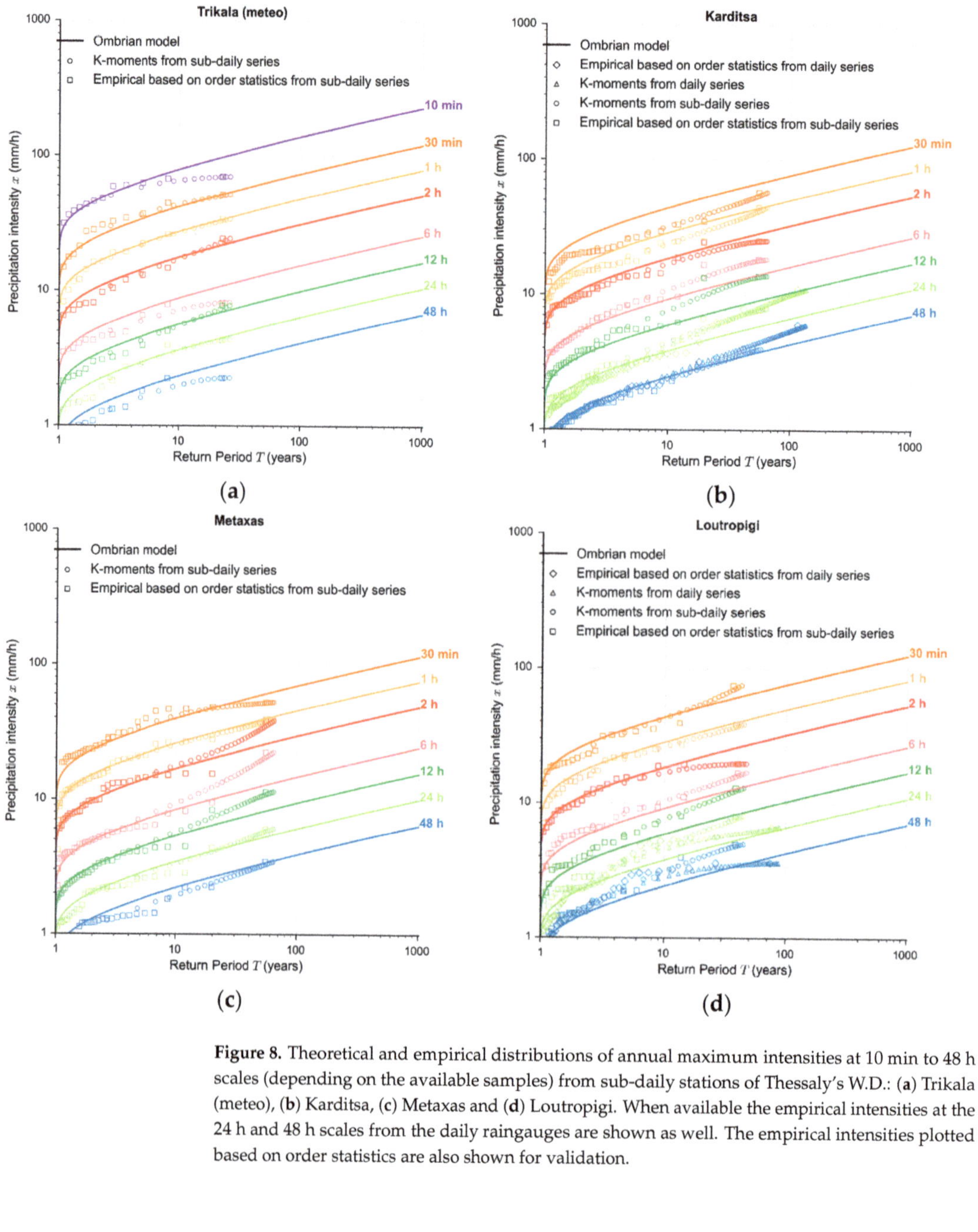

Figure 8. Theoretical and empirical distributions of annual maximum intensities at 10 min to 48 h scales (depending on the available samples) from sub-daily stations of Thessaly's W.D.: (**a**) Trikala (meteo), (**b**) Karditsa, (**c**) Metaxas and (**d**) Loutropigi. When available the empirical intensities at the 24 h and 48 h scales from the daily raingauges are shown as well. The empirical intensities plotted based on order statistics are also shown for validation.

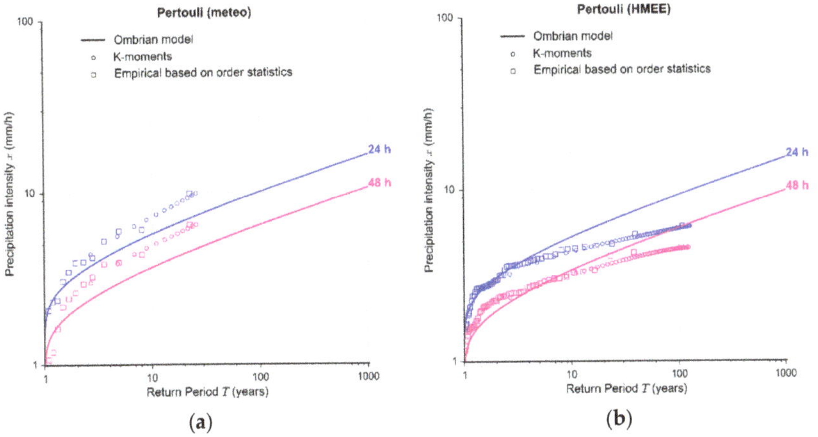

Figure 9. Example of the model fitting to two neighboring stations (from (**a**) meteo and (**b**) HMEE) at the Pertouli area with significant deviations in the stations' recordings.

5. Discussion and Conclusions

Ombrian curves have been around in hydrological engineering for approximately a century and are considered a standard task for at-site modeling. In this work, we address the more complex problem of constructing ombrian curves at regional scales that are of practical interest to the hydrologist in regional flood studies and in the common case that single-site data are either not available for the catchment of interest or the catchment is too large to characterize based on single-site data, e.g., [51]. Even more, constructing the curves by regional fitting is a powerful approach to limit estimation/sampling uncertainty resulting from short-length single records. As such, regional rainfall modeling has been an active research field in hydrological literature. The approach devised herein aims to create a framework for regional ombrian curves that incorporates recent advances in the field of regional frequency analysis and regionalization approaches within a theoretically consistent formulation of ombrian curves. The approach is tested in a challenging case study of the Thessaly WD in Greece, which shows a large variability of rainfall patterns stemming from its complex topography and large extent (~13,700 km^2).

The curves are constructed following the newly revisited mathematical formulation of single-site curves by Koutsoyiannis [2] coupled with a new regionalization approach that is developed herein. Four common parameters are identified, and one spatially varying scale parameter is employed. The site-specific scale parameter is explicitly modeled by a spatial smoothing model (BSSE) that employs elevation as an additional explanatory variable and produces a continuous 2D surface for the average 24 h annual maxima regime. This 2D surface suffices to model the spatial heterogeneity of the curves without involving cluster analysis for delineation of homogenous regions and avoiding related discontinuities and abrupt changes in the parameter space. The result is a model explicitly describing maximum rainfall at any point in a given space, which simplifies hydrological design. The BSSE model is selected for being more interpretable and less involved in parametric choices than common geostatistical software, while its robustness in cases of high spatial uncertainty has been documented [15]. We note that a map of the average maximum regime could also be used instead if already available. Still, the explicit incorporation of a surface smoothing model into the framework guarantees the consistency of the final spatial estimates to the point data.

The approach is also based on recent advances in the analysis of extremes, namely the use of reliable high-order moment estimators that account for the effect of the spatial dependence structure in assigning return periods (K-moments; [2]). This enables a rigorous fitting procedure aiming at minimizing the estimation uncertainty while respecting spatial dependence. To our knowledge, this is the first time that such a high number of moments (42 for estimation and 2305 for validation) is estimated from a sample of spatially correlated data with a provision for space dependence bias.

Results show that the model efficiently captures the spatial variability of extreme rainfall in the area covering scales from 5 min to 48 h, and its estimates are robust even under increased spatial uncertainty due to inconsistencies among the point data, which were present in a few cases.

A few modifications to the present approach would be required if one were to generalize the model over greater time scales, above the order of a few days [2]. While this task is not of direct use to flood estimation and typical applications, it is to be considered in view of a multi-purpose rainfall model at the regional scale. Further research is also required on the effect that spatial dependence exerts on the estimation of high return periods and the accompanying uncertainty bounds. This task is demanding as results are expected to depend on the assumed type and magnitude of the spatial dependence structure. Yet this research represents a first step toward significantly increasing the number of moments that can be justifiably employed in regional analyses of extremes.

Supplementary Materials: The following supporting information can be downloaded at: https://www.mdpi.com/article/10.3390/hydrology9050067/s1, Table S1: Properties of the daily raingauges (coordinates, elevation, source and record length) used in the analysis, Table S2: Properties (coordinates, elevation, source and record length) of the sub-daily raingauges (tipping-buckets) used in the analysis.

Author Contributions: Conceptualization, D.K. and T.I.; methodology, D.K., T.I., N.M.; software, T.I., N.M.; validation, T.I., N.M. and D.K.; formal analysis, T.I.; investigation, T.I.; resources, T.I.; data curation, T.I.; writing—original draft preparation, T.I.; writing—review and editing, T.I., N.M., D.K.; visualization, T.I.; supervision, D.K.; project administration, D.K. All authors have read and agreed to the published version of the manuscript.

Funding: T.I. and D.K. worked on, and were funded by, a related study for floods in Thessaly conducted by G.T.B. Anodos S.A., which included collection, preprocessing and preliminary analysis of data used in this paper.

Institutional Review Board Statement: Not applicable.

Informed Consent Statement: Not applicable.

Data Availability Statement: The rainfall data analyzed in this study were obtained, in the frame of a study for floods in Thessaly, from the Public Power Corporation of Greece, the Hellenic Ministry of Environment and Energy, the Hellenic Ministry of Agricultural Development and Food, the Hellenic Ministry of Agriculture, the Hellenic National Meteorological Service and the meteo network of the National Observatory of Athens. Requests to access these datasets should be directed to the respective parties. The SRTM elevation data used are publicly available from http://srtm.csi.cgiar.org (accessed on 27 November 2021).

Acknowledgments: We thank Nikos Mamassis, Antonis Koukouvinos and Andreas Efstratiadis for their help and discussions on preliminary analyses, and Paola Mazzoglio for her helpful feedback and suggestions on the manuscript. We also acknowledge comments by anonymous (Greek for nameless, unspeakable, inglorious or, in more modern terms, masked) reviewers on a previous version of the manuscript submitted elsewhere (cf. [52]) that motivated us to strengthen the paper against their criticism and highlight its contribution. We thank the Guest Editor Aristoteles Tegos for inviting us to submit this work to the Special Issue "Modern Developments in Flood Modelling". We are grateful to Alonso Pizarro and an anonymous reviewer of the current submission for encouraging comments and suggestions that improved the manuscript.

Conflicts of Interest: We declare no conflict of interest. We believe that anonymous reviewing may hide possible conflicts of interest and hence we favor eponymous reviewing.

References

1. Lesher, J.H. Saphêneia in Aristotle: "Clarity", "Precision", and "Knowledge". *Apeiron* **2010**, *43*, 143–156. [CrossRef]
2. Koutsoyiannis, D. *Stochastics of Hydroclimatic Extremes—A Cool Look at Risk*; Kallipos: Athens, Greece, 2021; 333p, ISBN 978-618-85370-0-2. Available online: https://repository.kallipos.gr/handle/11419/6522 (accessed on 1 April 2021).
3. Koutsoyiannis, D.; Iliopoulou, T. Ombrian curves advanced to stochastic modelling of rainfall intensity. In *Rainfall: Modeling, Measurement and Applications*; Elsevier: Amsterdam, The Netherlands, 2022; pp. 261–284. [CrossRef]
4. Iliopoulou, T.; Koutsoyiannis, D. PythOm: A python toolbox implementing recent advances in rainfall intensity (ombrian) curves. *Eur. Geosci. Union Gen. Assem.* **2021**, EGU21-389. [CrossRef]
5. Papalexiou, S.M.; Koutsoyiannis, D. Ombrian curves in a maximum entropy framework. *Eur. Geosci. Union Gen. Assem.* **2008**, *10*, 00702.
6. Sherman, C.W. Frequency and intensity of excessive rainfalls at Boston, Massachusetts. *Trans. Am. Soc. Civ. Eng.* **1931**, *95*, 951–960. [CrossRef]
7. Bernard, M.M. Formulas for rainfall intensities of long duration. *Trans. Am. Soc. Civ. Eng.* **1932**, *96*, 592–606. [CrossRef]
8. Koutsoyiannis, D.; Kozonis, D.; Manetas, A. A mathematical framework for studying rainfall intensity-duration-frequency relationships. *J. Hydrol.* **1998**, *206*, 118–135. [CrossRef]
9. Veneziano, D.; Furcolo, P. Multifractality of rainfall and scaling of intensity-duration-frequency curves. *Water Resour. Res.* **2002**, *38*, 42-1–42-12. [CrossRef]
10. Langousis, A.; Veneziano, D. Intensity-duration-frequency curves from scaling representations of rainfall. *Water Resour. Res.* **2007**, *43*, 1306. [CrossRef]
11. Svensson, C.; Jones, D.A. Review of rainfall frequency estimation methods. *J. Flood Risk Manag.* **2010**, *3*, 296–313. [CrossRef]
12. Hershfield, D.M. *Rainfall Frequency Atlas of the United States for Durations From 30 Minutes to 24 Hours and Return Periods From l to 100 Years*; US Weather Bureau Technical Paper 40; U.S. Weather Bureau: Washington, DC, USA, 1961; Volume 40, pp. 1–61.
13. Hogg, W.D.; Carr, D.A.; Routledge, B. *Rainfall Intensity-Duration Frequency Values for Canadian Locations*; Environment Canada, Atmospheric Environment Service: Downsview, ON, Canada, 1985.
14. Borga, M.; Vezzani, C.; Dalla Fontana, G. Regional rainfall depth–duration–frequency equations for an alpine region. *Nat. Hazards* **2005**, *36*, 221–235. [CrossRef]
15. Malamos, N.; Koutsoyiannis, D. Field survey and modelling of irrigation water quality indices in a Mediterranean island catchment: A comparison between spatial interpolation methods. *Hydrol. Sci. J.* **2018**, *63*, 1447–1467. [CrossRef]
16. Oliver, M.A.; Webster, R. A tutorial guide to geostatistics: Computing and modelling variograms and kriging. *Catena* **2014**, *113*, 56–69. [CrossRef]
17. Li, J.; Heap, A.D. Spatial interpolation methods applied in the environmental sciences: A review. *Environ. Model. Softw.* **2014**, *53*, 173–189. [CrossRef]
18. Malamos, N.; Koutsoyiannis, D. Bilinear surface smoothing for spatial interpolation with optional incorporation of an explanatory variable. Part 1: Theory. *Hydrol. Sci. J.* **2016**, *61*, 519–526. [CrossRef]
19. Malamos, N.; Koutsoyiannis, D. Bilinear surface smoothing for spatial interpolation with optional incorporation of an explanatory variable. Part 2: Application to synthesized and rainfall data. *Hydrol. Sci. J.* **2016**, *61*, 527–540. [CrossRef]
20. Hosking, J.R.M.; Wallis, J.R. *Regional Frequency Analysis: An Approach Based on L-Moments*; Cambridge University Press: Cambridge, UK, 1997. [CrossRef]
21. Hosking, J.R. L-moments: Analysis and estimation of distributions using linear combinations of order statistics, J.R. *Stat. Soc. Ser. B Stat. Methodol.* **1990**, *52*, 105–124. [CrossRef]
22. Hosking, J.R.M.; Wallis, J.R. The effect of intersite dependence on regional flood frequency analysis. *Water Resour. Res.* **1988**, *24*, 588–600. [CrossRef]
23. Koutsoyiannis, D. Knowable moments for high-order stochastic characterization and modelling of hydrological processes. *Hydrol. Sci. J.* **2019**, *64*, 19–33. [CrossRef]
24. Weinmann, P.E.; Nandakumar, N.; Siriwardena, L.; Mein, R.G.; Nathan, R.J. Estimation of rare design rainfalls for Victoria using the CRC-FORGE methodology. In *Water 99: Joint Congress; 25th Hydrology & Water Resources Symposium, 2nd International Conference on Water Resources & Environment Research; Handbook and Proceedings*; Institution of Engineers: Barton, Australia, 1999; p. 284.
25. Claps, P.; Ganora, D.; Mazzoglio, P. Rainfall regionalization techniques. In *Rainfall: Modeling, Measurement and Applications*; Elsevier: Amsterdam, The Netherlands, 2022; pp. 327–350. [CrossRef]
26. Hailegeorgis, T.T.; Thorolfsson, S.T.; Alfredsen, K. Regional frequency analysis of extreme precipitation with consideration of uncertainties to update IDF curves for the city of Trondheim. *J. Hydrol.* **2013**, *498*, 305–318. [CrossRef]
27. Aron, G.; Wall, D.J.; White, E.L.; Dunn, C.N. Regional rainfall intensity-duration-frequency curves for Pennsylvania 1. *JAWRA J. Am. Water Resour. Assoc.* **1987**, *23*, 479–485. [CrossRef]
28. Trefry, C.M.; Watkins, D.W., Jr.; Johnson, D. Regional rainfall frequency analysis for the state of Michigan. *J. Hydrol. Eng.* **2005**, *10*, 437–449. [CrossRef]
29. Dalrymple, T. *Flood-Frequency Analyses. Manual of Hydrology: Part 3. Flood-Flow Techniques*; USGPO: Washington, DC, USA, 1960; Volume 80. [CrossRef]
30. Burn, D.H. An appraisal of the "region of influence" approach to flood frequency analysis. *Hydrol. Sci. J.* **1990**, *35*, 149–165. [CrossRef]
31. Faulkner, D.S.; Prudhomme, C. Mapping an index of extreme rainfall across the UK. *Hydrol. Earth Syst. Sci.* **1998**, *2*, 183–194. [CrossRef]

32. Deidda, R.; Hellies, M.; Langousis, A. A critical analysis of the shortcomings in spatial frequency analysis of rainfall extremes based on homogeneous regions and a comparison with a hierarchical boundaryless approach. *Stoch. Environ. Res. Risk Assess.* **2021**, *35*, 2605–2628. [CrossRef]
33. Dimitriadis, P.; Koutsoyiannis, D.; Iliopoulou, T.; Papanicolaou, P. A global-scale investigation of stochastic similarities in marginal distribution and dependence structure of key hydrological-cycle processes. *Hydrology* **2021**, *8*, 59. [CrossRef]
34. Glynis, K.G.; Iliopoulou, T.; Dimitriadis, P.; Koutsoyiannis, D. Stochastic investigation of daily air temperature extremes from a global ground station network. *Stoch. Environ. Res. Risk Assess.* **2021**, *35*, 1585–1603. [CrossRef]
35. Koutsoyiannis, D.; Dimitriadis, P. Towards generic simulation for demanding stochastic processes. *Sci* **2021**, *3*, 34. [CrossRef]
36. Koutsoyiannis, D. An entropic-stochastic representation of rainfall intermittency: The origin of clustering and persistence. *Water Resour. Res.* **2006**, *42*, W01401. [CrossRef]
37. Koutsoyiannis, D. Statistics of extremes and estimation of extreme rainfall: II. Empirical investigation of long rainfall records/Statistiques de valeurs extrêmes et estimation de précipitations extrêmes: II. Recherche empirique sur de longues séries de précipitations. *Hydrol. Sci. J.* **2004**, *49*, 591–610. [CrossRef]
38. Koutsoyiannis, D.; Papalexiou, S.M. Extreme rainfall: Global perspective. In *Handbook of Applied Hydrology*; McGraw-Hill: New York, NY, USA, 2017; pp. 74.1–74.16.
39. Iliopoulou, T.; Koutsoyiannis, D.; Montanari, A. Characterizing and modeling seasonality in extreme rainfall. *Water Resour. Res.* **2018**, *54*, 6242–6258. [CrossRef]
40. Koutsoyiannis, D. Broken line smoothing: A simple method for interpolating and smoothing data series. *Environ. Model. Softw.* **2000**, *15*, 139–149. [CrossRef]
41. Malamos, N.; Koutsoyiannis, D. Broken line smoothing for data series interpolation by incorporating an explanatory variable with denser observations: Application to soil-water and rainfall data. *Hydrol. Sci. J.* **2015**, *60*, 468–481. [CrossRef]
42. Wahba, G.; Wendelberger, J. Some New Mathematical Methods for Variational Objective Analysis Using Splines and Cross Validation. *Mon. Weather. Rev.* **1980**, *108*, 1122–1143. [CrossRef]
43. Jarvis, A.; Reuter, H.I.; Nelson, A.; Guevara, E. Hole-Filled SRTM for the Globe Version 4. 2008. Available online: http://srtm.csi.cgiar.org (accessed on 27 November 2021).
44. Greenwood, J.A.; Landwehr, J.M.; Matalas, N.C.; Wallis, J.R. Probability weighted moments: Definition and relation to parameters of several distributions expressable in inverse form. *Water Resour. Res.* **1979**, *15*, 1049–1054. [CrossRef]
45. Mimikou, M.; Koutsoyiannis, D. Extreme floods in Greece: The case of 1994. In Proceedings of the US-ITALY Research Workshop on the Hydrometeorology, Impacts, and Management of Extreme Floods, Perugia, Italy, 13–17 November 1995.
46. Bathrellos, G.D.; Skilodimou, H.D.; Soukis, K.; Koskeridou, E. Temporal and spatial analysis of flood occurrences in the drainage basin of pinios river (thessaly, central greece). *Land* **2018**, *7*, 106. [CrossRef]
47. Koutsoyiannis, D.; Mamassis, N.; Efstratiadis, A.; Zarkadoulas, N.; Markonis, Y. *Floods in Greece, in Changes of Flood Risk in Europe*; Wallingford—International Association of Hydrological Sciences; IAHS Press: Wallingford, UK, 2012; pp. 238–256.
48. Loukas, A.; Vasiliades, L. Probabilistic analysis of drought spatiotemporal characteristics in Thessaly region, Greece. *Nat. Hazards Earth Syst. Sci.* **2004**, *4*, 719–731. [CrossRef]
49. Lagouvardos, K.; Kotroni, V.; Bezes, A.; Koletsis, I.; Kopania, T.; Lykoudis, S.; Mazarakis, N.; Papagiannaki, K.; Vougioukas, S. The automatic weather stations NOANN network of the National Observatory of Athens: Operation and database. *Geosci. Data J.* **2017**, *4*, 4–16. [CrossRef]
50. Hersfield, D.M.; Wilson, W.T. Generalizing of rainfall-intensity-frequency data. *AIHS. Gen. Ass. Tor.* **1957**, *1*, 499–506.
51. Pasculli, A.; Longo, R.; Sciarra, N.; Di Nucci, C. Surface Water Flow Balance of a River Basin Using a Shallow Water Approach and GPU Parallel Computing; Pescara River (Italy) as Test Case. *Water* **2022**, *14*, 234. [CrossRef]
52. Koutsoyiannis, D. An open letter to the Editor of Frontiers. *Researchgate* **2021**. [CrossRef]

hydrology MDPI

Article

Evaluation of Various Resolution DEMs in Flood Risk Assessment and Practical Rules for Flood Mapping in Data-Scarce Geospatial Areas: A Case Study in Thessaly, Greece

Nikolaos Xafoulis [1,*], Yiannis Kontos [2,*], Evangelia Farsirotou [1], Spyridon Kotsopoulos [3], Konstantinos Perifanos [4], Nikolaos Alamanis [5], Dimitrios Dedousis [6] and Konstantinos Katsifarakis [2]

1 Laboratory of Ecohydraulics & Inland Water Management, Department of Ichthyology and Aquatic Environment, University of Thessaly, 38446 Volos, Greece; efars@uth.gr
2 School of Civil Engineering, Aristotle University of Thessaloniki, 54124 Thessaloniki, Greece; klkats@civil.auth.gr
3 Department of Energy Systems, University of Thessaly, 41500 Larissa, Greece; kotsopoulos@uth.gr
4 Department of Language and Linguistics, National and Kapodistrian University of Athens, 10679 Athens, Greece; kperifanos@phil.uoa.gr
5 Department of Agrotechnology, University of Thessaly, 41500 Larissa, Greece; alam@uth.gr
6 Department of Civil Engineering, University of Thessaly, 38334 Volos, Greece; ddedousis@uth.gr
* Correspondence: nxafoulis@uth.gr (N.X.); ykontos@civil.auth.gr (Y.K.)

Abstract: Floods are lethal and destructive natural hazards. The Mediterranean, including Greece, has recently experienced many flood events (e.g., Medicanes Zorbas and Ianos), while climate change results in more frequent and intense flood events. Accurate flood mapping in river areas is crucial for flood risk assessment, planning mitigation measures, protecting existing infrastructure, and sustainable planning. The accuracy of results is affected by all simplifying assumptions concerning the conceptual and numerical model implemented and the quality of geospatial data used (Digital Terrain Models—DTMs). The current research investigates flood modelling sensitivity against geospatial data accuracy using the following DTM resolutions in a mountainous river sub-basin of Thessaly's Water District (Greece): (a) open 5 m and (b) 2 m data from Hellenic Cadastre (HC) and (c) 0.05 m data from an Unmanned Aerial Vehicle (UAV) topographical mission. RAS-Mapper and HEC-RAS are used for 1D (steady state) hydraulic simulation regarding a 1000-year return period. Results include flood maps and cross section-specific flow characteristics. They are analysed in a graphical flood map-based empirical fashion, whereas a statistical analysis based on the correlation matrix and a more sophisticated Machine Learning analysis based on the interpretation of nonlinear relationships between input–output variables support and particularise the conclusions in a quantifiable manner.

Keywords: hydraulic simulation; flood maps; digital elevation model; random forests; UAV mapping; DEM sensitivity; DEM errors; HEC-RAS; flood extent; flood risk assessment

Citation: Xafoulis, N.; Kontos, Y.; Farsirotou, E.; Kotsopoulos, S.; Perifanos, K.; Alamanis, N.; Dedousis, D.; Katsifarakis, K. Evaluation of Various Resolution DEMs in Flood Risk Assessment and Practical Rules for Flood Mapping in Data-Scarce Geospatial Areas: A Case Study in Thessaly, Greece. *Hydrology* **2023**, *10*, 91. https://doi.org/10.3390/hydrology10040091

Academic Editors: Aristoteles Tegos, Alexandros Ziogas and Vasilis Bellos

Received: 20 March 2023
Revised: 5 April 2023
Accepted: 9 April 2023
Published: 12 April 2023

1. Introduction

Floods are natural disasters that can have severe impacts on human lives, infrastructure, and the environment [1,2]. Floods can occur due to various reasons, such as heavy rainfall, river overflow, coastal storm surges, or tsunamis. The response of mountain basins to intense rainfall is rapid, mainly due to large slopes, while precipitation is spatially and temporally variable [3]. Mountain basin floods are often flashy [4], allowing limited time for warnings. Flash floods usually occur in mountain river catchments draining less than 1000 km^2 [5]. They constitute a common, extremely dangerous natural hazard and they are responsible for many deaths [6,7]. Their impacts on various socioeconomic activities are extremely diverse [8,9]. Around 40% of flood-related deaths in Europe between 1950 and 2006 are linked to flash floods [10]; still, there is a lack of relevant data, especially reliable discharge estimates [4]. The Mediterranean region is one of the most flood-prone areas in the world due to its unique geographic, climatic, and environmental conditions, with floods occurring on average every two years; floods are one of the most lethal and destructive natural hazards [11] there. Greece is also affected by floods that are mainly caused by heavy rainfall (e.g., flood in the city of Karditsa caused by Medicane Ianos [12]), river overflow, and flash floods (e.g., flood in Mandra with 24 fatalities caused by Medicane Numa-Zenon [13]). Greece has also experienced many flood events during the last decades [7].

One of the most effective ways to reduce the impacts of floods is to develop and implement flood management plans that include prevention, preparedness, response, and recovery measures. Flood mapping is a crucial tool for flood management planning as it enables the a priori identification of flood-prone areas, and the estimation of flood extent, flow depth and characteristics, and flood frequency [14]. There are several challenges in modelling floods using informed modelling, such as data quality issues, uncertainty in input parameters, and the need for improved model calibration. Flood model parameters are of grey-box nature and their global use is not suggested but rather should be carefully adopted [13]. Moreover, advanced modelling approaches supported by detailed spatial information are not always the answer. They are extremely computationally and data-greedy in order to overcome uncertainties [15]. Most of the time, a compromise between simulation accuracy and time defines the simulation scheme/model used. Hence, 1D flood modelling can be implemented, especially in data-scarce areas. In such cases, open real-world datasets can improve flood modelling [12]. The use of forensic hydrology, reconstructing flood events through field observations, hydrological and hydraulic modelling, and geomorphological analysis, is an established method to overcome the lack of data and provide valuable insights into past events and inform future flood risk management strategies [12].

An essential tool for successful flood risk management is accurate flood mapping. This requires the use of high resolution and accurate Digital Elevation Models (DEMs) [16]. High resolution does not always guarantee DEM accuracy, especially when dense vegetation and canopy are involved and the mapping is based on orthophotos. In such cases, the accuracy also depends on the vegetation filtering techniques used.

Current research investigates flood modelling sensitivity against geospatial data accuracy, in a case study concerning a part of the mountainous Enipeas river basin of Thessaly's Water District (Greece). The methodology that is implemented in the current research is graphically presented step-by-step in Figure 1. In particular, the following DEMs for the study area concerning flood modelling (flood area) are tested: (a) open 5 m resolution DEM data (DEM_5 m) from the Hellenic Cadastre (HC; [17]), (b) open 2 m resolution DEM data (DEM_2 m) from the HC, and (c) sub-meter (0.05 m) resolution DEM data (DEM_0.05 m) from a research team's own designated Unmanned Aerial Vehicle (UAV) topographical mission. The US Army Corps of Engineers' software [18] RAS-Mapper [19] and HEC-RAS 1D [20] are used for 1D (steady state) hydraulic simulations (Sims) regarding a 1000-year return period for the three different DEMs (DEM_5 m = Sim 1; DEM_2 m = Sim 2; DEM_0.05 m = Sim 3). Results include 2D flood maps graphically presenting spatial flood extents and flow depths, as well as flow characteristics for every cross-section of the hydrographic network, such as Froude number, hydraulic radius, and flood extent. In the absence of an ideal terrestrial mapping mission using land-based topographical instruments of the studied hydrographic network, DEM_0.05 m and the resulting hydraulic simulation (Sim 3) are assumed to be the "ground truth". Thus, the investigation focuses on the comparison of the results of the two open data-based simulations, Sim 1 and Sim 2, against the closer to the truth Sim 3 results.

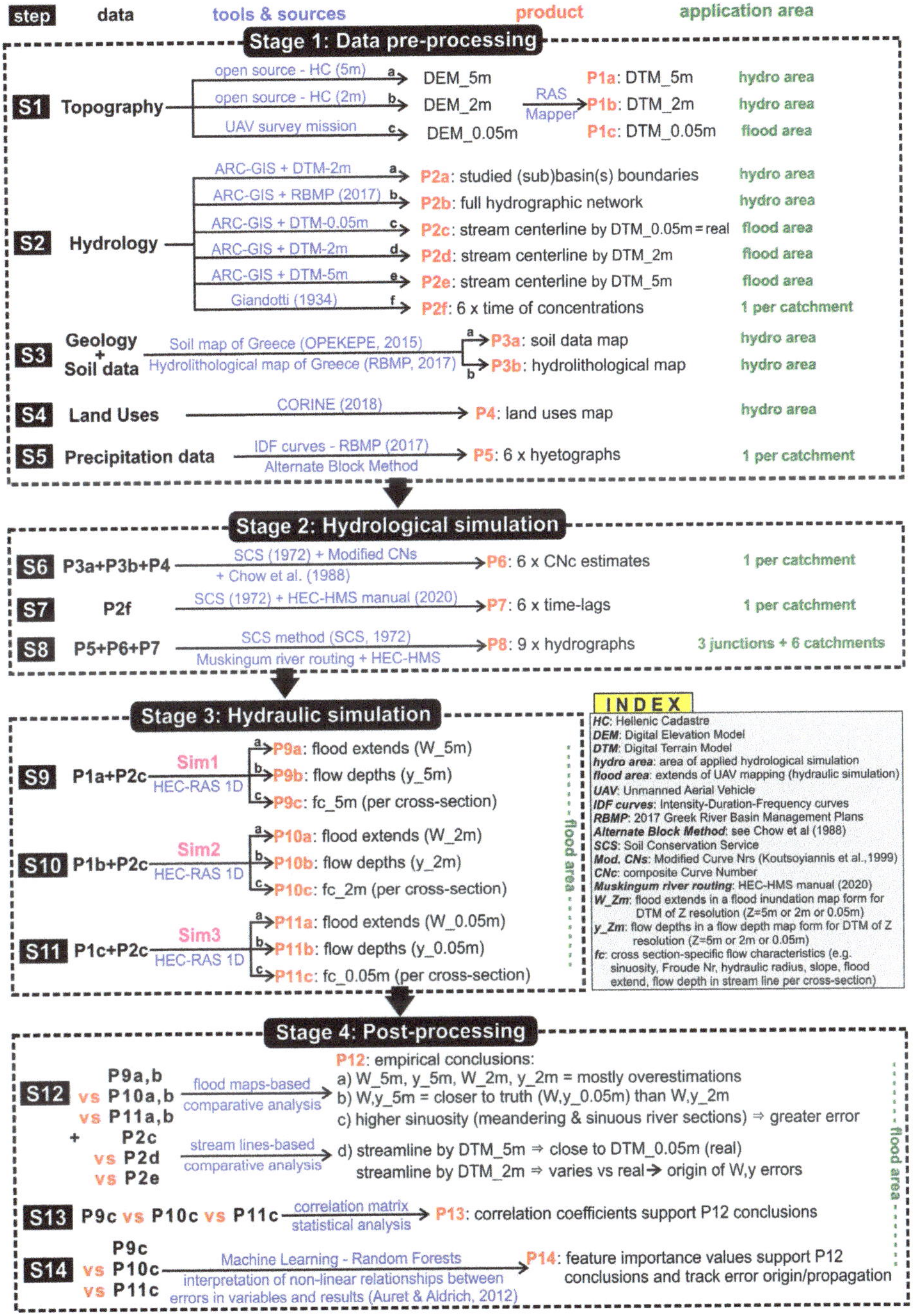

Figure 1. Graphical abstract of the methodology implemented to investigate the sensitivity of flood risk mapping via 1D hydraulic simulations vs. various DEM resolutions. Aim: conclude on study area features that render the use of more accurate but costly and time-consuming UAV mapping imperative and decide on the next best free alternative option in Greek reality [29,42].

The flood modelling results are analysed in a graphical flood map-based empirical fashion, whereas a statistical analysis, based on the correlation matrix, and a more sophisticated Machine Learning (ML) analysis, support and particularise the conclusions in a quantifiable manner; the ML-assisted analysis is based on the interpretation of the non-linear relationships between input–output variables (i.e., DEM, sinuosity, slope vs. flow extents, flow depths, flow characteristics). The goal is to track the errors in the simulation results and trace them back to the initial features that generated them, in relation to the selected DEM; this way, one can conclude on the features that render the use of the more accurate, but costly and time-consuming, UAV mapping imperative, while deciding on the next best free alternative DEM (5 m or 2 m) option in Greek reality.

The practical aim of this paper is to produce practical rules for optimal hydraulic simulation of a river basin, in terms of minimization of in situ topographical mapping costs without compromising the hydraulic simulation accuracy. This way, one can decide on which hydrographic network sections (if any) of any river basin demand UAV or other accurate mapping or not and what the free alternative is. This requires a transparent and detailed presentation of the implemented and proposed methodology so that the conclusions can be generalized.

2. Materials and Methods

This section presents the study area and all stages of the implemented methodology, with their discrete steps, as presented in the graphical abstract (Figure 1), mentioning all data sources and methods along the way. Stage 1 (see Section 2.2) refers to data pre-processing, including procurement and processing/manipulation of topography, hydrology, geology, soil, land uses, and precipitation data. Stage 2 (see Section 2.3) refers to hydrological simulations, including calculation of hydrological parameters and production of hydrographs, required as input for the hydraulic simulations included in Stage 3. Stage 3 (see Section 2.4) produces 2D flood maps and cross-section-specific results that are analysed in Stage 4 (see Section 2.5) via three different approaches: (a) empirical, (b) statistical, and (c) interpretation of non-linear relationships using Machine Learning.

2.1. Study Area

The Thessaly Water District (EL08; [21]) includes two main river basins: (a) the Pinios river basin and (b) the Almyros-Pilion basin. The study area is located in the district's southern section, being part of the Pinios basin. It specifically lies in the north-west of Mount Othris, being a part of the mountainous river basin of Enipeas river (code GR00080004002203). Figure 2 presents the location of the study area, and the wider study area (where the hydrological simulations are conducted; shortly referred to as "hydro area") as a part of the Thessaly Water District and the Enipeas river basin. Figure 3 presents the specific study area where the hydraulic simulations are conducted (shortly referred to as "flood area"), as a part of the wider "hydro area".

2.2. Data Pre-Processing (Methodology Stage 1)

2.2.1. Topographic Data (Step 1)

The geospatial data utilised in this research come from HC [17] and a private UAV mission. DEM_5 m is actually the "Digital Elevation Model-DEM-LSO (5 m)" dataset series, as presented by HC [22], which *"is a 5 m pixel size grid compilation (1:5000 cadastral tile distribution), deriving from the Large Scale Orthophotos project. It is a homogenous systematic point grid which refers to terrain elevation and creates an Earth Elevation Model"*. RAS-Mapper [19] is used to convert DEM_5 m into DTM_5 m (research product P1a; see Figure 1); the respective map is provided as Appendix A File SM1 (see Appendix A for details).

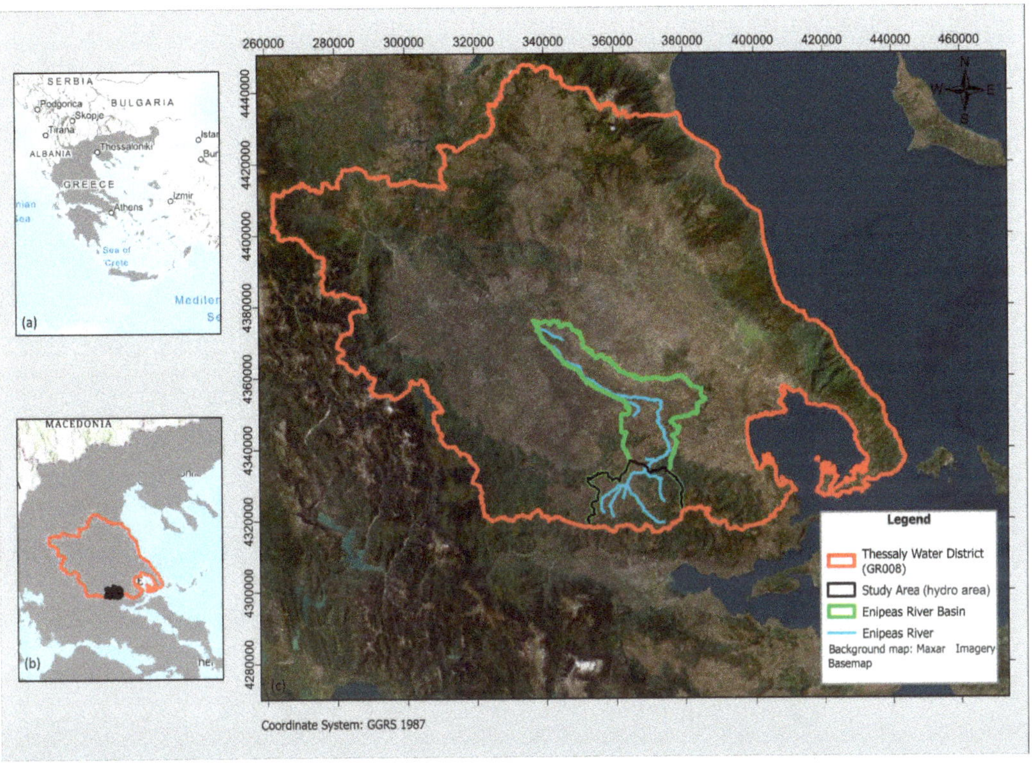

Figure 2. (**a**) Map of Greece, (**b**) Thessaly Water District, and (**c**) Enipeas river basin and wider studied catchment (hydro area) with the studied hydrographic network.

DEM_2 m is actually the "Digital Elevation Model-DEM-LSO25" dataset series as presented by HC [23], which "is a 2 m pixel size grid compilation (1:2500 cadastral tile distribution), for the entire country from airphotos taken between 2014 and 2016, deriving from the Large Scale Orthophotos 25 cm (LSO25) project. It is a homogenous systematic point grid which refers to terrain elevation and creates an Earth Elevation Model". RAS-Mapper [19] is used to convert DEM_2 m into DTM_2 m (P1b_flood for "flood area" and P1b_hydro for "hydro area"; see SM2a and SM2b, respectively).

DEM_0.05 m is produced by the research team's own designated UAV topographical mission. Structure-from-Motion (SfM) photogrammetry using photographs obtained by UAVs is increasingly being utilised for producing high resolution DEMs. The UAV used is WingtraOne GEN II. The flight took place on 3 March 2022 and lasted about 20 min to survey the "flood area" of approximately 1.05 km^2. The DEMs are interpolated from point clouds that represent entire landscapes, including terrain, vegetation, and infrastructure [24]. In the current research, the vegetation filtering is conducted with the standard method of Agisoft Metashape software application [25]. RAS-Mapper [19] is used to convert DEM_0.05 m into DTM_0.05 m (P1c; see SM3). DTM_0.05 is contained within the boundaries of the UAV mapping (DEM_0.05 m) where the hydraulic simulations are conducted (flood area), as presented in Figure 3.

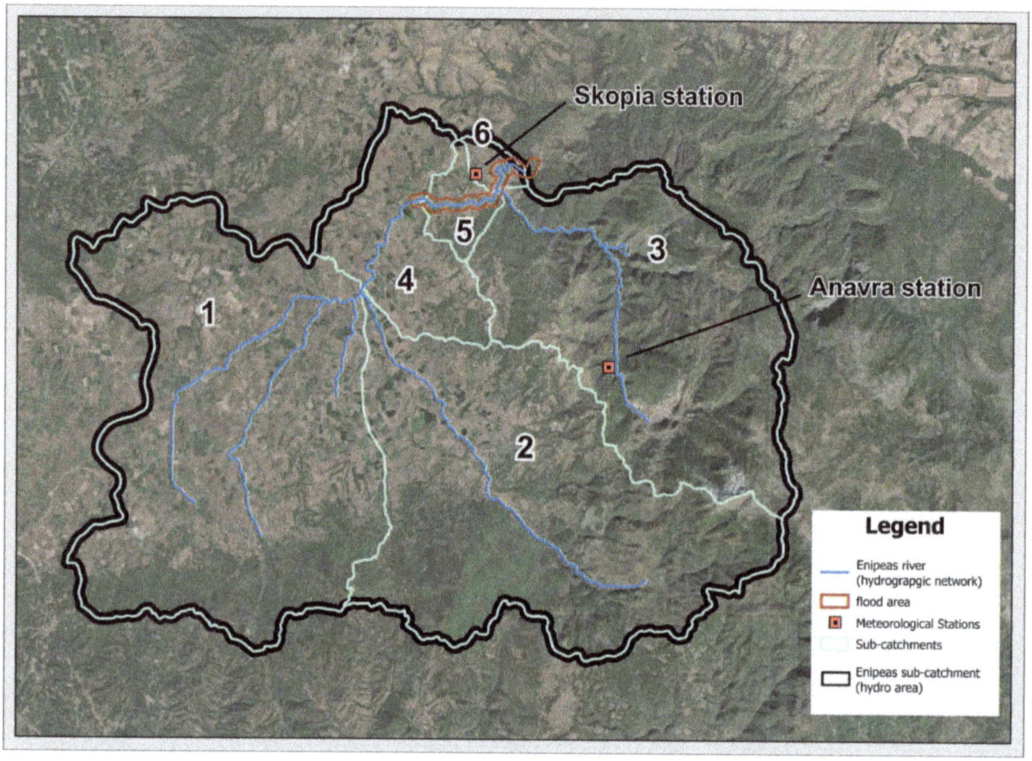

Figure 3. Study area of Enipeas river basin where the hydraulic simulations are conducted (flood area) as a part of the wider study area where the hydrological simulations are conducted (hydro area), together with the hydrographic network, the sub-catchments, and available meteorological stations.

2.2.2. Hydrology-Related Data and Calculations (Step 2)

Using the software ArcGIS Pro v3.1.0 [26], the "hydro" studied area is divided into sub-basins/catchments (P2a; see Figure 3 and SM4) based on DTM_2 m in conjunction with orthophotos derived from HC [23] and satellite images. The finer of the two resolutions available in the wider "hydro" study area DTM_2 m is used, as it allows for a finer representation of the low- and very low-slope areas of the study area, as well as various technical structures such as embankments. Figure 4 presents the "hydro" area, together with the hydrographic network. The latter derives from the River Basin Management Plans of the Water District of Thessaly [27], in conformation with the Water Framework Directive (2000/60/EC; [28]) and comprises the main stream sections of Enipeas hydrographic network (P2b; see SM5).

A more accurate stream centerline (part of the hydrographic network in the "flood" area) is also produced in this step (P2c and P2c_DTM; see SM6a,b), derived from the most accurate DTM_0.05 m. This will be used for the hydraulic simulation. Moreover, although not needed for the simulations, the respective stream centerlines derived from DTM_5 m (P2d and P2d_DTM; see SM7a,b) and DTM_2 m (P2e and P2e_DTM; see SM8a,b) are produced to be used for comparison and deduction of conclusions in the last section of the paper. All stream lines are produced using ArcGIS Pro. All of the geomorphological characteristics of the sub-catchments of the "hydro" area are presented in Table 1. Step 2 ultimately features use of the Giandotti methodology [29] for the calculation of the six concentration times (Table 1), one per sub-catchment (P2f; see SM9).

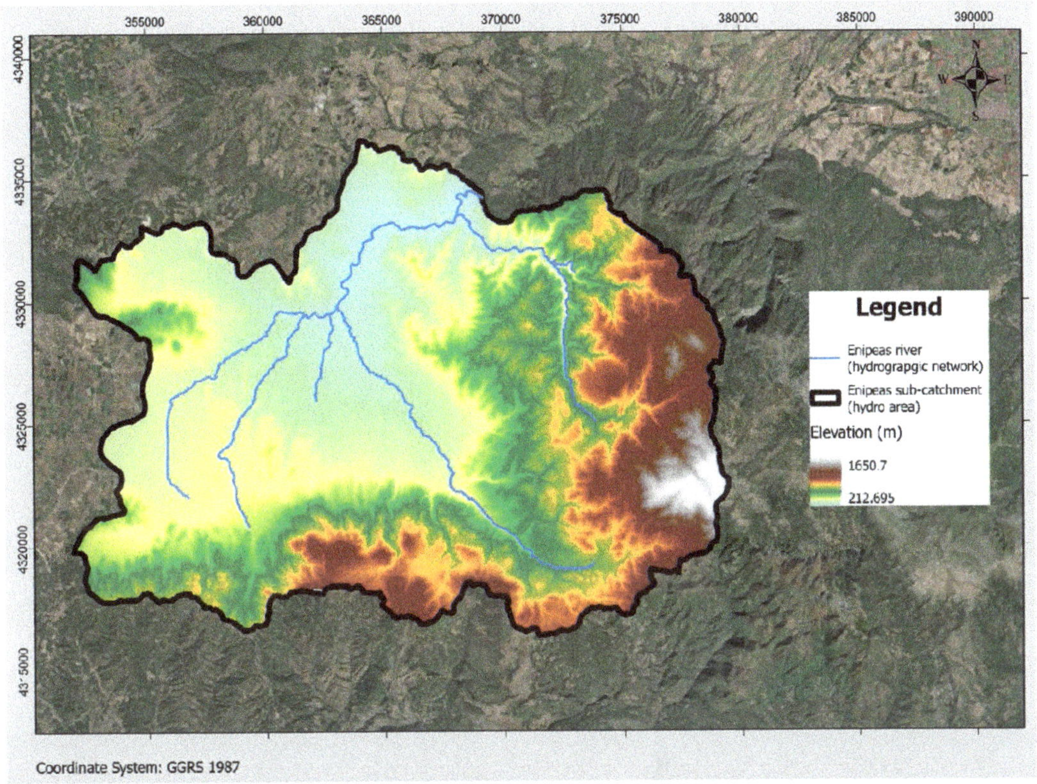

Figure 4. Digital Terrain Model of 2 m × 2 m resolution (DTM_2 m) of the study area ("hydro") used for the delineation of the sub-catchments and hydrographic network (see SM2a and SM5).

Table 1. The geomorphological features of all sub-catchment areas, together with the respective concentration times, lag times, and Areal Reduction factor (ARF).

Sub-Catchment	Area (km^2)	Length of Main River (km)	Mean Sub. Elevation (m)	Outlet Elevation (m)	Time of Concentration (h)	Lag Time (h)	Areal Reduction Factor	CNc
Sub 1	141.21	14.339	538.95	375.00	6.74	4.04	0.89	71
Sub 2	121.68	17.867	731.95	375.00	4.69	2.81	0.88	71
Sub 3	96.26	13.416	842.28	344.73	3.33	2.00	0.87	55
Sub 4	30.30	8.362	426.47	360.23	5.31	3.19	0.92	74
Sub 5	6.72	3.156	416.9	344.72	2.22	1.33	0.93	63
Sub 6	3.19	2.287	397.29	327.23	1.58	0.95	0.94	55

2.2.3. Geological and Soil Data (Step 3)

The most reliable sources for geological/soil data and, consequently, hydrolithological data are the European Soil Data Center (ESDAC) [30–33], the Soil Map of Greece by the Greek Payment Authority of Common Agricultural Policy (OPEKEPE; [34]), and the River Basin Management Plan for the Water District of Thessaly (RBMP-EL08) [27]. The main source of the soil data is OPEKEPE; the available separate soil map tiles are scanned and georeferenced on the "hydro" area using ArcGIS Pro (P3a; see SM10). The soil map does not cover the full extent of the study area. The missing data are drawn by the hydrolithological map provided by RBMP-EL08 [27]. The available data from ESDAC successfully validate the other sources. The soil types of OPEKEPE are linked to the respective hydromorphy and are categorised into classes (A: high; B: moderate; C: low; D and E: very low infiltration rates), whereas the RBMP map also categorises the soil types concerning the hydrolithological characteristics. The resulting merged hydrolithological map is presented in Figure 5 (P3b; see SM11).

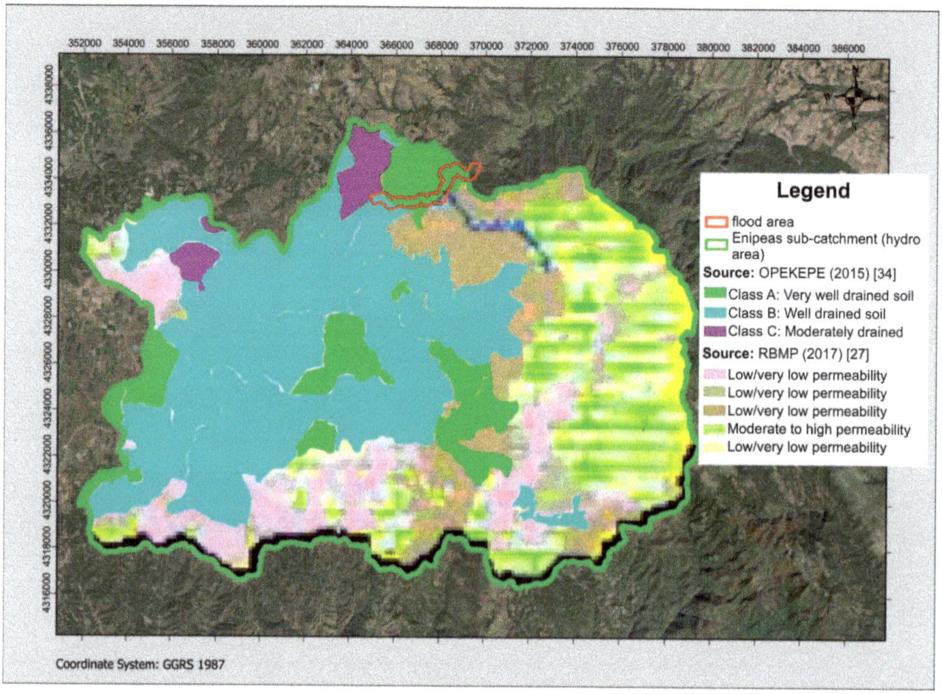

Figure 5. Hydrolithological map of the study area by merged OPEKEPE [34] and RBMP [27] data (see SM11).

2.2.4. Land Use Data (Step 4)

The determination of the land cover was based on CORINE land cover data [35]. The land use map is presented in Figure 6 (P4; see SM12). All information regarding CORINE land cover classes and the respective land cover area per sub-catchment are presented in SM14 (P6).

2.2.5. Precipitation Data Hyetograph Production (Step 5)

The only available meteorological stations in the study area (hydro) are those located at Anavra and Skopia, as presented in Figure 3. In order to be on the safe side and investigate the worst-case scenario regarding rainfall intensity, the higher-elevation Anavra station is selected; it always provides greater precipitation heights. Under these data-scarce conditions, following the methodology by Koutsoyiannis et al. [36], the Intensity Duration Frequency (IDF) curve was designed, using the proposed equation:

$$\mathrm{i(t,T)} = \frac{\lambda' \cdot \left(\mathbf{T}^{\kappa} - \psi'\right)}{\left(1 + \frac{t}{\theta}\right)^{n}}, \quad (1)$$

where i is the max point rainfall intensity of duration t for a return period of T; θ and η are parameters to be estimated, with $\theta \geq 0$ (in time units) and $0 < \eta < 1$; $\kappa > 0$ is the shape parameter; $\lambda' > 0$ is the scale parameter; and ψ' is the location parameter.

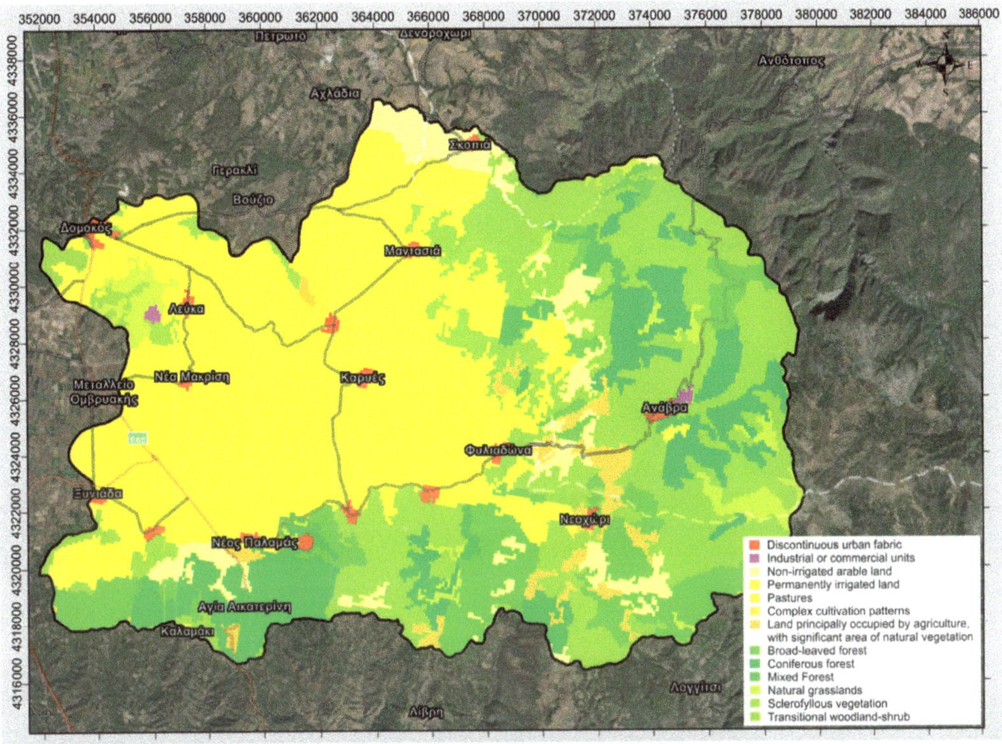

Figure 6. Land use map derived from CORINE [35] (see SM12).

The parameters proposed by the IDF Report v4 [37] (supporting documents of [27]) for the Anavra station and for a return period of 1000 years are presented in Table 2. The resulting function and IDF curve are presented in Figure 7.

Table 2. Parameters used in the IDF curve equation [36], regarding the Anavra station and a 1000-year return period.

Water District Code	Station ID	Station Name	X (m)	Υ (m)	Z (m)	κ	λ′	ψ′	θ	n
GR08	355	Anavra	372,326.71	4,327,100.77	208	0.092	592.3	0.768	0.042	0.639

Next, the point rainfall intensity is transformed to areal rainfall intensity using the respective Areal Reduction Factor (ARF) per sub-catchment, calculated by [36,37]:

$$\varphi = \max\left(1 - \frac{0.048 \cdot \mathrm{A}^{0.36-0.01\cdot\ln \mathrm{A}}}{\mathrm{d}^{0.35}}, 0.25\right), \tag{2}$$

where A is the river basin area (km^2) and d is the rainfall duration (h).

The ARF values per sub-catchment are presented in Table 1 (sources: [38–41]). The respective design hyetographs per sub-catchment are produced (P5; see SM13) based on the IDF curve using the Alternate Block Method [42,43].

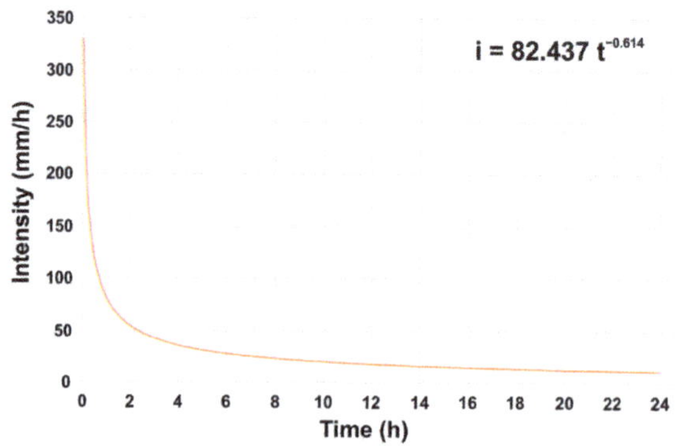

Figure 7. The IDF function and curve (point rainfall intensity vs. time) for the Anavra station for a return period of 1000 years.

2.3. Hydrological Simulation (Methodology Stage 2)

All current hydrological simulations are conducted using the Hydrologic Engineering Center—Hydrologic Modelling System (HEC-HMS), developed by the US Army Corps of Engineers [18]. HEC-HMS is a widely used and established tool that can simulate all hydrological processes of a watershed, including precipitation, infiltration, evaporation, snowmelt, and runoff. It is used for the design and management of water resource infrastructure, such as dams, reservoirs, and water supply systems. The transformation of precipitation into runoff for every sub-catchment was conducted using the Soil Conservation Service—Curve Number (SCS-CN) [44] unit hydrograph method, whereas losses are estimated using the SCS-CN model.

2.3.1. Curve Numbers (Step 6)

SCS-CN is a simple, widely used, and efficient method for determining the amount of runoff from rainfall, even in a particular area. A Curve Number (CN) [42] expresses the percentage of precipitation that will runoff as a function of the area's hydrologic soil group, land use, treatment, and hydrologic condition. Based on Chow et al. (1988; [42]), Koutsoyiannis and Xanthopoulos (1999; [45]) provided updated CN values, used in current research. There are various land uses in every sub-catchment, hence the weighted Curve Number (composite) value CNc is calculated [46]. The estimation process is presented in SM14, whereas the calculated CNc values are presented in Table 1.

2.3.2. Lag Time Estimation (Step 7)

Following the SCS methodology [44] and the HEC-HMS Technical Reference manual [47], the lag times for each sub-catchment are estimated (Table 1). Lag time refers to the delay between the occurrence of rainfall and the peak discharge of a river. It depends on size and shape of the catchment, soil type, and vegetation cover.

2.3.3. Hydrograph Production (Step 8)

Following the SCS methodology [44], hydrographs are produced (P8; see SM15) using HEC-HMS, applying the well-established Muskingum river routing method to all reaches. Parameters such as reach length and slope are estimated using topographical data, namely DTM_2 m. The nine hydrographs refer to the three junctions and six sub-catchments created by the studied Enipeas basin (hydro area) model, as presented in Figure 8.

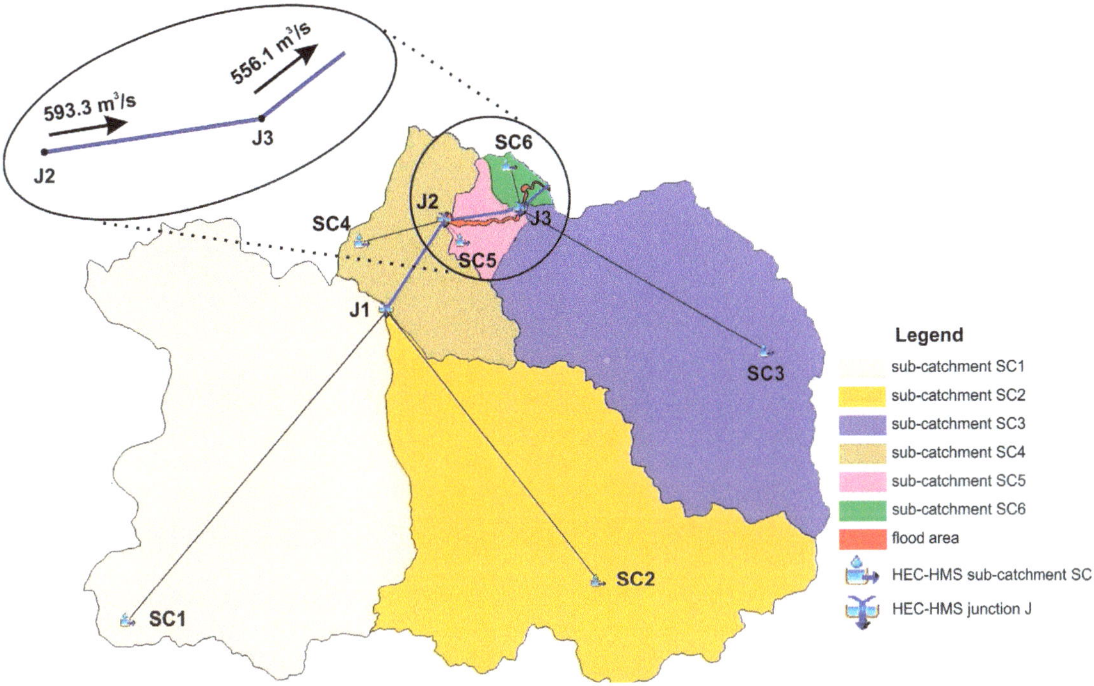

Figure 8. HEC-HMS hydrological model of the studied (hydro) area, featuring its six sub-catchments and three junctions, with detail of the calculated design discharges used in HEC-RAS.

2.4. Hydraulic Simulation (Methodology Stage 3; Steps 9, 10, 11)

All current hydraulic simulations are conducted using the Hydrologic Engineering Center—River Analysis System (HEC-RAS 1D) developed by the US Army Corps of Engineers [48]. It is a powerful tool for hydraulic modelling, based on a 1D numerical system, able to simulate steady/unsteady flow conditions. HEC-RAS can be used for a variety of hydraulic analyses, including floodplain mapping, bridge and culvert design, dam safety evaluations, sediment transport studies, and water quality modelling. The model solves equations for conservation of mass and momentum to calculate water surface elevation, flow velocity, and other hydraulic characteristics. It exhibits a user-friendly interface and features tools such as RAS-Mapper [19], supporting a range of data input formats, including DEMs, topographic maps, and surveyed cross-sections and post-processing utilities for displaying and analysing simulation results.

Although there is only one shared hydrological simulation applied in the wider "hydro" area, there are actually three hydraulic simulations, namely 1D implementations of the HEC-RAS 1D model, corresponding to the three input DTMs applied in the "flood" area: (a) Sim 1 uses DTM_5 m, (b) Sim 2 uses DTM_2 m, and (c) Sim 3 uses DTM_0.05 m. All three Sims share the river centerlines (Step 2; products P2c, P2d, P2e), as well as the bank lines, produced by DTM_0.05 m, which is assumed to be the "ground truth". Sims 1–3 also share 270 cross-sections positioned in an interval of approximately 20 m along the river centerline (see SM23). For all Sims, two Manning coefficient values are used, one for the main channel (n = 0.08) and a different one for the overbanks (n = 0.07). The values were decided after observations of satellite images and orthophotos and in situ inspection, based on Chow et al. (1988; [42]) and the HEC-RAS hydraulic reference manual [48]. The values were adjusted, as [42] suggests, based on river irregularities, variation in channel cross-section, obstructions, vegetation, and meandering. The contraction and expansion coefficients for all Sims are assumed to be equal to 0.1 and 0.3, respectively [48]. The maximum flow (sic.) values in the respective HEC-HMS-produced hydrographs of junctions J2 (539.3 m^3/s; see SM15, page 1) and J3 (556.1 m^3/s; see SM15, page 3) are used as steady flow discharges in all hydraulic simulations (Figure 8). Concerning upstream and downstream boundary conditions, the energy grade line slope values are assumed to be equal to the riverbed slope values [48].

Step 9 in Stage 2 of the applied methodology features Sim 1, namely HEC-RAS 1D simulation of the hydraulic model based on DTM_5 m, producing the respective W_5 m flood extent map (P9a; see SM16), the respective y_5 m flow depths map (P9b; see SM17), and calculating the flow characteristics. The fc_5 m flow characteristics selected to be investigated for their involvement in the propagation of errors generated by the lower resolution DEMs are presented per cross-section in SM18, including meter mark, hydraulic radius, Froude number, and flood extent.

Step 10 in Stage 2 of the applied methodology features Sim 2, namely HEC-RAS 1D simulation of the hydraulic model based on DTM_2 m, producing the respective W_2 m flood extent map (P10a; see SM19) and the respective y_2 m flow depths map (P10b; see SM20), and calculating the fc_2 m flow characteristics (P10c; see SM18).

Step 11 in Stage 2 of the applied methodology features Sim 3, namely HEC-RAS 1D simulation of the hydraulic model based on DTM_0.05 m produces the respective W_0.05 m flood extent map (P11a; see SM21) and the y_0.05 m flow depths map (P11b; see SM22), and calculating the fc_0.05 m flow characteristics (P11c; see SM18).

2.5. Post-Processing and Analysis (Methodology Stage 4)

Stage 4 features the post-processing of the results of the three hydraulic simulations (Sim 1–3) via (a) empirical (Step 12) and (b) statistical (Step 13) analysis as well as (c) use of Machine Learning for the interpretation of non-linear relationships of variables vs. results (Step 14).

The production of conclusions concerning the free DEMs and the UAV-produced DEM in relation to the hydraulic simulation results requires an initial conventional comparative analysis of the results of the three simulations (Sim 1–3). This demands the post-processing of the results, including graphical representation in the form of comparative maps (i.e., flood extend maps, flow depths maps, and maps with centerlines produced from the various DTMs), for the empirical but informed conclusion production. These are presented and discussed in the "Results and discussion" section (Section 3.1).

These initial conclusions must be supported by a more quantitative statistical analysis of cross-section-specific variables and results. Specifically, the correlation coefficient (Pearson product-moment correlation coefficient), a measure of linear association between two variables, is calculated between all involved input data and variables (Sim/DTM-specific or common for all Sims/DTMs) and results (flood extents, flow depths, flow characteristics). The calculated correlation matrix can be used to support the empirical conclusions of Step 12, but also help explore other patterns, namely the factors that magnify and propagate errors originating from lower resolution DEMs in various sections.

As the investigation process is actually a root cause analysis concerning the impact of the various DEM resolutions on the hydraulic simulation, a more sophisticated method to interpret the nonlinear dependencies between variables and results is needed. A well-documented methodology from Machine Learning, Random Forest (RF) importance [49], is utilised. An interval (t_lower, t_upper) is defined, where the residuals are reasonably small, and RF is fitted on the residuals that are larger than t_upper and lower than t_lower, observing the mean node impurity of the forest as a feature importance proxy. The key insight here is to include a gaussian noise "dummy" variable, uncorrelated with the target variable, as an additional feature, which is known to have no impact on the target output. By using the importance of this variable as a baseline, conclusions can be drawn on the importance of the other variables that have larger importance compared to the "dummy" variable, also including the error bars. This implementation should overcome the theoretical weakness of the statistically limited dataset, due to the study of a geospatially data-scarce area.

3. Results and Discussion

3.1. Comparative Analysis Based on Produced Maps (Step 12)

The production of empirical conclusions is mainly based on the comparative analysis of flood extent and flow depth maps. Figure 9 presents the flood extents for all three Sims projected on the same map (see SM24). It includes the five identified river sections exhibiting different sinuosity ratios (SR) in order to investigate the possible correlation between SR and simulation errors in lower resolution Sims 1 and 2. For a finer, more detailed comparison of the flood extent maps' differences, comparative maps for all combinations are available as Appendix A (SM25: W_5 m vs. W_0.05 m; SM26: W_2 m vs. W_0.05 m; SM27: W_5 m vs. W_2 m). The red line, representing the UAV-produced flood extents of Sim 3 (W_0.05 m) is assumed to be the "ground truth", namely the closest to the truth available flood extents (inundated area = 0.7 km^2). Although Sim 1 and Sim 2 flood extents (W_5 m and W_2 m, respectively) constitute reliable simulations and provide satisfactory approximations, there are many errors, as presented in Tables 3 and 4. Inundation areas produced by Sim 1 and Sim 2 are 0.836 km^2 and 0.896 km^2, respectively. Most of the errors are overestimations rather than underestimations (Table 3), at least being on the safe side.

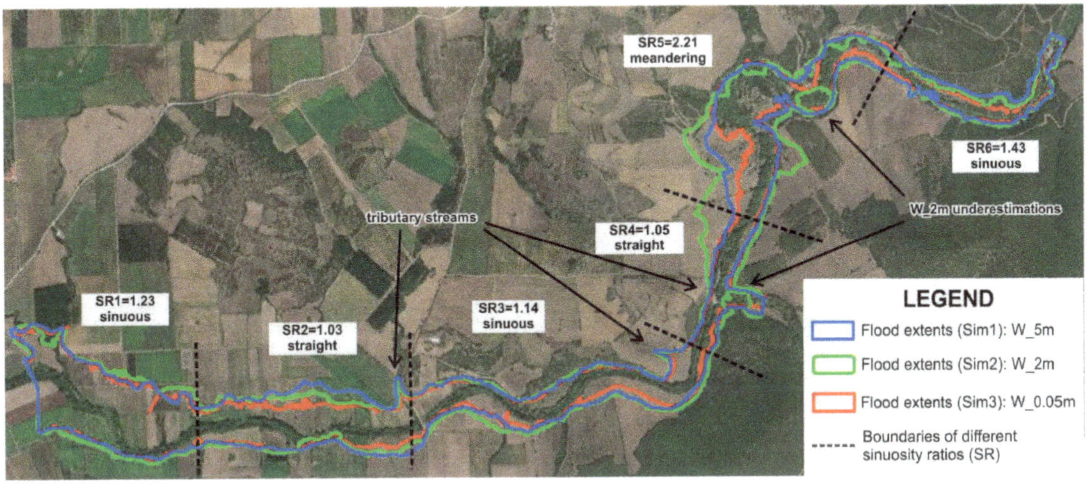

Figure 9. Flood extents of the three hydraulic simulations (Sim 1 = W_5 m; Sim 2 = W_2 m; Sim 3 = W_0.05 m) projected on the same map (separate maps in SM16, SM19, SM21).

Table 3. Area differences (km^2) of flood inundated regions between Sim 1 and Sim 3 (W_5 m–W_0.05 m) and Sim 2 and Sim 3 (W_2 m–W_0.05 m) as overall error, overestimations, and underestimations.

Flood Extent Error	Overall Error Area (km^2)	Overestimation Area (km^2)	Overestimation Area (%)	Underestimation Area (km^2)	Underestimation Area (%)
W_5 m vs. W_0.05 m	0.144	0.143	99.31%	0.001	0.69%
W_2 m vs. W_0.05 m	0.242	0.219	90.50%	0.023	9.50%

Table 4. Flood extent and flow depth statistics for the three simulations.

Sim	Flood Extent			Flow Depth	
	Area (km^2)	Algebraic Error (vs. Sim 3; km^2)	Algebraic Error (vs. Sim 3; %)	Min (m)	Max (m)
1	0.836	0.144	17.22%	0.002	5.979
2	0.896	0.242	27.01%	0.002	11.474
3	0.700	-	-	0.002	11.397

As presented in Table 3, the overall flood extent error of free DEM-based Sims (1 and 2) vs. the "ground truth" (Sim 3), expressed as the total area (km^2) of overestimation and underestimation, is 70% larger for Sim 2 (0.242 km^2 or 27.01% of the respective inundation area) compared to Sim 1 (0.242 km^2 or 17.22% of the respective inundation area). This is rather counter-intuitive: W_5 m is closer to the "ground truth", whereas W_2 m exhibits extended overestimations (0.219 km^2). Moreover, the most alarming result is that, while Sim 1's error is 99.31% overestimations, and only 0.69% underestimations (area-wise), Sim 2 exhibits 0.023 km^2 underestimations, approximately 10% of all its errors (Table 3). Examples of serious underestimations are indicated on Figure 9, e.g., in river section 5 (SR = 2.2). These underestimation errors are possibly risky, as they are not on the safe side. Overall, it is observed that there seem to be more errors in the sinuous and especially meandering sections of the river. Particular consideration should be given in areas where tributary streams converge to the main river (see Figure 9); these areas can be mistakenly considered as of greater than the actual importance flood-wise, due to their simulated extensive flooding.

Figure 10 presents the flow depth maps for all Sims side-by-side, whereas Figure 11 presents the flow depth differences between Sim 1 and Sim 3, and Sim 2 and Sim 3, respectively. Whereas Sim 2's range of flow depths (0.002 m–11.474 m) generally matches the respective range of Sim 3 (0.002 m–11.397 m; Table 4), a closer inspection reveals an inconsistency in their spatial distribution (Figure 10). With that in mind, Sim 1, though generally underestimating the flow depths, is in principle closer to Sim 3, and hydraulically more accurate. This is apparent in Figure 10, as the higher flow depths are positioned on the river centerline in Sim 1 (just like Sim 3). This is not the case for Sim 2, where the flow seems inconsistent and does not follow the real river centerline. Sim 1's flow depth errors (compared to Sim 3 "ground truth") range from −9.89 m to +5.85 m, whereas Sim 2's errors range from −7.47 m to +11.47 m. Sim 1 generally tends to underestimate flow depths up to 33% more than Sim 2, compared to the "ground truth", whereas Sim 2 tends to overestimate them up to 96% more than Sim 1. Considering the spatial distribution of flow depth errors, Sim 1 overestimates flow depths in a smaller area than Sim 2, while underestimating them in a larger area (Figure 11).

In search of the root of the errors in flood extents and flow depths, the meandering section of the river is selected to be scrutinised, as it is observed to exhibit extreme differences. Figure 12a presents the elevation differences between DTM_5 m and DTM_0.05, whereas Figure 12c presents the resulting flood extents of Sim 1 and Sim 3 (W_5 m and W_0.05 m) together with the flow depth differences between Sim 1 (y_5 m) and Sim 3 (y_0.05 m). In a similar fashion, Figure 12b presents DTM_5 m vs. DTM_0.05, whereas Figure 12d presents the resulting W_5 m and W_0.05 m, together with y_5 m vs. y_0.05 m. Figure 13 presents the locations of the two selected cross-sections, featured in detail in Figure 14.

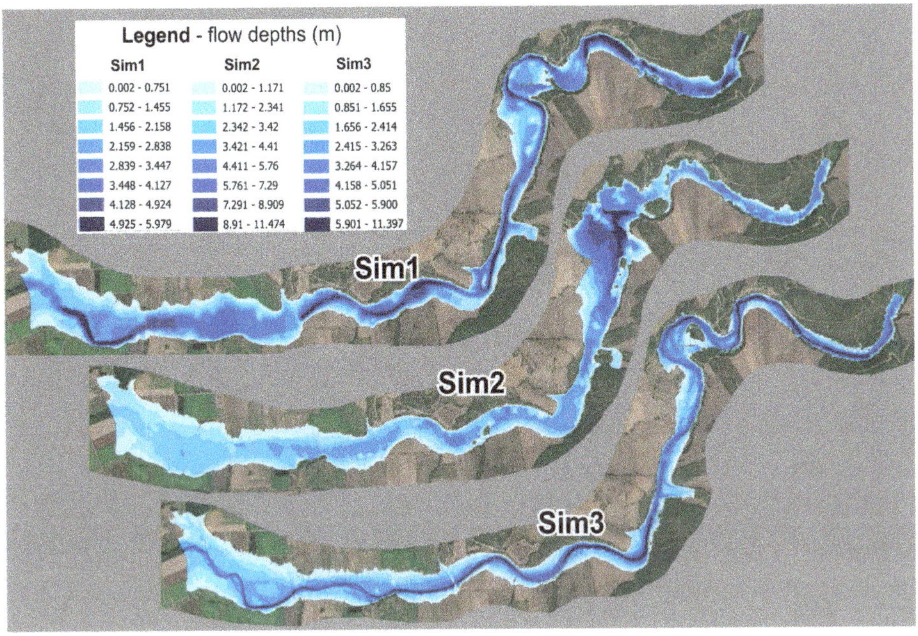

Figure 10. Flow depths of the three hydraulic simulations (Sim 1 = y_5 m; Sim 2 = y_2 m; Sim 3 = y_0.05 m; for separate hi-res maps see SM17, SM20, SM22).

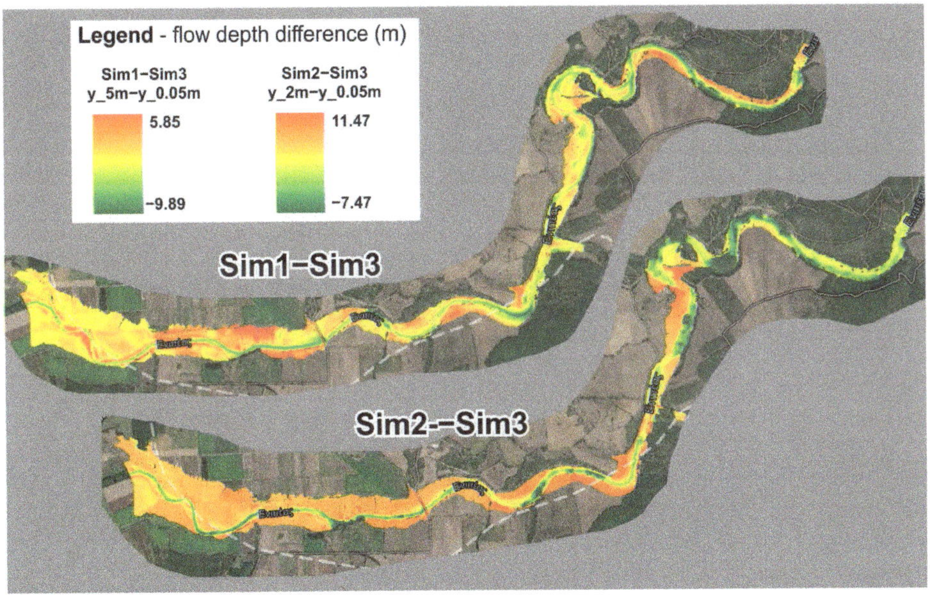

Figure 11. Flow depth differences between Sim 1 and Sim 3 (y_5 m−y_0.05 m) and Sim 2 and Sim 3 (y_2 m−y_0.05 m), presented only for the intersection of the respective inundated areas.

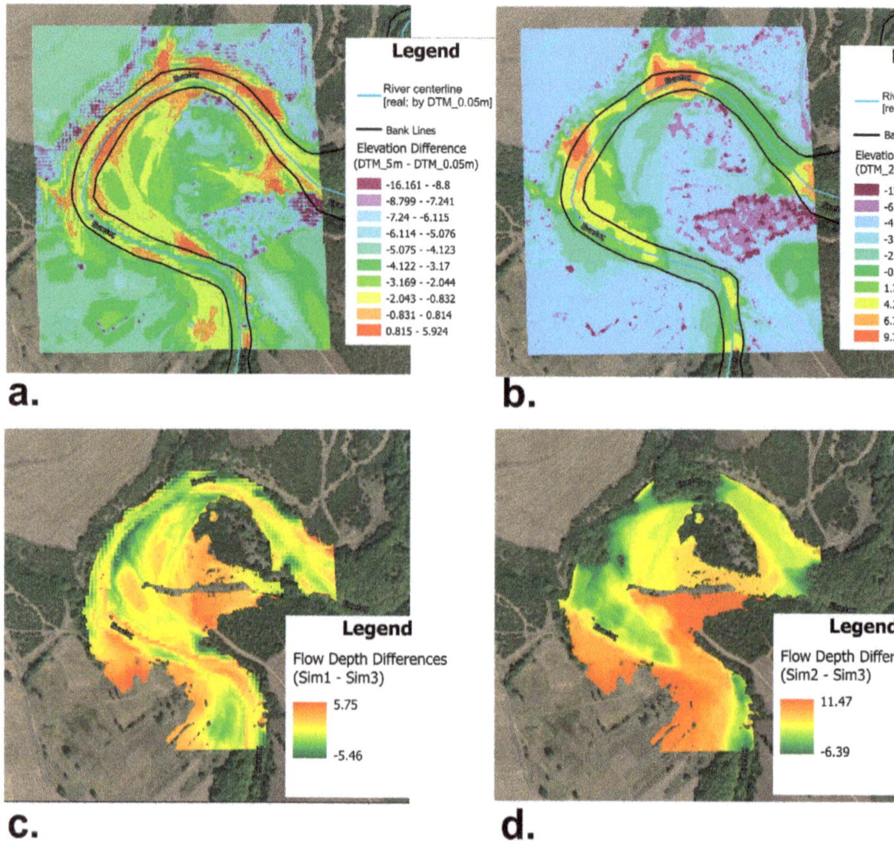

Figure 12. A section of the "flood area" (meandering) presenting differences in elevation (**a**) DTM_5 m vs. DTM_0.05 m, (**b**) DTM_2 m vs. DTM_0.05 m and flow depths for (**c**) Sim 1–Sim 3 and (**d**) Sim 2–Sim 3.

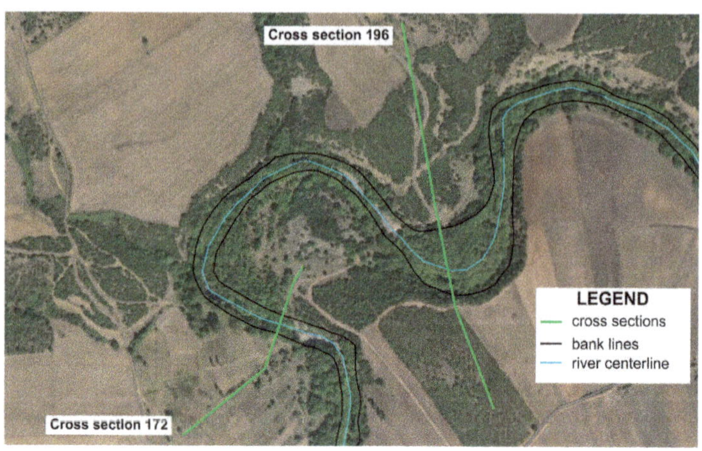

Figure 13. Exact locations of two selected cross-sections (172 and 196; see SM23) in the meandering part of the river in the "flood area", featured in detail in Figure 14.

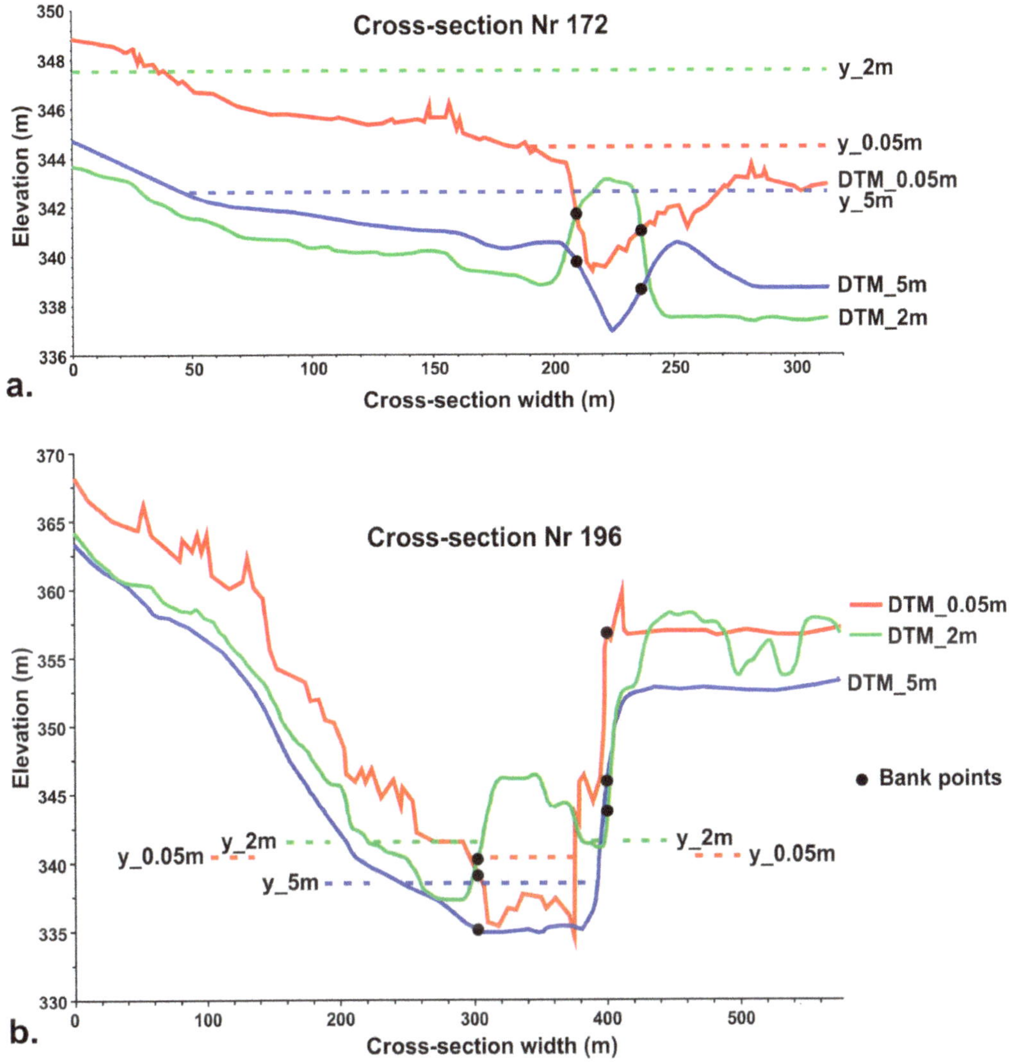

Figure 14. The geometry of the three versions of (**a**) cross-section 172 and (**b**) cross-section 196; (see SM23) as derived from DTM_5 m (blue), DTM_2 m (green), and DTM_0.05 m (red). DTM_5 m captures the general geometry, whereas DTM_2 m fails to delineate the main channel.

It is apparent that DTM_5 m exhibits elevation underestimations in most of the surveyed area (Figure 12a). The areas where it overestimates elevation are limited and marginally inside the banks, never on the centerline, hence managing to capture the geometry of the cross-sections (Figure 14). On the other hand, DTM_2 m generally underestimates elevation and overestimates near and inside the banks, even on the centerline, hence being unable to capture the true geometry of the cross sections (Figure 14). This practically results in the altering of the river pathways, as explicitly delineated in Figure 15, that simultaneously presents the river centerlines, as automatically produced based on DTM_5 m, DTM_2 m, and DTM_0.05 m, respectively. Whereas the DTM_5 m-derived centerline is a good approximation of the "ground truth" DTM_0.05 m, the DTM_2 m-derived centerline exhibits serious deviations, especially along the meandering sections. These errors in the initial topography data propagate and are the root causes of the errors in flood extent and flow depth results (Figure 12c,d). This is also obvious in Figure 14, where the water surface elevations (flow depths; y) vary significantly. A closer look reveals a strong connection of the errors in DTMs with the mapping of dense vegetation and canopy, especially in DTM_2 m, despite the fact that its resolution is in principle higher than DTM_5 m.

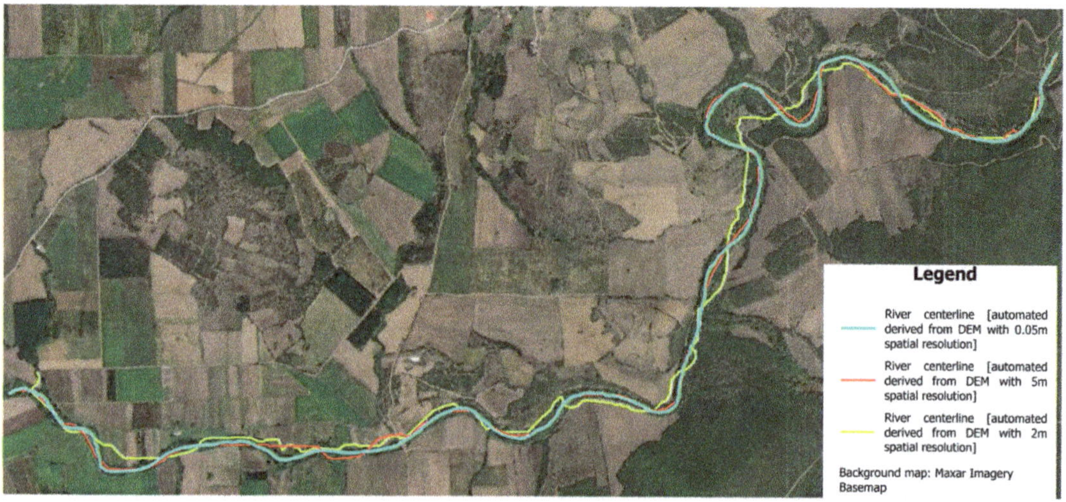

Figure 15. River centerlines derived from the 3 DTMs projected on the same map (see SM30) for comparison purposes. All simulations are based on "ground truth" centerline (by DTM_0.05 m).

3.2. Statistical Analysis Based on Correlation Matrix (Step 13)

The correlation matrix presenting correlation coefficient values between all involved input data and variables (Sim/DTM-specific or common for all Sims/DTMs) and results (flood extents, flow depths, flow characteristics) are presented in Figure 16. The values of interest are highlighted and discussed.

There is an almost linear relationship between riverbed elevations of all DTMs (in relation to the real centerline derived by DTM_0.05 m) and the cross-section ID (consecutive) numbers. The strong linearities are explained by the fact that rivers flow downhill and riverbeds exhibit a positive slope in the vast majority of their length. Hence, the larger the ID number of a cross-section, the lower the respective elevation. Although this is expected, the correlation coefficient values (Cc) are (stronger to weaker correlations) Cc_0.05 m = −0.99, Cc_5 m = −0.98, and Cc_2 m = −0.95. The small variations support the previous findings, indicating better riverbed elevation approximation (compared to the "ground truth" DTM_0.05 m) by DTM_5 m, rather than DTM_2 m.

Another interesting correlation is the relationship between riverbed elevations (Z) and the respective cross-section-specific flood extents (top widths; W). Again, the "ground truth" correlation between Z_0.05 m and W_0.05 m is the highest in value (Cc_0.05 m = +0.71), followed by Cc_5 m = 0.6 and finally Cc_2 m = 0.55. The absolute value of any of the aforementioned Cc is rather random. A finding worth mentioning is the variation in Cc that supports the claim that the closest fit between DTM_5 m and DTM_0.05 m, compared to DTM_2 m, also results in a closest fit between W_5 m and W_0.05 m, compared to W_2 m. This pattern continues in the relationship between the hydraulic radius values (R), as well as Froude number (Fr), and the respective W for each Sim. Sim 1 results are closer to Sim 3 compared to Sim 2. Specifically, the variation of the impact of R on W concerning Sim 2 is extreme: while Cc_0.05 m = −0.7 and Cc_5 m = −0.53, Cc_2 m is positive and equal to +0.39. Finally, the correlations between the flood extents themselves support the main argument: Cc_5 − 0.05 = +0.82, Cc_5 − 2 = +0.81, and Cc_2 − 0.05 = +0.71.

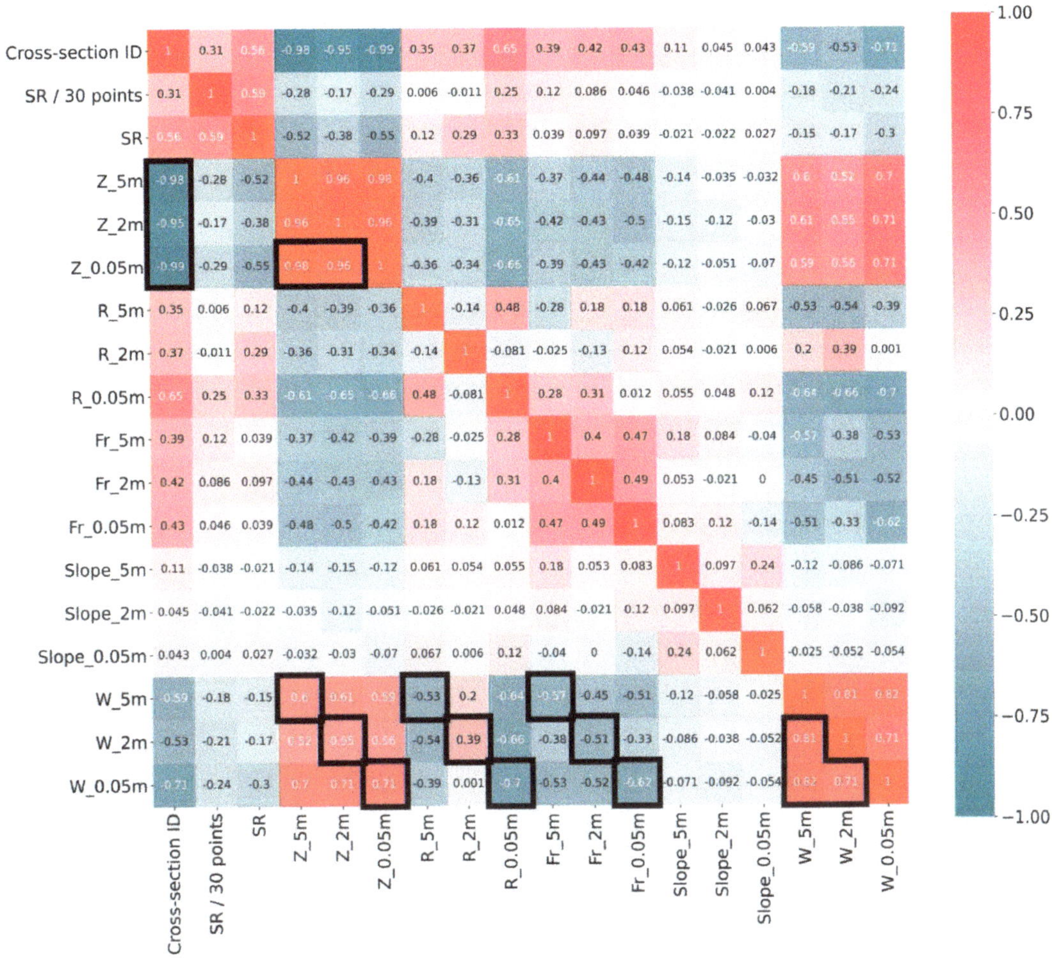

Figure 16. Correlation matrix presenting correlation coefficient values between all involved input data and variables (Sim/DTM-specific or common for all Sims/DTMs) and results (flood extents, flow depths, flow characteristics). Values of interest are highlighted.

3.3. Machine Learning for Interpretation of Nonlinear Relationships (Random Forests)

The histograms of Figure 17 present the distribution of errors of Sim 1 and Sim 2 flood extent errors, compared to Sim 3. The distributions indicate larger errors for Sim 2 compared to Sim 1, and are skewed, also indicating nonlinearities in the error generation and propagation. This is why the method of feature importance calculation using Mean Decrease in Impurity (MDI) is implemented with Random Forest. After a series of tests, the interval (t_lower, t_upper) is empirically selected as (−10, +1), so that the remaining negative and positive value sets of the distribution are split equally.

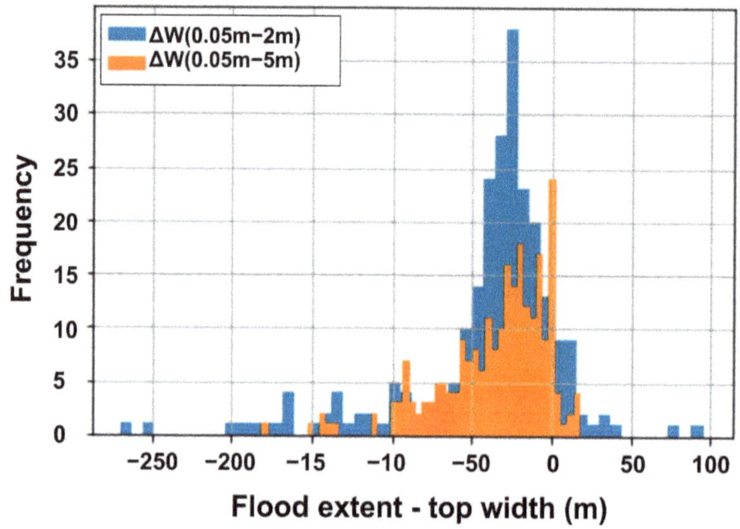

Figure 17. Histograms of the errors of Sim 1 and Sim 2 vs. "ground truth" Sim 3, indicating skewed distributions and nonlinear dependencies.

Following the already presented methodology, using a random noise feature as the "dummy" variable as a means of comparison, the feature importance of the following variables and variable errors regarding their impact on flood extent error ΔW are tested: sinuosity (SR; fixed, see Figure 9), rolling sinuosity (SR/30 sections), error of riverbed elevations (ΔZ), error of hydraulic radii (ΔR), and error of Froude number values (ΔFr). Figure 18 presents the MDI for the aforementioned features. Specifically, Figure 18a presents results for Sim 1 vs. Sim 3 (W_0.05 m–W_5 m) and the feature importance of SR, SR/30 sections, Z_5 m–W_0.05 m, R_5 m–R_0.05 m, and Fr_5 m–Fr_0.05 m. Figure 18b presents results for Sim 2 vs. Sim 3 (W_0.05 m–W_2 m) and the feature importance of SR, SR/30 sections, Z_2 m–W_0.05 m, R_2 m–R_0.05 m, and Fr_2 m–Fr_0.05 m.

As far as Sim 1 flood extent errors are concerned, SR, ΔR, and ΔFr seem to be equally important. This can be interpreted as follows: the DEM_5 m intrinsic errors propagate up to the flood extent results in the sections of increased sinuosity, driven especially by the resulting errors in the hydraulic radius and Froude number values calculation. On the other hand, Sim 2 flood extent errors' origin and root cause are different. The feature that stands out is the hydraulic radii of the cross-sections; their importance is far higher than the respective features related to Sim 1 errors. These results fully support the earlier conclusions drawn by the flood maps-based comparative analysis (Step 12) and the correlation matrix-based statistical analysis (Step 13). The hydraulic radii per cross-section are distorted in the main channel in Sim 2, due to the DEM_2 m production process failing to filter dense vegetation.

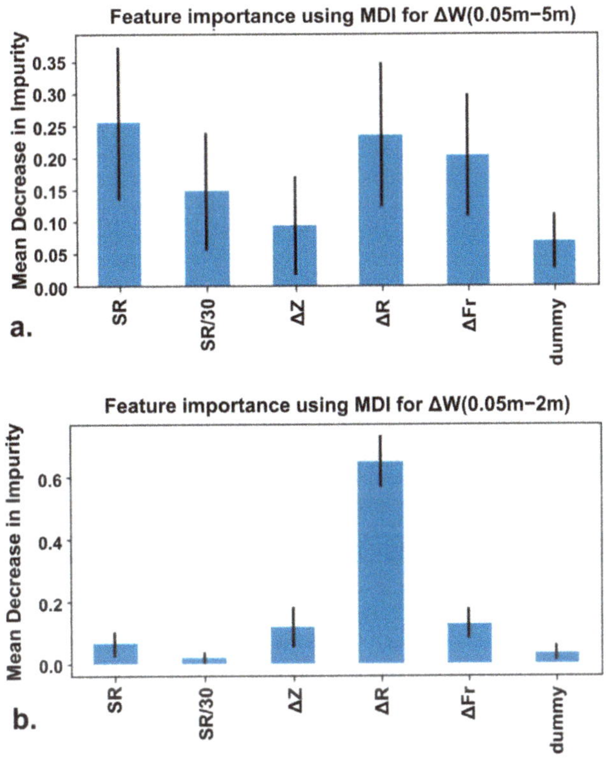

Figure 18. Random Forest-calculated feature importance for fixed sinuosity ratio SR, rolling SR per 30 cross-sections, ΔZ (0.05 m–5 m), ΔR (0.05 m–5 m), ΔFr (0.05 m–5 m), and a dummy variable (random noise) using Mean Decrease in Impurity (MDI) for (**a**) ΔW (0.05 m–2 m) and (**b**) ΔW (0.05 m–5 m).

4. Conclusions

This paper investigates flood modelling sensitivity against two sets of open access geospatial elevation data (5 m and 2 m resolution, respectively), derived from the Hellenic Cadastre, and an own designated Unmanned Aerial Vehicle topographical mission (0.05 m resolution). A case study is used concerning a part of the mountainous Enipeas river basin of Thessaly's Water District (Greece).

The first step of the proposed methodology includes a flood maps-based comparative analysis so that experts can empirically draw conclusions on the specific studied river catchment. In the current case study, most of the flood extent errors are overestimations rather than underestimations, at least being on the safe side. Though counter-intuitive, the DEM_5 m-derived (Sim 1) flood extents are closer to the "ground truth", whereas DEM_2 m-derived (Sim 2) extents are extensively overestimated, while also exhibiting relatively alarmingly high underestimations, which are not on the safe side and can have potentially catastrophic implications if used for design purposes. The sections of increased sinuosity ratio, especially the meandering river sections, seem more prone to flood modelling errors. The same applies for junctions of the main channel with modelled, or not, tributary streams. Concerning flow depth results, Sim 1 generally underestimates them and is, in principle, closer to DEM_0.05 m-derived Sim 3, and hydraulically more accurate. The reason is that, although Sim 2 range of flow depths is generally correct, their spatial distribution is inconsistent, as is the flow that does not follow the real river centerline.

The root of the errors concerning flood extents and flow depths lies in the topography data used. DEM_5 m mostly underestimates elevation, but manages to capture the geometry of the cross-sections, whereas DEM_2 m generally underestimates, but overestimates inside the critical zone of the main channel, near and inside the banks, even on the centerline; it is unable to capture the true geometry of the cross sections, practically altering the river pathways. This is more intense in the meandering river sections. The map-based analysis indicates that the true cause is the inability of DEM_2 m to capture the elevation of the ground in areas of dense vegetation and canopy, usually being wider in the meandering river sections. In absence of detailed information concerning the surveying process followed by the Hellenic Cadastre, and specifically the classification and filtering of the vegetation (tree removal), one can only speculate on the true source of this error. On the other hand, DEM_5 m, though constituting a lower resolution product, seems more suitable for approximating the real terrain.

The flood modelling results are also analysed via a statistical analysis, based on the correlation matrix presenting linear relationships between input data variables (i.e., elevation, slope, sinuosity ratio) and cross section-specific results, including flow characteristics (i.e., Froude number, hydraulic radius), flood extents, and flow depths. The correlation results indicate strong linearities where expected (riverbed elevations vs. cross-section ID numbers), and weaker where expected (i.e., riverbed elevations and hydraulic radii and Froude number vs. flood extents). Nevertheless, the important finding the statistical analysis has to offer is the quantifiable proof of the superiority of DEM_5 m-derived Sim 1 compared to the DEM_2 m-derived Sim 2 results, which supports the preceding empirical analysis conclusions. This is suggested by the fact that correlations of the analysed variables and flood extent results constantly follow the classification (stronger-to-weaker) Sim 3 > Sim 1 > Sim 2. The simple comparison between the correlation of the cross section-specific flood extents of Sim 1–Sim 3 and Sim 2–Sim 3 also supports the argument.

As the conventional approaches fail to identify the nonlinear dependencies of the root cause analysis and error propagation tracking side of the research problem, the proposed methodology finally implements a more sophisticated Machine Learning (ML) analysis, specifically Random Forest importance. The ML approach results further support and solidify the earlier conclusions drawn by the flood maps-based comparative analysis and the correlation matrix-based statistical analysis. The failure of the DEM_2 m production process to map the terrain in areas of dense vegetation and wide canopy leads to unreal cross-section geometries and inserts critical errors in the respective hydraulic radii, really important at least in 1D hydraulic analyses. These errors further propagate to the flood extent results, as the RF importance approach robustly indicates.

As far as the general proposed methodology is concerned, for deciding the best available alternative DEM of an accurate but costly UAV-based or in situ ground survey-based DEM, no step is redundant. The flood map-based comparative analysis by experts is the main and key evaluation tool and cannot be replaced by a statistical or even a more sophisticated Machine Learning-based analysis. Machine Learning methods can interpret nonlinear dependencies but depend on the way they are implemented and are susceptible to parameter errors. Nevertheless, they provide further insight on the root and cause of the error and the propagation mechanism, while identifying additional error patterns. The proposed stages and steps should be implemented as an integrated methodology.

The conclusions of the current paper and related research can be summed up as steps of a suggested procedure for the optimal hydraulic simulation of a river basin, in terms of minimisation of in situ topographical mapping costs without compromising the hydraulic simulation accuracy:

1. Approximate the real river centerline, as accurately as possible, utilising any available source and technique possible. A realistic approach would be the use of the most recent and high resolution open-source DEM available, in order to automatically produce an approximate river centerline, calibrated by recent satellite imagery (e.g., google earth) and orthophotos (e.g., Hellenic Cadastre in Greek reality), supported by in situ inspection if possible or necessary.

2. In river areas exhibiting one or more of the following characteristics: (a) dense vegetation or/and wide canopy, (b) meandering sections, (c) junctions of the main channel with tributary streams, or/and generally (d) centerline-bank lines derived from the available DEM diverging from the real centerline-bank lines, proceed to UAV-based mapping (combined with an a-posteriori vegetation removal process) or in situ topographical ground surveying.
3. In areas not belonging to the previous category, if more than one open-source DEM is available, the implemented and proposed methodology of this paper should be followed in order to select the one that inserts and propagates less errors in the hydraulic simulations. In Greek reality, just use the DEM_5 m by Hellenic Cadastre, as it is proven to be more efficient for, at least 1D, hydraulic simulations, rather than the higher resolution DEM_2 m.
4. Future research should include more case studies of different terrain characteristics, such as sinuosity, lush and weak vegetation areas, etc., combined with real in situ topographical ground surveying and validated flood extents and flow depth measurements. The larger datasets will also provide further credibility to the Machine Learning-based analysis. In the proposed methodology, 2D hydraulic simulations (e.g., using HEC-RAS 2D) should be tested to check their sensitivity vs. the various spatial resolution DEMs. Finally, more sophisticated ML techniques, such as Gradient Boost Algorithms, Fuzzy Cognitive Maps, and Self Organizing Maps, can be used for root cause analysis.

Author Contributions: Conceptualization, N.X., Y.K. and E.F.; methodology, N.X., Y.K., E.F. and K.P.; software, N.X., Y.K., E.F. and K.P.; validation, N.X., Y.K., E.F. and K.P.; investigation, N.X., Y.K., E.F. and K.P.; resources, N.X., Y.K. and E.F.; data curation, N.X., Y.K. and E.F.; writing—original draft preparation, N.X., Y.K., E.F. and K.P.; writing—review and editing, N.X., Y.K., E.F., S.K., N.A., D.D. and K.K.; visualization, N.X., Y.K. and K.P.; supervision, E.F., S.K., N.A. and K.K.; project administration, N.X., Y.K. and E.F. All authors have read and agreed to the published version of the manuscript.

Funding: This research received no external funding.

Data Availability Statement: Data supporting the reported results can be found at: https://drive.google.com/drive/folders/1L6_Whis8PSO5ArZjJTT375ClNRMYPMNt?usp=share_link (accessed on 26 February 2023).

Conflicts of Interest: The authors declare no conflict of interest.

Appendix A

The supporting information presented in detail in Table A1 of Appendix A, can be downloaded at: https://drive.google.com/drive/folders/1L6_Whis8PSO5ArZjJTT375ClNRMYPMNt?usp=share_link (accessed on 26 February 2023).

Table A1 presents the list and relevant information concerning the Appendix A referred to in the text.

Table A1. List of Appendix A and relevant information.

Nr	Product	Filename	Description
SM1	P1a	SM1-(P1a) DTM5 m flood.pdf	Map of the DTM_5 m covering only the "flood area", derived from DEM_5 m by Hellenic Cadastre open data, only covering the "flood area".
SM2a	P1b_hydro	SM2a-(P1b_hydro) DTM2 m hydro.pdf	Map of the DTM_5 m covering the full "hydro area" derived from DEM_2 m by Hellenic Cadastre open data.

Table A1. *Cont.*

Nr	Product	Filename	Description
SM2b	P1b_flood	SM2b-(P1b_flood) DTM2 m flood.pdf	Map of the DTM_5 m covering only the "flood area", derived from DEM_2 m by Hellenic Cadastre open data.
SM3	P1c	SM3-(P1c) DTM0.05 m.pdf	Map of the DTM_0.05 m derived from a research team's UAV survey mission in the "flood area".
SM4	P2a	SM4-(P2a) subcatchments.pdf	Map of the 6 sub-catchments' delineation.
SM5	P2b	SM5-(P2b) hydrographic network.pdf	Map of the hydrographic network in the "hydro area" derived by RBMP [27].
SM6a	P2c	SM6a-(P2c) river centerline by DTM0.05 m.pdf	Map of the river centerline, derived from DEM_0.05 m, in the delineated "flood area".
SM6b	P2c_DTM	SM6b-(P2c_DTM) river centerline by DTM0.05 m on DTM.pdf	Map of the river centerline, derived from DEM_0.05 m, on the respective DTM in the delineated "flood area".
SM7a	P2d	SM7a-(P2d) river centerline by DTM2 m.pdf	Map of the river centerline, derived from DEM_2 m, in the delineated "flood area".
SM7b	P2d_DTM	SM7b-(P2d_DTM) river centerline by DTM2 m on DTM.pdf	Map of the river centerline, derived from DEM_2 m, on the respective DTM in the delineated "flood area".
SM8a	P2e	SM8a-(P2e) river centerline by DTM5 m.pdf	Map of the river centerline, derived from DEM_5 m, in the delineated "flood area".
SM8b	P2e_DTM	SM8a-(P2e_DTM) river centerline by DTM5 m on DTM.pdf	Map of the river centerline, derived from DEM_5 m, on the respective DTM in the delineated "flood area".
SM9	P2f + P7	SM9-(P2f + P7) geomorphology + time conc. + time lags.xlsx	An excel file presenting the sub-catchment characteristics and the respective calculations.
SM10	P3a	SM10-(P3a) soil data map	A soil data map presenting the available soil data concerning the "hydro area", derived from [34].
SM11	P3b	SM11-(P3b).pdf hydrolithological map.pdf	A hydrolithological map presenting the drainage characteristics of the "hydro area", derived from the respective soil data (SM11; [34]), where available, and [27] in the remaining areas.
SM12	P4	SM12-(P4) land uses map.pdf	A land uses map of the "hydro area", derived from CORINE [35].
SM13	P5	SM13-(P5) 6 hyetographs.xlsx	An excel file presenting the hyetographs per sub-catchment based on the Anavra station IDF curve using the Alternate Block Method [42,43].
SM14	P6	SM14-(P6) Land cover and CN per subcatchment.xlsx	An excel file presenting the CNc estimations per sub-catchments with the respective calculations.
SM15	P8	SM15-(P8) 9 hydrographs.pdf	A pdf file with the 9 hydrographs (3 hydrographs for the 3 junctions and 6 for the 6 sub-catchments) produced during Step 8 of Stage 2.
SM16	P9a	SM16-(P9a) flood extents map W_5 m.pdf	Map of the flood extents simulated by Sim 1 with the automatically derived centerline by the respective DTM_5 m.
SM17	P9b	SM17-(P9b) flow depths map y_5 m.pdf	Map of the flow depths simulated by Sim 1 with the automatically derived centerline by the respective DTM_5 m.

Table A1. *Cont.*

Nr	Product	Filename	Description
SM18	P9c + P10c + P11c	SM18-(P9c + P10c + P11c) flow characteristics.xlsx	An excel file presenting the selected flow characteristics for all Sims (fc_5 m, fc_2 m, fc_0.05 m).
SM19	P10a	SM19-(P10a) flood extents map W_2 m.pdf	Map of the flood extents simulated by Sim 2 with the automatically derived centerline by the respective DTM_2 m.
SM20	P10b	SM20-(P10b) flow depths map y_2 m.pdf	Map of the flow depths simulated by Sim 2 with the automatically derived centerline by the respective DTM_2 m.
SM21	P11a	SM21-(P11a) flood extents map W_0.05 m.pdf	Map of the flood extents simulated by Sim 3 with the automatically derived centerline by the respective DTM_0.05 m.
SM22	P11b	SM22-(P11b) flow depths map y_0.05 m.pdf	Map of the flow depths simulated by Sim 3 with the automatically derived centerline by the respective DTM_0.05 m.
SM23	-	SM23-Map with cross-sections.pdf	A pdf file presenting all 270 cross-sections of the flood area created in an interval of 20 m
SM24	P9a + P10a + P11a	SM24-(P9a + P10a + P11a) flood extents maps.pdf	Map of all simulated flood extents (W_5 m, W_2 m, W_0.05 m).
SM25	P9a vs. P11a	SM25-(P9a + P11a) flood extents maps.pdf	Map of simulated flood extents of Sim 1 and Sim 3 (W_5 m vs. W_0.05 m).
SM26	P10a vs. P11a	SM26-(P10a + P11a) flood extents maps.pdf	Map of simulated flood extents of Sim 2 and Sim 3 (W_2 m vs. W_0.05 m).
SM27	P9a vs. P10a	SM27-(P9a + P10a) flood extents maps.pdf	Map of simulated flood extents of Sim 1 and Sim 2 (W_5 m vs. W_2 m).
SM28	-	SM28-DTM5 m-DTM0.05 m.pdf	Map of elevation differences between DTM_5 m and DTM_0.05 m in a selected meandering section of the river.
SM29	-	SM29-DTM2 m-DTM0.05 m.pdf	Map of elevation differences between DTM_2 m and DTM_0.05 m in a selected meandering section of the river.
SM30	-	SM30-all river centerlines.pdf	Map of all three centerlines derived from the respective DEM.
SM31	-	SM31-y_5 m-y_0.05 m.pdf	Map of flow depth differences between Sim 1 and Sim 3 (y_5 m-y_0.05 m).
SM32	-	SM32-y_2 m-y_0.05 m.pdf	Map of flow depth differences between Sim 2 and Sim 3 (y_2 m-y_0.05 m).

References

1. Talbot, C.J.; Bennett, E.M.; Cassell, K.; Hanes, D.M.; Minor, E.C.; Paerl, H.; Raymond, P.A.; Vargas, R.; Vidon, P.G.; Wollheim, W.; et al. The Impact of Flooding on Aquatic Ecosystem Services. *Biogeochemistry* **2018**, *141*, 439–461. [CrossRef] [PubMed]
2. Yu, Q.; Wang, Y.; Li, N. Extreme Flood Disasters: Comprehensive Impact and Assessment. *Water* **2022**, *14*, 1211. [CrossRef]
3. Rotunno, R.; Houze, R.A. Lessons on Orographic Precipitation from the Mesoscale Alpine Programme. *Q. J. R. Meteorol. Soc.* **2007**, *133*, 811–830. [CrossRef]
4. Borga, M.; Stoffel, M.; Marchi, L.; Marra, F.; Jakob, M. Hydrogeomorphic Response to Extreme Rainfall in Headwater Systems: Flash Floods and Debris Flows. *J. Hydrol.* **2014**, *518*, 194–205. [CrossRef]
5. Lumbroso, D.; Gaume, E. Reducing the Uncertainty in Indirect Estimates of Extreme Flash Flood Discharges. *J. Hydrol.* **2012**, *414–415*, 16–30. [CrossRef]
6. Ashley, S.T.; Ashley, W.S. Flood Fatalities in the United States. *J. Appl. Meteorol. Climatol.* **2008**, *47*, 805–818. [CrossRef]
7. Diakakis, M.; Deligiannakis, G. Flood Fatalities in Greece: 1970–2010. *J. Flood Risk Manag.* **2015**, *10*, 115–123. [CrossRef]

8. Diakakis, M.; Deligiannakis, G.; Katsetsiadou, K.; Antoniadis, Z.; Melaki, M. Mapping and Classification of Direct Flood Impacts in the Complex Conditions of an Urban Environment. The Case Study of the 2014 Flood in Athens, Greece. *Urban Water J.* **2017**, *14*, 1065–1074. [CrossRef]
9. Merz, B.; Kreibich, H.; Schwarze, R.; Thieken, A. Review Article "Assessment of Economic Flood Damage". *Nat. Hazards Earth Syst. Sci.* **2010**, *10*, 1697–1724. [CrossRef]
10. Barredo, J.I. Normalised Flood Losses in Europe: 1970–2006. *Nat. Hazards Earth Syst. Sci.* **2009**, *9*, 97–104. [CrossRef]
11. Gaume, E.; Bain, V.; Bernardara, P.; Newinger, O.; Barbuc, M.; Bateman, A.; Blaškovičová, L.; Blöschl, G.; Borga, M.; Dumitrescu, A.; et al. A Compilation of Data on European Flash Floods. *J. Hydrol.* **2009**, *367*, 70–78. [CrossRef]
12. Tegos, A.; Ziogas, A.; Bellos, V.; Tzimas, A. Forensic hydrology: A complete reconstruction of an extreme flood event in data-scarce area. *Hydrology* **2022**, *9*, 93. [CrossRef]
13. Bellos, V.; Kourtis, I.; Raptaki, E.; Handrinos, S.; Kalogiros, J.; Sibetheros, I.A.; Tsihrintzis, V.A. Identifying Modelling Issues through the Use of an Open Real-World Flood Dataset. *Hydrology* **2022**, *9*, 194. [CrossRef]
14. Jongman, B. Effective Adaptation to Rising Flood Risk. *Nat. Commun.* **2018**, *9*, 1986. [CrossRef] [PubMed]
15. Dimitriadis, P.; Tegos, A.; Oikonomou, A.; Pagana, V.; Koukouvinos, A.; Mamassis, N.; Koutsoyiannis, D.; Efstratiadis, A. Comparative evaluation of 1D and quasi-2D hydraulic models based on benchmark and real-world applications for uncertainty assessment in flood mapping. *J. Hydrol.* **2016**, *534*, 478–492. [CrossRef]
16. Hawker, L.; Bates, P.; Neal, J.; Rougier, J. Perspectives on Digital Elevation Model (DEM) Simulation for Flood Modeling in the Absence of a High-Accuracy Open Access Global Dem. *Front. Earth Sci.* **2018**, *6*, 233. [CrossRef]
17. Hellenic Cadastre. Available online: https://www.ktimatologio.gr (accessed on 28 January 2023).
18. US Army Corps of Engineers. Available online: https://www.usace.army.mil (accessed on 10 January 2023).
19. Ras Mapper User's Manual. Available online: https://www.hec.usace.army.mil/confluence/rasdocs/rmum/latest (accessed on 25 January 2023).
20. HEC-RAS 1D—Hydrologic Engineering Center—River Analysis System. Available online: https://www.hec.usace.army.mil/software/waterquality/hec-ras.aspx (accessed on 20 December 2022).
21. Official Government Gazette (O.G.G) of the Hellenic Republic. *Approving the River Basin Management Plan of the River Basins of Thessalia Water District*; No. 2561/B/25.09.14; National Printing Office of Greece: Athens, Greece, 2014.
22. Digital Elevation Model (DEM) for the LSO Project (5m). Available online: http://gis.ktimanet.gr/geoportal/catalog/search/resource/details.page?uuid=%7B456CB655-B899-450A-87BF-8322B8FB8370%7D (accessed on 28 January 2023).
23. Digital Elevation Model (DEM) for the LSO25 Ortho Mapping Production (2m). Available online: https://www.ktimanet.gr/geoportal/catalog/search/resource/details.page?uuid=%7B86CABAFD-00B8-41D4-B36A-2FE8E2DCF958%7D (accessed on 28 January 2023).
24. Anders, N.; Valente, J.; Masselink, R.; Keesstra, S. Comparing Filtering Techniques for Removing Vegetation from UAV-Based Photogrammetric Point Clouds. *Drones* **2019**, *3*, 61. [CrossRef]
25. Agisoft Metashape. Available online: https://www.agisoft.com/downloads/user-manuals/ (accessed on 2 February 2023).
26. ESRI ArcGIS Pro. Available online: https://www.esri.com/en-us/arcgis/products/arcgis-pro/resources (accessed on 10 October 2022).
27. River Basin Management Plan (RBMP)—EL08 Water District of Thessaly. Available online: http://wfdver.ypeka.gr/en/management-plans-en/approved-management-plans-en/gr08-approved-en/ (accessed on 10 October 2022).
28. European Union. Water Framework Directive (WFD) 2000/60/EC, 2000. Available online: https://environment.ec.europa.eu/topics/water/water-framework-directive_en (accessed on 26 February 2023).
29. Giandotti, M. *Previsione Delle Piene e Delle Magre dei Corsi D'acqua*; Istituto Poligrafico dello Stato: Rome, Italy, 1934; pp. 107–117.
30. European Soil Data Centre (ESDAC). European Commission, Joint Research Centre. Available online: https://esdac.jrc.ec.europa.eu (accessed on 12 December 2022).
31. Panagos, P.; Van Liedekerke, M.; Jones, A.; Montanarella, L. European Soil Data Centre: Response to European Policy Support and Public Data Requirements. *Land Use Policy* **2012**, *29*, 329–338. [CrossRef]
32. Toth, G.; Jones, A.; Montanarella, L. *LUCAS Topsoil Survey. Methodology, Data and Results*; JRC Technical Reports; EUR26102—Scientific and Technical Research series; Publications Office of the European Union: Copenhagen, Denmark, 2013.
33. Jones, A.; Fernandes-Ugalde, O.; Scarpa, S.; Eiselt, B. *LUCAS 2022, EUR 30331 EN*; Publications Office of the European Union: Copenhagen, Denmark, 2022.
34. OPEKEPE—Payment and Control Agency for Guidance and Guarantee Community Aid. *Soil Map of Greece, Scale 1:30,000*; Aristotle University of Thessaloniki: Thessaloniki, Greece, 2015.
35. Corine Land Cover. Available online: https://land.copernicus.eu/pan-european/corine-land-cover (accessed on 10 October 2022).
36. Koutsoyiannis, D.; Markonis, Y.; Koukouvinos, A.; Papalexiou, S.; Mamassis, N.; Dimitriadis, P. Hydrological Study of Severe Rainfall in the Kephisos Basin, Greece, Study of the Management of Kephisos, Commissioner: General Secretariat of Public Works—Ministry of Environment, Planning and Public Works, Contractors: Exarhou Nikolopoulos Bensasson, Denco, G. Karavokiris; et al., 154 Pages, Athens. 2010. Available online: http://www.itia.ntua.gr/el/docinfo/970/ (accessed on 26 February 2023).
37. Intensity Duration Frequency Curves (IDF) Report. Available online: https://floods.ypeka.gr/egyFloods/IDF/IDF_Report_V4.pdf (accessed on 12 December 2022).

38. Koutsoyiannis, D.; Kozonis, D.; Manetas, A. A mathematical framework for studying rainfall intensity-duration-frequency relationships. *J. Hydrol.* **1998**, *206*, 118–135. [CrossRef]
39. Veneziano, D.; Langousis, A. The areal reduction factor: A multifractal analysis. *Water Resour. Res.* **2005**, *41*, W07008. [CrossRef]
40. Koutsoyiannis, D. *Statistical Hydrology*, 4th ed.; National Technical University of Athens: Athens, Greece, 1997; 312p.
41. Collier, C.G.; Hardaker, P.J. Estimating Probable Maximum Precipitation Using a Storm Model Approach. *J. Hydrol.* **1996**, *183*, 277–306. [CrossRef]
42. Chow, V.T.; Maidment, D.R.; Mays, L.W. *Applied Hydrology*; McGraw-Hill Professional: New York, NY, USA, 1988.
43. Efstratiadis, A.; Koussis, A.D.; Koutsoyiannis, D.; Mamassis, N. Flood Design Recipes vs. Reality: Can Predictions for Ungauged Basins Be Trusted? *Nat. Hazards Earth Syst. Sci.* **2014**, *14*, 1417–1428. [CrossRef]
44. Soil Conservation Service. *SCS National Engineering Handbook, Section 4, Hydrology*; Govt. Print. Off.: Washington, DC, USA, 1972.
45. Koutsoyiannis, D.; Xanthopoulos, T.H. *Engineering Hydrology*, 3rd ed.; National Technical University of Athens: Athens, Greece, 1999; 418p.
46. Wanielista, M.; Kersten, R.; Eaglin, R. *Hydrology: Water Quantity and Quality Control*; Wiley: New York, NY, USA, 1997.
47. Brunner, G.W. *HEC-Ras River Analysis System: User's Manual*; US Army Corps of Engineers, Hydrologic Engineering Center: Davis, CA, USA, 2021.
48. Brunner, G.W. *HEC-Ras River Analysis System: Hydraulic Reference Manual*; US Army Corps of Engineers, Institute for Water Resources, Hydrologic Engineering Center: Davis, CA, USA, 2021.
49. Auret, L.; Aldrich, C. Interpretation of nonlinear relationships between process variables by use of random forests. *Miner. Eng.* **2012**, *35*, 27–42. [CrossRef]

hydrology

Article

Forensic Hydrology: A Complete Reconstruction of an Extreme Flood Event in Data-Scarce Area

Aristoteles Tegos [1,2,*], Alexandros Ziogas [3], Vasilis Bellos [4] and Apostolos Tzimas [3]

1 Laboratory of Hydrology and Water Resources Development, School of Civil Engineering, National Technical University of Athens, Heroon Polytechneiou 9, 15780 Zographou, Greece
2 Ryan Hanley Ltd. Ireland, 170/173 Ivy Exchange, Granby Pl, Parnell Square W, D01 N938 Dublin, Ireland
3 EMVIS S.A., Consultant Engineers-Environmental Services Research Information Technology & Services, 21, Paparrigopoulou Str., 15343 Agia Paraskevi, Greece; aziogas@emvis.gr (A.Z.); atzimas@emvis.gr (A.T.)
4 Laboratory of Ecological Engineering and Technology, Department of Environmental Engineering, School of Engineering, Democritus University of Thrace, Vas. Sofias 12, 67100 Xanthi, Greece; vbellos@env.duth.gr
* Correspondence: tegosaris@itia.ntua.gr

Abstract: On 18 September 2020, the Karditsa prefecture of Thessaly region (Greece) experienced a catastrophic flood as a consequence of the IANOS hurricane. This intense phenomenon was characterized by rainfall records ranging from 220 mm up to 530 mm, in a time interval of 15 h. Extended public infrastructure was damaged and thousands of houses and commercial properties were flooded, while four casualties were recorded. The aim of this study was to provide forensic research on a reconstruction of the flood event in the vicinity of Karditsa city. First, we performed a statistical analysis of the rainfall. Then, we used two numerical models and observed data, either captured by satellites or mined from social media, in order to simulate the event a posteriori. Specifically, a rainfall–runoff CN-unit hydrograph model was combined with a hydrodynamic model based on 2D-shallow water equations model, through the coupling of the hydrological software HEC-HMS with the hydrodynamic software HEC-RAS. Regarding the observed data, the limited available gauged records led us to use a wide spectrum of remote sensing datasets associated with rainfall, such as NASA GPM–IMREG, and numerous videos posted on social media, such as Facebook, in order to validate the extent of the flood. The overall assessment proved that the exceedance probability of the IANOS flooding event ranged from 1:400 years in the low-lying catchments, to 1:1000 years in the upstream mountainous catchments. Moreover, a good performance for the simulated flooding extent was achieved using the numerical models and by comparing their output with the remote sensing footage provided by SENTINEL satellites images, along with the georeferenced videos posted on social media.

Keywords: IANOS; medicane; Karditsa; HEC-HMS; HEC-RAS; remote sensing; SENTINEL

Citation: Tegos, A.; Ziogas, A.; Bellos, V.; Tzimas, A. Forensic Hydrology: A Complete Reconstruction of an Extreme Flood Event in Data-Scarce Area. *Hydrology* **2022**, *9*, 93. https://doi.org/10.3390/hydrology9050093

Academic Editor: Andrea Petroselli

Received: 18 April 2022
Accepted: 18 May 2022
Published: 20 May 2022

1. Introduction

Floods are among the most destructive natural hazards, and are caused by river overflows, flash floods of ephemeral streams, pluvial floods in the cities, floods in the coastal zone, and floods due to a potential dam or a levee failure, and with several time scales, ranging from large-scale to flash floods. Having identified a growing concern that the flood risk is increasing in Europe and globally, joint scientific efforts are necessary for establishing a reliable flood risk management framework [1]. The latter, associated with the increasing stress to the system due to urbanization and the changing climate, led the European Union to set in force the new Flood Directive 2007/60, which aims to provide a thorough investigation of the flooding risk in vulnerable areas with the use of advanced hydrological and hydrodynamic environmental approaches, and minimizing the flooding risk with structural and non-structural measures [2].

In accordance with Koutsoyiannis et al., 2012 [3], the main categories of potential flooding areas in Greece are associated with large rivers with insufficient capacity to route the natural flood and floods caused in ephemeral streams, whose cross-section dimensions have been significantly reduced by anthropogenic activities (land use change, urban sprawl). The latter, in conjunction with the limited available gauge network in the Greek catchments, offers a great challenge to the geoscience scientific community, to gather all the publicly available sources and use them within a consistent flood management framework. Flood risk management is associated with complex uncertainty sources related to hydrological, hydraulic, environmental, and social phenomena, and these sources significantly influenced the recent flooding event in Karditsa city [4].

IANOS was an intense medicane formed over the warm Mediterranean Sea. Following a path of approximately 1900 km, the medicane affected Greece, resulting in four casualties in Karditsa prefecture and devastating damage in the western and central parts of Greece. Analysis of the available observations showed that IANOS was the most intense medicane ever recorded in the Mediterranean [5]. The 15-h rainfall records ranged from 220 mm in Karditsa city up to 530 mm at the Plastiras dam rainfall gauge station. The flooding event was followed by extended public infrastructure deterioration, bridge collapses, soil sliding, and high debris flow, which were well documented by Zekkos et al., 2020 and Lolli et al., 2022 [6,7].

Several reconstructions of extreme flood events can be found in the literature, such as Borga et al. (2007) [8] and Costabile et al. (2013) [9]. Recently, similar works have been presented, providing coupled hydrological and hydraulic modelling of severe past flood events in Greece [10–14]. In this work, we followed the forensic hydrology framework, as proposed by Ramirez and Herrera (2016) [15], in which a reproduction of an extreme event has the following phases: (a) information gathering and integration; (b) hydrometeorological and hydrological analysis; (c) hydraulic analysis; (d) integrative analysis; and (e) final diagnosis.

According to this framework, first we used the full spectrum of available remote and gauge information to inform the spatial variability of the rainfall depths over the catchment study area. In addition, we collected observed data using new technologies, such as a remote sensing (SENTINEL platform), which indicated the flood inundation area, and crowdsourcing (videos uploaded to Facebook), which indicated the arrival time of flooding and the water depths. Then, we estimated the return period of the event and applied a rainfall–runoff hydrological model in the Kaletzis catchment (with runoffs at Karditsa city), using the HEC-HMS software, and having as an input, findings from the previous phase. The result of this phase was the derivation of the flood hydrographs that hit the greater area of Karditsa city. This is the next phase input, namely the hydraulic analysis. For this phase, a 2D hydrodynamic model was used for the flood propagation through the urban and peri-urban areas. The results of this phase were validated against the data collected during the first phase.

To our knowledge, this study is the first integrated hydrological–hydrodynamic analysis implemented for the greater area of Karditsa city, and in which a plausible check is performed regarding the results derived by numerical modelling, using a wide spectrum of satellite datasets and crowdsourced data, aiming to tackle the main challenge, which is the lack of flood-related data and measurements.

2. Materials and Methods

2.1. Study Area

Karditsa city is located in the south-western part of the Thessaly region, Karditsa prefecture, Greece (Figure 1). It has a population of approximately 42,000 people, based on the 2011 population census. The city lies within the Kaletzis river catchment, and two rivers are drained south and east of the city, which are named Gavrias and Karampalis, respectively (Figure 1c). The Kaletzis catchment has an area of 653.8 km^2, while the average elevation of the watershed is 254.8 m. The maximum river length is estimated to be 66.11 km. The main sub-catchments of Kaletzis relevant to the study (marked in Figure 1c) are the (a) upstream and downstream Karampalis stream catchments, which runoff at the south-eastern part of Karditsa city; and (b) three upstream catchments of the Gavrias stream, which runoff into the south-western part of the city, and through a man-made canal are conveyed to the Karampalis river. The determination of the land cover was based on CORINE land cover data. The majority of the watershed is covered by agricultural areas and forest–mountainous areas.

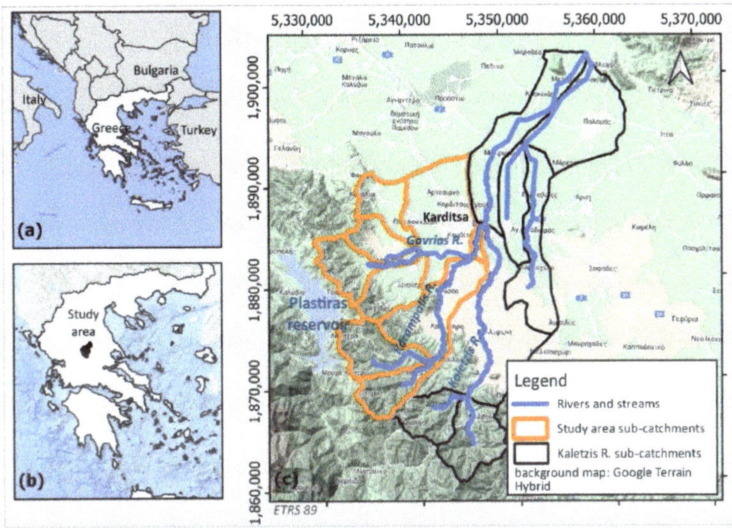

Figure 1. (**a**) Greece, (**b**) study area location within Greece and (**c**) catchments and rivers of the study area.

On 18 September 2020, the city of Karditsa was hit by the extreme medicane IANOS. An intense rainfall that lasted approximately 15 h had as a consequence severe economic losses and damage to public assets, transportation networks, buildings, and agricultural areas, including four human losses. The city of Karditsa was flooded by extreme river overflows, mainly from the Karampalis river, and most of the urban city area remained flooded for over two days. Rainfall records from local gauges and remote sensing datasets are detailed in the forensic analysis presented herein. It should be noted that the city has a sewer stormwater gravity system, which, however, failed before the river's flooding. Although the pluvial flood component contributed to the flooding in the city area, it was considered a small part of the fluvial flooding volume of the most intense medicane ever recorded in the Mediterranean, as the extreme precipitation magnitude recorded in the mountainous area suggests. The focus of the present work was given to the fluvial components that led to the extreme flooding of the greater Karditsa area, and the pluvial mechanism was excluded from further analysis.

2.1.1. Remote Sensing Flooding Records

The catchment area is ungauged, without the presence of flow and level gauge stations along the river system, and a reliable flood simulation is a challenging scientific task, since there are no records to validate the numerical results. Some rainfall gauges are operated by various authorities, such as the Public Power Company, National Observatory, and Minister of Public Works. In our analysis, due to the absence of a monitoring system, remote sensing dataset has been used to estimate the rainfall patterns of the IANOS event, which is in line with previous studies on event-based flood hydrological modelling [16–18]. Specifically, the following remote sensing rainfall has been considered for further use:

- MSWEP Multi-Source Weighted-Ensemble Precipitation
- GPM-IMREG (Global Precipitation Measurement) NASA
- CMORPH, Climate Prediction Center National Weather Service National Oceanic and Atmospheric Administration (NOAA)
- PERSIANN-CCS (Precipitation Estimation from Remotely Sensed Information using Artificial Neural Networks- CHRS Irvine)

GPEM-IMREG products and PERSIANN-CCS were selected based on record event availability.

Two re-analysis products were also collected for investigating the daily rainfall spatial extents over the catchments, namely:

- ERA5 land (ECMWF) (ERA5-Land hourly data from 1981 to present)
- MERRA land (Modern-Era Retrospective analysis for Research and Application) Global Modeling and Assimilation Office NASA.

More details on the/reliability of the remote sensing rainfall dataset for reproducing the real rainfall pattern are presented below (see Section 3.2).

In order to investigate the flooding event and support the modeling efforts, nine flooding remote sensing recordings were collected by SENTINEL-1 and SENTINEL-2 associated with two delineation products and seven grading products, produced by the Copernicus Emergency Management Service (EMS). Remote sensing products are shown in Figure 2. Records are available for 20 September 2020 and 24 September 2020, two and five days after the main event respectively. A third post-processing map was used, and it is associated with the maximum combined flood extent captured by SENTINEL and provided by Zekkos et al., 2020 (see Figure 2c) [6].

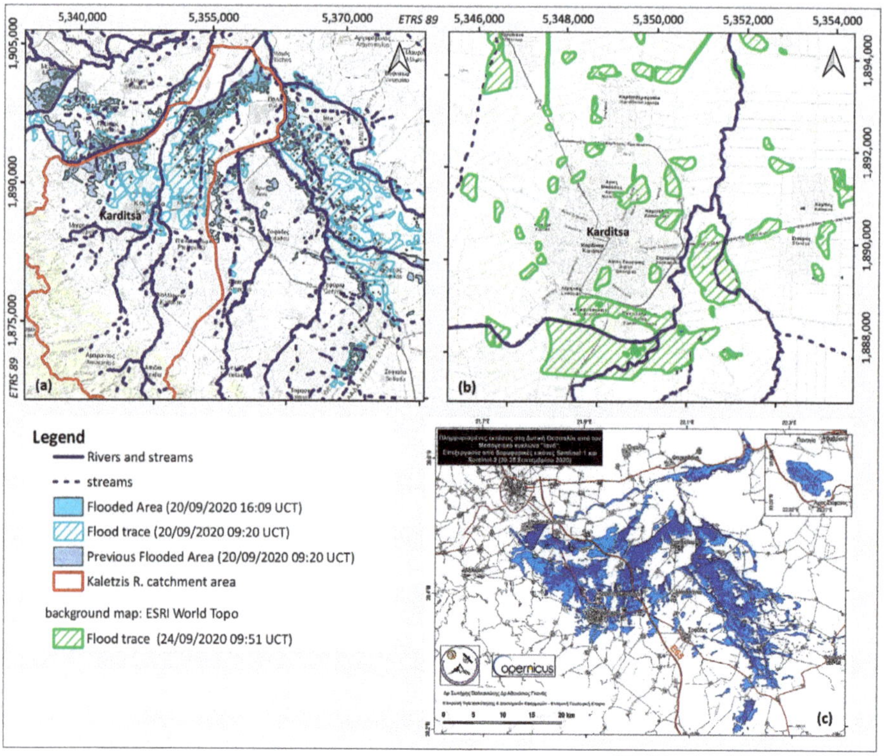

Figure 2. Remote sensing flooding footage products: (**a**,**b**) Copernicus EMS–Mapping products and (**c**) Copernicus Sentinel-1 and Sentinel-2 in map by Zekkos et al. [6].

2.1.2. Crowd Sourcing Data

Public evidence is critical for the confirmation of a flood wave propagation [19], and therefore numerous videos posted on social media along with photos were gathered to use in the validation of the flooding extent and evolution. Nine videos were collected, which were well-distributed along the floodplain and highlight the importance of relevant information in analyzing complex flooding events. All of the aforementioned information was used in combination with Google Maps to identify (when that was possible) the location of each element-record (Figure 3). Based on the characteristics of the damage estimation records, the use of this dataset was restricted to the validation of the flood extent.

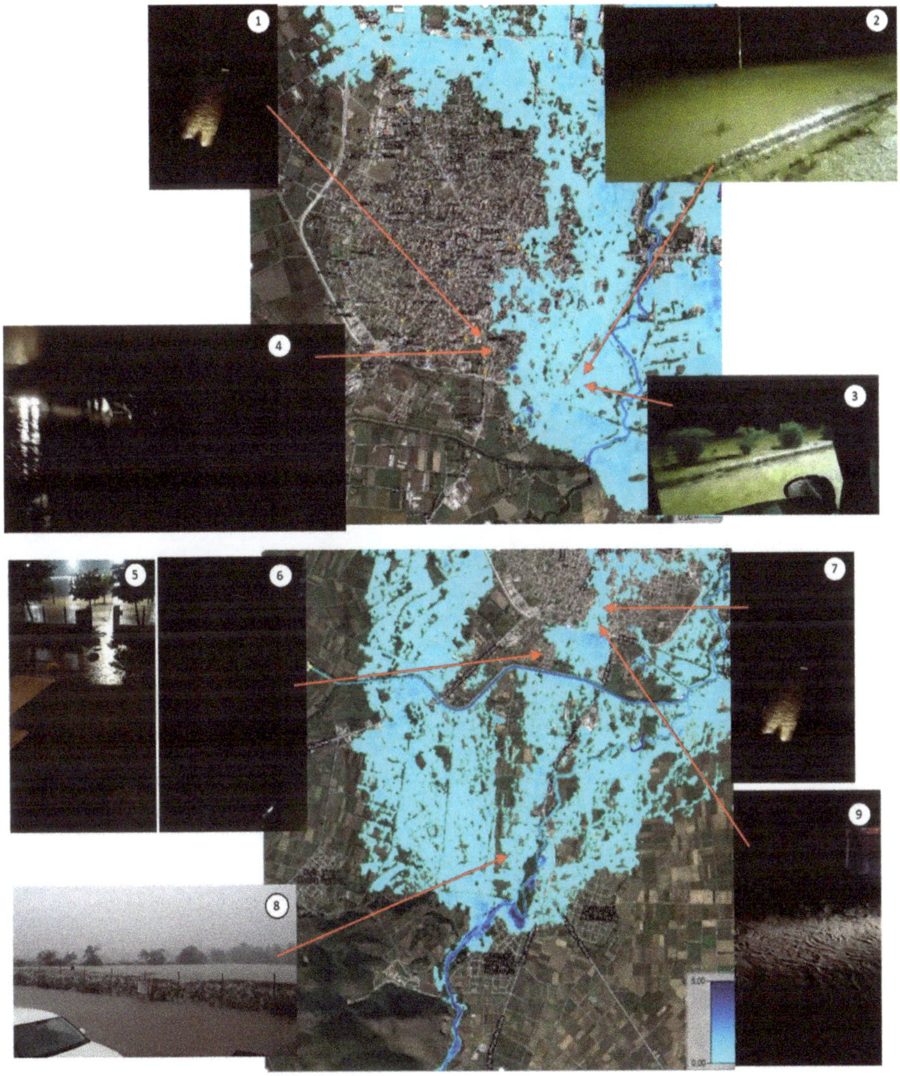

Figure 3. Spatial distribution of the ΙΑΝΟΣ flood data (event of 18 September 2020) collected by social media videos in comparison to a simulated flood map (refer also to Section 3.4 for more photo records) at 9 positions.

2.2. Hydrological Model Set-Up

The Hydrologic Engineering Center—Hydrologic Modelling System (HEC-HMS) software was developed by the US Army Corps of Engineers and incorporates several rainfall–runoff models, for the determination of the hydrological response of the catchments of a river. It is one of the foremost and most globally well-known software tools for the hydrological simulation of flood relief schemes, drainage, flood early warning systems, dam design, debris flows simulation, etc. [13,20,21]. The Kalentzis catchment was delineated and divided into 16 sub-basins using the HEC-GeoHMS extension in ArcGIS. Terrain preprocessing and basin processing tools were used to generate the basin model file containing the drainage network and delineated catchment. Precipitation was estimated using the Thiessen polygon method and suitable weights for each subbasin were defined to create a meteorological model. Remote sensing rainfall datasets and gauged rainfall records were compared to identify the most suitable rainfall pattern. More details on the aforementioned comparison are provided below. As an outcome, gauged precipitation data from the Karditsa (sub-hourly) and the Plastiras dam (15-min) gauging station were selected.

In general, HEC-HMS allows for the separate modelling of hydrological processes; loss, transformation, baseflow, and routing, with several models for each process. The selection of the model for each process should be based on the catchment characteristics, data availability, and whether the simulation is event-based or continuous. The SCS-CN unit hydrograph method was selected to simulate the transformation, in conjunction with the deficit and constant loss method, and the recession baseflow model. The lag time t_p, which is a parameter required by the methodology, was defined as the time period between the centroid of excess rainfall and the peak discharge. The later was calculated for each sub-basin using the Giandiotti equation [22], in order to derive the time of concentration. Average soil-moisture condition was selected considering the active irrigation period for the extended irrigation areas in low-lying catchment areas, and a small rainfall of about 10 mm occurred before the main flooding event.

Regarding the routing method, the well-known Muskingum–Cunge model was applied to all reaches, with a Manning n coefficient equal to 0.040. Parameters such as reach length and slope were estimated using topographical data. The overall hydrological schematization follows the technical specification of the Flood Directive implementation in Greece, and more details are provided by Papaioannou et al. [23].

2.3. Hydrodynamic Model Set-Up

The HEC-RAS software was used for the hydraulic routing simulation of the flood hydrograph through the Karditsa town stream network. It is a well-known software developed by the Hydrologic Engineering Center (HEC) of the U.S Army Corps of Engineers and used for river flood modelling and floodplain management [4,11]. Since the data availability was limited (lack of detailed river surveys, extended low-lying flood plains) and as the urban and peri-urban area of Karditsa city is quite complex, having multiple hydraulic directions due to a low-lying surface, we selected the two-dimensional (2D) mode of the HEC-RAS software. The latter mode is either based on the full form of the 2D shallow water equations (2D-SWE) or in 2D diffusion wave equations, which were selected after a sensitivity analysis, with respect to numerical accuracy and computational time. The latter approach was successfully applied in similar projects in the past, especially in data-scarce areas [11,13,24]. It should be mentioned that the hydrodynamic model setup was developed in line with the Flood Directive contracts, as previously outlined by Papaioannou et al. [23].

The importance of DEM accuracy has been highlighted by several authors, especially in two-dimensional hydraulic–hydrodynamic modelling applications [25–27]. To meet these requirements, a DEM with a horizontal resolution of 5 × 5 m generated from aerial images collected from 2007 to 2009 and provided by the National Cadastre and Mapping Agency S.A. (NCMA) was used in this study.

A critical parameter in hydrodynamic modelling applications is the selection of the roughness coefficient in the entire computational area [28]. In this work, we coupled the corresponding values proposed in Greek Flood Management Plans [29] informed by LAND COVER maps.

One of the modelling challenges was the representation of various hydraulic structures, such as bridges, culverts, weirs, etc., with respect to the cell size and their false description within the DEM. A non-detailed DEM spatial resolution, in combination with the appearance of natural or artificial structures close to the flood mitigation works and hydraulic structures, can lead to distortions of the elevation and by extension to their false representation within the DEM. To overcome this problem, a sensitivity analysis was carried out at the locations where structures exist. In order to represent the structure's roughness and potential blockage effect during the flood event, a local increase of the Manning coefficient was justified. A sensitivity analysis was also carried out, regarding the optimal computational grid size, which influences the modelling accuracy [27] and the computational time substantially. Based on this analysis, we selected a squared grid-size of 20 m.

3. Results

3.1. Extreme Statistical Analysis of Plastiras Reservoir Annual Runoff

As previously described, there is lack of long-term reliable gauge records in the riverine system, which would offer the capability to understand the complex transformation from rainfall to real runoff. To overcome this problem and in order to assess the exceedance probability of the IANOS flooding event, an annual extreme statistical analysis was carried out for 12-years of the annual max daily water level of the Plastiras reservoir located in the west of the study area. The reservoir is a multipurpose reservoir operating for 70-years and having irrigation, water supply, and tourism uses [30,31]. HYDROGNOMON software was used to fit numerous suitable statistical distributions [32].

EV2-max was selected as an appropriate statistical distribution, after applying the Kolmogorov–Smirnov test. Figure 4 visualizes that the daily 3-m reservoir level rise approximately corresponds to a 1-in-200-year event (according to the theoretical distribution). The statistical analysis is shown herein only for indicative purposes, since the 12-year time record was insufficient for demonstrating a reliable extreme statistical analysis.

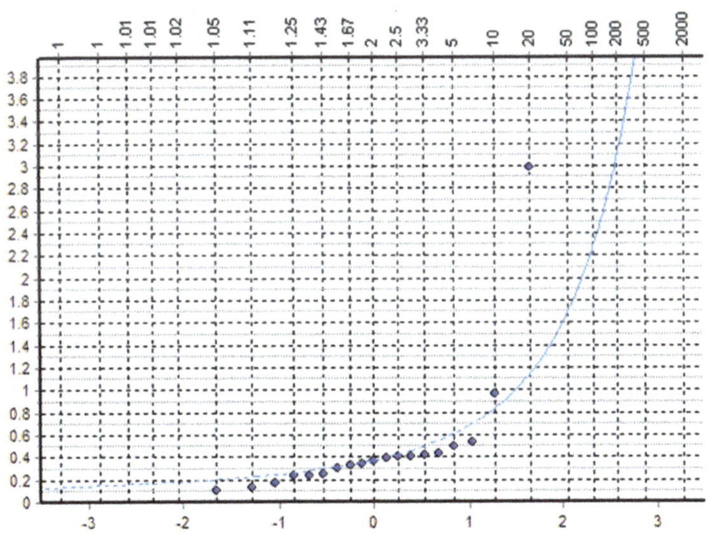

Figure 4. EV2-max statistical distribution on annual the max reservoir level (y-axis daily reservoir raising level in m, x-axis normal distribution).

3.2. Rainfall–Runoff Analysis

A comparative analysis was performed for the spatial variability of the rainfall regime, using a suite of the available remote sensing dataset rainfall records and rainfall gauge records at fine time scale. For the hydrometeorological formation of the extreme phenomenon, the readers are encouraged to study the work presented by Karagiannidis et al., 2021 [5].

As outlined above, three NASA GPM–IMREG products were assessed, named according to the initial assessment of the satellite records, and specifically the "Early", the "Late", and the "Final" products, which include further post-processing based on the ground meteorological observations. The "Final" product has significant differences in comparison with the corresponding "Early" and "Late" versions, due to the use of a climate change adjustment factor. It significantly underestimates the magnitude of the precipitation and presents a "buffered" and smooth temporal distribution, contrary to the data captured by local, ground measurements. Therefore, it fails to reproduce the IANOS rainfall spatial event. It seems that the "Late" product is the most suitable for describing the spatial variability of the extreme meteorological event, since it presents a better fit for the ground measurements of the representative ground stations for the study area, in terms of total precipitation, as well as the temporal distribution of precipitation. Figure 5 depicts the NASA satellite products for different time records. The higher rainfall records in the vicinity of the study area for both Early and Late products can be seen.

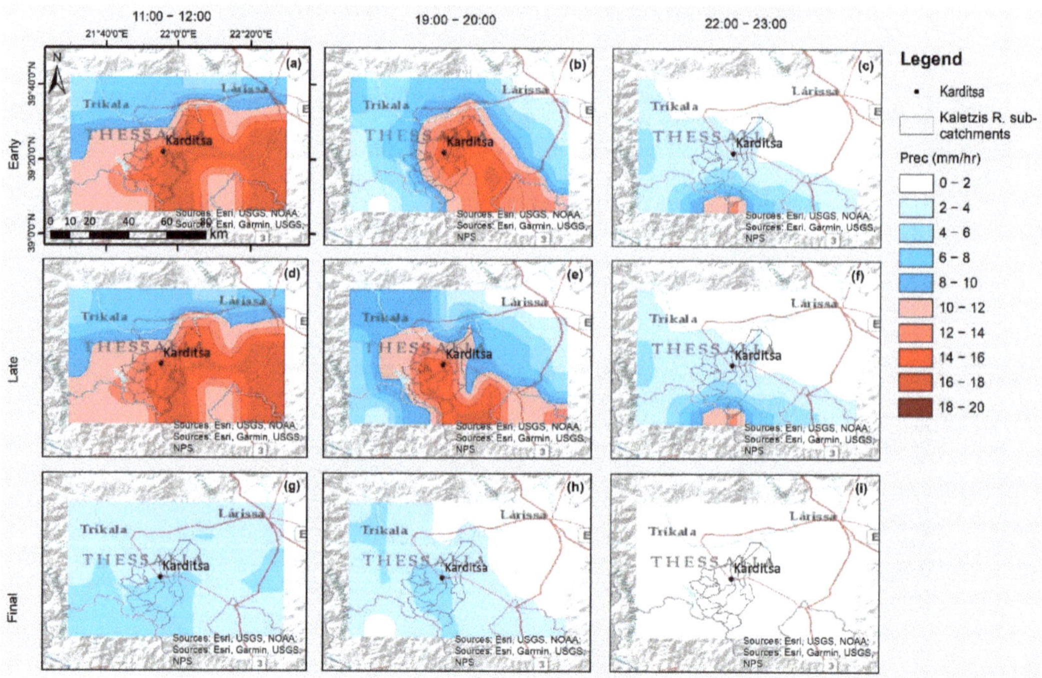

Figure 5. Records of different remote sensing NASA GPM-IMREG datasets (Early, Late, Final) for three time intervals on 18 September 2020: (**a**) Early; 11:00–12:00, (**b**) Early; 19:00–20:00, (**c**) Early; 22:00–23:00, (**d**) Late; 11:00–12:00, (**e**) Late; 19:00–20:00, (**f**) Late; 22:00–23:00, (**g**) Final; 11:00–12:00, (**h**) Final; 19:00–20:00, (**i**) Final; 22:00–23:00.

Except for the above, two reanalysis remote sensing products were also gathered and analyzed in conjunction with three NASA GPM–IMREG estimates: the ERA5 land-CNR and the MERRA land-NASA reanalysis products. Both of them underestimate the rainfall in comparison with the NASA-GPM–IMREG, as observed in Figure 6.

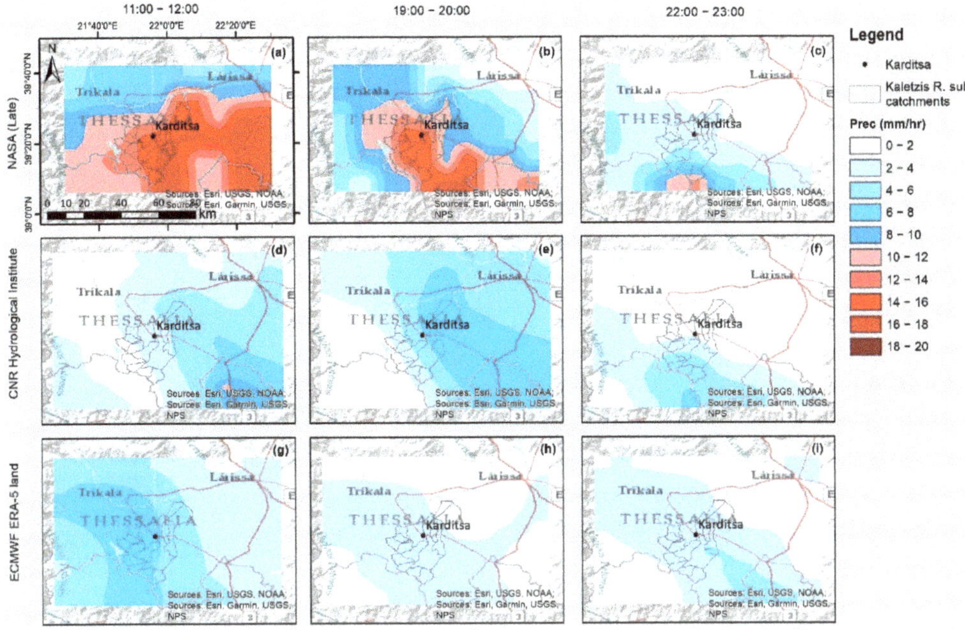

Figure 6. Records of different remote sensing datasets for three time intervals on 18 September 2020: (**a**) NASA (Late); 11:00–12:00, (**b**) NASA (Late); 19:00–20:00, (**c**) NASA (Late); 22:00–23:00, (**d**) CNR Hydrological Institute; 11:00–12:00, (**e**) CNR Hydrological Institute; 19:00–20:00, (**f**) CNR Hydrological Institute; 22:00–23:00, (**g**) ERA-5 land; 11:00–12:00, (**h**) ERA-5 land; 19:00–20:00, (**i**) ERA-5 land; 22:00–23:00.

Following on from the above comparative analysis, it can be concluded that NASA's and CNR's remote products show better agreements and greater rainfall depths in the south and southeast of the study area. In contrast, ERA5 land (ECMWF) shows higher rainfall depths in the southwest of the study area. Figure 7 exhibits the daily cumulative rainfall depths retrieved by the aforementioned remote sensing products. The NASA GPM–IMREG Late product shows a higher daily rainfall depth of up to 200 mm and a better performance for the greater area of the city of Karditsa. ERA-5 and CNR products present substantially lower records, of up to 100 mm.

All the remote sensing products failed to provide accurate rainfall records in comparison with the gauged rainfall records. Specifically, the 15-min records from the Plastiras dam gauge station west of the study area exhibit a relatively higher record of approximately 530 mm in the 15-h time period on the 18 September 2020. Karditsa's rainfall gauge station provides a lower record of 220 mm for the same record period (Figure 8).

Although the remote sensing precipitation products underestimated the magnitude and intensity of the phenomenon, they provided useful insights regarding the evolution and spatial variation of the phenomenon. The spatial information revealed, indicates that the available gauging stations captured the spatial variability of the event's precipitation and could be used for the hydrological investigation. In the light of the above analysis, the gauge rainfall records were the most suitable for the flooding event analysis; and following a Thiessen analysis, the point records of Plastiras dam station and Karditsas station were mapped over the sub-catchment of the study area. The mountainous sub-catchments have a significantly higher rainfall input, influenced by the Plastiras dam record representing 30% of the total catchment study area. The low-lying catchments were impacted by Karditsa's station rainfall record.

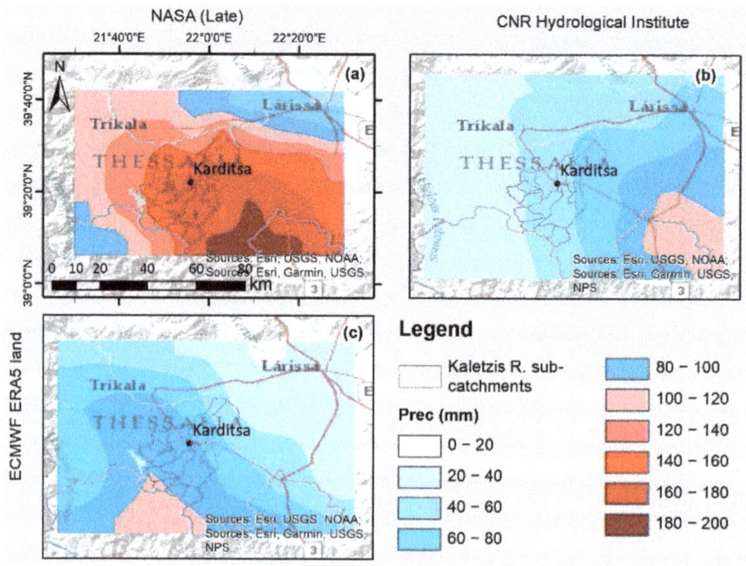

Figure 7. Cumulative remote daily records for the three different gridded precipitation products: (**a**) NASA (Late), (**b**) CNR Hydrological Institute and (**c**) ECMWF ERA-5 land.

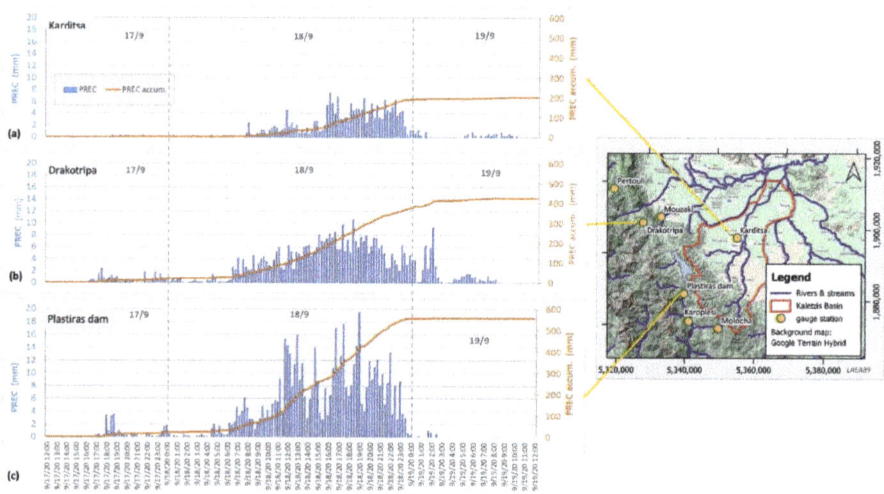

Figure 8. Rainfall gauge records: (**a**) Karditsa, (**b**) Drakotripa and (**c**) Plastiras dam.

The final output of the HEC-HMS software is depicted in Figure 9. Specifically, three flood hydrographs are presented at the junctions of interest. For the J5 junction of the Gavrias river (which is located in the southwest part of the city), the maximum peak flow was simulated as about 630 m^3/s, while the time to peak was estimated as about 12 h. The Karampalis upstream junction J6 peak flow was estimated as about 600 m^3/s, while Karampalis downstream junction peak flow was estimated as about 1400 m^3/s. At the latter junction, the Gavrias river and upstream Karampalis are connected. The latter provided a specific event discharge ranging from 5.06 to 10.5 $m^3/s*km^2$. It is worth mentioning that the time to flooding responses of the Gavrias and upstream Karampalis sub-catchments coincided, leading to a very high peak flow in the east part of the city.

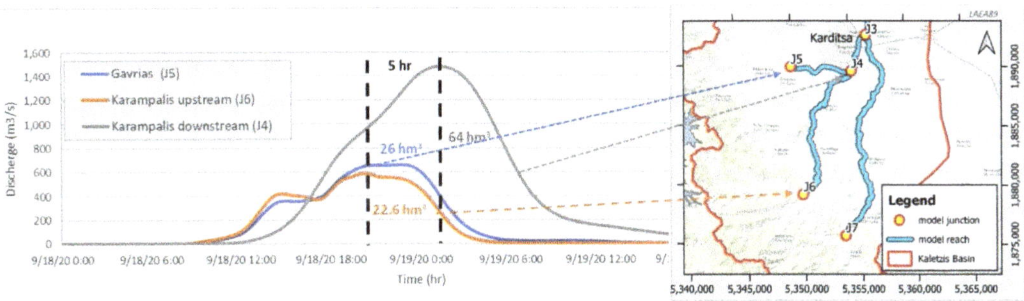

Figure 9. Generated hydrographs at river junctions (hm^3: cubic hectometres).

In order to quantify the return period of the IANOS event, three rainfall scenarios were developed using the intensity–duration–frequency curves of Karditsa station, namely T = 50 years, 100 years, and 1000 years. The latter was developed as part of the national implementation of the Flood European Directive [33]. A recent framework presents more insights into the regional implementation of the intensity–duration–frequency in the study area [34], and it is recommended for defining the extreme rainfall depths for different exceedance probabilities.

Figure 10 shows the plots of the estimated return period with respect to the flows (a and b), as well the flooding volumes (c and d), which are the equivalent volumes extracted by the flow event hydrographs. It was calculated that the IANOS flooding event's return period is around 400 years for the downstream low-lying catchments and about 1000 years for the upstream mountainous catchments. Most interestingly, the flooding volume return periods were estimated as being 1000 years for the low-lying areas, and for the upstream catchments reaching 10,000 years, highlighting the catastrophic nature of the flooding event.

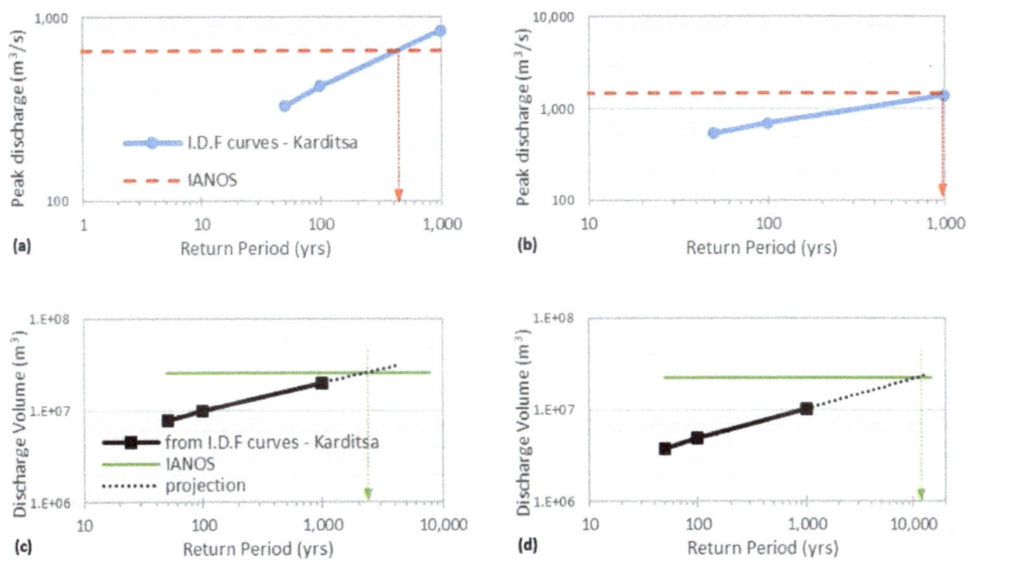

Figure 10. Estimated return period event: (**a**) flows at Gavrias junction (J5), (**b**) flows at Karambalis downstream junction (J4), (**c**) volumes at Gavrias junction (J5), (**d**) volumes at Karambalis upstream junction (J6).

3.3. Flood Mapping

Figure 11 presents the maximum water depths and the maximum flood extent simulated by the HEC-RAS software, having as an input the HEC-HMS output for each junction point, namely J4, J5, and J6. An extensive overflow of the upstream Karampalis river was observed in the south part of the city areas. The Garvias river was flooded in the northwest part of the city. Its overflow was directed to the north, while the remaining flows were routed through the man-made canal in the north part of the town as well. It is worth mentioning that the extreme flows from the upstream Karampalis catchments had as a consequence the overtopping of the Gavrias river and the railway rail embankments. These overflows were propagated from the north floodplains to the city center through the complex urban stream network. The most severe overflows that impacted the city were observed in the east Karampalis river, downstream of the junction with the Gavrias river.

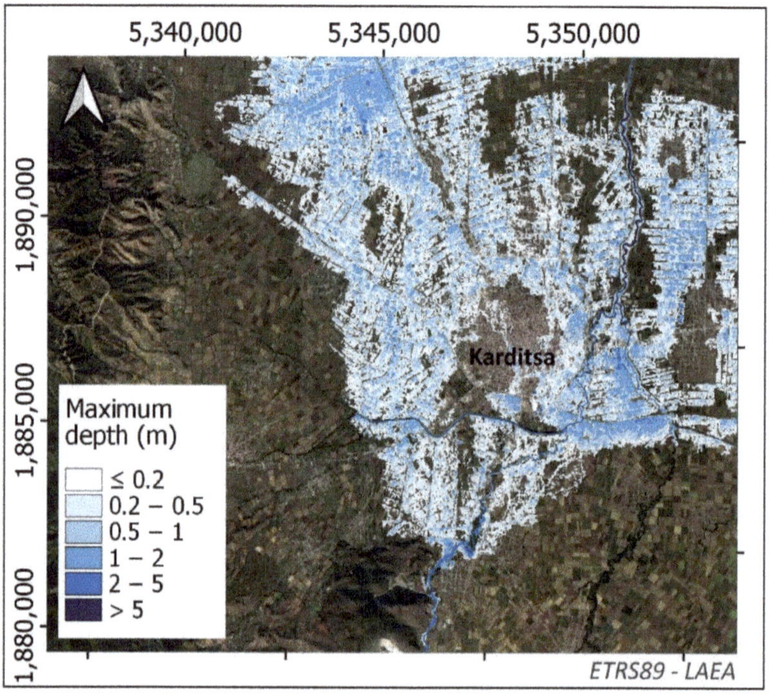

Figure 11. IANOS flood extent map.

For the main channel of the river, the maximum water depth was simulated as about 5 m, while the corresponding maximum water depth in floodplain was about 1 m. All the simulations demonstrated that the city was hit by a significant flooding wave coming from the west, east, and north.

3.4. Validation

Since remote flooding footage is critical for validating the performance of our simulations, three satellite flooding footages were used to compare the simulated flood extent with the observed flood extend [35]. In this vein, two SENTINEL satellites images were acquired from the Copernicus EMS service: (a) the first refers to approximately 35 h (date 20 September 2020) after the estimated peak flow of the flood event; (b) the second refers to the flooding extent 5 days after the event. In addition to this dataset, a post-event flood image was acquired by Zekkos et al., 2020 [6] and is also presented here. Post-event satellite images of the same day or the next day after the flood were sought, in order to supplement the analysis with an as recent as possible flood extent delineation. However, the cloud coverage was dense for these dates, significantly reducing the potential of those images for useful flood extent delineation.

Figure 12 shows the simulated flood extent, in comparison to the observed flood extent provided by the satellites. The performance, in terms of the main flooding extent and the capturing of the main, overland flood pathways, was satisfactory. As expected, the simulated flood extent was greater than the three observed extents. The observed footage is a snapshot, referring to significantly later observation times, varying from 1.5 to 5 days post-event; while the simulated flood represents the maximum flood extent during the development of the phenomenon. It can be safely assumed that, in the period following the flood, the overflows were drained through the town's sewer network.

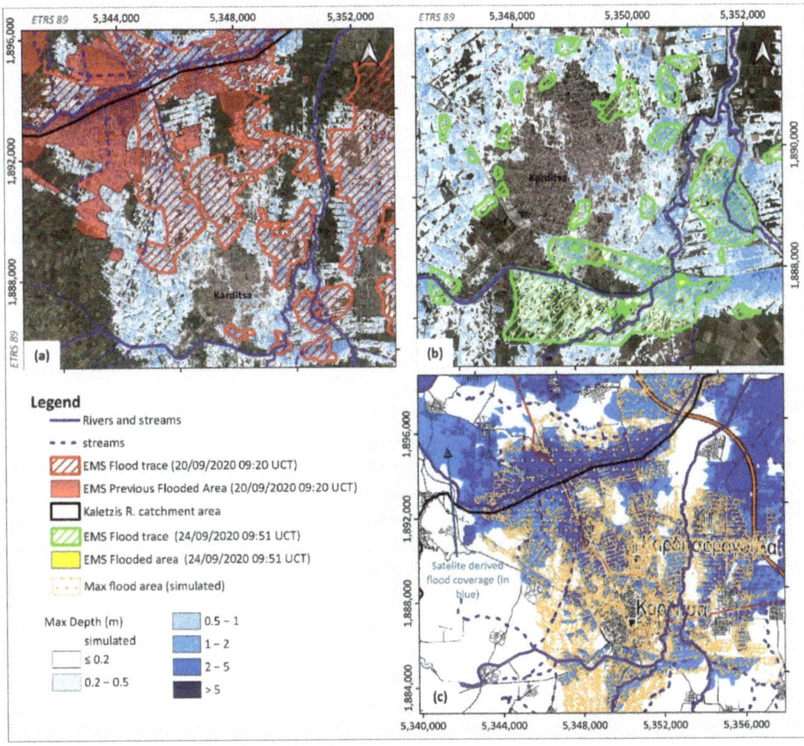

Figure 12. Flood extent map—simulation results comparison with: (**a**) EMS Flood capture of 20 September 2020, (**b**) EMS Flood capture of 24 September 2020 and (**c**) combined flood extend assessment by Zekkos et al., 2020 [6].

In addition to the remote sensing data, crowdsourced photos and videos captured by social media were used, in order to perform a plausible check of the simulated flood maps and evolution of the simulated flood. Numerous georeferenced photos were posted on Facebook approximately 12 h after the peak flow. Some of these are depicted in Figure 13. Furthermore, several posted videos exist, which are provided in the appendix. As mentioned in the previous section, the town's sewer system was submerged during and after the flood event. As the gauging records reveal, extreme precipitation was recorded in the mountainous part of the basin. Although the pluvial flooding component contributed to the flooding in the city area, it was considered a small part of the fluvial flooding volumes. Therefore, we focused our analysis on the fluvial flood component, coming from the Gavrias, upstream Karampalis, and downstream Karampalis rivers.

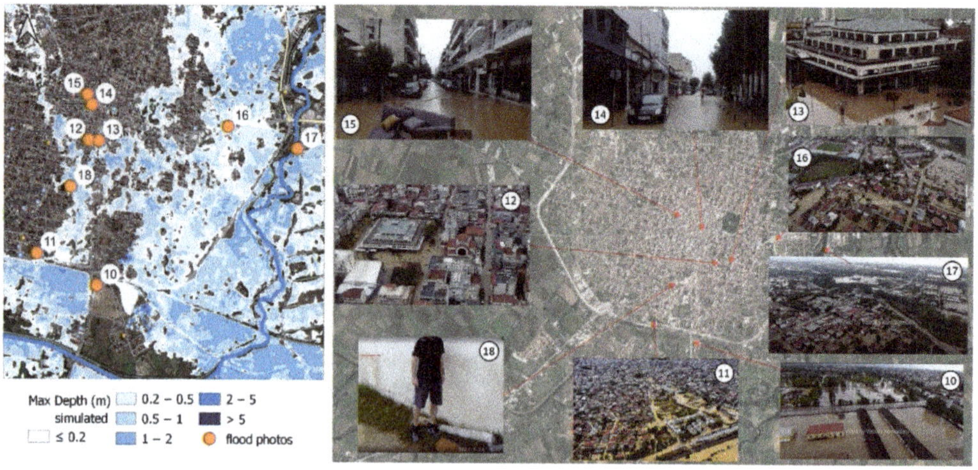

Figure 13. Simulated flood map vs. photo records in 9 positions (continuous numbering from Figure 3).

Flood depth estimations were made based on the crowdsourced data. Two sets of estimations were carried out. Photos of the next day (19 September 2020, see Figure 13) were utilized to estimate the depth that the flood reached (positions 10 to 18). These depths were compared with the maximum simulated depth in those positions. Furthermore, videos capturing the evolution of the flood were utilized to estimate the depth at the time of the video capture (see Figure 3, points 1 to 9). The depth estimation was conducted based on expert judgment and knowledge of the area. In order to validate the model, the simulated flood depth was calculated for the corresponding observation points (1 to 18). Due to the uncertainty of the depth estimations based on the photos and videos, as well as the fact that the areas captured in the photos do not represent a single, specific point, the corresponding modelling depth was calculated for an area of 1 cell radius (when larger areas were depicted in photos, larger areas were used from the model results), as a range of values. The range of the simulated depth values was compared to the estimated ones in Figure 14.

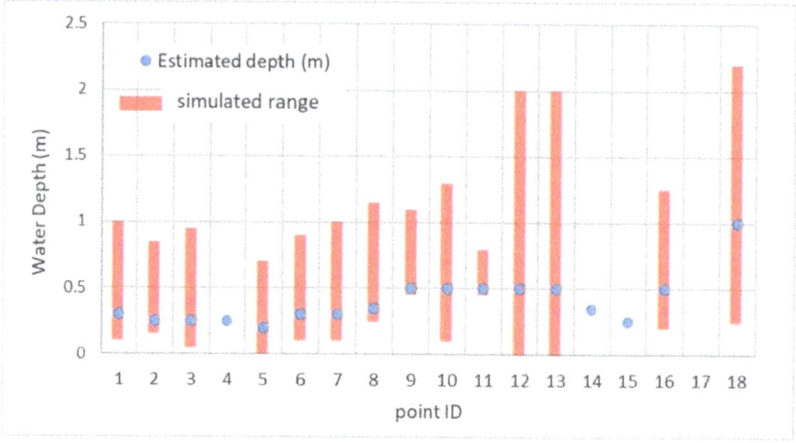

Figure 14. Modelled vs observed water depths based on the crowdsourced dataset.

Based on this comparison, the estimated observed values lie within the range of the simulated depths. The modelled (averages) depths overestimated the estimated actual depths by about 0.25 m or less in the majority of positions considered. Additionally, the spatial (between observation) variation of the estimated observed values is consistent with the variation of the corresponding modelled (averages) depths. Based on the above, and taking into account the various factors of uncertainty, the agreement between modelled and observed values is considered satisfactory.

4. Discussion

Herein, we discuss how our analysis can contribute to existing knowledge, in order to reconstruct past flooding events using a suite of remote sensing datasets, limited gauged records, rainfall-runoff modelling, and 2D-hydrodynamic modelling. The main findings and issues for further consideration are presented below:

1. Both the rainfall spatial and temporal variability is of high importance in order to reconstruct a past flood event with reliability, especially in small catchments with high complexity (Bellos et al., 2020 [10]). Topography in smaller scales (such as local horographic phenomena) is a factor that can significantly affect the meteorological conditions of the atmosphere. For this reason, the scientific community has turned its attention to meteorological satellite and meteorological radar products, in order to derive distributed rainfall data, both in space and time. However, in our study the remote sensing rainfall dataset at fine time scale (namely 30 min records) seemed to underestimate the order of magnitude of the storm, in comparison with the ground meteorological stations. This is probably due to the fact that the space step of the satellite product is rather large and smoothens the extreme rainfall intensity, which was (fortunately) captured by the ground observations. Needless to say, this is not a global conclusion, but an ad hoc remark based on our study. In the end, there are no doctrines for reconstructing a flood event: all the available rainfall products should be evaluated and used.
2. The latter agrees with the recent global analysis presented by Pradhan et al., 2022 [36], an analysis of an extreme event in Mexico [37], highlighting the requirement for further improvements to achieve a higher rainfall estimate accuracy [38]. Similar results were presented by a comparative evaluation of GPM IMERG Early, Late, and Final hourly precipitation products over the Sichuan Basin, China [39]. Most interestingly, our findings, where the Early-run product was found to have a better performance than the Late and Final IMERG products, agrees with recent research on the Evaluation of IMERG GPM products during Tropical Storm Imelda [40]. An additional source of rainfall data is also the meteorological radars, which can achieve a better accuracy and denser resolution for both space and time. Finally, the importance of expansion of the ground meteorological network should not be underestimated. In the end, ground observations are the only "reality", compared with the proxy data provided by satellite and radar products.
3. Integrated catchment modelling is the most significant tool for reconstructing a past flood event. In our case study, we linked a rainfall–runoff model with a 2D hydrodynamic model and compared the model output with crowdsourced data. Regarding rainfall–runoff modelling, caution should be taken when assessing the time of concentration of different sub-catchments and assuming the pre soil moisture conditions of the catchment. All these parameters are very sensitive for reliable flow estimates in ungauged catchments and have already been referred to in previous studies [41–43].
4. The question raised by Apel et al., 2009 [44] is still valid, in respect to how detailed we need to be in flood modelling. In our case, the 2D hydrodynamic modelling of an extended low-lying irrigated area exhibited a satisfactory performance in reproducing an extreme event; however, 2D-modelling has some shortcomings (e.g., of blockage bridge assessment under high debris rates). It is recommended that the modelling analysis should be always considered in conjunction with the available survey information, as well datasets for validation. We should highlight that we cannot exclude

a priori any modelling option (2D or full 1D hydrodynamic modelling) and this is subject of the data availability in each case study.

5. Our analysis introduced a data-driven integrated hydrological–hydrodynamic assessment of a major past fluvial event, including several datasets for validating our model approaches. It was based on a deterministic approach, which is the current practice for natural hazards and exhibited satisfactory results herein. However, given the complexity and uncertainty associated with the hydrological and hydrodynamic components, probabilistic flood mapping approaches [2,4], coupled hydrological–hydrodynamic physically-based numerical modelling [45], and the recent hybrid-stochastic approaches are strongly recommended [46], especially when we deal with real-world engineering design.
6. It is generally accepted that flood studies suffer from a lack of data. The majority of the basins are ungauged, and in gauged basins an extreme and violent event, such as a flood, can destroy the monitoring system. Forensic hydrology gives a framework in which proxy data are mined from several sources, such as human observations (crowdsourced data). Recent technological advances, namely cell phones with good cameras, widespread internet access, and social media platforms, let us to derive this kind of data for flood studies more easily. A novel part of our study was the use of distributed public information posted on Facebook. This information seems to be a treasure trove for validating complex flooding events in data-scarce areas with unavailable gauged records. This has already been documented by several recent studies [47,48] and is strongly recommended for similar future studies.

5. Conclusions

The IANOS hurricane was an extreme hydrometeorological event, which caused catastrophic flooding in the Karditsa prefecture, with four casualties and extended infrastructure damage. The aim of this study was to present a combined approach of hydrological and hydrodynamic analysis with remote sensing and crowdsourced data analysis for the reconstruction of this flood event in the vicinity of Karditsa city, which was flooded by overtopping flows from the surrounding river system. The data availability was rather limited: there were few rainfall gauges, while there was no monitoring system for either flow or water level stages in the rivers of the area. First, an analysis of the rainfall spatially variability over the catchment was carried out using numerous freely available remote rainfall datasets, along with data captured by rainfall gauges. The analysis showed that all the remote sensing datasets underestimated the rainfall depths, and a rainfall–runoff analysis was performed using gauged rainfall along with a rainfall–runoff CN approach for assessing the river flows at representative river nodes. Although the examined remote sensing datasets were not used as input data, they provided useful insights regarding the evolution and spatial variation of the phenomenon, proving to be an asset of added value for the study. The flows were estimated as being 1:400 years in the low-lying catchments and 1:1000 years using as a design scenario the IDF curve of the Karditsa gauge station. Investigation of the severity of the event was supplemented by a statistical analysis of the annual maximum water levels in a reservoir adjacent to the study area. The flows were then mapped using the HEC-RAS software and a validation was made using remote sensing footage, photos, and videos posted on social media. The overall modelling performance was satisfactory and highlighted the importance of gathering all the available records for revealing past flood events.

In addition, the advantages of using remote sensing datasets are unique in flood modelling, underlining the need for introducing new concepts and frameworks in flood risk management analysis.

Author Contributions: A.T.: methodology, data mining, analysis and modelling, draft reporting A.Z.: methodology, data mining, analysis and modelling, draft reporting, V.B.: draft reviewing/editing A.T.: draft reviewing, supervision, administration. All authors have read and agreed to the published version of the manuscript.

Funding: This research was carried out as part of the project "Hydrology and Hydraulic analysis of the flooding event in Karditsa city caused by Medicane Ianos" funded by Thessaly Regional Government.

Institutional Review Board Statement: Not applicable.

Informed Consent Statement: Informed consent was obtained from all subjects involved in the study.

Acknowledgments: The manuscript is an invited paper as part of the Special Issue "Modern Developments in Flood Modelling" organized by the Hydrology journal. We are thankful to the three anonymous reviewers for the constructive comments, which helped us to improve our manuscript substantially. A detailed hydraulic animation along with georeferenced videos is presented in the publication "Forensic hydrology and hydraulic study for the reproduction of extreme flooding event caused by Medicane IANOS Available online: https://www.researchgate.net/publication/352993775_Forensic_hydrology_and_hydraulic_study_for_the_reproduction_of_extreme_flooing_event_caused_by_Medicane_IANOS/related (accessed on 18 May 2022). We are grateful to the people from Karditsa who shared event-posts on social media that supported us carrying out this research. We, finally, thank the composer Konstantis Papakonstantinou for license to use the music of "Archaeologist" in our video animation created as part of this research.

Conflicts of Interest: The authors declare no conflict of interest.

References

1. Hall, J.; Arheimer, B.; Borga, M.; Brázdil, R.; Claps, P.; Kiss, A.; Kjeldsen, T.R.; Kriaučiūnienė, J.; Kundzewicz, Z.W.; Lang, M.; et al. Understanding flood regime changes in Europe: A state-of-the-art assessment. *Hydrol. Earth Syst. Sci.* **2014**, *18*, 2735–2772. [CrossRef]
2. Dimitriadis, P.; Tegos, A.; Petsiou, A.; Pagana, V.; Apostolopoulos, I.; Vassilopoulos, E.; Gini, M.; Koussis, A.D.; Mamassis, N.; Koutsoyiannis, D.; et al. Flood Directive implementation in Greece: Experiences and future improvements. *Eur. Water* **2017**, *57*, 35–41.
3. Koutsoyiannis, D.; Mamassis, N.; Efstratiadis, A.; Zarkadoulas, N.; Markonis, Y. Floods in Greece. In *Changes of Flood Risk in Europe*; AHS Press: Wallingford, UK, 2012; pp. 238–256.
4. Dimitriadis, P.; Tegos, A.; Oikonomou, A.; Pagana, V.; Koukouvinos, A.; Mamassis, N.; Efstratiadis, A. Comparative evaluation of 1D and quasi-2D hydraulic models based on benchmark and real-world applications for uncertainty assessment in flood mapping. *J. Hydrol.* **2016**, *534*, 478–492. [CrossRef]
5. Karagiannidis, A.; Dafis, S.; Kalimeris, A.; Kotroni, V. Ianos-A hurricane in the Mediterranean. *Bull. Am. Meteorol. Soc.* **2021**, *1*, 1–31.
6. Zekkos, D.; Zalachoris, G.; Alvertos, A.E.; Amatya, P.M.; Blunts, P.; Clark, M.; Dafis, S.; Farmakis, I.; Ganas, A.; Hille, M.; et al. The September 18–20 2020 Medicane Ianos Impact on Greece—Phase I Reconnaissance Report; GEER-068; Geotechnical Extreme Events Reconnaissance Association: 2020. Available online: https://www.researchgate.net/publication/347240663_The_September_18-20_2020_Medicane_Ianos_Impact_on_Greece_Phase_I_Reconnaissance_Report#:~{}:text=On%20September%2017%2D20%202020,areas%20the%20mean%20annual%20precipitation (accessed on 18 May 2022).
7. Loli, M.; Mitoulis, S.A.; Tsatsis, A.; Manousakis, J.; Kourkoulis, R.; Zekkos, D. Flood characterization based on forensic analysis of bridge collapse using UAV reconnaissance and CFD simulations. *Sci. Total Environ.* **2022**, *822*, 153661. [CrossRef]
8. Borga, M.; Boscolo, P.; Zanon, F.; Sangati, M. Hydrometeorological Analysis of the 29 August 2003 flash flood in the Eastern Italian Alps. *J. Hydrometeorol.* **2007**, *8*, 1049–1067. [CrossRef]
9. Costabile, P.; Costanzo, C.; Macchione, F. A storm event watershed model for surface runoff based on 2D fully dynamic wave equations. *Hydrol. Processes* **2013**, *27*, 554–569. [CrossRef]
10. Bellos, V.; Papageorgaki, I.; Kourtis, I.; Vangelis, H.; Kalogiros, I.; Tsakiris, G. Reconstruction of a flash flood event using a 2D hydrodynamic model under spatial and temporal variability of storm. *Nat. Hazards* **2020**, *101*, 711–726. [CrossRef]
11. Sarchani, S.; Tsanis, I. Analysis of a Flash Flood in a Small Basin in Crete. *Water* **2019**, *11*, 2253. [CrossRef]
12. Diakakis, M.; Andreadakis, E.; Nikolopoulos, E.I.; Spyrou, N.I.; Gogou, M.E.; Deligiannakis, G.; Katsetsiadou, N.; Antoniadis, Z.; Melaki, M.; Georgakopoulos, A.; et al. An integrated approach of ground and aerial observations in flash flood disaster investigations. The case of the 2017 Mandra flash flood in Greece. *Int. J. Disaster Risk Reduct.* **2019**, *33*, 290–309. [CrossRef]
13. Kastridis, A.; Kirkenidis, C.; Sapountzis, M. An integrated approach of flash flood analysis in ungauged Mediterranean watersheds using posFlood surveys and unmanned aerial vehicles. *Hydrol. Processes* **2020**, *34*, 4920–4939. [CrossRef]
14. Sapountzis, M.; Kastridis, A.; Kazamias, A.P.; Karagiannidis, A.; Nikopoulos, P.; Lagouvardos, K. Utilization and uncertainties of satellite precipitation data in flash flood hydrological analysis in ungauged watersheds. *Glob. Nest J.* **2021**, *23*, 388–399.
15. Ramirez, A.I.; Herrera, A. Forensic hydrology. In *Forensic Analysis—From Death to Justice*; Shetty, B.S.K., Padubidri, J.R., Eds.; IntechOpen: London, UK, 2016. [CrossRef]

16. Gilewski, P.; Nawalany, M. Inter-comparison of rain-gauge, radar, and satellite (IMERG GPM) precipitation estimates performance for rainfall-runoff modeling in a mountainous catchment in Poland. *Water* **2018**, *10*, 1665. [CrossRef]
17. Kazamias, A.P.; Sapountzis, M.; Lagouvardos, K. Evaluation of GPM-IMERG rainfall estimates at multiple temporal and spatial scales over Greece. *Atmos. Res.* **2022**, *269*, 106014. [CrossRef]
18. Varlas, G.; Anagnostou, M.N.; Spyrou, C.; Papadopoulos, A.; Kalogiros, J.; Mentzafou, A.; Michaelides, S.; Baltas, E.; Karymbalis, E.; Katsafados, P. A Multi-Platform Hydrometeorological Analysis of the Flash Flood Event of 15 November 2017 in Attica, Greece. *Remote Sens.* **2019**, *11*, 45. [CrossRef]
19. Chaudhary, P.; D'Aronco, S.; Moy de Vitry, M.; Leitão, J.P.; Wegner, J.D. Flood-Water Level Estimation from Social Media Images. In Proceedings of the 2019 ISPRS Geospatial Week 2019, Enschede, The Netherlands, 10–14 June 2019.
20. Dhanapala, L.; Gunarathna, M.H.J.P.; Kumari, M.K.N.; Ranagalage, M.; Sakai, K.; Meegastenna, T.J. Towards Coupling of 1D and 2D Models for Flood Simulation—A Case Study of Nilwala River Basin, Sri Lanka. *Hydrology* **2022**, *9*, 17. [CrossRef]
21. Tedla, M.G.; Cho, Y.; Jun, K. Flood Mapping from Dam Break Due to Peak Inflow: A Coupled Rainfall–Runoff and Hydraulic Models Approach. *Hydrology* **2021**, *8*, 89. [CrossRef]
22. Giandotti, M. *Previsione Delle Piene e Delle, Magre dei Corsi D'acqua*; Istituto Poligrafico dello Stato: Rome, Italy, 1934; pp. 107–117.
23. Papaioannou, G.; Efstratiadis, A.; Vasiliades, L.; Loukas, A.; Papalexiou, S.M.; Koukouvinos, A.; Tsoukalas, I.; Kossieris, P. An Operational Method for Flood Directive Implementation in Ungauged Urban Areas. *Hydrology* **2018**, *5*, 24. [CrossRef]
24. Papaioannou, G.; Varlas, G.; Terti, G.; Papadopoulos, A.; Loukas, A.; Panagopoulos, Y.; Dimitriou, E. Flood inundation mapping at Ungauged basins using coupled Hydrometeorological–hydraulic Modelling: The catastrophic case of the 2006 flash flood in Volos City, Greece. *Water* **2019**, *11*, 2328. [CrossRef]
25. Bellos, V. Ways for flood hazard mapping in urbanised environments: A short literature review. *Water Util.* **2012**, *4*, 25–31.
26. Apel, H.; Thieken, A.H.; Merz, B.; Blöschl, G. Flood risk assessment and associated uncertainty. Nat. *Hazards Earth Syst. Sci.* **2004**, *4*, 295–308. [CrossRef]
27. Psomiadis, E.; Tomanis, L.; Kavvadias, A.; Soulis, K.X.; Charizopoulos, N.; Michas, S. Potential dam breach analysis and flood wave risk assessment using HEC-RAS and remote sensing data: A multicriteria approach. *Water* **2021**, *13*, 364. [CrossRef]
28. Papaioannou, G.; Vasiliades, L.; Loukas, A.; Aronica, G.T. Probabilistic flood inundation mapping at ungauged streams due to roughness coefficient uncertainty in hydraulic modelling. *Adv. Geosci.* **2017**, *44*, 23–34. [CrossRef]
29. Shustikova, I.; Domeneghetti, A.; Neal, J.C.; Bates, P.; Castellarin, A. Comparing 2D capabilities of HEC-RAS and LISFLOOD-FP on complex topography. *Hydrol. Sci. J.* **2019**, *64*, 1769–1782. [CrossRef]
30. Efstratiadis, A.; Hadjibiros, K. Can an environment-friendly management policy improve the overall performance of an artificial lake? Analysis of a multipurpose dam in Greece. *Environ. Sci. Policy* **2011**, *14*, 1151–1162. [CrossRef]
31. Tyralis, H.; Aristoteles, T.; Delichatsiou, A.; Mamassis, N.; Koutsoyiannis, D. A perpetually interrupted interbasin water transfer as a modern Greek drama: Assessing the Acheloos to Pinios interbasin water transfer in the context of integrated water resources management. *Open Water J.* **2017**, *4*, 11.
32. Kozanis, S.; Christofides, A.; Mamassis, N.; Efstratiadis, A.; Koutsoyiannis, D. Hydrognomon–open source software for the analysis of hydrological data. In Proceedings of the European Geosciences Union (EGU) General Assembly, Vienna, Austria, 2–7 May 2010.
33. Efstratiadis, A.; Papalexiou, S.M.; Markonis, I.; Mamassis, N. Ombrian curves. In *Flood Risk Management Plan of River Basin District of Thessaly (GR08)–Phase A*; Special Secretariat for Water, Ministry of Environment and Energy (SSW-MEE): Athens, Greece, 2016.
34. Iliopoulou, T.; Malamos, N.; Koutsoyiannis, D. Regional Ombrian Curves: Design Rainfall Estimation for a Spatially Diverse Rainfall Regime. *Hydrology* **2022**, *9*, 67. [CrossRef]
35. Zotou, I.; Bellos, V.; Gkouma, A.; Karathanassi, V.; Tsihrintzis, V. Using Sentinel-1 Imagery to Assess Predictive Performance of a Hydraulic Model. *Water Resour. Manag.* **2020**, *34*, 4415–4430. [CrossRef]
36. Pradhan, R.K.; Markonis, Y.; Godoy, M.R.V.; Villalba-Pradas, A.; Andreadis, K.M.; Nikolopoulos, E.I.; Papalexiou, S.M.; Rahim, A.; Tapiador, F.J.; Hanel, M. Review of GPM IMERG performance: A global perspective. *Remote Sens. Environ.* **2022**, *268*, 112754. [CrossRef]
37. Tang, G.; Clark, M.P.; Papalexiou, S.M.; Ma, Z.; Hong, Y. Have satellite precipitation products improved over last two decades? A comprehensive comparison of GPM IMERG with nine satellite and reanalysis datasets. *Remote Sens. Environ.* **2020**, *240*, 111697. [CrossRef]
38. Mayor, Y.G.; Tereshchenko, I.; Fonseca-Hernández, M.; Pantoja, D.A.; Montes, J.M. Evaluation of Error in IMERG Precipitation Estimates under Different Topographic Conditions and Temporal Scales over Mexico. *Remote Sens.* **2017**, *9*, 503. [CrossRef]
39. Tang, S.; Li, R.; He, J.; Wang, H.; Fan, X.; Yao, S. Comparative Evaluation of the GPM IMERG Early, Late, and Final Hourly Precipitation Products Using the CMPA Data over Sichuan Basin of China. *Water* **2020**, *12*, 554. [CrossRef]
40. Sakib, S.; Ghebreyesus, D.; Sharif, H.O. Performance Evaluation of IMERG GPM Products during Tropical Storm Imelda. *Atmosphere* **2021**, *12*, 687. [CrossRef]
41. Efstratiadis, A.; Koussis, A.D.; Koutsoyiannis, D.; Mamassis, N. Flood design recipes vs. reality: Can predictions for ungauged basins be trusted? *Nat. Hazards Earth Syst. Sci. Discuss.* **2013**, *1*, 7387–7416. [CrossRef]
42. Michailidi, E.M.; Antoniadi, S.; Koukouvinos, A.; Bacchi, B.; Efstratiadis, A. Timing the time of concentration: Shedding light on a paradox. *Hydrol. Sci. J.* **2018**, *63*, 721–740. [CrossRef]
43. Grimaldi, S.; Petroselli, A.; Tauro, F.; Porfiri, M. Time of concentration: A paradox in modern hydrology. *Hydrol. Sci. J.* **2012**, *57*, 217–228. [CrossRef]

44. Apel, H.; Aronica, G.T.; Kreibich, H.; Thieken, A.H. Flood risk analyses—How detailed do we need to be? *Nat. Hazards* **2009**, *49*, 79–98. [CrossRef]
45. Bellos, V.; Tsakiris, G. A hybrid method for flood simulation in small catchments combining hydrodynamic and hydrological techniques. *J. Hydrol.* **2016**, *540*, 331–339. [CrossRef]
46. Efstratiadis, A.; Dimas, P.; Pouliasis, G.; Tsoukalas, I.; Kossieris, P.; Bellos, V.; Sakki, G.-K.; Makropoulos, C.; Michas, S. Revisiting Flood Hazard Assessment Practices under a Hybrid Stochastic Simulation Framework. *Water* **2022**, *14*, 457. [CrossRef]
47. Fohringer, J.; Dransch, D.; Kreibich, H.; Schröter, K. Social media as an information source for rapid flood inundation mapping. *Nat. Hazards Earth Syst. Sci.* **2015**, *15*, 2725–2738. [CrossRef]
48. Brouwer, T.; Eilander, D.; Van Loenen, A.; Booij, M.J.; Wijnberg, K.M.; Verkade, J.S.; Wagemaker, J. Probabilistic flood extent estimates from social media flood observations. *Nat. Hazards Earth Syst. Sci.* **2017**, *17*, 735–747. [CrossRef]

hydrology MDPI

Article

Identifying Modelling Issues through the Use of an Open Real-World Flood Dataset

Vasilis Bellos [1,*], Ioannis Kourtis [2], Eirini Raptaki [2], Spyros Handrinos [3], John Kalogiros [4], Ioannis A. Sibetheros [5] and Vassilios A. Tsihrintzis [2]

1 Laboratory of Ecological Engineering and Technology, Department of Environmental Engineering, Democritus University of Thrace, 12 Vas. Sofias St., 67100 Xanthi, Greece
2 Centre for the Assessment of Natural Hazards and Proactive Planning, Laboratory of Reclamation Works and Water Resources Management, School of Rural, Surveying and Geoinformatics Engineering, National Technical University of Athens, 9 Heroon Polytechniou St., Zografou, 15780 Athens, Greece
3 MSc Civil Engineering Program, Department of Civil Engineering and Geosciences, Delft University of Technology, Stevinweg 1, 2628 Delft, The Netherlands
4 Institute for Environmental Research and Sustainable Development, National Observatory of Athens, I. Metaxas & Vassilis Pavlou St., Old Penteli, 15236 Athens, Greece
5 Department of Civil Engineering, School of Engineering, University of West Attica, 250 Petrou Ralli & Thivon St., Egaleo, 12244 Athens, Greece
* Correspondence: vbellos@env.duth.gr

Abstract: The present work deals with the reconstruction of the flood wave that hit Mandra town (Athens, Greece) on 15 November 2017, using the framework of forensic hydrology. The flash flood event was caused by a huge storm event with a high level of spatial and temporal variability, which was part of the Medicane Numa-Zenon. The reconstruction included: (a) the post-event collection of 44 maximum water depth traces in the town; and (b) the hydrodynamic simulation employing the HEC-RAS and MIKE FLOOD software. The derived open dataset (which also includes additional data required for hydrodynamic modeling) is shared with the community for possible use as a benchmark case for flood model developers. With regards to the modeling issues, we investigate the calibration strategies in computationally demanding cases, and test whether the calibrated parameters can be blindly transferred to another simulator (informed modeling). Regarding the calibration, it seems that the coupling of an initial screening phase with a simple grid-search algorithm is efficient. On the other hand, the informed modeling concept does not work for our study area: every numerical model has its own dynamics while the parameters are of grey-box nature. As a result, the modeler should always be skeptical about their global use.

Keywords: forensic hydrology; flood modeling; open dataset; HEC-RAS; MIKE FLOOD

Citation: Bellos, V.; Kourtis, I.; Raptaki, E.; Handrinos, S.; Kalogiros, J.; Sibetheros, I.A.; Tsihrintzis, V.A. Identifying Modelling Issues through the Use of an Open Real-World Flood Dataset. *Hydrology* **2022**, *9*, 194. https://doi.org/10.3390/hydrology9110194

Academic Editor: Asaad Y. Shamseldin

Received: 4 October 2022
Accepted: 25 October 2022
Published: 31 October 2022

Publisher's Note: MDPI stays neutral with regard to jurisdictional claims in published maps and institutional affiliations.

1. Introduction

Floods occur in both rural and urban environments and are among the most destructive natural hazards, creating huge economic losses and casualties at global scale [1,2]. Impervious surfaces pose a major effect on watershed hydrology [2], since urban sprawl affects total runoff volumes, peak flow rates, and catchment response times. Moreover, discharges, associated with storm events with high and/or low probability of occurrence before development, increase after urban development takes place [3]. The combined effects of urban development and climatic variability may affect the urban water cycle [1]. Moreover, extreme urbanization and the projected climate change have led the scientific community to focus more on urban flood risk, urban flood dynamics and flood mitigation measures, estimation of return periods of extreme events through extreme value analysis, in both quantitative and qualitative terms, and the update of intensity–duration–frequency curves [4–12].

In our era, physically based simulators are the main tools used for simulating the flow dynamics driven by urbanization, land use/land cover (LULC) changes, and climatic variability in the urban environment [7]. The main approach employed is the 1D approach. However, 1D-1D, 1D-2D, and 2D approaches are also applied [13]. Various uncertainties, both aleatory and epistemic, are associated with these models and this is mainly because of the model structure, initial and boundary conditions, input and forcing data. Moreover, the rational estimation of the various parameters, incorporated in these models, requires a calibration procedure [14]. For example, parameters regarding the proper coupling of 1D flow in the sewer system with 2D flow on the surface of the catchment include the proper representation of buildings and other obstacles to the flow, the estimation of mass losses due to infiltration or interception, or the estimation of energy losses due to friction, among others [14].

Accurate simulation of flood events is extremely difficult. The main reason for this is associated with the absence of a proper dataset which can be used for model calibration-validation purposes, especially in urban catchments which are partially gauged or completely ungauged. It should be noted herein that even in a gauged urban catchment the monitoring system is often destroyed during a flood. The framework of forensic hydrology, as proposed by [15], which comprises five steps (i.e., information gathering; hydrometeorological and hydrological analysis; hydrodynamic analysis; integrative analysis; and final diagnosis), pays special attention to data collection for a proper reconstruction of the event. Some recent studies propose the post-event data collection using some proxies as measurements, which are mainly the maximum flood depths at several observational points [16,17], while other studies are based on data derived by flood event reconstruction using physical modelling [18,19]. Other strategies with potential merit are the crowd sourcing of data derived by social media, also focusing on water depths at specific time moment [20], and finally, remote sensing data, providing flood extents at various time points [21].

This paper is divided in two parts. In the first part, we describe and share a full dataset, deposited in Zenodo platform, acquired using the forensic hydrology framework at the site. The flood dataset is associated with the catastrophic flood event that hit Mandra town in Athens, Greece, on 15 November 2017, causing 24 casualties. The purpose of this open dataset is to provide a complete dataset for benchmarking flood simulators. In the second part, we focus on the second and third steps of forensic hydrology framework. Specifically, we reconstructed the flood event with the assistance of commercial, physically based simulators, and we investigate some practical issues regarding flood modeling, which can be summarized as follows: (a) the strategy for calibrating the required parameters with respect to the computational burden; (b) the potential transferability of the calibrated parameters from one software to the other, in order to identify if the calibrated parameters are of global or grey-box nature.

2. Materials and Methods

2.1. Study Area

The town of Mandra is located in Attica, in the western part of the greater metropolitan area of Athens, Greece (Figure 1). It is built at the outlet of two catchments, namely Agia Aikaterini and Soures catchments (Figure 1), which are part of the greater river system of Sarantapotamos, and extends along the eastern-southeastern foothills of Mt. Pateras (1130 m).

The storm event under study caused the severe Mandra flood, which occurred between the 14 November 2017 at 23:00 UTC and the 15 November 2017 at 12:00 UTC and was part of the Medicane Numa-Zenon. It was a highly localized phenomenon with extreme spatial and temporal variability. According to the National Observatory of Athens (NOA), which recorded the rainfall field with a mobile X-band polarimetric weather radar (XPOL), the total rainfall on Mt. Pateras, above Nea Peramos and Mandra, exceeded 200 mm in depth during the 6-h main storm event, with instant rainfall intensities reaching peak values of up to 120–140 mm/h, while the accumulated rainfall in 10 h reached nearly 300 mm (Figure 2, left).

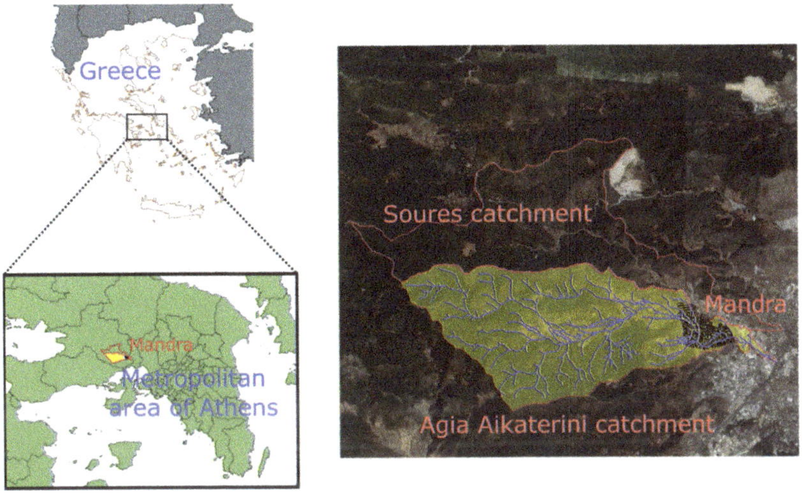

Figure 1. General location of Mandra town and the Agia Aikaterini and Soures catchments.

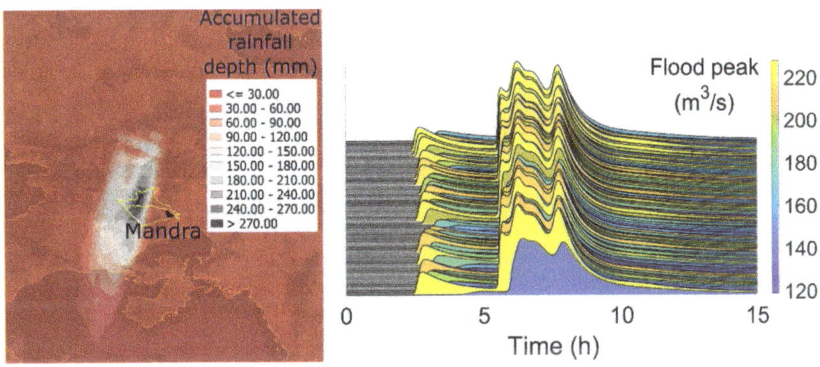

Figure 2. Accumulated rainfall depth recorded between 14 November 2017 at 22:30 and 15 November 2017 at 08:30 GMT (left); Joyplot of the 100 input hydrographs computed at Mandra town starting from 14 November 2017 at 22:30 until 15 November 2017 at 13:30 (right).

The response of the storm was a flash flood wave, which hit the town during the night, with catastrophic consequences. It is one of the biggest multi-fatality flood events recorded in Greece. Several researchers reconstructed the event, simulating the flood hydrographs at the inlet of the town [8,22–28].

In this work, the input hydrographs were taken from a previous study [29] which reconstructed the flood dynamics of Agia Aikaterini catchment using a 2D-SWE-based simulator named FLOW-R2D [23]. A parametric and input data uncertainty analysis was performed. Therefore, the output was an ensemble of 100 hydrographs at the outlet of the catchment and at the inlet of the town (Figure 2, right). The uncertainty band of the flood discharge peak ranges between 120 m^3/s and 220 m^3/s, while the median was about 180 m^3/s. This peak is in accordance with the post event rough estimation of the flood peak made by [24].

2.2. Data Collection

The research team visited the field four days after the flood event, and specifically, on 19 November 2017. During the visit, they derived a dataset of 44 maximum water depths using the dry mud or leaf footprints on several building walls as a proxy indication. The depth was measured with a measuring tape while the exact coordinates were recorded with a hand-held GPS. Figure 3 depicts a general view of the observation points while Figures 4–7 depict some indicative photos taken in the field. The majority of the observation points are along the two main streets of Mandra, since the major flood impact was observed in these roads. Specifically, the roads are Vaggeli Koropouli str. and Str. Nik. Rokka str., which are the extensions of Agia Aiakterini str (which in fact is the extension of the ephemeral Agia Aikaterini stream).

Figure 3. Location of observation points across Mandra town.

Figure 4. Indicative observation points in Mandra town (in roads): the observed maximum depths are 0.64 m (up & left), 1.01 m (up & right), 0.54 m (down & left), 1.70 (down & right).

Figure 5. Indicative observation points in Mandra town (inside the buildings): the observed maximum depths are 1.15 m (left), 1.90 m (middle), 1.06 m (right).

Figure 6. Uncertainty issues at the observation points: the finally selected values for the observed maximum depths are 0.80 m (left), 1.01 m (right).

Figure 7. Example of the of the urban drainage system failure.

Specifically, Figure 4 is a representative photo of flood traces observed in the roads while Figure 5 is a representative photo of flood traces observed inside buildings. At some points, local inhabitants had marked with a green line the flood maxima. This information was used carefully, employing engineering judgement and after cross checking. It should be noted that the guidance of the locals in the field was substantial in order to address the uncertainties of the measurements and provided and increased reliability of the dataset. For example, the building in Figure 6 (left) has two mud traces: the lower trace denotes the real flood depth maximum, while the higher denotes mud splashing as a result of turbulence from obstacles. The flood mark depicted in Figure 6 (right) is also another point which is characterized by uncertainties. All these issued were addressed doing forensic research and with the assistance provided by the local community in the form of small discussions, interviews, etc.

Finally, Figure 7 depicts an example of the sewer system failure. We should mention here that the main storm sewer system is underground and was designed for a maximum flood capacity of 10 m^3/s. The main pipe of the latter system is below the Agia Aikaterinin str. and its extensions (Vaggeli Koropouli str. and Str. Nik. Rokka str.) crossing the town. Based on the field visit and the interviews with the locals, it seems that the sewer system did not work (it was destroyed or filled with debris) and the flow inside the system was negligible compared to the flood wave that hit the town. For this reason, we did not include urban drainage flow in our hydrodynamic analysis.

2.3. *Hydrodynamic Simulators*

For the hydrodynamic simulations, we used two well-known 2D hydrodynamic models, namely the HEC-RAS [30] and the MIKE-FLOOD [31,32] pieces of software. Both simulators are based on the 2D Shallow Water Equations (2D-SWE), either in their full form or in the simpler diffusion wave form, while the equations are solved using the finite volume method. In this study, we exploited the 2D diffusion wave mode for both software since the full form of 2D-SWEs numerical instabilities were observed.

We used the same input data, parameters, and boundary conditions in both models. Specifically, the digital terrain model (DTM), the friction coefficients, the internal boundaries defined by the buildings, and the forcing 100 hydrographs as defined from the above-mentioned uncertainty analysis were introduced in both software. It should be noted that only the inflow of Agia Aikaterini catchment (and not Soures catchment inflow) was considered, since it was the only driver for the flood observed inside the town.

The computational area consists of a polygon of 1.6 km^2 area. The digital terrain model (DTM) which was used for geographical information was derived from the National Cadastre & Mapping Agency of Greece with a spatial resolution of 5 m. The internal boundaries for the buildings were manually designed with the assistance of satellite images in the Google Earth platform. For both pieces of software, the lateral boundaries of the computational domain were assumed to have a no-slip condition (solid boundaries) in order to preserve the water volume, while for the downstream boundaries the open boundaries mode was selected.

Regarding the HEC-RAS software, we manually inserted (233) flow breaklines around urban blocks, and then modified the mesh by generating orthogonal cells around the breaklines, thus improving the computational speed and accuracy of the model. For the whole mesh, 121,583 cells were generated with the average cell area being about 13 m^2. For the upstream boundaries, we used the mode of the input hydrograph (which requires no additional parameters), while for the downstream boundaries we selected the open boundaries, which are based on the Manning equation and require an extra parameter, namely the energy slope S_f. The time step was selected to be equal to 10 s.

The Manning coefficient of the computational domain n_r was assumed to lie in a plausible range of values and was calibrated. Since HEC-RAS software is not capable of including internal solid boundaries, buildings are represented with a local increase of roughness, a common methodology in the relative literature [33]. Therefore, we used two more Manning coefficients, n_h and n_l (high and low), for dense and less dense urban blocks, respectively.

Regarding the MIKE software, we performed a similar procedure in order to generate the mesh, but with triangular elements. Specifically, we generated a mesh comprising 22.264 elements, with an average area of 70 m^2, while the smallest permitted triangular angle was 23°. For the upstream boundaries, we also used the mode of the input hydrograph, while for the downstream boundaries, we selected the open boundaries as well which required no additional parameters. A constant eddy viscosity coefficient was selected with a value equal to 0.1, while the time step ranged between 0.001 and 0.002 s. Finally, the wet-dry threshold was given a value equal to 0.005 m.

In contrast to HEC-RAS, MIKE FLOOD has the option to represent the buildings with a free-slip boundary condition. Therefore, there was no need to introduce new parameters in order to simulate the flood in a built-up area. Figure 8 depicts the computational area, the building footprints, and the upstream/downstream boundaries, while Figure 9 depicts the mesh generation for both pieces of software in an indicative part of the town.

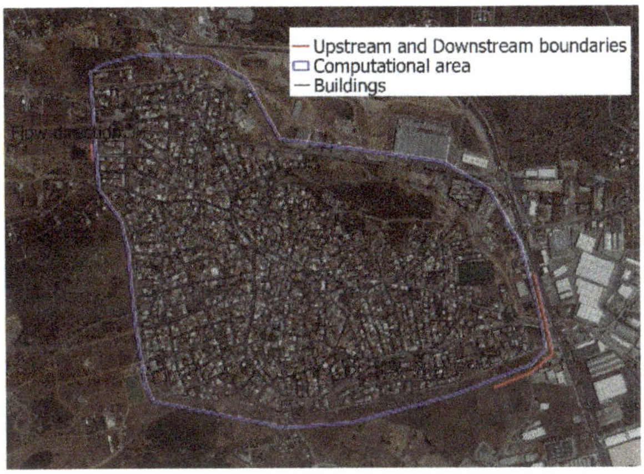

Figure 8. Computational area, upstream/downstream boundaries and buildings footprint used for HEC-RAS and MIKE FLOOD software.

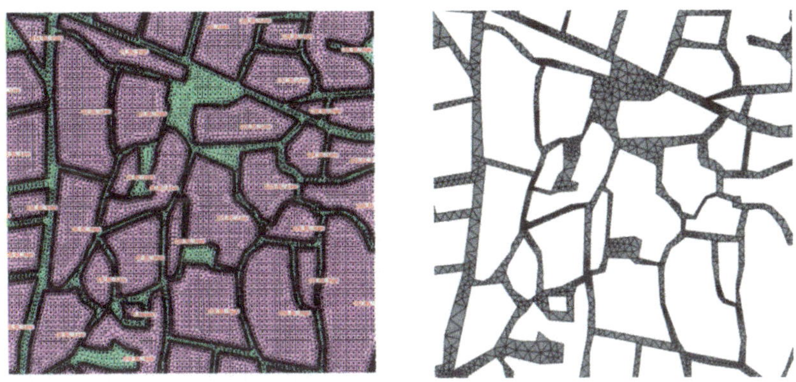

Figure 9. Mesh generation with HEC-RAS (left) and MIKE FLOOD (right) software.

2.4. Calibration Strategy

Although 2D hydrodynamic simulators have a strong physical base, they incorporate several grey-box parameters whose values, instead of being unique, adopted from handbooks or good practice guidance, lie in a plausible range. It is recommended that these parameters should be calibrated against relevant on-site measurements [14,34]. In theory, the latter simulators shall be calibrated against both water depths and flow velocities. However, the difficulty in finding observed velocities led us to limit the calibration only using maximum water depths. Moreover, the use of this kind of simulator for flood simulation has as a consequence an extremely high computational burden, which is exacerbated in cases where procedures, such as calibration, optimization, and uncertainty quantification, are employed. For example, each HEC-RAS simulation required approximately 2 h while each MIKE simulation required 4 h in an Intel(R) i7-4790 CPU/3.60 GHz.

Overall, it is not necessary or feasible to incorporate all the parameters of a model in such processes if we aim at an efficient optimization. To this end, sensitivity analysis, a process aiming to screen the most influential parameters on model results, can help modelers and practitioners identify a model's key input parameters, and thus reduce computational cost.

The two most common classifications of sensitivity analysis methods are the global sensitivity analysis (GSA) and the local sensitivity analysis (LSA). It is not in the scope of the present work to describe in detail sensitivity analysis approaches and for further information the reader is referred to [35]. In recent literature, a wide variety of sensitivity analysis methods are presented (e.g., local methods, global methods, qualitative, quantitative methods) from different scientific fields [36,37]. Two of the most widely used sensitivity analysis methods are Sobol's sensitivity analysis method [38] and the screening technique proposed by [39]. The latter is one of the most widely used sensitivity analysis methods as it is easy to implement and is not time consuming [7,36].

Specifically, Morris [39] introduced the concept of elementary effects and proposed a sensitivity analysis method which is based on the computation of the mean and standard deviation of elementary effects in order to determine the effects of input parameter on the final result. According to [35], the input parameters may have: (i) negligible; (ii) linear and additive; or (iii) nonlinear or involved in interactions with other parameters. The mean of the elementary effects is used to assess the overall effect of the input parameter on the final result, while the standard deviation is used as a metric for the interactions with other parameters. However, [40] proposed to use the absolute mean instead of the mean in order to not introduce type II errors. In the present work, the sampling strategy proposed by [39] was used. Overall, the Morris elementary effects method can be categorized as a GSA approach, and it is simple, robust, and has low computational burden compared to other GSA methods (e.g., Sobol's method, GLUE etc.).

In the present study, we performed the calibration phase of the simulation only with HEC-RAS and we divided it in two stages: (a) First, we performed a GSA for the required model parameters using the SAFE toolbox [41] in order to reduce the dimensional space through parametric screening. Then, (b) we performed a grid-search calibration. The required five parameters were: a) the Manning coefficient for the roads, n_r; (b) the Manning coefficient for the urban blocks (high value), n_H; (c) the Manning coefficient for the urban blocks (low value) n_L; (d) the confidence interval of the uncertainty band of the upstream hydrograph, CI; and (e) the downstream energy slope, S_f. The range of values assigned to each of the five parameters are presented in Table 1.

Table 1. Parametric range.

Parameter	Lower Limit	Upper Limit
n_r [s/m$^{1/3}$]	0.03	0.06
n_H [s/m$^{1/3}$]	40	60
n_L [s/m$^{1/3}$]	15	25
CI [%]	10	90
S_f [-]	0.01	0.03

3. Results

3.1. Flood Dataset

Table 1 presents the maximum water depths measured at the 44 observation points, with their coordinates in both the Greek Geodetic Reference System 1987 (GGRS87) and the World Geodetic System 84 (WGS84).

In this link (https://zenodo.org/record/7140750, accessed on 3 October 2022), the reader can also download the full dataset on the Zenodo platform. The dataset consists of the following:

1. The maximum water depths recorded at the 44 points after the event with their coordinates in both GGRS87 and WGS84 systems, as shown in Table 2.
2. The digital terrain model (DTM) of the greater area with a resolution of 5 m, as provided by the National Cadastre and Mapping Agency of Greece.
3. A shape file with the boundaries of the computational area used for both HEC-RAS and MIKE FLOOD software, which coincide with the boundaries of the town of Mandra.
4. A shape file with the Mandra urban blocks footprint, as were manually drawn on Google Earth platform.
5. The ensemble of 100 hydrographs as depicted in Figure 2, which serve as the inflow from Agia Aikaterini catchment to Mandra town. They are derived by implementing the FLOW-R2D hydrodynamic simulator at the catchment scale, having as an input the rainfall field captured by the weather radar during the Mandra flood event [23].
6. The rainfall field of the greater area with a spatial resolution of 200 m and a temporal resolution of 2 min, recorded by the X-band weather radar of the National Observatory of Athens.

Table 2. Maximum water depths measured at the field.

Gauge	X, Y (GGRS87)	φ, λ (WGS84)	Depth (m)
1	455668.2588, 4214030.3444	38.075555095412, 23.496248974378	3.00
2	455714.0308, 4213859.4439	38.074017108644, 23.496781377525	1.41
3	455618.1325, 4214089.9909	38.076090202058, 23.495673789261	1.91
4	455648.7464, 4214045.8128	38.075693548981, 23.496025555384	2.60
5	455715.7245, 4213941.2321	38.074754298656, 23.496795637049	2.00
6	455714.7635, 4213840.7551	38.073848713598, 23.496790884875	2.40
7	455708.8934, 4213757.7429	38.073100288440, 23.496729087177	1.90
8	455718.1527, 4213696.6548	38.072550190317, 23.496838421089	2.60
9	455711.4275, 4213686.7258	38.072460378003, 23.496762362771	2.40
10	455693.3690, 4213585.9232	38.071551023452, 23.496562711634	3.30
11	455683.1461, 4213546.1162	38.071191767747, 23.496448625272	2.44
12	455694.2735, 4213521.9724	38.070974717633, 23.496576973349	2.00
13	455724.7554, 4213510.6998	38.070874612326, 23.496925173983	2.80
14	455815.0195, 4213431.7319	38.070167321865, 23.497959080654	2.43
15	455893.3008, 4213392.0692	38.069813674021, 23.498853945292	1.17
16	455950.5524, 4213396.1206	38.069852967810, 23.499506375731	1.28
17	456148.2251, 4213517.7609	38.070958813396, 23.501752448480	1.56
18	456149.9273, 4213543.6000	38.071191768069, 23.501770274694	0.88
19	456244.9914, 4213542.7794	38.071188961397, 23.502854094564	1.55
20	456349.9641, 4213582.0807	38.071548216307, 23.504048435197	1.19
21	456405.9467, 4213583.0278	38.071559443195, 23.504686605580	1.68
22	456498.3168, 4213604.9584	38.071761523636, 23.505738338489	3.00
23	456555.1662, 4213651.3694	38.072182522132, 23.506383639558	2.47
24	456597.3211, 4213639.6230	38.072078676069, 23.506864940258	1.19
25	456662.0088, 4213690.0413	38.072536158243, 23.507599370013	1.87
26	456690.6165, 4213659.3707	38.072261108064, 23.507927367552	1.06
27	455950.3622, 4213418.8560	38.070057859161, 23.499502811367	1.15
28	455722.5295, 4213734.9355	38.072895404700, 23.496885956174	2.00
29	455761.5685, 4213956.7620	38.074896496719, 23.497317344483	1.70
30	455764.4553, 4213970.1369	38.075017177319, 23.497349431868	1.30
31	455770.2635, 4213945.5041	38.074795460034, 23.497417170265	0.81
32	455783.0573, 4213940.4525	38.074750556643, 23.497563343334	1.10
33	455811.7063, 4213918.1875	38.074551291492, 23.497891341098	1.16
34	456228.1013, 4213834.3580	38.073815970735, 23.502643747389	1.01
35	456335.9814, 4213774.6124	38.073282717248, 23.503877306043	0.28
36	456306.8247, 4213760.7545	38.073156420710, 23.503545742317	0.61
37	456333.7110, 4213759.3652	38.073145194065, 23.503852350022	1.21

Table 2. *Cont.*

Gauge	X, Y (GGRS87)	φ, λ (WGS84)	Depth (m)
38	456346.2390, 4213821.5813	38.073706513488, 23.503991391677	1.01
39	456344.8884, 4213802.9037	38.073538118443, 23.503977130532	0.62
40	456422.4147, 4213797.8190	38.073496019748, 23.504861299648	0.54
41	456529.0244, 4213792.8917	38.073456726769, 23.506077031385	0.46
42	456528.4916, 4213751.4763	38.073083448814, 23.506073467053	0.54
43	456557.3220, 4213703.9877	38.072656842958, 23.506405030212	0.91
44	456652.5360, 4213731.8215	38.072912246041, 23.507488849352	0.78

3.2. Calibration Phase

As previously described, we first performed the GSA in order to reduce the number of parameters to be calibrated, and therefore, reduce the dimensional space. Based on engineering judgement and similar studies [23], we assumed that the trajectory number was equal to 15. Hence, the required number of HEC-RAS simulations was equal to 90.

For each simulation, we calculated the root mean square error (RMSE) between the simulated and the observed maximum water depths. Based on this value, the SAFE toolbox calculates the mean and the standard deviation of the elementary effects.

Figure 10 depicts the results of the sensitivity analysis. It seems that the *CI* is by far the most influential parameter regarding the RMSE. The impact of the forcing driver in the model output is in accordance with similar flood studies [42]. The second most influential parameter is the Manning coefficient of the computational domain n_r, while the remaining Manning coefficients and the energy slope required at the downstream boundaries seem to have a negligible impact on the RMSE.

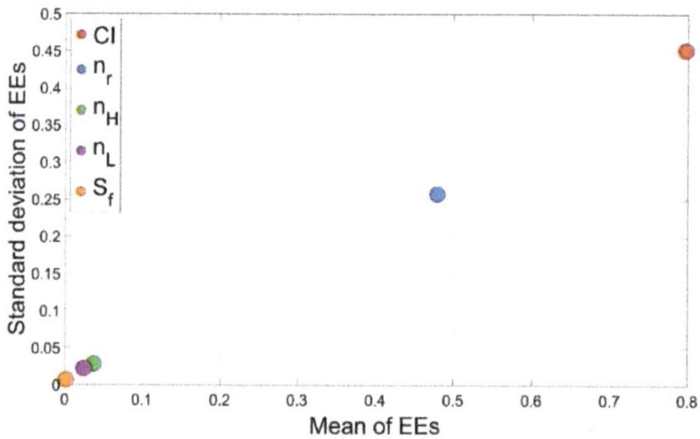

Figure 10. Sensitivity analysis results.

Based on the previous analysis, we then focused on the pair of the most influential parameters (*CI*, n_r). Therefore, we implemented a grid-search calibration, in order to find the optimal combination of the pair values. Assuming a step of 10% for the *CI* (which according to Table 1 ranges from 10% to 90%) and 0.01 s/m$^{1/3}$ for the n_r (which according to Table 1 ranges from 0.03 s/m$^{1/3}$ to 0.06 s/m$^{1/3}$), we produced 9 × 4 = 36 scenarios with different combinations of *CI* and n_r, while the other parameters (n_H, n_L and S_f) were assigned the average values 50 s/m$^{1/3}$, 20 s/m$^{1/3}$ and 0.02, respectively, according to Table 1.

Then, we performed again the hydrodynamic analysis for these scenarios, and we defined as an objective function the RMSE of the simulated maximum flood depths against the observed data. Figure 11 (up, left) depicts the dimensional space of the objective function. Since the target was the combination of the *CI* and n_r values for which the RMSE is minimized, it can be deduced that these values are 40% and 0.06 s/m$^{1/3}$, respectively (denoted by red star in the figure), while the RMSE value is equal to 0.70 m.

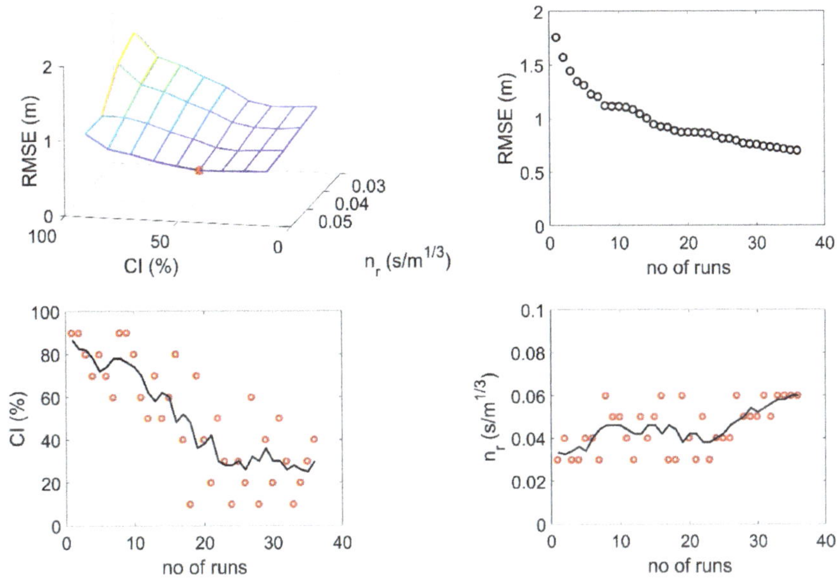

Figure 11. Dimensional space for the optimization function used for the grid-search calibration (up & left); Runs are sorted from maximum to minimum RMSE (up & right); Confidence Interval *CI* in respect to the run number with 5-period moving average denoted by the black line (down & left); Manning coefficient n_r in respect to the run number with 5-period moving average denoted by the black line (down & right).

It should be noted that the optimization procedure is trapped in the boundaries, which means that the dimensional space is not sufficiently explored. If the runs are sorted from the maximum until the minimum RMSE, it can be observed (Figure 11, up, right) that the objective function can be further minimized, but not that much. Taking into account the computational cost, we did not perform additional runs. Finally, Figure 11 (down, left and down, right) depicts the general trend of the calibrated parameters *CI* and n_r in respect of the run number, as previously sorted. It seems that the parameters tend to reach their optimal values as the calibration is in progress. This is not a proof that we achieved the global optimum, but it is a strong indication that we avoided the equifinality issue.

3.3. Calibrated HEC-RAS vs. Informed MIKE FLOOD

Since we assumed that we identified the optimal combination for the HEC-RAS parameter values, we then tried to identify whether we can inform MIKE FLOOD with these values. With this blind test, we aimed to investigate whether these calibrated values have a global nature or are model-specific.

As previously described, MIKE FLOOD includes the option for representing the buildings using a free-slip condition, which in general is better practice than the other two available methodologies in the literature, namely the local increase of the elevation or of the Manning coefficient [33]. Therefore, there is no need for estimating the Manning coefficient for the urban blocks. Besides, there is also no need for estimating a parameter for the downstream open boundaries, in contrast with HEC-RAS which requires the energy slope S_f. In order to make a meaningful comparison, MIKE FLOOD configuration (computational area, boundary conditions, bathymetry, boundaries of buildings) was the same as in HEC-RAS. Furthermore, for the MIKE-FLOOD simulations, the forcing driver was the hydrograph with the optimal value of $CI = 40\%$, as described in the previous section, while the Manning coefficient of the computational domain n_r was given the optimal value of 0.06 $s/m^{1/3}$. It was found that the RMSE metric was significantly bigger than the corresponding RMSE derived by HEC-RAS, which was equal to 1.63 m

Figure 12 depicts the comparison of the observed maximum flood depths vs. the simulated results derived from HEC-RAS and MIKE FLOOD. The inundation maps shown in Figures 13 and 14 are derived with the calibrated HEC-RAS and the informed MIKE FLOOD, respectively, and depict the maximum water depths. Figure 15 depicts the differences between the calibrated HEC-RAS and the informed MIKE FLOOD, for maximum water depths, while Figure 16 depicts the distribution of the residuals.

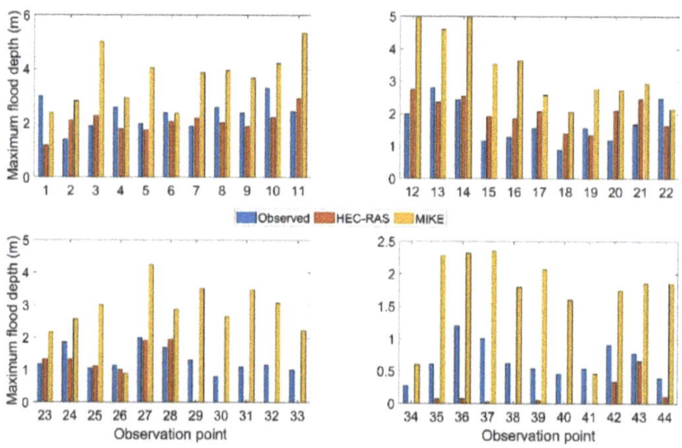

Figure 12. Comparison of the maximum flood depths derived by field observations: HEC-RAS simulation and MIKE FLOOD simulation.

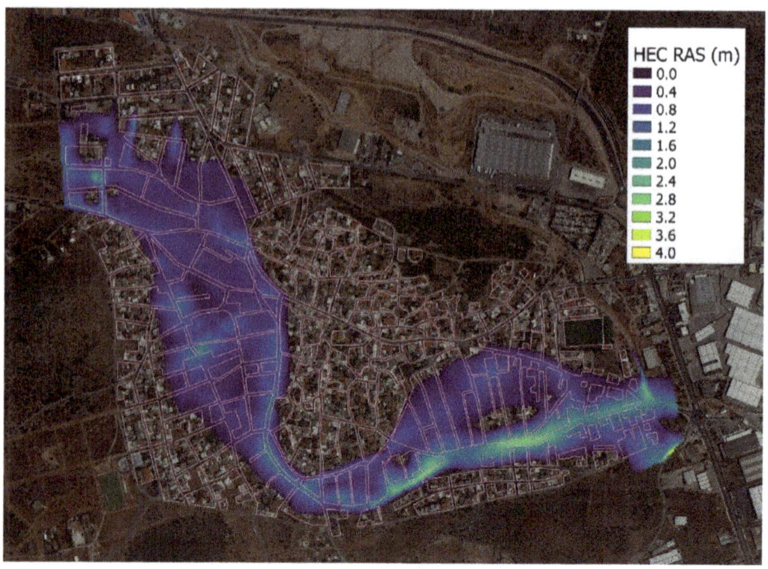

Figure 13. Results of the calibrated HEC-RAS for maximum water depths.

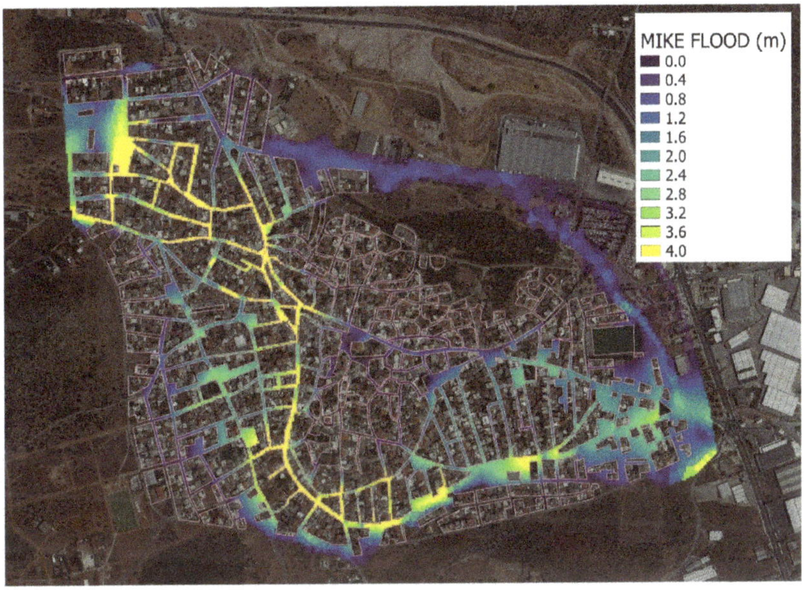

Figure 14. Results of the informed MIKE FLOOD for maximum water depths.

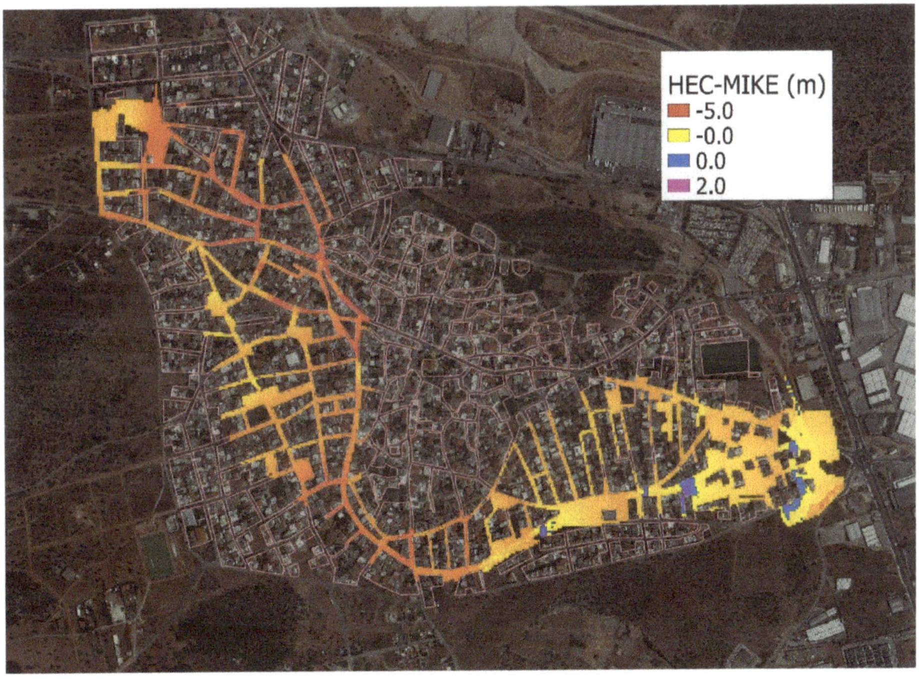

Figure 15. Calibrated HEC-RAS minus informed MIKE FLOOD for maximum water depths.

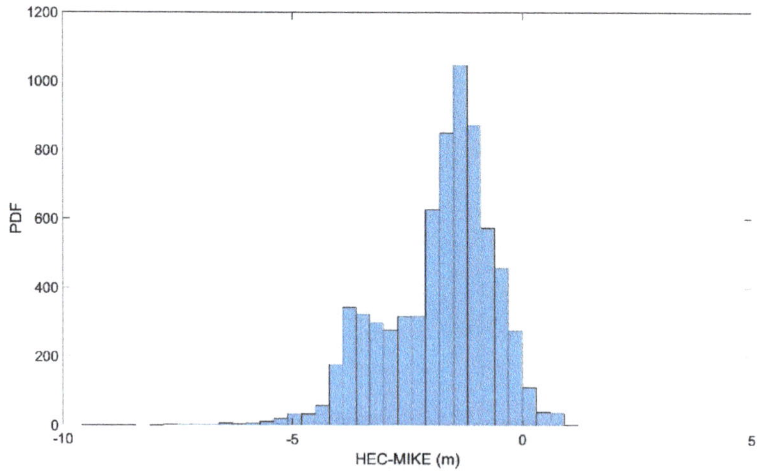

Figure 16. Distribution of the residuals HEC minus MIKE.

There is an area of the city where both models failed to reproduce the flood sufficiently, either overestimating or underestimating the maximum observed flood depths (gauges 29–41). This is probably due to each model's structure and the simplifying assumptions regarding the validity of the diffusion wave model. For the remaining observation points, the calibrated HEC-RAS seems to capture the flood characteristics reasonably well. On the other hand, the informed MIKE FLOOD seems to overestimate flood depths in a systematic way. The residual (HEC-RAS minus MIKE FLOOD results) shows that an overestimation of flood depths by MIKE FLOOD compared to the ones by HEC-RAS occurs in the upstream and the main computational area, while in some parts of the downstream computational area we observe the opposite. The residual distribution ranges from −5 m (MIKE FLOOD is overestimating compared to HEC-RAS) to 1 m (HEC-RAS is overestimating compared to MIKE FLOOD). It should be noted that the abstraction is performed just for the intersection between these two flood extents.

4. Discussion and Concluding Remarks

Our work had a dual objective: First, to perform forensic research, collecting post-event flood data in the field, and second, to investigate several modelling issues regarding the use of well-known flood simulators in this kind of complex case studies. More specifically, in this paper we presented and shared a flood dataset for the Mandra flood event, which occurred in the greater metropolitan area of Athens, Greece, on 15 November 2017. This real-world flood dataset was used to calibrate flood simulations by the HEC-RAS software and used to inform the MIKE FLOOD software. The open flood dataset should be welcome given the scarcity of this kind of data. Since many simulators are usually verified with numerical experiments and analytical or simplified physical models, this dataset can potentially contribute to benchmarking robust flood simulators, tested in real world case studies.

The major findings of our work consist of: (a) the lessons learned from a post-event data collection in the field; (b) the calibration issues raised from a computationally demanding simulator; (c) the answer to the research question regarding the global use of the calibrated parameters.

As far as the first finding is concerned, it seems that a post-event forensic research is feasible, in terms of resources and equipment. Since urban flooding is characterized by mud flows, there are lot of clues for the maximum water depth observed in several points. The drawback is the absence of data indicating the time evolution of the phenomenon and flow velocity data. The participation of the locals in the field survey was of high importance in order to derive a reliable and representative flood dataset. Small interviews and chats could provide answers for several unclear points.

Regarding the second finding, the proposed two-phase calibration procedure, including the parametric screening and the grid-search methodology, seems to be an efficient way of reducing the computational demand of the simulator. Although the calibration procedure seems to be trapped in the boundaries of the dimensional space, there are strong indications that the optimized pair of parameters gives the optimal result for these model structures, without equifinality problems.

Finally, the concept of the informed modeling does not seem to work. Possible reasons for this are the differences regarding the way in which buildings are represented, as well as the different form of the downstream open boundaries and the mesh structure. However, we strongly highlight that the systematic overestimation of MIKE FLOOD against the observed data does not indicate that one software is better than the other, but that every software has its dynamics, and the transferability of parameter values cannot be performed in blind trust, while the direct calibration of model input parameters is a must. This reinforces the belief that the flood model parameters are of grey-box nature and their global use should be avoided or adopted with the utmost care.

Author Contributions: Conceptualization, V.B., I.K. and V.A.T.; methodology, V.B. and I.K.; software, E.R. and S.H.; validation, V.B. and I.K.; investigation, V.B., I.K., E.R. and S.H.; data curation V.B. and J.K.; writing—original draft preparation, V.B. and I.K.; writing—review and editing, J.K., I.A.S. and V.A.T.; visualization, V.B., I.K. and S.H.; supervision, V.B., I.A.S. and V.A.T. All authors have read and agreed to the published version of the manuscript.

Funding: This research received no external funding.

Data Availability Statement: The data presented in this study are openly available in Zenodo at https://doi.org/10.5281/zenodo.7140750 (accessed on 3 October 2022).

Acknowledgments: We thank Moussoulis of DHI in Athens for the educational license of MIKE FLOOD.

Conflicts of Interest: The authors declare no conflict of interest.

References

1. Kourtis, I.M.; Tsihrintzis, V.A. Adaptation of urban drainage networks to climate change: A review. *Sci. Total Environ.* **2021**, *771*, 145431. [CrossRef]
2. Paprotny, D.; Sebastian, A.; Morales-Nápoles, O.; Jonkman, S.N. Trends in flood losses in Europe over the past 150 years. *Nat. Commun.* **2018**, *9*, 1985. [CrossRef] [PubMed]
3. Shuster, W.D.; Bonta, J.; Thurston, H.; Warnemuende, E.; Smith, D.R. Impacts of impervious surface on watershed hydrology: A review. *Urban Water J.* **2007**, *2*, 263–275. [CrossRef]
4. Jha, A.K.; Bloch, R.; Lamond, J. *Cities and Flooding: A Guide to Integrated Urban Flood Risk Management for the 21st Century*; The World Bank: Washington, DC, USA, 2012.
5. Mimikou, M.A.; Baltas, E.A.; Tsihrintzis, V.A. *Hydrology and Water Resource Systems Analysis*; CRC Press: Boca Raton, FL, USA, 2016.
6. Kourtis, I.M.; Tsihrintzis, V.A.; Baltas, E. A robust approach for comparing conventional and sustainable flood mitigation measures in urban basins. *J. Environ. Manag.* **2020**, *269*, 110822. [CrossRef]
7. Kourtis, I.M.; Bellos, V.; Kopsiaftis, G.; Psiloglou, B.; Tsihrintzis, V.A. Methodology for holistic assessment of grey-green flood mitigation measures for climate change adaptation in urban basins. *J. Hydrol.* **2021**, *603*, 126885. [CrossRef]
8. Theodosopoulou, Z.; Kourtis, I.M.; Bellos, V.; Apostolopoulos, K.; Potsiou, C.; Tsihrintzis, V.A. A Fast Data-Driven Tool for Flood Risk Assessment in Urban Areas. *Hydrology* **2022**, *9*, 147. [CrossRef]
9. Eregno, F.E.; Nilsen, V.; Seidu, R.; Heistad, A. Evaluating the trend and extreme values of faecal indicator organisms in a raw water source: A potential approach for watershed management and optimizing water treatment practice. *Environ. Process.* **2014**, *1*, 287–309. [CrossRef]
10. Yannopoulos, S.; Eleftheriadou, E.; Mpouri, S.; Giannopoulou, I. Implementing the requirements of the european flood directive: The case of ungauged and poorly gauged watersheds. *Environ. Process.* **2015**, *2*, 191–207. [CrossRef]
11. Ullah, H.; Akbar, M. Drought risk analysis for water assessment at gauged and ungauged sites in the low rainfall regions of Pakistan. *Environ. Process.* **2021**, *8*, 139–162. [CrossRef]
12. Razmi, A.; Mardani-Fard, H.A.; Golian, S.; Zahmatkesh, Z. Time-varying univariate and bivariate frequency analysis of nonstationary extreme sea level for New York City. *Environ. Process.* **2022**, *9*, 8. [CrossRef]

13. Bellos, V.; Kourtis, I.M.; Moreno-Rodenas, A.; Tsihrintzis, V.A. Quantifying Roughness Coefficient Uncertainty in Urban Flooding Simulations through a Simplified Methodology. *Water* **2017**, *9*, 944. [CrossRef]
14. Bellos, V.; Nalbantis, I.; Tsakiris, G. Friction Modeling of Flood Flow Simulations. *J. Hydraul. Eng.* **2018**, *144*, 04018073. [CrossRef]
15. Ramirez, A.I.; Herrera, A. Forensic hydrology. In *Forensic Analysis—From Death to Justice*; Shetty, B.S.K., Padubidri, J.R., Eds.; IntechOpen: London, UK, 2016.
16. Alcrudo, F.; Mulet, J. Description of the Tous dam break case study (Spain). *J. Hydraul. Res.* **2007**, *45* (Suppl. 1), 45–57. [CrossRef]
17. Diakakis, M.; Andreadakis, E.; Nikolopoulos, E.I.; Spyrou, N.I.; Gogou, M.E.; Deligiannakis, G.; Katsetsiadou, N.K.; Antoniadis, Z.; Melaki, M.; Georgakopoulos, A.; et al. An integrated approach of ground and aerial observations in flash flood disaster investigations. The case of the 2017 Mandra flash flood in Greece. *Int. J. Disaster Risk Reduct.* **2019**, *33*, 290–309. [CrossRef]
18. Testa, G.; Zuccalà, D.; Alcrudo, F.; Mulet, J.; Soares-Frazão, S. Flash flood flow experiment in a simplified urban district. *J. Hydraul. Res.* **2007**, *45*, 37–44. [CrossRef]
19. Haltas, I.; Tayfur, G.; Elci, S. Two-dimensional numerical modeling of flood wave propagation in an urban area due to Ürkmez dam-break, İzmir, Turkey. *Nat. Hazards* **2016**, *81*, 2103–2119. [CrossRef]
20. Tegos, A.; Ziogas, A.; Bellos, V.; Tzimas, A. Forensic Hydrology: A Complete Reconstruction of an Extreme Flood Event in Data-Scarce Area. *Hydrology* **2022**, *9*, 93. [CrossRef]
21. Zotou, I.; Bellos, V.; Gkouma, A.; Karathanassi, V.; Tsihrintzis, V.A. Using Sentinel-1 imagery to assess predictive per-formance of a hydraulic model. *Water Resour. Manag.* **2020**, *34*, 4415–4430. [CrossRef]
22. Varlas, G.; Anagnostou, M.N.; Spyrou, C.; Papadopoulos, A.; Kalogiros, J.; Mentzafou, A.; Michaelides, S.; Baltas, E.; Karymbalis, E.; Katsafados, P. A multi-platform hydrometeorological analysis of the flash flood event of 15 November 2017 in Attica, Greece. *Remote Sens.* **2019**, *11*, 45. [CrossRef]
23. Bellos, V.; Papageorgaki, I.; Kourtis, I.; Vangelis, H.; Kalogiros, I.; Tsakiris, G. Reconstruction of a flash flood event using a 2D hydrodynamic model under spatial and temporal variability of storm. *Nat. Hazards* **2020**, *101*, 711–726. [CrossRef]
24. Diakakis, M.; Deligiannakis, G.; Andreadakis, E.; Katsetsiadou, K.N.; Spyrou, N.I.; Gogou, M.E. How different surrounding environments influence the characteristics of flash flood-mortality: The case of the 2017 extreme flood in Mandra, Greece. *J. Flood Risk Manag.* **2020**, *13*, e12613. [CrossRef]
25. Spyrou, C.; Varlas, G.; Pappa, A.; Mentzafou, A.; Katsafados, P.; Papadopoulos, A.; Anagnostou, M.N.; Kalogiros, J. Implementation of a Nowcasting Hydrometeorological System for Studying Flash Flood Events: The Case of Mandra, Greece. *Remote Sens.* **2020**, *12*, 2784. [CrossRef]
26. Mitsopoulos, G.; Panagiotatou, E.; Sant, V.; Baltas, E.; Diakakis, M.; Lekkas, E.; Stamou, A. Optimizing the Performance of Coupled 1D/2D Hydrodynamic Models for Early Warning of Flash Floods. *Water* **2022**, *14*, 2356. [CrossRef]
27. Mitsopoulos, G.; Diakakis, M.; Panagiotatou, E.; Sant, V.; Bloutsos, A.; Lekkas, E.; Baltas, E.; Stamou, A.I. How would an extreme flood have behaved if flood protection works were built? the case of the disastrous flash flood of November 2017 in Mandra, Attica, Greece. *Urban Water J.* **2022**, *19*, 1–11. [CrossRef]
28. Bournas, A.; Baltas, E. Investigation of the gridded flash flood Guidance in a Peri-Urban basin in greater Athens area, Greece. *J. Hydrol.* **2022**, *610*, 127820. [CrossRef]
29. Tsakiris, G.; Bellos, V. A Numerical Model for Two-Dimensional Flood Routing in Complex Terrains. *Water Resour. Manag.* **2014**, *28*, 1277–1291. [CrossRef]
30. Brunner, G.W. *HEC-RAS, River Analysis System Hydraulic Reference Manual, Version 6.0*; Report Number CPD-69, Hydrologic Engineering Center; US Army Corps of Engineers: Davis, CA, USA, 2020.
31. DHI Water & Environment. *MIKE 11-A Modelling System for Rivers and Channels Reference Manual*; DHI Water & Environment: Hørsholm, Denmark, 2009.
32. DHI Water &Environment. *MIKE 21 Flow Model-Hydrodynamic Model User Guide*; DHI Water & Environment: Hørsholm, Denmark, 2011.
33. Bellos, V.; Tsakiris, G. Comparing Various Methods of Building Representation for 2D Flood Modelling in Built-Up Areas. *Water Resour. Manag.* **2014**, *29*, 379–397. [CrossRef]
34. Tscheikner-Gratl, F.; Zeisl, P.; Kinzel, C.; Leimgruber, J.; Ertl, T.; Rauch, W.; Kleidorfer, M. Lost in calibration: Why people still do not calibrate their models, and why they still should—A case study from urban drainage modelling. *Water Sci. Technol.* **2016**, *74*, 2337–2348. [CrossRef] [PubMed]
35. Saltelli, A.; Ratto, M.; Andres, T.; Campolongo, F.; Cariboni, J.; Gatelli, D.; Saisana, M.; Tarantola, S. *Global Sensitivity Analysis: The Primer*; John Wiley & Sons, Ltd.: Hoboken, NJ, USA, 2008; pp. 1–292. [CrossRef]
36. Christelis, V.; Hughes, A.G. Metamodel-assisted analysis of an integrated model composition: An example using linked surface water—Groundwater models. *Environ. Model. Softw.* **2018**, *107*, 298–306. [CrossRef]
37. Bournas, A.; Baltas, E. Increasing the Efficiency of the Sacramento Model on Event Basis in a Mountainous River Basin. *Environ. Process.* **2021**, *8*, 943–958. [CrossRef]
38. Sobol, I.M. Sensitivity analysis for non-linear mathematical models. *Math. Model. Comput. Exp.* **1993**, *1*, 407–414. Available online: http://www.andreasaltelli.eu/file/repository/sobol1993.pdf (accessed on 24 August 2022).
39. Morris, M.D. Factorial Sampling Plans for Preliminary Computational Experiments. *Technometrics* **1991**, *33*, 161. [CrossRef]
40. Campolongo, F.; Cariboni, J.; Saltelli, A. An effective screening design for sensitivity analysis of large models. *Environ. Model. Softw.* **2007**, *22*, 1509–1518. [CrossRef]

41. Pianosi, F.; Sarrazin, F.; Wagener, T. A Matlab toolbox for Global Sensitivity Analysis. *Environ. Model. Softw.* **2015**, *70*, 80–85. [CrossRef]
42. Bellos, V.; Tsihrintzis, V. Uncertainty aspects of 2D flood modelling in a benchmark case study. In Proceedings of the 17th Conference on Environmental Science and Technology (CEST2021), Athens, Greece, 1–4 September 2021.

hydrology

Article

CoastFLOOD: A High-Resolution Model for the Simulation of Coastal Inundation Due to Storm Surges

Christos Makris *, Zisis Mallios, Yannis Androulidakis and Yannis Krestenitis

Laboratory of Maritime Engineering and Maritime Works, Division of Hydraulics and Environmental Engineering, School of Civil Engineering, Aristotle University of Thessaloniki, 54124 Thessaloniki, Greece; zmallios@civil.auth.gr (Z.M.); iandroul@civil.auth.gr (Y.A.); ynkrest@civil.auth.gr (Y.K.)

* Correspondence: cmakris@civil.auth.gr

Abstract: Storm surges due to severe weather events threaten low-land littoral areas by increasing the risk of seawater inundation of coastal floodplains. In this paper, we present recent developments of a numerical modelling system for coastal inundation induced by sea level elevation due to storm surges enhanced by astronomical tides. The proposed numerical code (CoastFLOOD) performs high-resolution (5 m × 5 m) raster-based, storage-cell modelling of coastal inundation by Manning-type equations in decoupled 2-D formulation at local-scale (20 km × 20 km) lowland littoral floodplains. It is fed either by outputs of either regional-scale storm surge simulations or satellite altimetry data for the sea level anomaly. The presented case studies refer to model applications at 10 selected coastal sites of the Ionian Sea (east-central Mediterranean Sea). The implemented regular Cartesian grids (up to 5 m) are based on Digital Elevation/Surface Models (DEM/DSM) of the Hellenic Cadastre. New updated features of the model are discussed herein concerning the detailed surveying of terrain roughness and bottom friction, the expansion of Dirichlet boundary conditions for coastal currents (besides sea level), and the enhancement of wet/dry cell techniques for flood front propagation over steep water slopes. Verification of the model is performed by comparisons against satellite ocean color observations (Sentinel-2 images) and estimated flooded areas by the Normalized Difference Water Index (NDWI). The qualitative comparisons are acceptable, i.e., the modelled flooded areas contain all wet area estimations by NDWI. CoastFLOOD results are also compared to a simplified, static level, "bathtub" inundation approach with hydraulic connectivity revealing very good agreement (goodness-of-fit > 0.95). Furthermore, we show that proper treatment of bottom roughness referring to realistic Land Cover datasets provides more realistic estimations of the maximum flood extent timeframe.

Keywords: coastal flooding; numerical modelling; storm surge; sea level elevation; inundation maps; Manning coefficient; raster grid

Citation: Makris, C.; Mallios, Z.; Androulidakis, Y.; Krestenitis, Y. CoastFLOOD: A High-Resolution Model for the Simulation of Coastal Inundation Due to Storm Surges. *Hydrology* **2023**, *10*, 103. https://doi.org/10.3390/hydrology10050103

Academic Editors: Aristoteles Tegos, Alexandros Ziogas and Vasilis Bellos

Received: 20 March 2023
Revised: 22 April 2023
Accepted: 27 April 2023
Published: 30 April 2023

1. Introduction

Storm surges, i.e., a (spatially) broad-scale and abnormal elevation of sea level in coastal areas due to severe weather events (storms, tropical cyclones, hurricanes, typhoons, etc.), threaten low-land littoral areas by increasing the risk of seawater inundation of coastal floodplains and low-lying urban environments [1]. This threat intensifies when high seas due to storm surges (meteorological residual of sea level rise) are combined with high astronomical tides (storm tides) [2]. The projected possible Mean Sea Level Rise (MSLR) due to probable future environmental changes in the climatic scale can also further stimulate the intensity of such phenomena on the coastal zone. Moreover, future projections of cyclone characteristics have shown that detrimental extreme events of marine storminess, such as heavy precipitation, windstorms, and storm surges, are strongly associated with each other and can drive coastal flood hazards in a combined way over the Mediterranean basin [3,4]. Thus, storms may affect the sea level elevation on the shoreline/waterfront in two ways: (a) by increasing the Sea Surface Height (SSH) due to the inverted (or inverse) barometer effect (low-barometric atmospheric pressure), and (b) by winds pushing seawater onshore. The floodwater in coastal areas can overtop physical obstacles or artificial barriers (e.g., dunes and knolls or seawalls, levees, embankments, and armoured slopes), and consequently inundate large parts of inland rural and urban areas. Coastal inundation is mainly responsible for land loss, erosion, damages to onshore infrastructure and properties, environmental degradation of coastal aquatic ecosystems, saltwater intrusion in coastal aquifers, and, occasionally, human casualties, etc.

1.1. Research Theme

The most prominent natural hazard induced by episodic bursts of SSH or (semi-)permanent long-term MSLR is coastal flooding and/or the inundation of littoral lowlands, with various significant implications for the coastal communities and environments [5]. Several studies have presented investigations of the coastal vulnerability due to the impact of MSLR and storm surges, related flood hazards, and damage assessment in the eastern and western Mediterranean littorals; e.g., Moroccan and Egyptian coasts [6–8]; NE Mediterranean coastal zone [9]; Ebro river delta in Spain (NW Mediterranean) [10]; coastal inundation risk assessment due to combined land subsidence and MSLR in southern Italy [11]; estimation of 49 cultural World Heritage Sites in low-lying Mediterranean coastal areas until 2100 [12]; potential MSLR-induced inundation in the central Mediterranean (Malta) for susceptibility assessment and risk assessment scenarios to lead policy action [13]. Hauer et al. [14] assessed the exposure of the U.S. population to coastal flooding due to MSLR, while Kulp and Strauss [15] showed that the latest developments in assessments and error corrections of Digital Elevation Models (DEM) have induced a rise in estimates of global vulnerability to MSLR and related coastal flooding. A robust model implementation for such phenomena producing realistic inundation hazard maps is crucial in terms of coastal management, the study of risk, flood hazard mitigation, first-level response to disaster, and decision support.

In this paper, we present recent developments of a numerical model for coastal inundation on littoral floodplains induced by sea level elevation due to storm surges (CoastFLOOD) [16], potentially enhanced by astronomical tides (and MSLR, not investigated in this paper). The model performs numerical simulations of hydraulic flood flow on inland coastal domains covering local-scale areas up to a few hundred km^2 [17,18]. The inundation model can be forced either with sea level observations (e.g., in situ measurements from tide-gauges and satellite-derived data) or with modelling outputs of regional-scale simulations for storm surges [17]. The High-Resolution Storm Surge (HiReSS) [19] has been used in operational forecast mode for short-term marine weather predictions (sea level and currents) [20,21], providing boundary conditions for CoastFLOOD simulations over adjacent coastal zones [17]. Furthermore, it has been applied as the Mediterranean Climatic Storm Surge (MeCSS) model in climatic studies for long-term hindcasts [9] or future projections of storm surge patterns in the Mediterranean Sea [22–24].

CoastFLOOD performs detailed modelling of the rather shallow and slow process of seawater uprush and flood routing due to episodic, mid- or long-term sea level elevation, i.e., induced by storm surges/tides. It is a very fine resolution, raster-based, 2-D horizontal, mass balance flood model for coastal inlands, following the simplified concept of a reduced complexity form of the Shallow Water Equations (SWEs) running on a storage-cell GIS domain [25–27]. Only the large-scale low-frequency phenomena of coastal inundation due to storm surges and tides are simulated by the model, which does not consider the high-frequency processes of coastal flooding due to wave run-up. The storm-induced SSH on the coastline feeds the seawater surge on the littoral floodplain via a set of 2-D decoupled Manning-type flow equations. The floodwater inundation on the coastal terrain is simulated on a very high resolution (dx = 2–5 m) ortho-regular Cartesian raster grid. Land elevation data are derived by the post-processing of available DEM datasets by the Hellenic Cadastre [28], available in 4600 × 3600 m^2 ground tiles by the projection of the

Hellenic Geodetic Reference System 1987 (HGRS87). The detailed features of the model are discussed in Section 2.

1.2. Literature Review

Numerous 2-D horizontal models exist for the simulation of the mid- to long-term coastal inundation due to storm tides with or without the influence of MSLR. The most representative and established flood inundation model suites have been developed for river flooding and fluvial inundation but can be also used for seawater inundation in coastal areas (Table 1).

Table 1. Representative 2-D inundation model suites for river flooding, also used in coastal areas.

Model	References	Concept and Applications
LISFLOOD-FP CAESAR-LISFLOOD	[29–34]	Reduced complexity inertial formulation of the SWEs leading to 2-D horizontally decomposed Manning-type volumetric flow equations mainly applied in coastal areas with rivers, optionally coupled to a Landscape Evolution Model (LEM) simulating the geomorphic development of flood basins.
MIKE FLOOD 2D	[35,36]	The well-known proprietary flood model suite combining different modules, MIKE URBAN, MIKE 11, and MIKE 21, for urban sewerage systems overflows, river/channel flood discharges, and coastal drivers applied in coastal cities.
HEC-RAS 2D	[37,38]	The classic non-commercial flood modelling system combining 1-D/2-D for river flood flow and fluvial plain inundation with the coastal floodplain extent due to sea level changes simulated with the ADCIRC hydrodynamic model.
SOBEK-2DFLOW (Overland Flow)	[39,40]	Based on complete Saint Venant equations; a fully hydrodynamic 2-D simulation engine for steep floodwater fronts, wetting and drying processes, subcritical and supercritical flow, including rainfall runoff; applications combine pluvial floods with storm surge influence in urban areas.
FLO-2D	[41]	Reduced complexity 2-D Manning-type volumetric flow storage cell simulator coupled to JMA storm surge model.
Multi-Scale Nested MSN-Flood model	[42,43]	High-resolution multi-scale modelling of coastal flooding due to tides, storm surges, and river flows specifically for urban coastal inundation.
FloodMap-Inertia	[44]	An urban flood inundation model neglecting the convective acceleration term in the momentum equation, coupled to ADCIRC for sea level on its coastal boundary, assuming that the floodplain is filled with water by an embankment-type of river-littoral boundaries essentially acting as a continuous, broad-crested weir, through which flow exchange occurs between channel and floodplain.
Floodity	[45]	An anisotropic dynamic mesh optimization (DMO) technique for 2-D double control-volume and a finite element adaptive mesh model for urban coincidental flood modelling.
Delft Flooding System Delft-FLS)	[46]	Overland flow simulation by the 2-D Saint-Venant equations on a rectangular, staggered grid with a finite difference method employing a shock-capturing numerical scheme suitable for rapidly modeling varying flows over rough terrains, including flow through defense breaches and around buildings (minimum depth of 0.01 m distinguishes "dry" from "flooded" cells).
Unstructured Tidal, Residual, Intertidal, Mudflat version 2 ($UnTRIM^2$)	[47]	A semi-implicit, Eulerian-Lagrangian finite difference/finite volume model, governed by 3-D SWEs with Boussinesq approximation solved for free surface elevation, water velocities (and salinity) in a Cartesian coordinate system on an unstructured orthogonal grid including both 3-D barotropic and baroclinic processes (tide, wind, and gravitationally-driven circulation).
Sea, Lake, and Overland Surges from Hurricanes (SLOSH)	[48,49]	A polar-grid storm surge model with gradually varying cell sizes covering a basin extending from the possibly flooded inland area up to deep water, with a dedicated computation scheme on a B-grid to simulate wetting and drying processes. Water surface elevation differences act as hydraulic load for floodwater propagation to the surrounding grid cells.
Stevens Estuarine And Coastal Ocean Model (sECOM)	[50,51]	A successor model to the Princeton Ocean Model (POM) family of models; a 3-D, free surface, hydrostatic, primitive equation estuarine and coastal ocean circulation model with a wetting-drying flood model approach along free moving boundaries.

Table 1. *Cont.*

Model	References	Concept and Applications
Xie-Pietrafesa-Peng (XPP) model	[52,53]	A HM-C mass-conserving inundation wetting/drying scheme coupled to POM-3D.
Cellular Automata (CA) modules	[54,55]	A simplified, grid-based, Saint-Venant equations, 2-D shallow hydrodynamic module discretized in time, space, and state, with local spatial interaction and temporal causality, also optionally running on a triangular finite element mesh.

Within this modelling framework, Hubbert and McInnes [56] introduced a storm surge inundation model through treatment of the coastal boundary configured to pass through the velocity grid points on the staggered grid in a stepwise manner and define wet/dry cells in inland areas based on a predefined threshold of local water depth on each cell. Nevertheless, many researchers have discussed the practical need for reduced physical complexity approaches [57] to adequately simulate 2-D flood inundation [58] compared to the full-scale 3-D hydrodynamic or 2-D SWE modelling of complex flood flow routing [59–66]. The latter mainly applies to 2-D river floodplain flows, but it is equally valid for 2-D coastal plain flooding either by waves or storm tides. Nevertheless, proper testing and validation of flood inundation models [67] intended for specific hydrodynamic, hydraulic, or hydrological processes dictate the concept of equifinality in model implementation [68]. Our case outlook is to adequately simulate (in terms of robustness and computational resources availability) the coastal inundation extents (including a fine 2-D horizontal local distribution of water heights) and the response times of coastal flood maxima within an oversimplified methodological framework minimizing the uncertainty of parametric analysis and dependence on unreliable or insufficient (topographic and land use) input [69].

1.3. Research Incentive

The proposed model follows the conceptual framework of reduced complexity flood inundation approaches on high-resolution computational grids in a way to balance between the reliability and practicality of applications in the coastal zone [41,44,70–72]. Hence, we introduce a recently developed in-house model (CoastFLOOD) specifically designed for fine-scale hydraulic flooding of seawater in littoral areas. It is specifically built to work in operational mode, meeting the need to be easily coupled to a coarser large-sale storm surge model (e.g., HiReSS) written in the same programming language and using similar coding modules and job execution tactics. Our goal was to further formulate proper and detailed input for spatially varying Manning roughness coefficients, especially fitted to 2-D coastal floodplains. This way we can uphold the physical properness, assist the calibration process and the robust performance of the model in a timely manner for operational forecasting, and engineer consulting purposes [73,74].

The scope of the study is to further evaluate the impact of detected sea level variations (either by modelling or monitoring procedures) on seawater inundation patterns over several characteristic regions of the Greek coastal zone. Kulp and Strauss [75] have discussed the necessity to minimize errors in DEMs to avoid underestimations of coastal vulnerability due to MSLR-induced flooding. Therefore, the CoastFLOOD model is tested in tandem with an updated dataset of land elevation derived from a DEM with a resolution of dx = 2–5 m that covers 10 selected lowland areas along the Ionian Sea coastline. These have been identified as highly impacted areas by intense flooding events in the past [17].

The model domains include various urban and suburban settlements, rural coastal plains, environmentally protected areas (lagoons, estuaries, wetlands, and aquatic habitats), touristic infrastructure areas, recreational coastal zones with sandy beaches, and coastal regions accumulating several activities (e.g., aquaculture, fisheries, navigational transportation, seaport commerce, etc.). Coastal inundation hazard maps are produced to estimate the littoral flooding variability over the Greek coastline. Model validation is performed for the operational forecast mode of CoastFLOOD simulations against fine-scale satellite observations (by Sentinel-2 images at 10-m resolution), producing the Normalized Difference Water Index (NDWI) [76,77]. Model results are also compared to a static level, enhanced "bathtub" inundation approach with "eight-side rule" hydraulic connectivity [78–82].

New model features are also presented, concerning the detailed surveying of terrain roughness and bottom friction, the expansion of Dirichlet boundary conditions for coastal currents (besides sea level), the enhancement of a wet/dry cell technique for flood front propagation over positive/negative steep terrain slopes, etc.

All of the methodological information regarding the model setup, parameterization features, numerical schemes, and computational grids are thoroughly described in Section 2. Case study characteristics and datasets for model validation are presented in Section 3. The results regarding coastal flooding are analysed in Section 4. A discussion of the study findings is presented in Section 5, followed by a section of concluding remarks (Section 6).

2. Methodology

2.1. *Conceptual Approach of Storm Surge Inundation*

The basic concept of our modelling approach refers to implementing a set of simplified continuity and momentum conservation equations for the simulation of rather shallow and slow inundation processes [16,18]. These are primarily (or solely) driven by the sea level elevation on the coastline and secondarily by the estimation of the barotropic coastal current as long as it has an onshore direction. Therefore, we simulate the sluggish seawater flooding on the low-lying coastal areas that is induced by a slow surface flow due to storm surge, unlike the fast-evolving undulating flows that are caused by swell and wind-wave action on the coast.

The model's advantageous feature is that it can be applied at very high spatial resolutions (e.g., dx = 1–5 m) for a geophysical-scale flow, while the feeding input of *SSH*, acting as the hydraulic head that defines the piezometric load on the boundary conditions, can be of wider scales (e.g., O(Dx) = 1–10 km) [17]. This allows for a practically efficient compromise between the validity of representation of the governing physics and operational model adequacy for hydraulic engineering problems in large-scale environmental flows. The chosen raster modelling approach adopts a (horizontally) decomposed uniform flow approximation for coastal floodplain flow, which is mainly dominated by gravity and friction to calculate the momentum balance [18]. This is a reasonable approximation for gradually evolving (laminar) flows over mild sloping floodplains in rural or natural areas; however, it may be an oversimplification for unsteady hydraulic flows in complex urban environments, where turbulent effects play a starker role in rapidly varying topographies. Neglecting pressure and/or inertial terms of the momentum equation may lead to erroneous representation of the floodwater flow characteristics in the built environment. Nevertheless, the assumed model approach has been shown in the past to be able to adequately predict the horizontal extents of inundation and the floodwater height in inland areas even if they lie in urban regions. The simplified kinematic scheme of the Manning-type hydraulic flow allows for numerical applications on regular gridded domains of large areas, typically incorporating up to 15×10^6 model grid cells, testing the limits of modern available computational resources.

2.2. *Numerical Model for Hydraulic Flow in Coastal Flooding*

CoastFLOOD [16–18] is an in-house numerical model built on a FORTRAN-95 code, that solves the depth-averaged, 2-D horizontal, mass balance, flood flow equations [25–27,29]. These have produced a series of 2-D floodplain applications [70,72,83] particularly implemented in coastal case studies [84–88]. The latest version of the model, presented herein, has been enhanced in terms of bottom roughness treatment to include cases in:

(a) Rural plains with agricultural zones and farmlands, wild flora or natural vegetated fields, forests, bare or stony lands, pastures, and grasslands, etc.;

(b) Wet inland areas, such as shores, estuaries, lagoons, river deltas, beaches, etc.;

(c) Urban and sub-urban areas with engineered coasts, built waterfronts, ports and coastal protection structures, roads, highways, railway networks, dense building constructions or open spaces and parks, mildly or highly developed built environments, etc.

The robustness of similar model approaches (e.g., LISFLOOD-FP, FLO-2D, Floodmap) has been validated and applied in 2-D floodplains in coastal areas or fluvial landforms [29,80,83,89]. CoastFLOOD also follows a simplistic finite difference scheme for hydraulic flow inundation, running on very fine resolution raster grids, able to reproduce the surge-induced 2-D flood on the coast [17,18]. Propagation of the floodwater front is decomposed in two horizontal Cartesian x- and y-directions, allowing for discrete zonal and meridional components of the flow, respectively, for inland flood routing [26,71,90].

The simplified form of the 2-D equations for conservation of mass (continuity) and momentum are discretized over an ortho-regular grid of rectangular cells (Figure 1a), in order to reproduce the evolution of a 2-D Manning-type flow between neighbouring cells over the entire floodplain [58]. The floodwater flow between adjacent cells is mainly driven by the hydraulic head created by the inter-cell difference of water surface height in all four cardinal directions of the horizon (Figure 1). Thus, the continuity equation relates the floodwater volume of an arbitrary cell to the volumetric flows in and out of it, during a typical timestep of the numerical solution. This is written in the form of generic volumetric (Equation (1)), (analytic) spatially discretized volumetric and piezometric head (Equations (2) and (3)), grid- and time-discretized (Equation (4)) equations, as:

$$\frac{\partial V}{\partial t} = Q_x^{in} - Q_x^{out} + Q_y^{in} - Q_y^{out} \tag{1}$$

$$\frac{\partial V_{i,j}}{\partial t} = Q_{xi-1/2,j} - Q_{xi+1/2,j} + Q_{yi,j-1/2} - Q_{yi,j+1/2} \tag{2}$$

$$\frac{\partial h_{i,j}}{\partial t} = \frac{Q_{xi-1/2,j} - Q_{xi+1/2,j} + Q_{yi,j-1/2} - Q_{yi,j+1/2}}{\partial x \cdot \partial y} \tag{3}$$

$$h_{i,j}^{t'} = h_{i,j}^{t} + dt \cdot \frac{Q_{xi-1/2,j}^{t} - Q_{xi+1/2,j}^{t} + Q_{yi,j-1/2}^{t} - Q_{yi,j+1/2}^{t}}{dx \cdot dy} \quad \text{or}$$
$$h_{i,j}^{t'} = h_{i,j}^{t} + dt \cdot \left(\left(\theta \cdot \frac{Q_{xi-1/2,j}^{t} - Q_{xi+1/2,j}^{t} + Q_{yi,j-1/2}^{t} - Q_{yi,j+1/2}^{t}}{dx \cdot dy} \right) + \left((1-\theta) \cdot \frac{Q_{x\,i-1/2,j}^{t'} - Q_{x\,i+1/2,j}^{t'} + Q_{y\,i,j-1/2}^{t'} - Q_{y\,i,j+1/2}^{t'}}{dx \cdot dy} \right) \right) \tag{4}$$

where, V is the volume with V_{ij} referring to cell (i,j), i and j being the x- and y-directions of the Cartesian grid; +1/2 in indexing denotes the intercell positioning of flow parameters; t is the time and dt the timestep of temporal discretization (hence, $t' = t + dt$ at the following timestep in the solution scheme); Q_x and Q_y are the volumetric flow rates between adjacent floodplain cells in the zonal x- and meridional y-directions of the Cartesian grid, respectively; Q^{in} and Q^{out} are the incoming and outgoing volumetric flow rates in a typical grid cell within the generic representation of the equations; h is the local floodwater height above each grid cell's land elevation, z; dx and dy are the cell dimensions in the zonal x- and meridional y-directions of the Cartesian grid, respectively; θ is a numerical weighting coefficient, which determines whether the equations are fully solved or partially implicitly for $\theta < 1$ or explicitly for $\theta = 1$ [58]. The explicit scheme is the norm, but both options are provided in the CoastFLOOD model. Note that the scalar magnitude of local water height, h, is calculated on each cell's centre or any adjacent cell's centre, e.g., $h_{i,j}$ or $h_{i+1,j}$ or $h_{i,j-1}$, while the vectorial magnitude of flow rate, Q, is calculated on either of the side faces of each cell or either of the side faces of any adjacent cell, e.g., $Q_{i-1/2,j}$ or $Q_{i,j+1/2}$; hence, practically rendering the solution scheme on a staggered grid (Figure 1b).

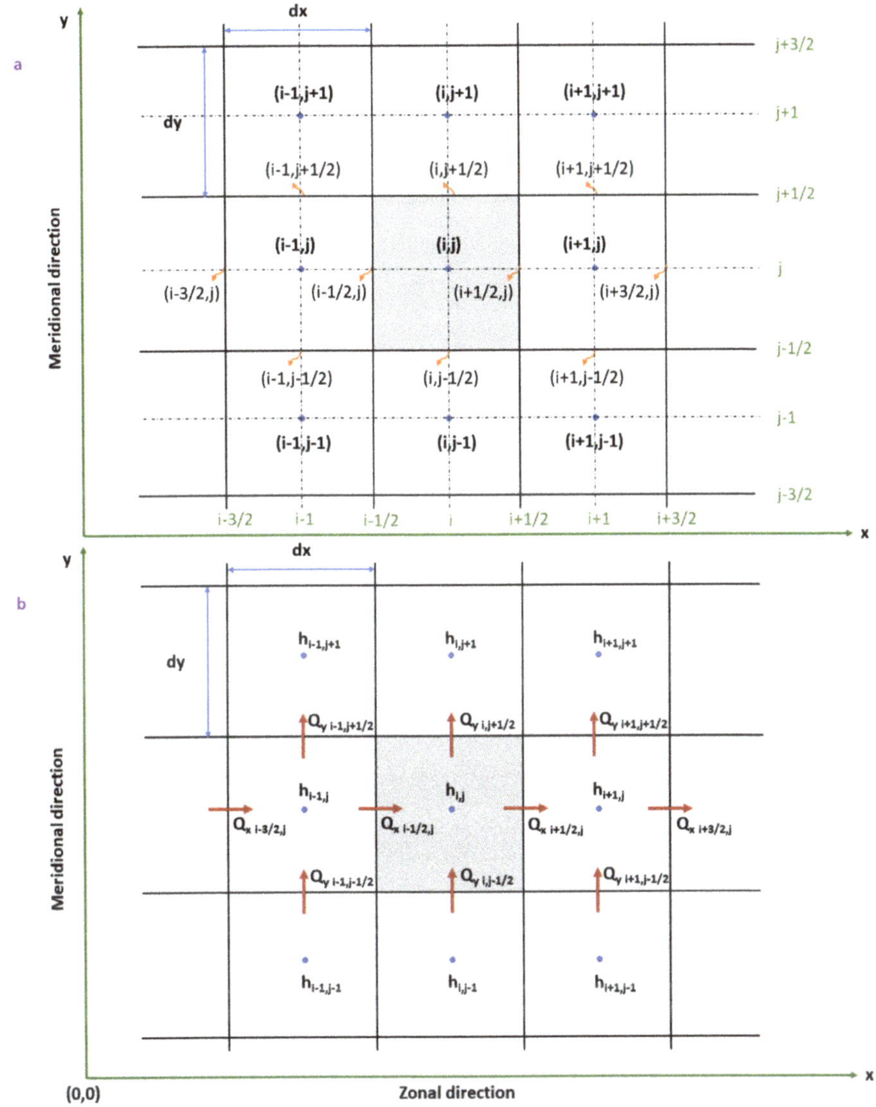

Figure 1. Depiction of the prototype Cartesian raster grid formulating a typical computational domain in CoastFLOOD model; (**a**) discretization of the staggered grid cells (at their centres and faces) with dx and dy dimensions, over an i−j coordinate system on the Cartesian x- and y-directions (zonal and meridional directions of the horizon); (**b**) notation of scalar parameter floodwater height h at the centres of the grid cells and decoupled vectorial parameter volumetric flow rate, Q_x and Q_y, between adjacent cells (at their interfaces). The shaded cell is the main cell of parametric numerical calculation at each timestep. Arrow directions represent the positive values of flow pathways between grid cells; i.e., from floodwater flow upstream areas to downstream ones.

This way, we allow for each floodplain grid element to function as an individual storage cell, letting a simplified formulation of the momentum equation derive inter-cell fluxes. Equations in x- and y-directions can be written in the form of an analytic kinematic function based on Manning's law, permitting the decomposed calculation of the flow rate in each grid cell, reading in generic form:

$$Q = \frac{h_{flow}^{5/3}}{n} \cdot \left(\frac{h_{upstream} - h_{downstream}}{CellWidth_{zonal}} \right)^{1/2} \cdot CellWidth_{meridional} \tag{5}$$

where, $CellWidth_{zonal} \equiv dx$ and $CellWidth_{meridional} \equiv dy$ are generic notations of cell dimensions in horizontal directions; indices $_{upstream}$ and $_{downstream}$ refer to generic representations of, e.g., (i − 1,j) and (i + 1,j) cells for (i,j) central element of numerical calculation at each timestep; n is the Manning's coefficient of roughness for bed friction inclusion; h_{flow} is the flow depth between two adjacent cells, i.e., defined as the difference of the highest floodwater surface elevation from Mean Sea Level (MSL), H, minus the maximum bed elevation, z, between two neighbouring cells (Figure 2).

The spatially discretized version of Equation (5) further reads:

$$\begin{aligned} Q_x^{in} &= \frac{h_{flow\,x,in}^{5/3}}{n} \cdot \left(\frac{h_{i-1,j} - h_{i,j}}{dx} \right)^{1/2} \cdot dy\,,\; Q_x^{out} = \frac{h_{flow\,x,out}^{5/3}}{n} \cdot \left(\frac{h_{i,j} - h_{i+1,j}}{dx} \right)^{1/2} \cdot dy \\ Q_y^{in} &= \frac{h_{flow\,y,in}^{5/3}}{n} \cdot \left(\frac{h_{i,j-1} - h_{i,j}}{dy} \right)^{1/2} \cdot dx\,,\; Q_y^{out} = \frac{h_{flow\,y,out}^{5/3}}{n} \cdot \left(\frac{h_{i,j} - h_{i,j+1}}{dy} \right)^{1/2} \cdot dx \end{aligned} \tag{6}$$

where, again, indices $_{in}$ and $_{out}$ denote incoming and outgoing flows.

The spatiotemporally discretized form of Equation (6) corresponding to placement on a typical model grid (Figure 1b) is written as:

$$\begin{aligned} Q_{xi-1/2,j}^t &= \frac{h_{flow\,xi-1/2,j}^{t\,5/3}}{n} \cdot \left(\frac{h_{i-1,j}^t - h_{i,j}^t}{dx} \right)^{1/2} \cdot dy,\; Q_{xi+1/2,j}^t = \frac{h_{flow\,xi+1/2,j}^{t\,5/3}}{n} \cdot \left(\frac{h_{i,j}^t - h_{i+1,j}^t}{dx} \right)^{1/2} \cdot dy \\ Q_{yi,j-1/2}^t &= \frac{h_{flow\,yi,j-1/2}^{t\,5/3}}{n} \cdot \left(\frac{h_{i,j-1}^t - h_{i,j}^t}{dy} \right)^{1/2} \cdot dx,\; Q_{yi,j+1/2}^t = \frac{h_{flow\,yi,j+1/2}^{t\,5/3}}{n} \cdot \left(\frac{h_{i,j}^t - h_{i,j+1}^t}{dy} \right)^{1/2} \cdot dx \end{aligned} \tag{7}$$

where t is the current time and can be also substituted by t' to represent the needed values in Equation (4), and h_{flow} in absolute discretised notation (Figure 2) can be calculated based on the equation:

$$h_{flow_{x_{i-1/2,j}}} = \left(max\{H_{i-1,j}, H_{i,j}\} - max\{z_{i-1,j}, z_{i,j}\} \right) \tag{8}$$

The exponent $h_{flow}{}^{5/3}$ refers to a Manning law approach for the flood propagation and can be used under the assumption of a uniform laminar flow over a flat rectangular cell (dx = dy wide grid element) of constant depth.

Equations (5)–(7) describe the reduced complexity versions of the momentum equations, which are typically based on a semi-analytical approach for hydraulic flows, such as the aforementioned Manning-type equation. Alternately, the user can choose to incorporate the 2-D finite difference approximation of a similar equation for diffusive waves [58]:

$$Q_{xi-1/2,j}^t = \frac{\frac{h_{flow\,xi-1/2,j}^{t\,5/3}}{n} \cdot \left(\frac{h_{i-1,j}^t - h_{i,j}^t}{dx} \right)^{1/2} \cdot dy}{\left(\left(\frac{h_{i-1,j}^t - h_{i,j}^t}{dx} \right)^{1/2} + \left(\frac{h_{i,j-1}^t - h_{i,j+1}^t}{dxdy} \right)^{1/2} \right)^{1/4}},\; Q_{xi+1/2,j}^t = \frac{\frac{h_{flow\,xi+1/2,j}^{t\,5/3}}{n} \cdot \left(\frac{h_{i,j}^t - h_{i+1,j}^t}{dx} \right)^{1/2} \cdot dy}{\left(\left(\frac{h_{i,j}^t - h_{i+1,j}^t}{dx} \right)^{1/2} + \left(\frac{h_{i,j-1}^t - h_{i,j+1}^t}{dxdy} \right)^{1/2} \right)^{1/4}} \tag{9}$$

$$Q_{yi,j-1/2}^t = \frac{\frac{h_{flow\,yi,j-1/2}^{t\,5/3}}{n} \cdot \left(\frac{h_{i,j-1}^t - h_{i,j}^t}{dy} \right)^{1/2} \cdot dx}{\left(\left(\frac{h_{i,j-1}^t - h_{i,j}^t}{dy} \right)^{1/2} + \left(\frac{h_{i-1,j}^t - h_{i+1,j}^t}{dxdy} \right)^{1/2} \right)^{1/4}},\; Q_{yi,j+1/2}^t = \frac{\frac{h_{flow\,yi,j+1/2}^{t\,5/3}}{n} \cdot \left(\frac{h_{i,j}^t - h_{i,j+1}^t}{dx} \right)^{1/2} \cdot dy}{\left(\left(\frac{h_{i,j}^t - h_{i,j+1}^t}{dx} \right)^{1/2} + \left(\frac{h_{i-1,j}^t - h_{i+1,j}^t}{dxdy} \right)^{1/2} \right)^{1/4}} \tag{10}$$

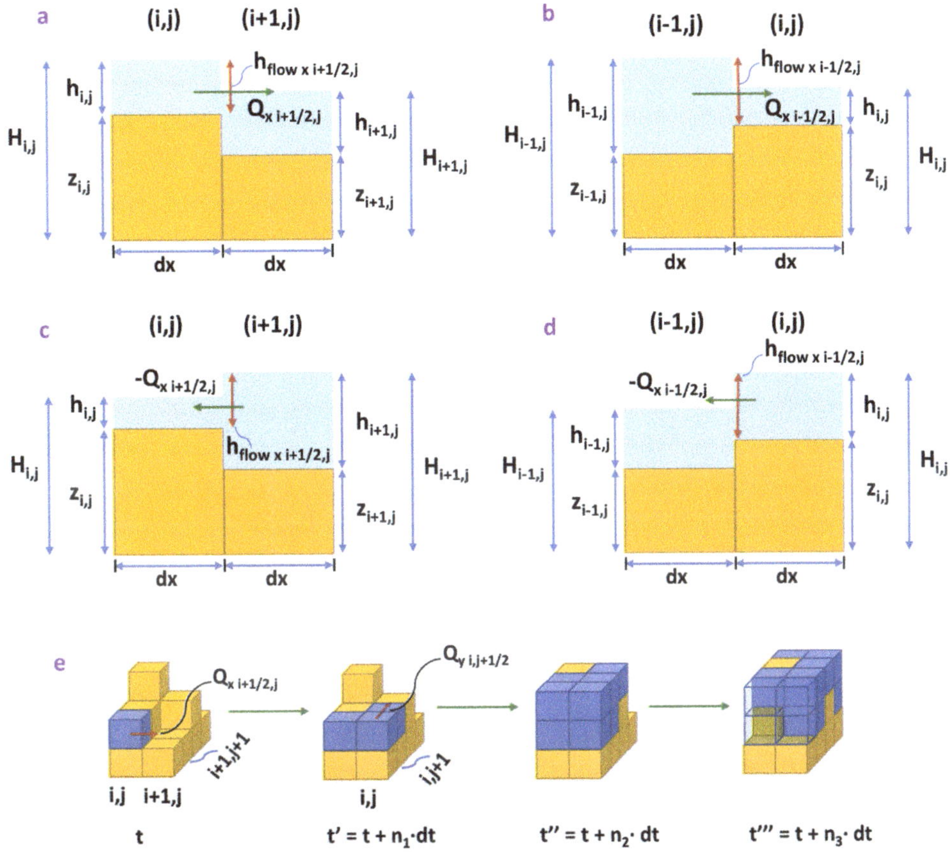

Figure 2. Depiction of flood front propagation over typical grid cells in the CoastFLOOD model's 2-D x-z plane (graphs (**a**,**b**)) and wet/dry cell expansion in pseudo-3-D projection (graph (**c**)). (**a**–**d**) Schematic representation of Q_x and h_{flow}, i.e., the flow depth between two adjacent cells, defined as the difference of the highest floodwater surface elevation from MSL zero-level, H, minus the maximum bed elevation, z, between two neighbouring cells either (i,j) and $(i+1,j)$ in graphs a and c, or $(i-1,j)$ and (i,j) in graphs (**b**,**d**). (**e**) Illustrative representation of progressive inundation front by discretized floodwater flow propagation and encroachment on an elevating model grid with explicitly modelled micro-topography at arbitrary $(n_1 - n_3 \cdot t)$ timesteps. Yellow-brown cube-cells refer to ground, while blue ones refer to floodwater.

2.3. Time Discretization—Numerical Schemes

The abovementioned discretized Equations (4) and (7)–(10) are solved with the use of appropriate boundary and initial conditions using certified numerical techniques. CoastFLOOD incorporates (user-identified) solvers that implement either an explicit ($\theta = 1$) forward-time and centered-space (FTCS) finite difference scheme or an implicit ($\theta < 1$) backward-time and centered-space (BTCS) algorithm to obtain predictions of Q_x, Q_y, and h at any given timestep. The choice of θ is a prerequisite from the CoastFLOOD user, resulting in different levels of solution complexity/stability and higher model runtimes for the implicit scheme. For $\theta = 1$, the Q and h, at t' can be explicitly computed by the known quantities at t (floodplain flows Q can be initially calculated by Equations (8)–(11)). Consequently, floodwater depths h can be updated by Equation (4a). Explicit algorithms are preferred for their coding simplicity and straightforward integration schemes on a staggered ortho-regular raster grid. Nevertheless, numerical stability is ensured by very small model timesteps, e.g., dt < 10 s, according to the Courant-Friedrichs-Lewy (CFL) criterion, C:

$$C = u_x dt/dx < 1 \tag{11}$$

e.g., for $C \leq 0.5$, the timestep should practically be $dt \leq (0.5h_{i,j}dx^2)/Q_x$, where $u_x = Q_x/A_x$ and $A_x = dx \cdot h_{i,j}$ in a typical grid cell. To ensure numerical stability, the following CFL condition, with α = 0.3–0.7, is proposed by [27,33]:

$$dt_{max} = a \cdot dx / \sqrt{gh_{ij}} < 1 \tag{12}$$

Practically, based on Equation (13), for values of, e.g., h = 0.001–1.5 m and dx = 5 m, the minimum achieved timestep should roughly range between $dt_{max} \approx$ 35–0.35 s, respectively (for corresponding α = 0.7–0.3). Nonetheless, the aforementioned dt values refer to an upper threshold value, while even lower timesteps may be needed in the course of cell-by-cell numerical solution. Previous studies have proposed the following adaptive timestep [71], based on the Von Neumann condition, especially for the diffusive wave case, as shown in Equations (9) and (10):

$$dt \leq \frac{dx^2}{4(1-\theta)} => dt = \frac{dx^2}{4} min\left(\frac{2n}{h_{flow x}^{5/3}} \left| \frac{\partial h}{\partial x} \right|^{1/2}, \frac{2n}{h_{flow y}^{5/3}} \left| \frac{\partial h}{\partial y} \right|^{1/2} \right) \tag{13}$$

This is supposed to eliminate "chequerboard" numerical oscillations, induced when dt becomes large, which essentially occurs for very low $h_{i,j}$ values and consequently low flood flow rates (and floodwater velocities). However, in the CoastFLOOD model, the practical lower/upper cut-off dt values are set to 0.5 s $\leq dt \leq$ 5 min (e.g., for dx = 5 m), allowing for reasonable computational times and the avoidance of lagging in the numerical solution, respectively. Likewise, to avoid further instabilities in the advancing iterations of the numerical solution (notably in high floodwater depths, $h_{i,j}$, or highly uneven elevation levels of adjacent cells), we adopt a flow rate limiter, especially for the most classic case of 2-D floodplain flow being controlled by momentum Equation (7). The flow limiter (minimum Q threshold) can also prevent instabilities in adjacent areas of very large differences in floodwater depth [25]:

$$\begin{aligned} Q_{xi-1/2,j} &= min\left\{ calculated\ Q_{xi-1/2,j}, \frac{dxdy\left(h_{i,j}^t - h_{i-1,j}^t\right)}{8dt} \right\} \\ Q_{xi+1/2,j} &= min\left\{ calculated\ Q_{xi+1/2,j}, \frac{dxdy\left(h_{i+1,j}^t - h_{i,j}^t\right)}{8dt} \right\} \end{aligned} \tag{14}$$

for a concomitant min/max limiter of floodwater velocity that reads 0.01 m/s $\leq u_x = Q_x/A_x \leq$ 5 m/s. Similar equations apply to the y-direction of the flow. With this numerical treatment, the user can actually prevent over- or under-shooting of the numerical solution. The flow limiter essentially ensures that floodwater depth change in an arbitrary cell at t is not adequately large to reverse the y flow entering or exiting the cell at t' [71]. Q values derived by Manning's equation are replaced, when overestimated, with values strictly determined by model domain parameters (dx and dt). If a small dt or large dx is chosen, the limiter is nearly eliminated. Therefore, the results of the CoastFLOOD model, like many other storage-cell codes for flood flows, are far from invariant with respect to dx and dt. Their optimal choice is a matter of experience, taking into account the extents of the entire case study domain and its low-lying areas, etc. Moreover, this approach may undermine the simulations in terms of correctly predicting the advance of flood fronts and the volume of floodwater in inundated areas [91]. The choice of smaller CFL numbers, $C << 0.5$; hence, smaller dt can address this discrepancy.

The coastal flooding phenomena, induced by storm surges, may last from several hours up to a few days, i.e., resulting in simulations of $2–4 \times 10^4$ to $2–5 \times 10^5$ timesteps, for a few hours up to 3 days duration of the studied flood event, given that $dt \leq 1$ s. Depending on the number of inland grid cells to be flooded (e.g., up to 40×10^6 elements), this means that the estimated computational times range from one hour up to more than half a day on a PC with a 10th generation 12-core Intel® i7-CPU, 10750H, @2.60 GHz, with 64 GB RAM and 1 TB SSD hard disk 860 QVO. For the case of the implicit scheme, where Q and h variables depend on unknown quantities at t', an iterative solution technique (e.g., finite difference Preissmann scheme [92]) adds even more computational burden and time. Of course, the implicit scheme allows for larger timesteps in the O(5–10) mins, given the slow evolution of flood events over inundated plains.

The meridional- and zonal-direction decompositions of the flood flow components allow the derived 1-D flow equations for overland seawater propagation to be numerically and separately calculated for each grid cell face on a typical 2-D raster [90]. This makes the calculation of flood routing an easy task, through the use of a simplistic nearest neighbour or quad-tree search algorithm for the downstream cells. The latter are defined as dry or wet (for $h_{ij} > 0.005$ m) and then they are saved and/or updated in a storage cell matrix at every simulation timestep. To this end, the effective water flow depth between two neighbouring cells, h_{flow}, which is defined by the difference between the highest possible water level in adjacent grid elements and their largest land elevation, z (Figure 2), is not allowed to exceed the maximum threshold of $h_{flow} \leq \max(h_{i,j}) = \mathrm{SSH}-z_{i,j}$. The x- and y-direction decoupling of flood flow propagation may not represent the diffusive nature of the inundation wave spreading on the floodplain; however, it has been shown [83] that more complicated treatments of floodplain flows have yielded no significant improvements compared to reduced complexity models [70] when evaluated against Synthetic Aperture Radars (SAR) data.

2.4. Computational Domain and Raster Grid

The numerical grid formulation (terrain discretization) for typical, reduced complexity models of coastal inundation by storm surges follows the trends in the development of high-resolution topographic gridded data. Namely, DEMs represent bare earth or ground surface topography, excluding trees, buildings, and any other surface objects, while Digital Surface Models (DSMs) capture the land surface, including vegetation and manmade structures, such as buildings and infrastructures. DEMs are used to construct the entire model domain (mainly focused on natural areas, rural environments, wild lands, etc.), whilst DSMs are implemented within urban and suburban areas to include the flow obstruction by the built environment.

To firstly identify the low-lying areas along the Greek coastal zone and secondly create the detailed topographical input for the storm tide inundation simulations with CoastFLOOD, the GIS datasets of land elevation were retrieved from the official Greek service for the comprehensive recording of real-estate and property metes-and-bounds [28]. There are two available high-resolution DEMs in coastal and inland regions with spatial resolution $dx = 2$ m and 5 m. The rectangular model domains were produced by post-processing of the available polar coordinate geospatial data in the World Geodetic System 1984 (WGS84) to HGRS87. The DEM's geometric accuracy is less than 0.70 m, while its absolute accuracy is less than 1.37 m with a 95% confidence level [17]. Similarly, the DSM's accuracy is less than 0.32 m, while its absolute accuracy is less than 1 m with a 95% confidence level. The DSM has an even finer resolution of $dx = 0.8$ m, and thus its datasets were extrapolated to fit the fixed model's computational domains of $dx = 2–5$ m.

To avoid the underestimation of the storm surge effect driving the flood flow from any possible convex or crooked part of the coastline (no matter how complex it might be or what orientation the shoreline has in the domain), a cross-type scan of the model grid

(N→S and S→N in the meridional direction; W→E and E→W in the zonal direction) is applied in every timestep (Figure 3). This way, the volumetric flow rates' signs (Figure 2c,d) are corrected, based on the propagation of the flood front from all directions of the horizon, and thus the wet/dry storage cell matrix is updated with every possible change in water level of each grid element in the model domain (Figure 3). This is a step forward from traditional coastal inundation modeling that considers flood propagation from only one boundary at a time.

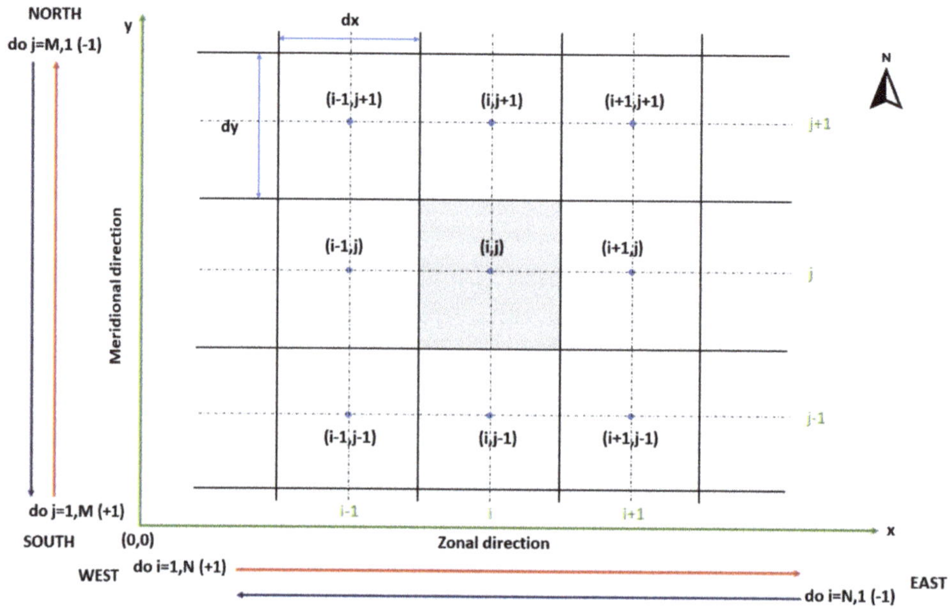

Figure 3. Depiction of the cross-type scanning process of the numerical grid by the computational domain in the CoastFLOOD model. Red and blue arrows represent the numerical propagation scan direction of the grid cells on zonal and meridional, x- and y-axis, for i = 1, N and j = 1, M (and reverse), respectively, applied at each timestep.

The discrepancies of the DEM/DSM are crucial factors of accuracy in CoastFLOOD simulations of flooded areas, even if the highest available resolution raster grid is used to include topographic details of the natural and urban parts of the coastal domain [93]. CoastFLOOD does not consider the effects of porous bed percolation and ground infiltration, flows in sewerage and drainage systems (e.g., conduits, bridge culverts, wells, shafts, etc.). However, this is not considered a crucial issue, as these constructions are usually saturated with fresh water or drainage/sewage waters from surface runoffs. Coastal inundation usually occurs within a compound flooding incident; i.e., concurrently to river flooding and/or urban flooding due to heavy rainfall and strong runoffs relevant to the storm event also driving the onshore sea surge [37].

2.5. Model Parameterization

Bottom friction is the main parameterization feature of reduced complexity flood inundation models. The calculation of hydraulic flows requires the specification of flow resistance or bed roughness in a parametric approach. As the typical model cell's dimensions and depth are assumed to be uniform for each grid element, an effective Manning's bottom roughness coefficient, n, at grid unit scale can be determined as a calibration parameter. Seenath [94] thoroughly discussed issues of achieved improvement in prediction modelling of coastal flooding (more in terms of inundated area extents) based on the fine representation of spatially distributed friction over the case study domain against a uniform n value all over the model grid.

The CoastFLOOD model incorporates both solutions, i.e., considering the friction effect of the floodplain terrain on the inundation flow either by defining a distributed, effective, grid-scale Manning's n on each cell of the model's raster domain or by proposing a representative "global" effective grid-scale n coefficient (on the entire domain or large homogenous parts of it). By integrating the relevant literature [74,80,94–101], we created a detailed collective ensemble of proposed Manning coefficient n values discretized at 36 increments (Table 2). These values are specifically fitted to 2-D coastal floodplain flows and refer to the most common and less likely types of (natural or artificial) ground material.

Beven [102] argued that a predetermination of bottom roughness parameters at each computational grid point was rarely possible due to scaling problems, i.e., differences between the in situ observation scale and the model grid scale, and other data availability constraints. However, the recent development of the CORINE Land Cover (CLC) inventory [103] provides a robust record of land cover in 44 classes for Europe. CLC uses a minimum mapping unit of 25 ha for areal phenomena and a minimum width of 100 m for linear phenomena; here, we use the latter. CLC is mainly produced on a country/state-level by visual interpretation of fine-resolution satellite imagery from Sentinel-2 and Landsat-8 (for gap filling) products, with the latest time consistency referring to 2017–2018.

Table 3 presents a detailed matching catalogue that we have created for all 36 discrete cases of CoastFLOOD's Manning coefficient listings in Table 2 to the CLC-2018 codes that refer to data of as many possible natural and manmade land cover types. CLC is available in both raster and vector formats; in our case studies, we used the second one, because it is easier to align the land cover data to the constructed model domains. Specifically, for each of the study areas, CLC data were retrieved in QGIS using its boundaries as a reference. Then, a Manning coefficient n was assigned to each vector polygon representing a specific land use, using the matching between n and land use from Table 2. Finally, a raster image with the same dimensions and spatial resolution as the Manning n matrix and the model grid was created. If no CLC are available, a parametric calibration of bottom roughness can be undertaken in order to identify empirical values for the Manning coefficient. Terrain heterogeneities on the sub-grid level can cause discrepancies in the representation of land cover texture, thus Manning's n is commonly used as a determinative calibration parameter rather than a physical factor of actual field friction.

Table 2. CoastFLOOD 2-D modified floodplain Manning coefficient list.

A/A	n	Description of Areas' Characteristics
1	0.001	open water
2	0.0115	concrete surfaces
3	0.010	rural driveways (dirt road and granules)
4	0.012	urban land uses (asphalt mixtures and other urban surface features: artificial stones, paving blocks, lightweight aggregate concrete), concrete rooftop, playground, yard, barren land
5	0.013	main asphalt roads (national, regional highway networks, autobahns, etc.)
6	0.015	brick terrain, unidentified high and low development urban environment, inland open waters (reservoirs, lakes, ponds, lagoons, estuaries)
7	0.017	city streets (asphalt, concrete, etc.)
8	0.018	unidentified/unclassified urban terrain
9	0.020	clean to gravelly earth pathways (pebbles with a small portion of cobbles), muddy/sandy open waters and sandy terrains, sea bottom (saturated wet sand or silt-sand) and channel beds
10	0.030	bare unidentified/unclassified soil

Table 2. *Cont.*

A/A	n	Description of Areas' Characteristics
11	0.022	bare land, stone paved road and ceramic sett, or paving sett pathways
12	0.029	stony cobble lands, pastures, and farmlands
13	0.025	manmade structures, gravel beds and pathways (pebbles with nominal diameter: d_{n50} = 4–64 mm, cobbles: d_{n50} = 64–256 mm)
14	0.0375	cultivated fields and pasture, grassland (including prairies, steppes, plains)
15	0.0425	isolated sand/gravel(mixed) pits, estuary channels, and uneven urban areas
16	0.029	emerged sloping sandy beaches, sand dunes
17	0.030	managed grasslands
18	0.0115	unclassified/unidentified rural areas
19	0.033	grass surfaces
20	0.035	short stiff grass areas
21	0.0575	weeds with or without structure
22	0.0555	heavy brush floodplains
23	0.040	arable land plains, heavy/coarse gravel (boulders: d_{n50} >= 256 mm) areas, unclassified grassland, and shrubs (including savannah, meadow, veldt, pampa, tundra)
24	0.050	unclassified trees, open development areas (containing parks, streets of rural character)
25	0.055	herbaceous wetlands
26	0.067	emerged barriers
27	0.140	hardwood woodland and cultivated woodland
28	0.086	unclassified wetlands (including watersheds, salt/fresh marshes, bottomland hardwood, swamps, mangrove swamps, seagrass flats, forest swamps)
29	0.100	forest land and unidentified forest trees evergreen forest, pasture, hay, crop, vegetation
30	0.120	deciduous forest, natural grassland, herbaceous lands
31	0.150	mixed forest, shrubs, scrub, emergent herbaceous wetlands
32	0.240	cultivated vegetation
33	0.300	unidentified densely built urbanized zones (uncharacterized structures)
34	0.320	very dense tall (long trunk) trees forest (jungles, etc.)
35	0.368	very dense and/or stiff grasslands (reedy bamboo, etc.)
36	0.400	very dense small forest trees and thick shrubs

Table 3. Matching of Table 2's A/A for Manning coefficient list to Corine Land Cover (CLC) data.

A/A	CLC Code	Description of CLC Label Areas' Characteristics
4–8	111, 112	Continuous urban fabric, Discontinuous urban fabric
10–8	121	Industrial or commercial units
5–7	122	Road and rail networks and associated land
4–2	123	Port areas
4–5	124	Airports
3	131	Mineral extraction sites
6–4	132, 133, 141	Dump sites, Construction sites, Green urban areas
4–7	142	Sport and leisure facilities
23–14	211, 212	Non-irrigated arable land, Permanently irrigated land

Table 3. *Cont.*

A/A	CLC Code	Description of CLC Label Areas' Characteristics
14	213	Rice fields
22	221	Vineyards
30	222	Fruit trees and berry plantations
29–30	223	Olive groves
12–14	231	Pastures
27	241	Annual crops associated with permanent crops
27–32	242	Complex cultivation patterns
21–29	243	Land principally occupied by agriculture, with areas of natural vegetation
29–32	244	Agroforestry areas
29–34	311	Broad-leaved forest
30–34	312	Coniferous forest
31–34	313	Mixed forest
19–30	321	Natural grasslands
22–30	322	Moors and heathland
32	323	Sclerophyllous vegetation
31	324	Transitional woodland-shrub
16–15	331	Beaches, dunes, sands
12–9	332	Bare rocks
29–32	333	Sparsely vegetated areas
10	334, 335	Burnt areas, Glaciers and perpetual snow
28	411, 412, 421, 422	Inland marshes, Peat bogs, Salt marshes, Salines
1–16	423	Intertidal flats
6	511, 512, 521, 522	Water courses, Water bodies, Coastal lagoons, Estuaries
1	523	Sea water

2.6. Input Data: Boundary and Initial Conditions, Simulation Time Limit

A basic assumption of the CoastFLOOD approach, except for the steady state forcing of the flood flow on the coastal boundary with smoothly varying sea level maxima, is the non-treatment of the floodwater ebbing phenomenon. The model considers the spatiotemporally local wetting and drying of individual cells during the numerical solution, yet the computations are ceased when floodwater reaches the farthest area from the coastline or the waterfront. Thus, the model is not allowed to simulate the large-scale drying phase of floodwater receding back to the sea after the storm surge begins to decrease on the marine coastal boundary.

The application of a flood inundation model to a specific coastal area requires the definition of boundary conditions (mainly shoreline sea level and optionally onshore currents), topographic features (land elevation), and local flow resistance (bottom friction) as model parameters that control the flow characteristics. If the SSH on the coastline exceeds the MSL, then Equations (4) and (7) or (8) are activated with a value of $h(t) \equiv \mathrm{SSH}(t)$ on the seaside boundary (ghost) cell, used to calculate the initial volume flux to all adjacent shoreland cells and then onto the floodplain cells. This implies that CoastFLOOD is driven by a Dirichlet-type boundary condition referring to local values of $h = \mathrm{SSH} - z$ (where z is the land elevation of a raster grid cell) [18], i.e., even for sea level timeseries SSH(t) varying in the tidal cycle on the seaward side of the computational domain [17]. These conditions should last for at least a few hours and up to 3 days, given that the storm-induced sea level does not abruptly change in time but follows the slow smooth variation of the tidal constituent. Furthermore, this approach is ideal for particular scenarios of long-term MSLR or Total Water Level (TWL) on the coastline [104,105].

Although this approach actually ignores the momentum exchange effects between neighbouring cells in the floodplain and therefore introduces a restricted physical interpretation of the flow characteristics, it can capture all of the dominant features of the shallow seawater onshore flow, which leads to the rather slow propagation process (thus, seawater flux may be neglected) of coastal inundation [26,30,57,83,94]. To include the barotropic current's effect on the momentum flux of the first land cell adjacent to the seawater cell, we added an impromptu $Q_{xs} = U_{cx} \cdot dy \cdot h_{flowx}$ (similar to Q_{ys}; where U_c is the storm surge-induced current velocity decoupled in Cartesian components U_{cx} and U_{cy}) added to the calculated Q_x, Q_y of Equations (7) and (9) or (10), only for the "first" dry shoreland cell. Its inclusion does not seem to drastically influence the inland flood inundation extent, but it is a step towards improvement of the physical representation of onshore seawater flow.

The storm tide (integrating surge and tide) levels can be extracted either from ocean modelling (Section 2.6.1) or from tide-gauge recordings and satellite altimetry (Section 2.6.2). The seawater elevation input can be entered as a boundary condition, representing the land-sea interface, on any cell in the computational domain.

2.6.1. Coupling with a Storm Surge Model

We coupled CoastFLOOD with the operational forecast model HiReSS, which simulates storm surges at both regional and local scales [17,19,21,106]. The latter is a 2-D horizontal SWE hydrodynamic circulation model for the simulation of sea level variations and depth-averaged currents, applied in large regional marine bodies and marginal seas [9,20,22–24], including several combined processes, such as:

- barotropic circulation hydrodynamics by momentum conservation and continuity SWEs;
- inverse barometer effect, i.e., the response of sea level to the atmospheric pressure gradient of large barometric systems;
- shear stresses of wind on the sea surface;
- Earth gravity and geostrophic effects (Coriolis force);
- interaction of surge-driven sea level and astronomical tides by a static model [107] approach based on the equilibrium theory of tides [108];
- ocean bottom friction;
- turbulence of horizontal eddies based on the eddy viscosity concept and the Smagorinsky model approach;
- interaction with coastal wave-induced currents by incorporating radiation stress terms in nearshore surf zones;

The model has been applied in operational forecast mode for short-term marine weather predictions and has been thoroughly validated, during the past 15 years, in the Mediterranean region against field data from in situ tide-gauge observations of storm-induced episodic SSH due to severe weather conditions or the derived Sea Level Anomaly (SLA; SLA = SSH−MSL) in inter-annual tidal cycles [2,17,20,21]. Its climatic mode counterpart, MeCSS, has also been evaluated for long-term historical simulations of mean and extreme storm surge patterns in the Mediterranean basin during (>30-year) reference periods [9,18,22–24,104,105]. Furthermore, the HiReSS model is the official numerical tool of the Operational Forecast Platform (OFP) Wave4Us, recently incorporated into the METEO.GR node managed by the National Observatory of Athens [109–111]. It is also advocated on a global scale by the Accu-Waves OFP [112] over several regional and marginal seas (e.g., Red Sea, Yellow Sea, Black Sea, Java Sea, NW Atlantic Ocean, etc.), gulfs, straits, and local aquatic bodies (e.g., Gulf of Finland, Osaka Gulf, Tokyo Gulf, Persian Gulf, English Channel, etc.), producing sea level forecasts for safer navigation in 50 important ports around the globe [113].

2.6.2. Boundary Conditions from Sea Level Observations

The model can also be forced by sea level observations on the study areas' coastlines, which are represented in the computational domain by marginal dry cells of the model grids during Still Water Level (SWL) conditions. Observations can be derived either from satellite altimetry (SLA) of the Copernicus Marine Service (CMS) that covers the last 30-years [114] or tide-gauges, located along the coastline [115]. The spatial resolution of the CMS product is 1/8° (~13 km), and it is provided in a daily step while the initial coverage of the dataset extends over all European seas. The Level L4 data are produced by merging observations of Topex/Poseidon, ERS1/2, Jason 1-2-3, Sentinel (3A/B and 6A), HaiYang-2A/B, Saral[-DP]/Altika, Cryosat-2, ENVISAT, and GFO altimetry missions. These satellite SLA fields have been previously used to evaluate the sea level variability in the Mediterranean Sea [116,117]. The tide-gauge observations can be derived from the Intergovernmental Oceanographic Commission (IOC) system. The IOC field data [115] have higher temporal resolution (e.g., 10-min step), but their spatial coverage along the coastline is coarser than the satellite or modelling data based on the locations of the tide-gauges, which are usually operated inside ports. Here, we focus on CoastFLOOD simulations forced by satellite-derived SLA data (see Section 4.2).

The CoastFLOOD simulations provide tide/surge-induced flooded areas due to satellite recordings or realistically modelled values of daily SLA or SSH values, respectively. From these, the timeseries' maxima are extracted, SLA_{max} or SSH_{max}, and are separately simulated together with several extreme case scenarios of onshore TWL, typical of the east-central Mediterranean and the Greek coastal zone; i.e., 0.5, 1, 1.5, and 2 m [23,104,105]. The latter produce reference values of Flooded Areas (*FA*) in regions prone to coastal inundation, assisting in the normalization of flood extents in different case studies.

3. Case Studies and Data for Model Validation

3.1. Case Study Areas

The CoastFLOOD model was tested at 10 selected case study areas of the western Greek coastal zone (Figure 4), which are rather frequently inundated by storm surges of the Ionian Sea. Similar to tropical storms, peculiar low-pressure atmospheric systems may form in the western and central Mediterranean (namely Medicanes) and propagate from the westernmost cyclogenesis centers of the basin towards the Ionian and Adriatic Seas, making landfall on the western shores of the Italian and Balkan Peninsulas [118–121]. These events are known to threaten the selected case study areas, located in coastal lowland regions prone to inundation (Figure 4a). Thus, the latter were chosen based on a series of recorded coastal (and/or compound) flooding events that were recently reported in mass media (i.e., some examples out of numerous documented flood inundation impacts in provincial and metropolitan Greek areas; Figure 4b–e):

- Manolada-Lechaina coastal zone (Area 1), east of Patra city, north-western Peloponnese, southeastern Ionian Sea, recorded during October 2021 storm Ballos [122] followed by incidents during December of the same year (December 2021).
- Vassiliki Bay (Area 2; Figure 4c) on the southern coast of Lefkada Island, northern Ionian Sea, recorded on 17 November 2017 [123] and on 18 September 2020 [124].
- Preveza coast (Area 3; Figure 4e), west-central Epirus, northern Ionian Sea, recorded on 30 November 2021 [125].
- Igoumenitsa port (Area 4), north-western Epirus (north Ionian Sea), recorded on 12 November 2017 [126].
- Cephalonia Island (including the Livadi coastal area in its southern bay; Area 5), central Ionian Sea, recorded on 18 September 2020 [127] during Ianos Medicane [17,120,121].
- Patra city (Area 10), broader metropolitan area in north-eastern Peloponnese, Rio town's flooded seafront during Ianos Medicane [17,120,121], on 18 September 2020 [128].

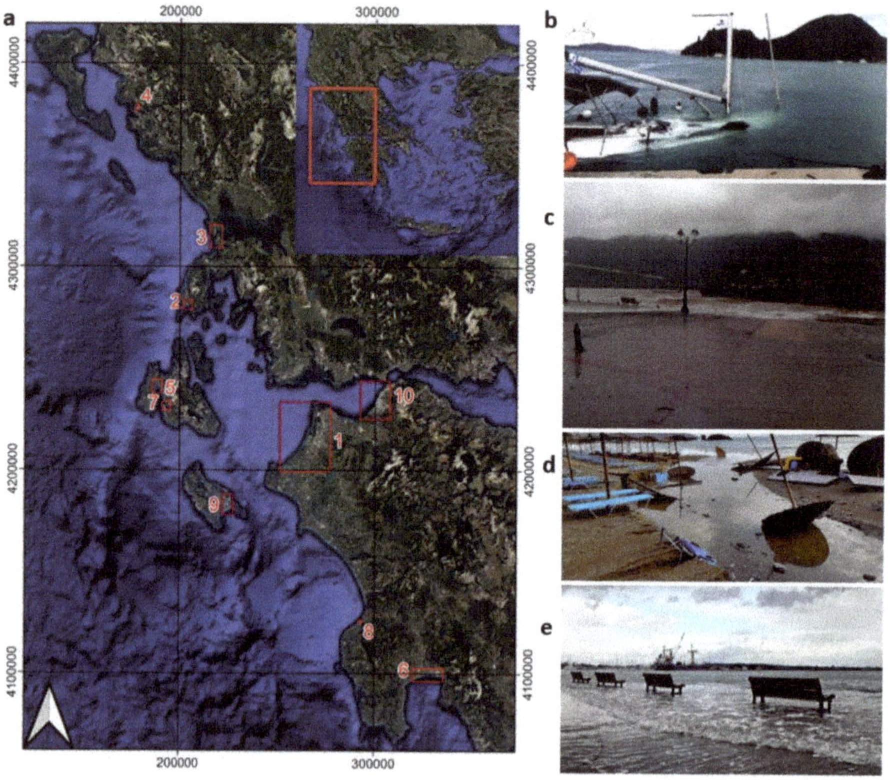

Figure 4. (**a**) Map of selected study areas to apply the CoastFLOOD model; Areas 1: Manolada-Lechaina, 2: Vassiliki bay, 3: Preveza coastal area, 4: Igoumenitsa port, 5: Livadi bay, 6: Kalamata, 7: Argostoli, 8: Kyparissia, 9: Laganas, 10: Patra city; (**b**) Depiction of boat wreck due to the passage of Ianos Medicane (September 2020) over Lefkada Island; (**c**) Storm seawater inundation of November 2017 in Area 2 (Vassiliki, Lefkada Island); (**d**) impact of "Ballos" (October 2021) storm on a touristic beach on Corfu Island; (**e**) storm surge coastal inundation at the seafront of Preveza (in November 2021).

Other interesting flood-prone areas (Figure 4a) frequently impacted by sea level elevation on the Ionian coastline comprise the towns of Kalamata (Messenia, southern Peloponnese; Area 6) and Argostoli (east Cephalonia Island, Ionian Sea; Area 7), the rural areas of Kyparissia (north-western Messenia, south-western Peloponnese; Area 8), and Laganas (southern Zakynthos Island, Ionian Sea; Area 9).

3.2. Observational Data for Model Evaluation

The coastal model validation was based on comparisons of simulation results against ocean colour images collected by the Sentinel-2 satellite with a spatial resolution of 10 m freely distributed by the Copernicus Data Space Ecosystem (CDSE) or Sentinel Hub [129,130]. To estimate the observed coastal inundation during stormy conditions, a remote sensing technique of Sentinel-2 raster images was used to compute the NDWI [131] on coastal areas affected by storm surges, shown to oversee any alterations in water content on the Earth's surface aquatic resources [132]. Several researchers have used NDWI in the past to assess flood extents due to hurricane-led storm surges, e.g., in the Gulf of Mexico, or to identify coastlines [133,134]. NDWI is computed based on *Band*3 and *Band*8 bands of the ocean colour images:

$$NDWI = \frac{Band3 - Band8}{Band3 + Band8} \tag{15}$$

where *Band*3 is the Visible Green Light (VGL) and *Band*8 is the Near-Infrared Radiation (NIR) of the spectrum.

Herein, we use NDWI to identify (wet) flooded areas on low coastal inlands following a storm surge event with two different procedures (see Section 4). In the first case, a satellite image taken on 14 December 2021 was used to calculate the NDWI > 0 on the raster grid, corresponding to "wet" cells of the study area (given that they are not all flooded by seawater, but by rainwater from precipitation or surface runoff as well). The second approach involved the estimation of flooded areas using two separate satellite images, the one before (15 September 2020) and the second after (20 September 2020) the recorded Ianos Medicane's storm surge that occurred on 17–18 September 2020 (due to the unavailability of datasets on these exact dates); the difference of the two calculated NDWIs was used to estimate the inundated area [17]. Specifically, after calculating the individual NDWI for each cell of the domain and both images, the difference in $NDWI_{dif}$ of post-storm NDWI minus pre-storm NDWI was calculated on each pixel of the raster grid. It was assumed that pixels with $|NDWI_{dif}| > 0.15$ corresponded to remaining wet ground (areas that were very likely flooded during the storm). To identify areas that were flooded likely due to the storm surge, the NDWI values of the second image were filtered to exclude the already wet cells before the storm surge. Notably, the areas identified as inundated by stormwater had $NDWI_{dif} > 0.5$ in many instances, confirming the result. To avoid misinterpretations, we mainly considered lowland areas close to the coastline (with hydraulic connectivity to the sea), nevertheless there is no safe method yet able to distinguish the source of floodwater (e.g., tidal surge, drainage or runoff, and rainfall) based on the NDWI technique.

An important limitation of the comparison with remote sensing NDWI fields is that satellite images are susceptible to the timeframe they refer to or are available in, namely due to the absence of satellites over the study regions during the storm event or due to cloud contamination, a process very common during storms, cyclones, and Medicanes. A second limitation of the NDWI method is that the water accumulation due to intense water precipitation or surface runoff from surrounding higher ground into bilged lowlands (e.g., cesspools, dugouts, sumps, pits, fosses, and cisterns) can contaminate the derived NDWI fields of humid surfaces or wetted areas, thus deregulating the coastal model validation procedure. Nevertheless, the NDWI method is essential for model performance testing of the occurrence of characteristic coastal hazard events.

3.3. *Enhanced Bathtub Module for Model Validation*

The CoastFLOOD model was compared with a static level "bathtub" approach inundation module [78,135]. This method easily identifies the flood-prone low-lying areas with ground elevation below a predefined threshold, e.g., an estimation of coastal seawater level maximum, $z <$ SSH or $z <$ TWL. The bathtub technique is known to be oversimplifying in terms of physical processes and can produce serious overestimations of coastal flood extents [34,136]. Therefore, an enhanced bathtub module with hydraulic connectivity (Bathtub-HC) was adopted [81,137,138]. To this end, we applied a nearest neighbour search algorithm following the 'eight-side rule' in order to identify the potential floodwater flow path between neighbouring raster-grid cells in both cardinal (cross-orthogonal) and ordinal (diagonal) directions of the horizon. This way, the unsubstantial excessive estimations of possible seawater inundation in coastal lowlands was restricted.

The Bathtub-HC method is known to provide fast and adequately robust estimations of flooded coastal area extents, yet they are practically more conservative than those by SWE models. Compared to the CoastFLOOD model, this method neglects the floodplain terrain sloping topography, the bottom friction effects, etc. Thus, it can predict the flooded areas, but it cannot account for flood duration, detailed floodwater height, and fluxes (velocities) that dynamically affect the onshore and overland floodwater flow. Hence, the Bathtub-HC results are usually only implemented as a reference level for potentially wet inland cells in evaluative assessments of reduced complexity numerical models [16–18]. Moreover, bathtub methods can collaterally identify and depict lowland bilge areas (e.g., pits, fosses, puddles, and cisterns) that can accumulate water from rainfall and surface runoff, unlike SWE coastal flooding models, which only account for seawater floods [17].

Convenient field data of coastal inundation based on in situ observations of floodwater height and extents are literally very rare, while their fitness for model verification is not always suitable due to several reasons [30]. There are no available floodwater level gauges in coastal areas (there are only a very few downstream of river embankments in fluvial floodplains), at least in Greek (scarce- or no-data) study areas of interest. Reduced complexity models of coastal flooding need field data for verification on geophysical scales (10–100 km wide) of observation and implementation. Therefore, only satellite data can serve as in situ references for impacted areas due to seawater flooding. The latter are susceptible to the timeframe they may be available in (e.g., the absence of satellite data during storm events, cloud contamination of satellite images, etc.). Uncertainty regarding the contribution of possible sources of recorded inundation besides storm surges (e.g., waves, rainfall, drainage) may obfuscate the derivation of inundated area coverage due to storm-induced floods.

4. Results

We examined the adequacy of the inundation model predictions under realistic severe storm surge conditions (Section 4.1) and simplified bathtub estimations (Section 4.2), on the Ionian Sea coasts of Greece. Idealized (extreme) and realistic (maxima from 2017–2021 period) scenarios of coastal flooding are also presented in Section 4.3.

4.1. Model Verification against Satellite Data during Severe Storm Surge Conditions

Two areas and events were used for qualitative verification of the coastal flooding model's performance due to a lack of imperative satellite data. In the Manolada-Lechaina study area, there was unfortunately no satellite data availability during October 2021, when storm Ballos hit. However, NDWI could be estimated on 14 December 2021, when another storm surge incident was traced based on the retrieved SLA datasets. These depictions serve as reference for qualitative comparisons with the modelled output of CoastFLOOD. Figures 5 and 6 present flood maps of model simulations overlaid by satellite-tracked wet regions. The CoastFLOOD results are driven on the coastal boundary of the Manolada-Lechaina study area by recorded SLA values on 14 December 2021 (see Section 2.6.2). The zoomed-in maps of Figure 6 depict the overlap of NDWI-identified wet areas by satellite images above flood inundation model output focusing on the mainly affected northern and southern parts of the study area. In general, the CoastFLOOD simulations seem to reproduce the coastal flooding mechanism in areas that are more-or-less affected (wetted) by stormy weather during the timeframe of analysis. Furthermore, model results may overpredict the momentary depiction of flood extents, as derived by the NDWI method based on the recorded image on 14 December 2021 at 09:24:01 (hh:mm:ss). However, there is no guarantee that the satellite data represent the actual situation of floodwater extents during the storm-induced high seas.

Figure 5. Map of estimated flooded areas as depicted by NDWI satellite data (purple colour) overlaid on CoastFLOOD simulation results driven by recorded SLA values on 14 December 2021 (blue colour) for the Manolada-Lechaina study area, north-western Peloponnese (western Greece). The flooded areas' extents are superimposed over a background of recent GoogleEarth satellite images.

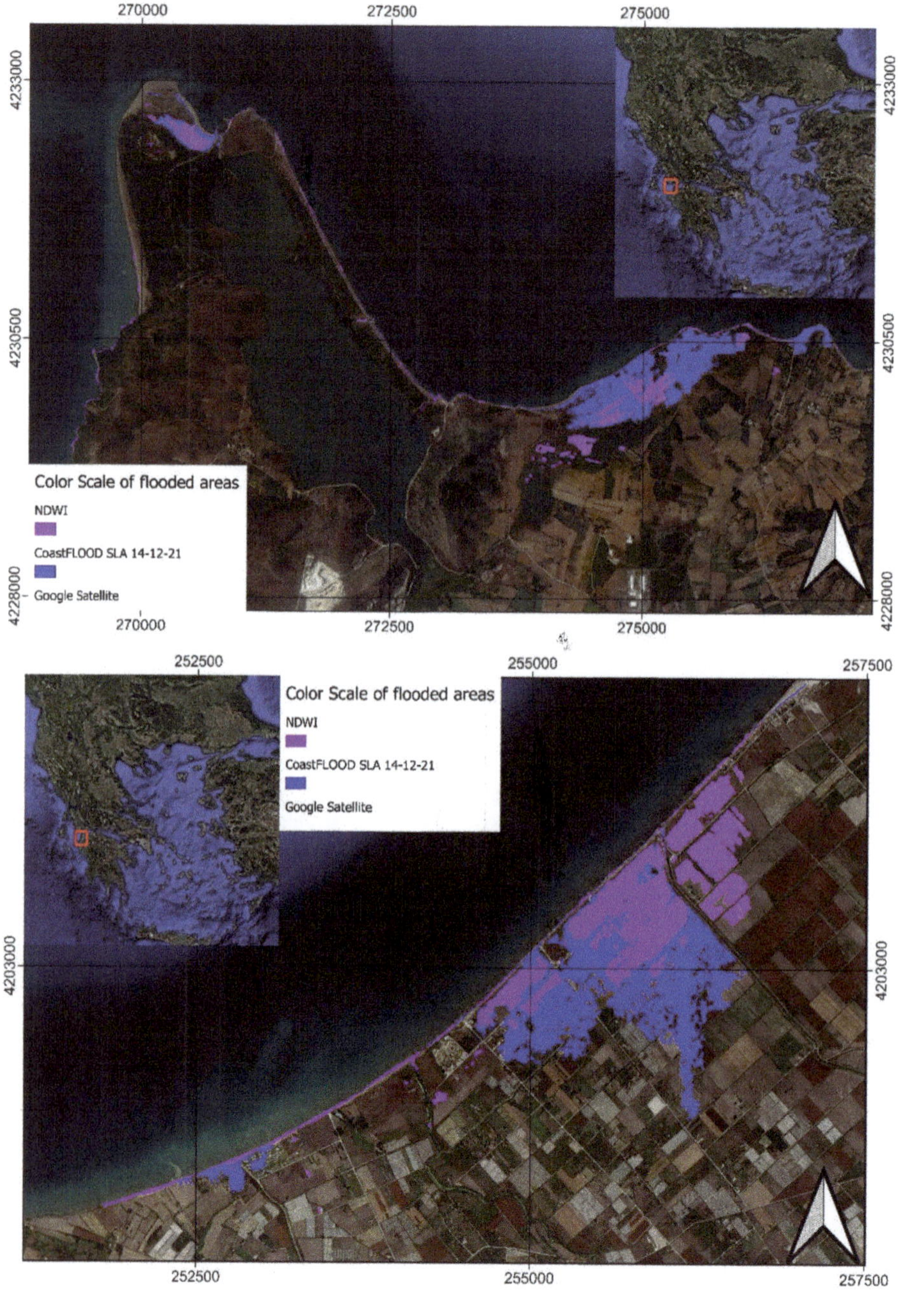

Figure 6. Zoom-in maps from the estimated flooded areas in Figure 5 as depicted by NDWI satellite data (purple colour) overlaid on CoastFLOOD simulation results driven by recorded SLA values on 14 December 2021 (blue colour) for the Manolada-Lechaina study area; upper map: northern part, lower map: southern part. The flooded areas' extents are superimposed over a background of recent GoogleEarth satellite images.

Figure 7 portrays the estimated flood map in the Livadi study area (Area 7; Figure 4a), based on a heuristic approach of NDWI differences before and after the landfall of Ianos Medicane on the study area [16,17,120,128], depicted on a 0.1–1.0 scale of values, overlaid on the CoastFLOOD simulation results. The latter were driven by HiReSS-modelled SSH (see Section 2.6.1) from operational forecasts by the WaveForUs system on 17 September 2020 [17,109]. The predicted flood extents (red patches) on the southern coastal zone of Cephalonia Island overlap and include the traced wet areas by remote sensing (purple patches). A large wet area (shown in shades of purple) in the northern part of the study area, set off of the model-predicted flooded area, is considered to be hydraulically detached from the impacted area due to the storm surge. These areas usually act as drainage bilges that are usually flooded with water originating from local intense rainfall and/or stormwater surface runoff from the surrounding hills and mountains. An extreme case scenario of TWL = 1 m, typical for a possible cumulative sea level increase due to the combined effects of surges and waves, is also provided (yellow patches) for comparison of the flood-prone littorals against the actually impacted touristic coastal areas. The extreme case flooded extents may reach a 250-m distance onshore in the southern part of the study area, occasionally reproduced by the model for the actually recorded SLA_{max}, too, while not along the entire beach stretch. An intrusion of floodwater around 167 m from the coastline, where the beach dunes are located parallel to the shoreline contour, is further plausibly reproduced by the model for both SLA_{max} and TWL cases on the north-western part of the coast. For the extreme TWL case, the model further predicts a 458-m inland flood extent on the northern part of the study area, but this is not reproduced by the SLA_{max} = 0.262 m simulations as the area is not hydraulically connected to the sea by an equally low land pathway.

4.2. Model Validation against the Bathtub-HC Approach

To validate the CoastFLOOD model's efficiency to reproduce the highest possible flood extent (on the safe-side in terms of engineering) in coastal plains, we implemented the performance metric goodness-of-fit, *GoF*, between the modelled (CoastFLOOD; subscript: mod_{CF}) and the estimated (Bathtub-HC; subscript: est_{BHC}) flooded area, *FA*, extents [26,30]:

$$GoF = \frac{FA_{mod_{CF}} \cap FA_{est_{BHC}}}{FA_{mod_{CF}} \cup FA_{est_{BHC}}} \tag{16}$$

where *FA* is defined by the amount of flooded grid cells by the CoastFLOOD model and Bathtub-HC estimations, respectively. The two predictions exactly overlap each other if *GoF* = 1 and no intersection of *FAs* occurs for *GoF* = 0 [94]. In the simulated test cases shown in Figures 8–11, the CoastFLOOD model agreement compared to Bathtub-HC was very high, i.e., *GoF* > 0.95 (see captions of Figures 8–11 for actual values), for several scenarios of SLA as a driver of coastal inundation, ranging from a recorded $SLA_{max} \approx 0.25$ m (minimum SLA_{max} recorded in Area 1) to extreme cases of TWL = 1.0–1.5 m. The model was able to evenly reproduce the estimated maximum inundation extent over lowland areas using the bathtub approach. As expected, it was slightly underestimated compared to the latter, yet, therefore, CoastFLOOD shows a more realistic perspective of littoral inundation, given the error of the retrieved DEM/DSM topography and the boundary conditions (SLA on the coastline) provided by satellite observations.

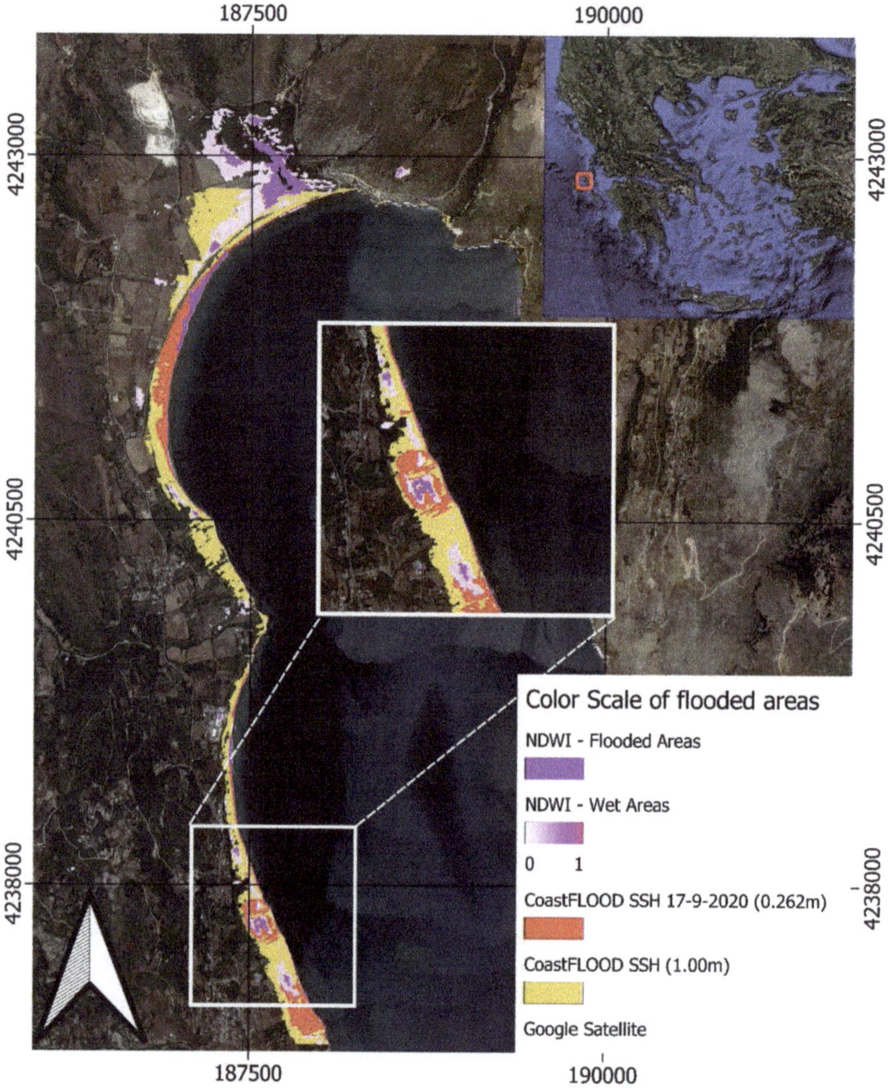

Figure 7. Map of estimated flooded and wet areas as depicted by NDWI differences by satellite data before and after Ianos Medicane passage from the study area (white-to-purple colour shift corresponding to 0–1 of NDWI values; method description in Section 3.2), overlaid on CoastFLOOD simulation results driven by HiReSS-modelled SSH from operational forecasts by the WaveForUs system, during the Ianos Medicane landfall on 17 September 2020 (red colour) flood extent magnitude, for the Livadi study area, on Cephalonia Island, in the Ionian Sea. Modelled flood area extents for an extreme case scenario of TWL = 1 m is also provided in yellow color. The insert map presents a zoomed-in depiction of the main impacted area corresponding to 17 September 2020, SLA_{max} = 0.262 m underlaid below the identified wet areas by the NDWI methodology (white-to-purple color).

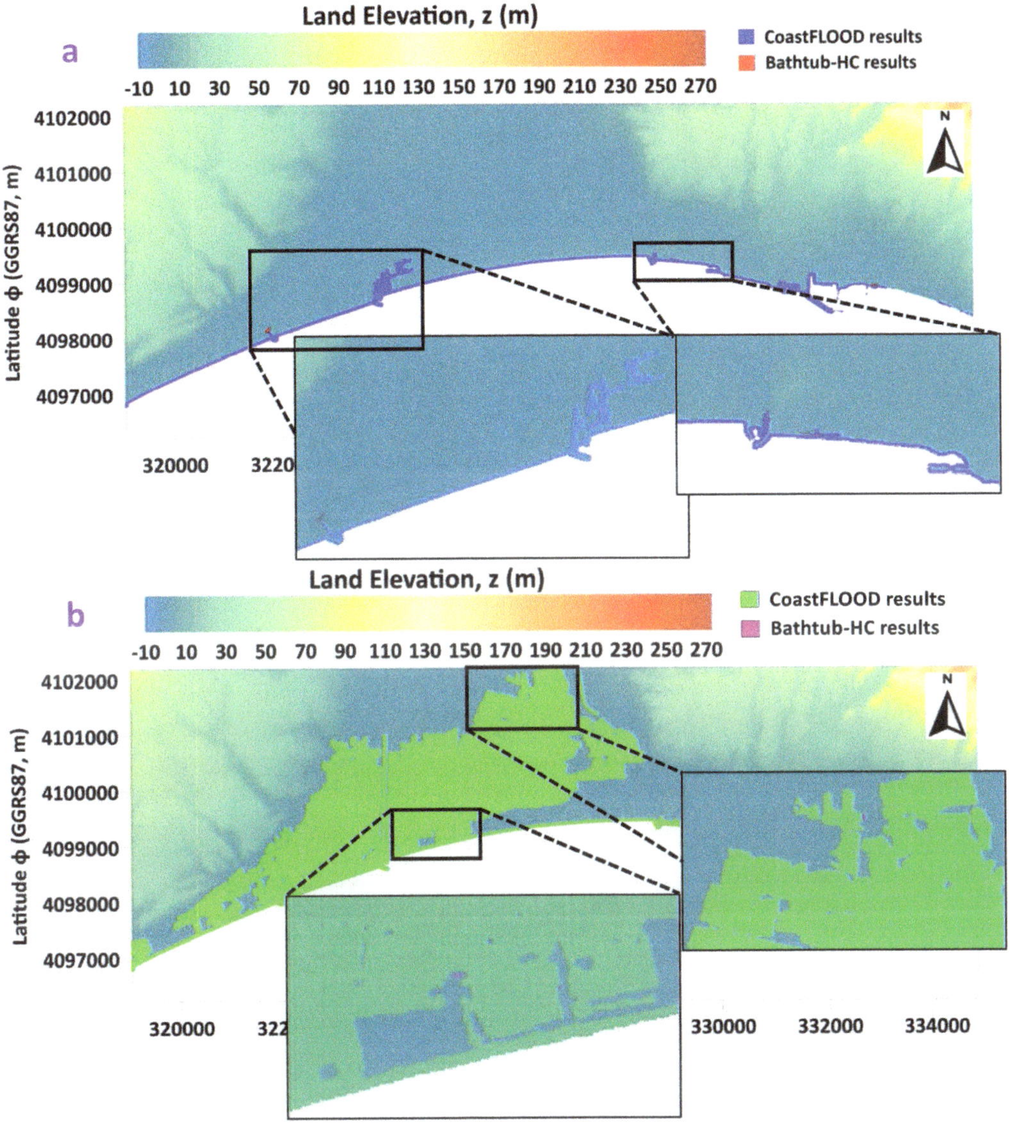

Figure 8. (**a**) Maps of estimated flooded areas as depicted by Bathtub-HC approach (red colour) and CoastFLOOD simulations (blue colour), driven by SLA_{max} = 0.253 m during December 2021, for the Kalamata coastal zone (Area 6), in Messenia of the southern Peloponnese. The insert maps present zoomed-in depictions of the main impacted areas showing the good agreement of the two methods and the superimposed discrepancies of flood extents on the boundaries of the floodwater "wet" regions (*GoF* = 0.972). (**b**) Maps of estimated flooded areas as depicted by Bathtub-HC approach (purple colour) and CoastFLOOD simulations (green colour), driven by an extreme scenario of TWL = 1.5 m, for the same study area, including respective zoom insert maps (*GoF* = 0.993). The two results overlap each other in such a way that Bathtub-HC red and purple areas are barely visible.

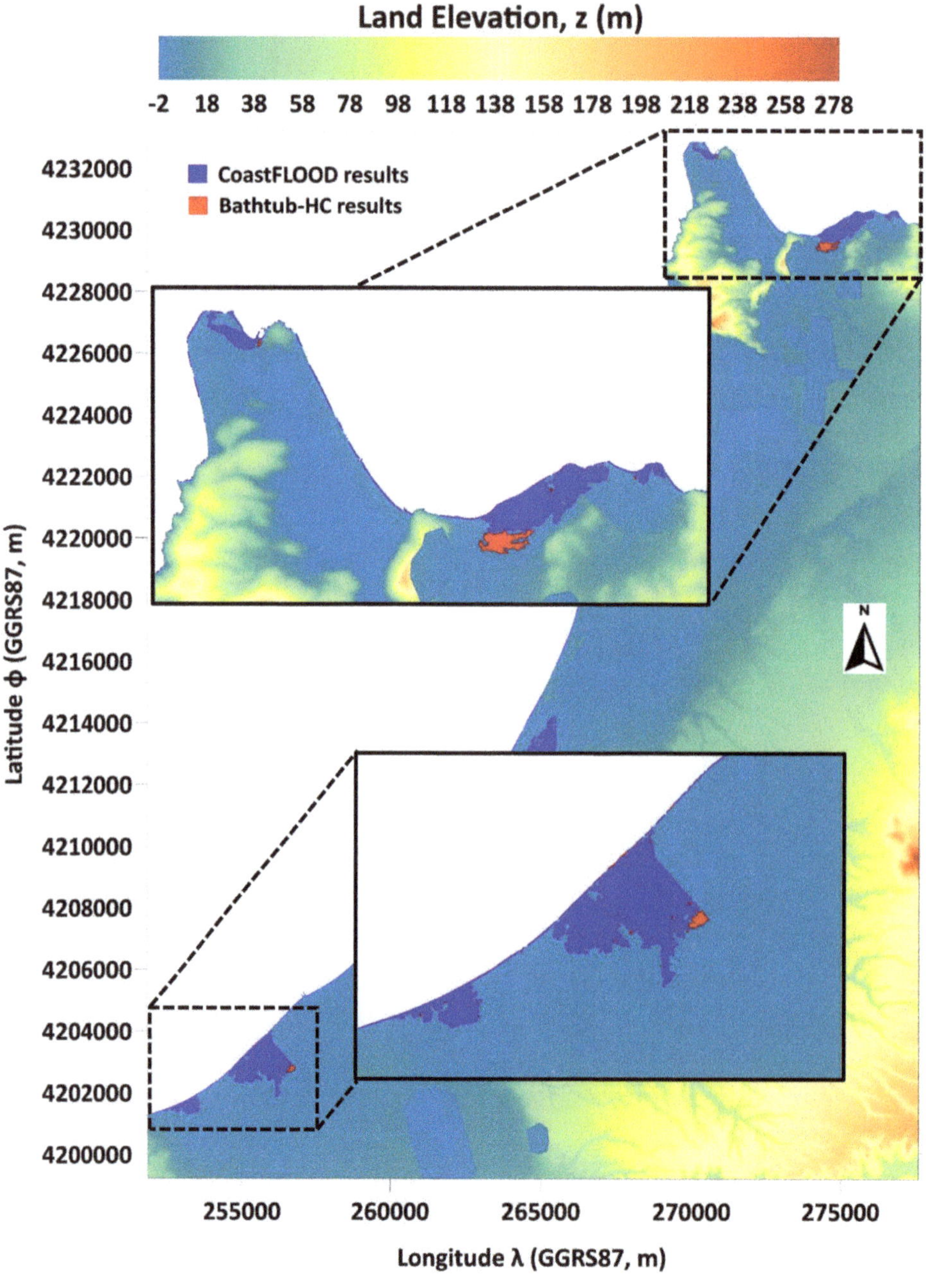

Figure 9. Maps of estimated flooded areas as depicted by Bathtub-HC approach (red colour) and CoastFLOOD simulations (blue colour), driven by $SLA_{max} = 0.25$ m during December 2021, for the Manolada-Lechaina coastal zone (Area 1), in north-western Peloponnese. The insert maps present zoomed-in depictions of the main impacted areas showing the good agreement of the two methods and the superimposed discrepancies of flood extents on the boundaries of the floodwater "wet" regions (*GoF* = 0.951).

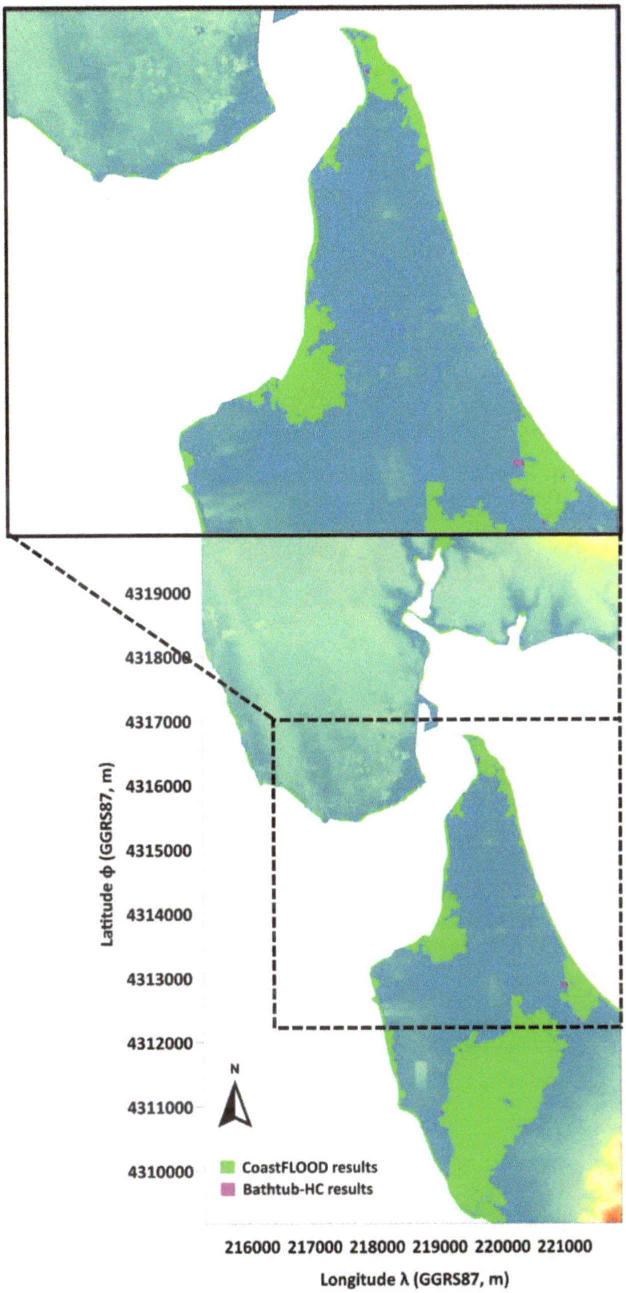

Figure 10. Map of estimated flooded areas as depicted by Bathtub-HC approach (purple colour) and CoastFLOOD simulations (green colour), driven by an extreme case scenario of TWL = 1.5 m, for the Preveza coastal case study (Area 3), in western Epirus. The insert map presents a zoomed-in depiction of the main impacted areas showing the good agreement of the two methods in tandem with the superimposed discrepancies of flood extents on the boundaries of the floodwater "wet" regions (*GoF* = 0.984).

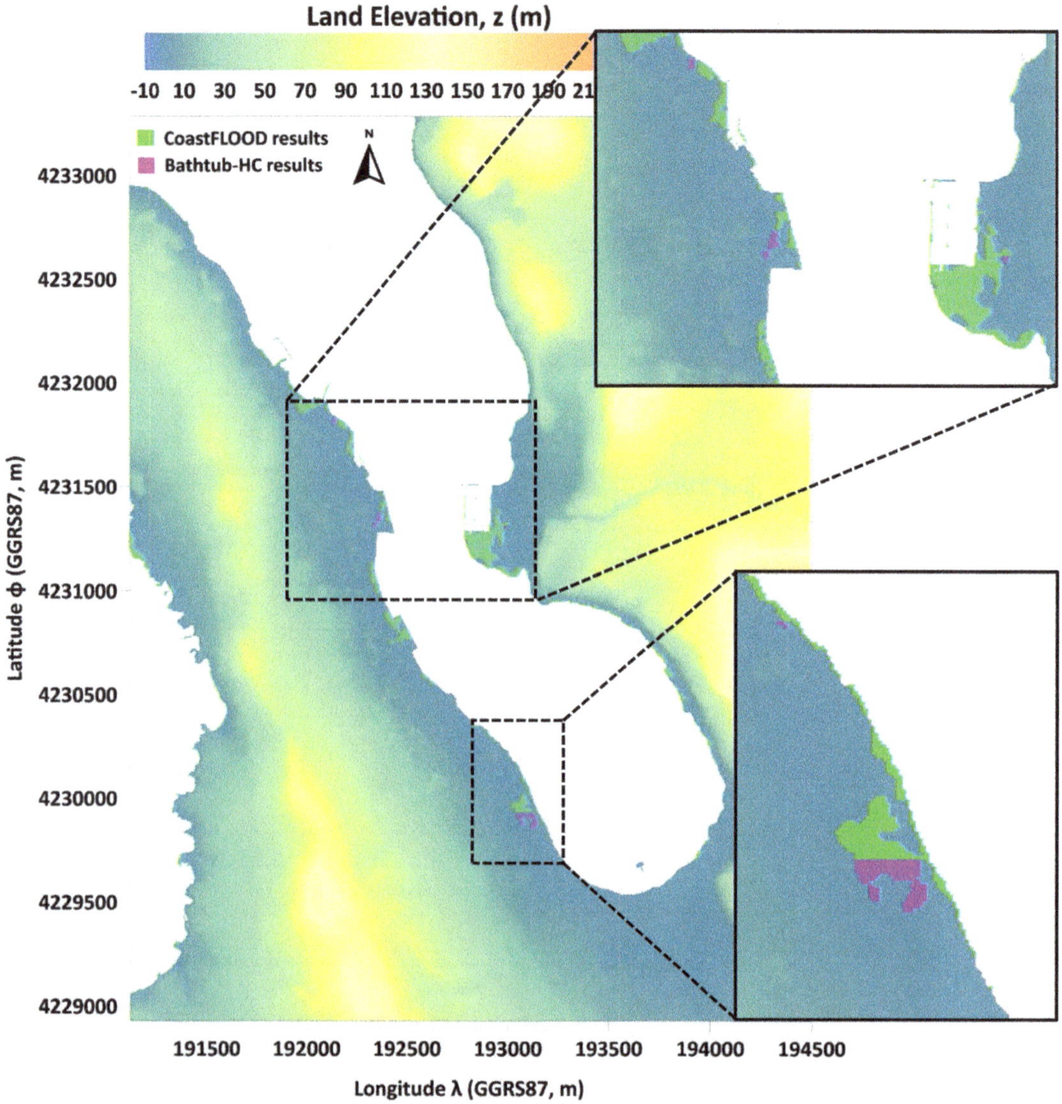

Figure 11. Map of estimated flooded areas as depicted by Bathtub-HC approach (purple colour) and CoastFLOOD simulations (green colour), driven by an extreme case scenario of TWL = 1 m, for the Argostoli coastal inlet (Area 7), in Cephalonia Island; the good agreement of the two methods is depicted in tandem with the superimposed discrepancies of flood extents on the boundaries of the floodwater "wet" regions (*GoF* = 0.96).

Figures 8 and 9 present maps of flooded areas driven by storm surge maxima of SLA > 0.25 m in Kalamata (Area 6) and Manolada-Lechaina (Area 1) with a plausible, nearly perfect overlap of the two methodologies, only showing "wet-area" differences (i.e., Bathtub-HC overestimations) in inland areas far away from the coastal boundary. Similar flood model behaviour is observed for an extreme case scenario of TWL = 1.5 m in one of these study areas. Figures 10 and 11 present maps of flooded areas driven by storm tide extremes of TWL $\geq$ 1 m in Preveza (Area 3) and Argostoli (Area 7), with an equally persuasive overlap of the two methodological results. The inland flood-prone areas identified through the Bathtub-HC approach are obviously located in inclined higher grounds on the maximum boundary of the floodwater extent modelled with CoastFLOOD. The model achieves similar performance in both the natural and urban settings.

4.3. Flooding Scenarios of Realistic and Extreme Sea Level Conditions

Within the framework of setting up an operational modelling platform for storm surge flooding in Greece towards a robust Early Warning System for coastal hazards [51,139–143], we presented CoastFLOOD outputs in the selected study areas.

Figure 12 shows a map of operationally modelled flooded areas, driven by a mild storm surge of SLA = 0.23 m (a maximum record in the 2017–2021 period) and additional extreme case scenarios of TWL = 0.5–2.0 m, for the coastal zone of Kyparissia (Area 8; north-western Messenia, south-western Peloponnese). The impacted areas associated with the satellite-derived sea level mainly refer to the first few tens of meters from the shoreline, which are more pronounced in the northern part of the Kyparissia coast. In general, the occurred storm surge maximum presented no serious impacts on the waterfront of the port area (marina and fishing harbour), but in the case of an extreme event (e.g., TWL $\geq$ 1.5 m), the flood expanse could locally reach up to 100–150 m onshore from the shoreline extending along the entire coastal stretch. In that case, the residential areas behind the port infrastructure can also be affected. The use of a global effective grid-scale Manning coefficient (n = 0.02 corresponding to A/A 9 of Table 2), compared to a properly distributed field of gridded n values based on CLC datasets in the area, does not highly affect the estimation of the flood extent and the location of impacted areas, but it drastically influences the calculation of the timespan for maximum flood reach, rendering it from almost half an hour to 49.2 min (0.82 h; Table 4), respectively.

Table 4. Timeframe for Maximum Flood Inundation Reach, t_{MIR}.

SLA (m)	0.2–0.3	0.5	1	1.5	2
Study Area			t_{MIR} (h)		
Laganas	4.25	3.61	4.45	6.40	8.87
Kyparissia	0.82	0.72	1.12	1.98	2.21
Kalamata	3.96	5.13	25.76	28.59	32.79
Patra	14.46	15.93	50.12	77.39	81.97
Vassiliki	0.18	0.45	1.11	4.21	8.90
Livadi	0.22	0.49	5.33	19.87	38.43
Igoumenitsa	0.20	0.32	0.93	3.76	5.28
Argostoli	0.67	1.57	6.97	9.23	10.18

* *The two highlighted rows correspond to exceptional cases of counterintuitively higher values of* t_{MIR} *for lower values of SLA = 0.2–0.3 m.*

Figure 13 presents the map of simulated flood extents, due to a recorded SLA_{max} = 0.266 m and four hypothetical extreme case scenarios of TWL = 0.5–2.0 m, for the coastal study areas on Zakynthos Island (Area 9) pertaining to Laganas beach (south) and the coastal town of Zante (north), the main port of the island. The southern beach of Laganas with the small fishing harbour on its south boundary cape is mainly impacted. The affected coastal stretch expands for several km along the entire Laganas bay with a cross-shore floodwater uprush of a maximum of 500 m inland for the extreme case of TWL = 2 m. The Zakynthos seaport in the northern part of the study area does not present any crucial impacts for regular SLA_{max} < 0.3 m, but in the case of extreme events (e.g., TWL $\geq$ 1.5 m), the leeward breakwater/jetty and parts of the secondary harbour's docks may be overtopped by high seas. The suburban coasts can be also affected by extreme sea levels, increasing the coastal flood risk for the adjacent coastal residencies.

The situation of storm surge impacts is similar for the coastal areas of Patra (Area 10), Vassiliki (Area 2), and Igoumenitsa (Area 4), presented in Figures 14–16. In the city of Patra, the town of Igoumenitsa, and their peri-urban coastal settings (Figures 14 and 16), the lowland shores can even be flooded by rather low values of storm surge maxima (e.g., SLA = 0.24–0.28 m); nonetheless, the impacts of inundation can be quite high with flood extents reaching hundreds of meters inland for the extreme cases of SLA or TWL > 1 m; i.e., by combining the tidal surge with the wave-induced run-up. In these two study areas, the urban spaces, where high-density populations and revenue-oriented assets are located, including the port-related infrastructure, open air locales, and road networks, are more exposed to surge-flood inundation. However, in Vassiliki bay (Area 2; Lefkada Island, Figure 15), the natural coastal sites and the surrounding touristic residencies may be more likely to be impacted by extreme seawater floods, rather than the small harbour in the north-eastern part of the bay.

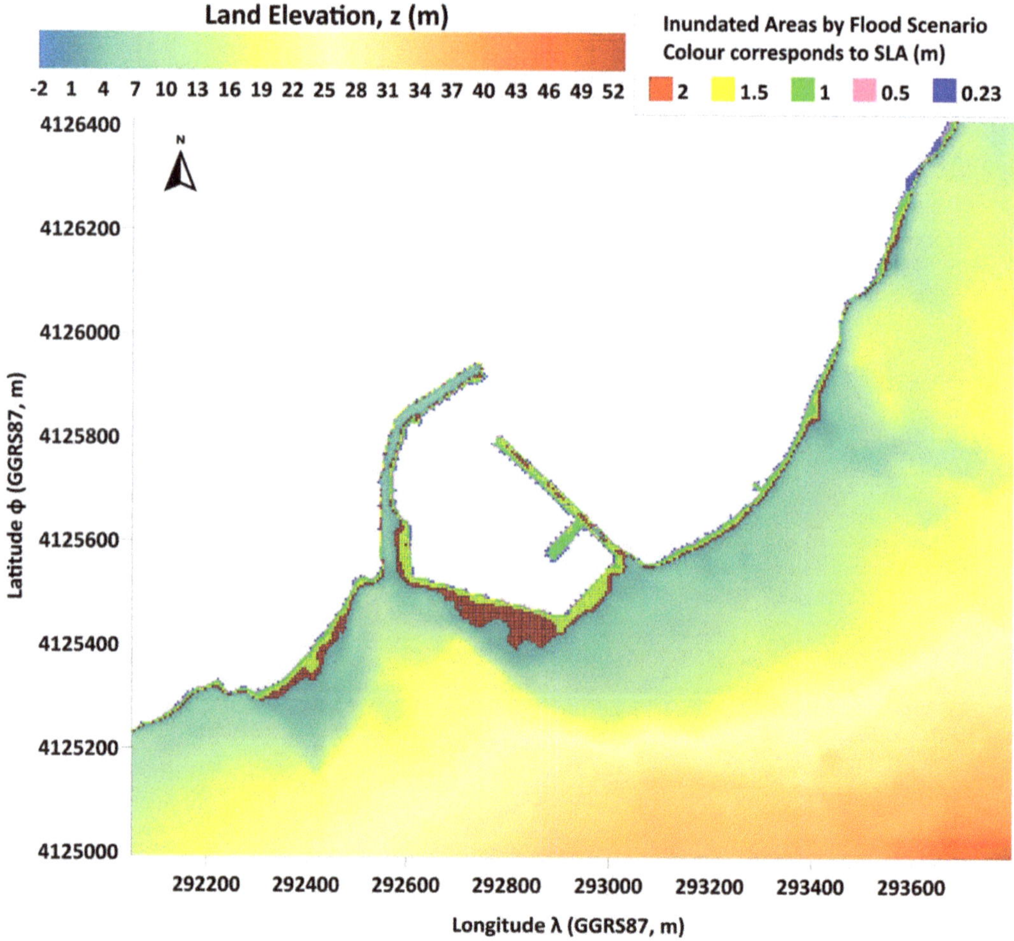

Figure 12. Map of estimated flooded areas as depicted by operational CoastFLOOD simulations, driven by an in situ recorded SLA = 0.23 m and four extreme case scenarios of TWL = 0.5–2.0 m, for the coastal study area of Kyparissia (Area 8; north-western Messenia, south-western Peloponnese), including a local marina harbour.

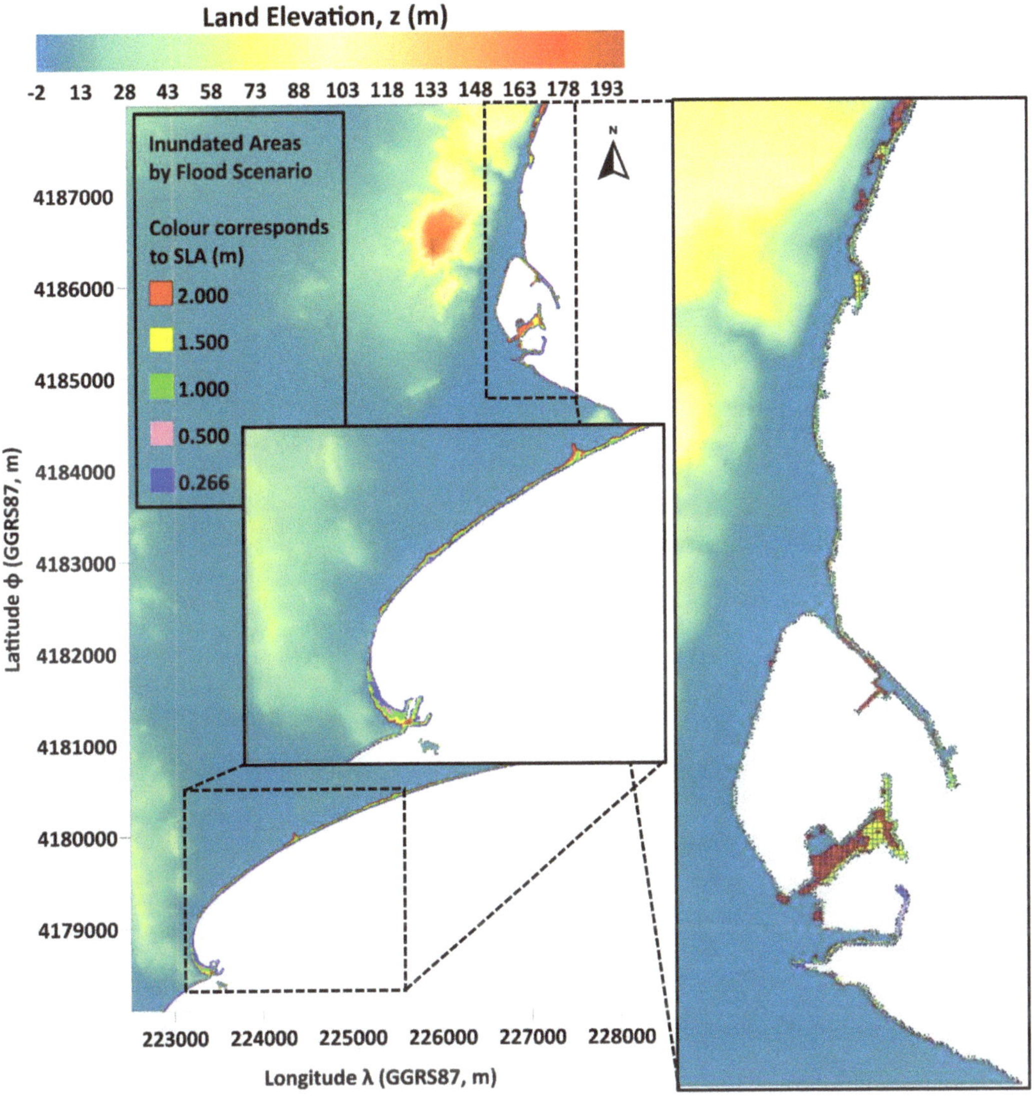

Figure 13. Map of estimated flooded areas as depicted by operational CoastFLOOD simulations, driven by an in situ recorded SLA = 0.266 m and four extreme case scenarios of TWL = 0.5–2.0 m, for the coastal study area of Laganas (Area 9; southern Zakynthos Island, Ionian Sea), also including Zakynthos' main port in the northern part.

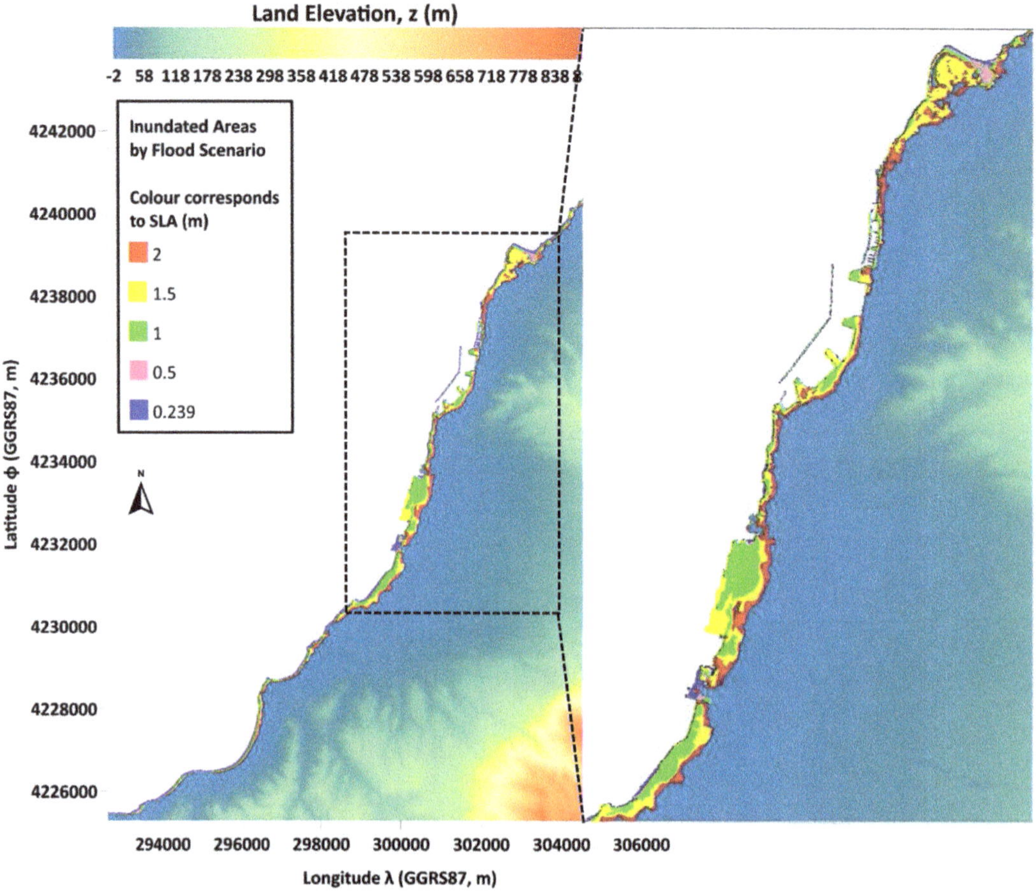

Figure 14. Map of estimated flooded areas as depicted by operational CoastFLOOD simulations, driven by an in situ recorded SLA = 0.239 m and four extreme case scenarios of TWL = 0.5–2.0 m, for the city of Patra (Area 10; north-eastern Peloponnese), also including the main port in the central part, the rural coastal areas of Achaia around the main urban settlement, and the town of Rio in the northern part of the graph.

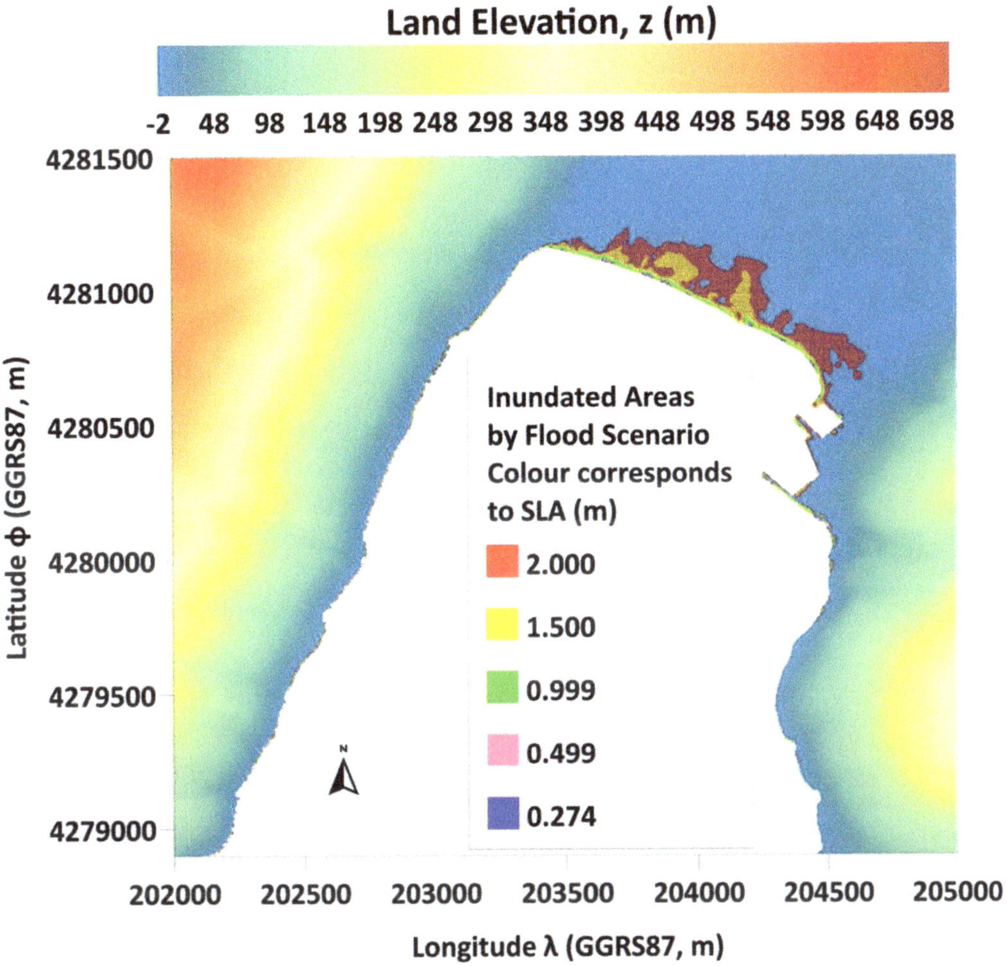

Figure 15. Map of estimated flooded areas as depicted by operational CoastFLOOD simulations, driven by an in situ recorded SLA = 0.274 m and four extreme case scenarios of TWL = 0.5–2.0 m, for the coastal study area of Vassiliki bay (Area 2; south-western Lefkada Island), also including a small fishing harbour port in the north-eastern part of the bay.

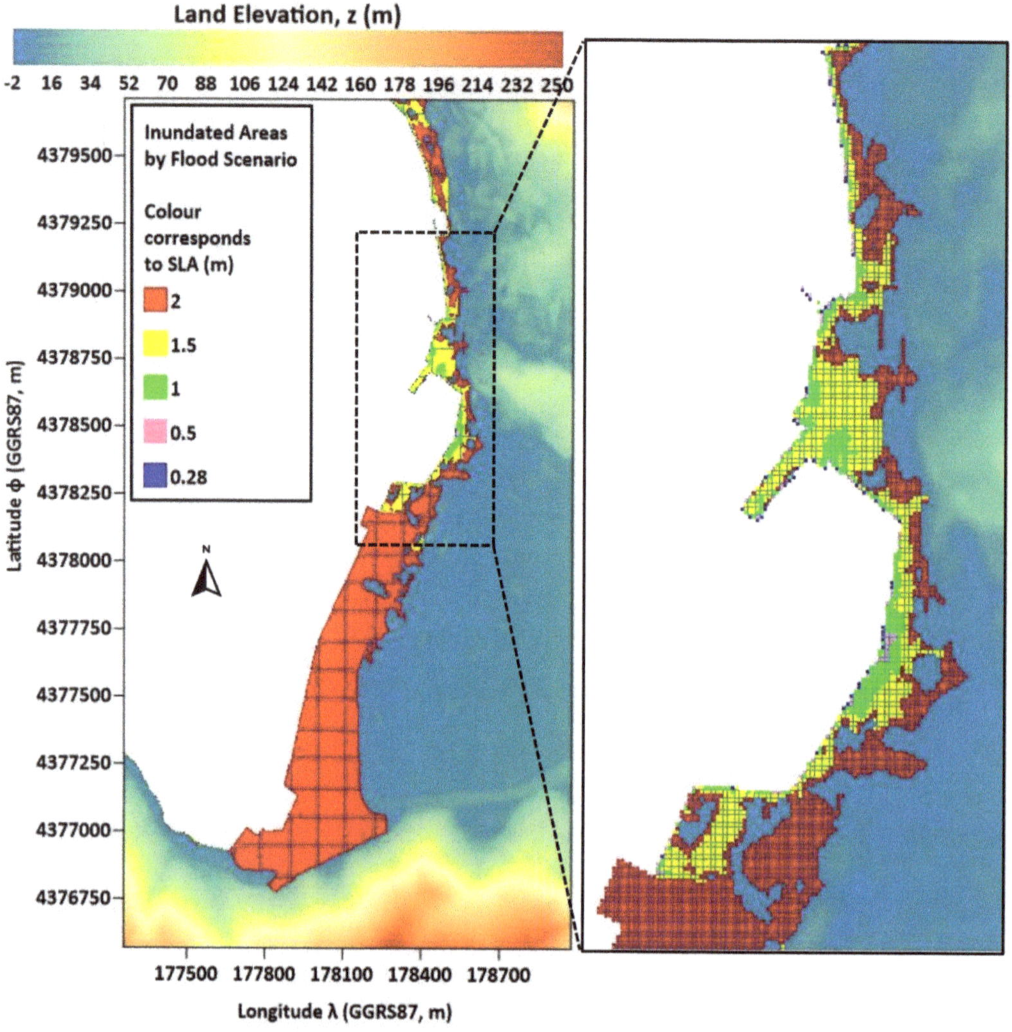

Figure 16. Map of estimated flooded areas as depicted by operational CoastFLOOD simulations, driven by an in situ recorded SLA = 0.28 m and four extreme case scenarios of TWL = 0.5–2.0 m, for the coastal town of Igoumenitsa port (Area 4; North-western Epirus) with its port. Insert map presents the zoomed-in depiction of the northern port area.

An interesting feature is the formulation of the timeframe for maximum flood inundation reach, t_{MIR}, in some study cases. The pattern of t_{MIR} is similar and, in general, increasing for the ascending values of SLA_{max} = 0.2–2 m, except from the Laganas and Kyparissia case studies (highlighted in Table 4) and the lower values of recorded SLA_{max} = 0.2–0.3 m, for which t_{MIR} is counterintuitively quite high; i.e., larger than the t_{MIR} of larger SLAs and consequent inundation extents. However, this is probably reasonable because lower SLA values on the coastline drive much slower inundation flows than larger storm surge levels, since shoreline SLA/SSH acts as the main formulation factor of the hydraulic head of the flood front propagation. The latter is valid given the peculiarities of the topographic formulation of the studied area. Nevertheless, this fact reveals that the CoastFLOOD model, with proper treatment of the bottom roughness (Manning coefficient n), can produce rather plausible estimations of the time evolution of flood inundation phenomena.

5. Discussion

During the last two decades, with the rise in available computational power and resources, the approach of reduced complexity in flood modelling has become the norm for the estimation of coastal inundation due to sea level increase and the lack of available field observations, in order to support Integrated Coastal Zone Management (ICZM) and strategic decision-making. ICZM requires spatiotemporally broad estimations in large-scale domains, $O(x) = 10–100$ km, yet with high-resolution modeling on grid cells with $O(dx) = 1$ m. High-frequency and/or robust updated field data (topographies, transient water areas, reliable DEM/DSM, etc.) in the highly changing coastal zone are hitherto rather limited, making it difficult to feasibly apply the multiple fine-resolution simulations needed at a regional scale in littoral areas. To this end, proper hydraulic models with quick solvers, that neglect secondary effects of turbulence, hypercritical flows, local acceleration terms in the momentum equations, water infiltration and percolation at the bed, etc., such as CoastFLOOD, presented herein, can provide a computationally viable alternative for modelling flood inundation in the coastal environment [26,83,94,99].

The x- and y-direction decoupling of the lower order semi-analytical flow equations in such models may undermine the reproduction of diffusive effects in the hydrodynamic flow of floodwater masses, but the proposed approach is rather simple and allows for easy numerical coding that is computationally robust and produces very similar results to more sophisticated models for flood wave propagation [58,60,83,100]. Thus, on each grid element, the mass and momentum conservation principles are translated into simplified semi-analytic hydraulic equations for continuity (based on floodwater depth and hydraulic head calculation) and volumetric flow rates (Manning-type flow driven by a hydrostatic approach for the piezometric load and bottom friction). These can be separately solved on the centre and faces of the grid cells of a finely discretized domain. The main advantage of such a method for flood routing is the easy use of a wet/dry cell storage module [58,91].

The main disadvantages of reduced complexity flood models are the oversight of sub-grid scale features of the flow (e.g., cavitation, recirculation, aeration, debris advection, and viscosity effects) [144] and fine-scale spatial features (e.g., drainage systems, sewers, conduits, bridge culverts, pools, and drillings). Nevertheless, if one needs to find spatially broad-scale information regarding the inundated areas' extents and the floodwater level in them, and not the full details of the transient flood hydrodynamics, then neglecting the aforementioned effects on the flow is plausible. The secondary fine-scale topographical features of small engineering structures (open canals and conduits, etc.) should play a role in properly modelling the flood flow only in the beginning of the inundation process, when these technical structures are empty and have adequate depth. After enough time, these open channel formations become filled either with rainwater or with seawater, allowing the floodwater to only flow above the hydraulic structures' crests, and this is what we approach herein. Another relevant issue is the exclusion of floodwater percolative interaction with the porous bed and the downward infiltration to the aquifer. However, these flows are usually very slow processes compared to the hydraulic propagation of flood fronts, and thus they cannot significantly influence the hydrodynamics of inundation (this might not be the case for extreme TWL > 1 m in Patra city, where floods that reach maximum duration might range between 2–3.5 days; Table 4). Moreover, the soil on which the floodwater propagates should probably be saturated with rainwater from the storm. Hence, seawater should flow as a runoff on the floodplain's saturated ground surface. Furthermore, inundation in coastal areas is apparently a combined result of river/watershed, precipitation, and ocean (compound) flooding. Therefore, there is a need to integrate fluvial floods with (pluvial) surface runoff and coastal water run-up in order to model flood inundation in littoral lowlands.

A matter that may cause uncertainties in coastal flood flow prediction in urban environments is the depiction of topographic details that are finer than the available DEM/DSM resolution or their vertical accuracy; i.e., the inclusion of outdoor microstructures (uneven pavements, sidewalk fringes, raised curbs, fences, roadblocks, bumps, and obstacles, etc.), stairs, gates, doors, and basement windows at ground level. These can either prevent the free flood flow or absorb floodwater, draining it inside buildings and basements. These effects cannot be taken into account by the model but seem to only be significant in densely built/populated urban spaces and not on coastal floodplains, such as the studied coastal zones of the Ionian Sea that were presented here (excluding Patra city; Area 10). The typical model grid cell should not exceed the upper thresholds of O(dx) = 10–100 m given that the characteristic flood flow depths range between 0.1–1.5 m and Manning's n fluctuates between 0.001–0.4 s·m$^{-1/3}$, respectively. The larger spatial discretization step used herein is dx = 5 m, which is considered quite fine. Consequently, the choice of a zero-inertia model that can reduce the complexity of floodplain hydraulics to an imperative minimum representation of the flow equations is acceptable for slow (big volumetric flow changes occur in timescales >> dt) and shallow (vertical changes in floodwater flow depths are practically a lot smaller than horizontal ones or the typical cell width dh << dx) flood flows [58]. Neglecting inertia terms can only play a local role, in the sense that the ability of 2-D reduced complexity models to reproduce flood propagation has been corroborated by several researchers in the past based on comparisons with available field data and other model approaches [71,83,90]. It is clear that the spatial resolution and the consequent timestep of the numerical solution are the most crucial factors in defining robust simulations for this kind of reductionistic modelling approaches. These issues are adequately addressed in the CoastFLOOD simulator, offering computational efficiency, ease of coding for GIS raster-based applications, broad-scale (regional flood reach) simulations, and repeatability from a pragmatic management perspective for engineers, scientists, and managing authorities.

The lack of field data for calibration and validation may be the major constraint in the further verification of reduced complexity flood inundation models for coastal areas. The recent evolution of remote sensing products and their available resolutions seems to partially address this issue in a qualitative manner. The inherent discrepancies to distinguish the source of floodwater (e.g., tidal surge, wave action, drainage or runoff, and rainfall) is a problem for the quantitative validation of coastal flooding modelling due to storm tides in tandem with MSLR [71,145]. Therefore, we also compared our hydraulic flood model results with a Bathtub-HC approach. However, when using the latter, one should consider issues arising from the omission of bottom friction leading the analysis by exaggerated flood vulnerability estimations. Several coastal managers have inferred that the latter can lead to overprotective engineering solutions, excessive defence schemes, and inflated investment against flood protection. Despite this, we believe that a Bathtub-HC method should always be applied to indicate low-lying flood-prone areas in the coastal zone to formulate an idea about potentially inundated areas and to direct the more focused (high-resolution) coastal flooding approaches under extreme sea level elevation in the future.

Finally, model implementations in areas that are too large might require rather large timestep values (given the available computational resources and timeframes, especially in operational mode), which may lead to chequerboard-type oscillations in the numerical solution, not easily suppressed or relaxed, especially in areas with small gradients of the floodwater free-surface and subsequent slow evolution of the flow. CoastFLOOD solves this issue with the use of a proper CFL criterion within an adaptive time-stepping algorithm [27,33,71,90,91].

Thus, CoastFLOOD has been recently upgraded to include very detailed depictions of bottom roughness (based on recently available land cover data), the influence of storm surge-led currents on the coastline boundary, fine-scale DEM/DSM, and the enhancement of wet/dry cell techniques for flood front propagation over steep water slopes. These techniques have been proposed by other researchers in the past, and we included them as options in this new updated code. An additional novelty is related to the very fine-scale DEM/DSM of $dx = 2$ m, providing high detail of the domain terrain. Moreover, a cross-type scan of the model grid (N→S/S→N in the meridional direction; W→E/E→W in the zonal direction) is now applied in every timestep, thus allowing for plausible estimations of the flood front propagation from any direction of the horizon or peripheral boundary, while some coastal inundation models still only allow one-way flood propagation; i.e., either from south/north or west/east.

Future research should include even finer scale simulations and comparisons with model formulations considering local acceleration terms in tandem with the proper depiction of details over and around coastal structures, port infrastructure, beach land formations, and rocky shores in the model grid. The treatment of sub-grid topographical features (weirs, drainage holes on embankments, drainage trenches and channels under bridges, sewerage networks, etc.) should also be included in future developments of the CoastFLOOD model. Incorporating a breaching mechanism for sand dunes and coastal embankments should also be implemented. An additional consideration is to combine a percolation and subterranean infiltration module to account for ground porosity effects on the floodplain, together with a simple approach for the evaporation of inundated seawater. The latter can always contribute to more long-term simulations that may result in different patterns of floodplain water storage, floodwater encroachment and conveyance, as well as possible backwater effects from flood flux blockage, etc.

Therefore, we believe that, although it presents no ground-breaking scientific novelty, it provides very much needed technical innovations, i.e., a first national-level OFP for surge-induced coastal floods established in Greece since the 1980's concepts of flood hydraulics for coastal (optionally combined with fluvial-deltaic) inundation by storm surges and sea level elevation in general.

6. Conclusions

In this study, we present applications using a new code (CoastFLOOD), developed in FORTRAN-95, for a classic modelling approach of 2-D hydraulic flood flow in coastal areas. CoastFLOOD is built on the concept of high-resolution, storage-cell, mass balance flood inundation for coastal lowlands, following the simplified approach of Manning-type flow equation, under a reduced complexity concept, running on a GIS raster-based domain. Although a detailed physical representation of turbulent floodwater hydrodynamics is overlooked, CoastFLOOD relies on computational efficiency and the delivery of stable simulations with robust results. The model's performance is evaluated for the case of predicted (i.e., ocean modelling) or observed (i.e., satellite altimetry) storm surges affected by tidal components of sea level elevation (also termed as storm tides). The proposed methodology and numerical model could be applied in operational applications as well as studies of long-term mean sea level rise or short-term extreme scenarios of total water levels, also considering an estimative mean condition for wave runup, but mainly excluding the high-frequency phenomena, such as the undulating sea surface uprush and backwater effects due to waves, etc.

The flood extent identification was based on the computation of the NDWI index derived from remote sensing ocean color data by Sentinel-2 satellite. The verification of the model was performed for two cases of recorded storm surges in the Ionian Sea; the first during a storm in December 2021 in the Manolada-Lechaina coastal zone (Area 1; north-western Peloponnese, western Greece), and the second in September 2020 during the Ianos Medicane landfall in Livadi bay (Area 5; southern Cephalonia Island [17]). The comparison of CoastFLOOD simulation results against NDWI-identified flooded areas show that our model can reproduce the coastal flooding mechanism in areas that are more-or-less affected (wetted) by stormy weather during the timeframe of analysis. The model results maybe overpredict the recorded flood extents because the satellite data are not totally accurate to represent the actual situation of floodwater extents during the storm surge, since the satellite does not usually coincide with the peak of the storm surge due to cloud contamination, and thus, it is not representative of the maximum flood reach. In the model's defence, the predicted flood extents on the southern coastal zone of Cephalonia Island (Area 5) definitely overlap and include the wet areas traced by remote sensing, and that is on the safe side in terms of engineering and coastal management. Moreover, some available soft data (visual proof and pictures from social and mass media reports) can also be used to corroborate the general performance of the model [30].

The validation of the CoastFLOOD model's efficiency to reproduce the highest possible flood extent in coastal plains was also tested against an efficient Bathtub-HC approach. The agreement between the two approaches is quite high with very high *GoF* [30,91,94] scores (>0.95) for both the realistic sea level and extreme scenario TWL cases. Furthermore, we show that proper treatment of the bottom roughness with spatially distributed Manning coefficients referring to realistic land cover datasets can formulate a more realistic estimation of the timeframe for reaching maximum flood inundation extents. Therefore, the bottom friction parameter is defined as the main calibration feature. The realistic reproduction of the flooded inland areas' roughness, based on different representations of the land cover information by CLC datasets, was investigated in detail. Specifically, we created a matching list of all CLC-2018 codes to a detailed set of discrete types for earth/ground material that correspond to a detailed list of different assigned Manning coefficient values in the CoastFLOOD model. The use of a horizontally distributed field of gridded Manning coefficient values (based on the CLC) compared to a global effective value of a grid-scale Manning coefficient did not highly affect the estimation of the flood extent and the location of impacted areas in agreement with previous studies [94]. However, it drastically influenced the calculation of the timespan for maximum flood reach. Moreover, it was shown that the latter heavily depends on the levels of the storm-induced sea level on the coastline, which acts as the hydraulic head of flood front propagation; i.e., lower storm surge heights may drive much slower inundation flows than larger ones. Hence, the proposed model also shows an intuitively correct sensitivity to realistic representations of floodplain friction, especially if it is applied in areas with complex topographies. The use of highly variable friction coefficients for coastal flood modelling should provide better predictions for the duration of an inundation event, which is crucial to first-level responders and coastal zone managers. Still, it is concluded that the detailed depiction of topography is the key constraint on robustly formulating and realistically simulating the floodwater flow for the accurate determination of the maximum flood extent.

The most probable explanation for any discrepancy in comparisons of modelled and observed flood extents in the coastal zone is the uncertainty of field data concerning the actually occurred flood rates. Thus, large uncertainties of the latter, mainly stemming from the sources of seawater inundation, except from storm tides, e.g., wind waves and swell, make it difficult to develop a definite benchmark case dataset with which to robustly test the performance of storm-induced coastal inundation models. Indeed, it has been argued [30] that for random coastal inundation events, storm surge flooding usually coincides with wave overtopping, making it very difficult to produce any reliable observation dataset capable of being used as a reference against competing coastal model formulations in a meaningful way. Hence, as a future research step, there is a need to incorporate a treatment of boundary conditions in the CoastFLOOD model as a varying timeseries of non-deterministic values in order to avoid substantial underestimations of coastal inundation and potentially relevant risk.

The Ionian Sea's coastal zone in Greece is eventually threatened by storm surge inundation in an annual cycle, with likely coastal flooding events occurring during mid-autumn (late September—early- to mid-October) and during December or early January, as also pointed out in [17,23,118–120]. The impacts are not very pronounced for usual storm surge levels (<0.3 m) but can be severe for extreme cases of total water levels (>0.6 m as found in future climatic projections along the Greek coastal zone [23,104,105]), e.g., in coastal urban areas (Igoumenitsa, Patra, and Kalamata).

Conclusively, we presented a robust, easy-to-use, numerical tool for coastal inundation due to storm surge/tide flooding, under the reduced complexity notion, imperatively needed for operational forecasts of storm impact. Nonetheless, it can hopefully be useful for both operational applications and projected climatic studies of coastal inundation under extreme scenarios to help coastal zone managers, policymakers, and involved stakeholders to better estimate the characteristics of coastal (or compound) flooding under conditions of environmental change.

Author Contributions: C.M.: Conceptualization, Formal analysis, Investigation, Methodology, Software, Validation, Visualization, Writing—original draft, Writing—review and editing; Z.M.: Data curation, Methodology, Software, Validation, Visualization, Writing—review and editing; Y.A.: Conceptualization, Data curation, Formal analysis Investigation, Methodology, Software, Writing—review and editing; Y.K.: Conceptualization, Investigation, Project administration, Resources, Supervision, Writing—review and editing. All authors have read and agreed to the published version of the manuscript.

Funding: This research received no external funding.

Data Availability Statement: The data presented in this study are available on request from the corresponding author. The data are not publicly available due to copyright restrictions of intellectual property produced within AUTh.

Conflicts of Interest: The authors declare no conflict of interest.

References

1. Lionello, P.; Barriopedro, D.; Ferrarin, C.; Nicholls, R.J.; Orlić, M.; Raicich, F.; Reale, M.; Umgiesser, G.; Vousdoukas, M.; Zanchettin, D. Extreme floods of Venice: Characteristics, dynamics, past and future evolution. *Nat. Hazards Earth Syst. Sci.* **2021**, *21*, 2705–2731. [CrossRef]
2. Krestenitis, Y.; Androulidakis, Y.; Kombiadou, K.; Makris, C.; Baltikas, V. Operational forecast system of storm tides in the aegean sea (Greece). In Proceedings of the 2015 ASLO Aquatic Sciences Meeting, Granada, Spain, 22–27 February 2015.
3. Mishra, A.K.; Jangir, B.; Strobach, E. Does Increasing Climate Model Horizontal Resolution Be Beneficial for the Mediterranean Region?: Multimodel Evaluation Framework for High-Resolution Model Intercomparison Project. *J. Geophys. Res. Atm.* **2023**, *128*, 2022JD037812. [CrossRef]
4. Reale, M.; Cabos Narvaez, W.D.; Cavicchia, L.; Conte, D.; Coppola, E.; Flaounas, E.; Giorgi, F.; Gualdi, S.; Hochman, A.; Li, L.; et al. Future projections of Mediterranean cyclone characteristics using the Med-CORDEX ensemble of coupled regional climate system models. *Clim. Dyn.* **2022**, *58*, 2501–2524. [CrossRef]
5. Nicholls, R.J.; Hoozemans, F.M.J. The Mediterranean: Vulnerability to coastal implications of climate change. *Ocean Coast. Manag.* **1996**, *31*, 105–132. [CrossRef]
6. Snoussi, M.; Ouchani, T.; Niazi, S. Vulnerability assessment of the impact of sea-level rise and flooding on the Moroccan coast: The case of the Mediterranean eastern zone. *Estuar. Coast. Shelf Sci.* **2008**, *77*, 206–213. [CrossRef]
7. Shaltout, M.; Tonbol, K.; Omstedt, A. Sea-level change and projected future flooding along the Egyptian Mediterranean coast. *Oceanologia* **2015**, *57*, 293–307. [CrossRef]
8. Refaat, M.M.; Eldeberky, Y. Assessment of coastal inundation due to sea-level rise along the Mediterranean Coast of Egypt. *Mar. Geod.* **2016**, *39*, 290–304. [CrossRef]
9. Krestenitis, Y.N.; Androulidakis, Y.S.; Kontos, Y.N.; Georgakopoulos, G. Coastal inundation in the north-eastern Mediterranean coastal zone due to storm surge events. *J. Coast. Conserv.* **2011**, *15*, 353–368. [CrossRef]
10. Alvarado-Aguilar, D.; Jiménez, J.A.; Nicholls, R.J. Flood hazard and damage assessment in the Ebro Delta (NW Mediterranean) to relative sea level rise. *Nat. Hazards* **2012**, *62*, 1301–1321. [CrossRef]
11. Aucelli, P.P.C.; Di Paola, G.; Incontri, P.; Rizzo, A.; Vilardo, G.; Benassai, G.; Buonocore, B.; Pappone, G. Coastal inundation risk assessment due to subsidence and sea level rise in a Mediterranean alluvial plain (Volturno coastal plain–southern Italy). *Estuar. Coast. Shelf Sci.* **2017**, *198*, 597–609. [CrossRef]
12. Reimann, L.; Vafeidis, A.T.; Brown, S.; Hinkel, J.; Tol, R.S. Mediterranean UNESCO World Heritage at risk from coastal flooding and erosion due to sea-level rise. *Nat. Commun.* **2018**, *9*, 1–11. [CrossRef]
13. Rizzo, A.; Vandelli, V.; Gauci, C.; Buhagiar, G.; Micallef, A.S.; Soldati, M. Potential Sea Level Rise Inundation in the Mediterranean: From Susceptibility Assessment to Risk Scenarios for Policy Action. *Water* **2022**, *14*, 416. [CrossRef]
14. Hauer, M.E.; Hardy, D.; Kulp, S.A.; Mueller, V.; Wrathall, D.J.; Clark, P.U. Assessing population exposure to coastal flooding due to sea level rise. *Nat. Commun.* **2021**, *12*, 6900. [CrossRef]
15. Kulp, S.A.; Strauss, B.H. New elevation data triple estimates of global vulnerability to sea-level rise and coastal flooding. *Nat. Commun.* **2019**, *10*, 1–12.

16. Makris, C.; Androulidakis, Y.; Mallios, Z.; Baltikas, V.; Krestenitis, Y. Towards an Operational Forecast Model for Coastal Inundation due to Storm Surges: Application during Ianos Medicane. In Proceedings of the 9th International Conference on Civil Protection & New Technologies, SafeThessaloniki, Thessaloniki, Greece, 29 September–1 October 2022; pp. 69–72.
17. Androulidakis, Y.; Makris, C.; Mallios, Z.; Pytharoulis, I.; Baltikas, V.; Krestenitis, Y. Storm surges and coastal inundation during extreme events in the Mediterranean Sea: The IANOS Medicane. *Nat. Hazards* **2023**, 1–40. [CrossRef]
18. Skoulikaris, C.; Makris, C.; Katirtzidou, M.; Baltikas, V.; Krestenitis, Y. Assessing the vulnerability of a deltaic environment due to climate change impact on surface and coastal waters: The case of Nestos River (Greece). *Environ. Model. Assess.* **2021**, *26*, 459–486. [CrossRef]
19. Makris, C.; Androulidakis, Y.; Baltikas, V.; Kontos, Y.; Karambas, T.; Krestenitis, Y. HiReSS: Storm Surge Simulation Model for the Operational Forecasting of Sea Level Elevation and Currents in Marine Areas with Harbor Works. In Proceedings of the International Scientific Conference DMPCO, Athens, Greece, 8–11 May 2019; Volume 1, pp. 11–15.
20. Krestenitis, Y.; Pytharoulis, I.; Karacostas, T.; Androulidakis, Y.; Makris, C.; Kombiadou, K.; Tegoulias, I.; Baltikas, V.; Kotsopoulos, S.; Kartsios, S. Severe weather events and sea level variability over the Mediterranean Sea: The WaveForUs operational platform. In *Perspectives on Atmospheric Sciences*; Springer Atmospheric Sciences; Karacostas, T., Bais, A., Nastos, P.T., Eds.; Springer: Cham, Switzerland, 2017; pp. 63–68. [CrossRef]
21. Makris, C.; Androulidakis, Y.; Karambas, T.; Papadimitriou, A.; Metallinos, A.; Kontos, Y.; Baltikas, V.; Chondros, M.; Krestenitis, Y.; Tsoukala, V.; et al. Integrated modelling of sea-state forecasts for safe navigation and operational management in ports: Application in the Mediterranean Sea. *Appl. Math. Model.* **2021**, *89*, 1206–1234. [CrossRef]
22. Androulidakis, Y.; Kombiadou, K.; Makris, C.; Baltikas, V.; Krestenitis, Y. Storm surges in the Mediterranean Sea: Variability and trends under future climatic conditions. *Dyn. Atmos. Ocean.* **2015**, *71*, 56–82. [CrossRef]
23. Makris, C.; Galiatsatou, P.; Tolika, K.; Anagnostopoulou, C.; Kombiadou, K.; Prinos, P.; Velikou, K.; Kapelonis, Z.; Tragou, E.; Androulidakis, Y.; et al. Climate change effects on the marine characteristics of the Aegean and the Ionian seas. *Ocean. Dyn.* **2016**, *66*, 1603–1635. [CrossRef]
24. Makris, C.; Tolika, K.; Baltikas, V.; Velikou, K.; Krestenitis, Y. The impact of climate change on the storm surges of the Mediterranean Sea: Coastal sea level responses to deep depression atmospheric systems. *Ocean. Model.* **2023**, *181*, 102149. [CrossRef]
25. Bates, P.D.; De Roo, A.P.J. A simple raster-based model for flood inundation simulation. *J. Hydrol.* **2000**, *236*, 54–77. [CrossRef]
26. Bates, P.D.; Dawson, R.J.; Hall, J.W.; Horritt, M.S.; Nicholls, R.J.; Wicks, J.; Hassan, M. Simplified two-dimensional numerical modelling of coastal flooding and example applications. *Coast. Eng.* **2005**, *52*, 793–810. [CrossRef]
27. Bates, P.D.; Horritt, M.S.; Fewtrell, T.J. A simple inertial formulation of the shallow water equations for efficient two-dimensional flood inundation modelling. *J. Hydrol.* **2010**, *387*, 33–45. [CrossRef]
28. Hellenic Cadastre. Available online: https://www.ktimatologio.gr/en (accessed on 19 March 2023).
29. Neal, J.; Schumann, G.; Fewtrell, T.; Budimir, M.; Bates, P.; Mason, D. Evaluating a new LISFLOOD-FP formulation with data from the summer 2007 floods in Tewkesbury, UK. *J. Flood Risk Manag.* **2011**, *4*, 88–95. [CrossRef]
30. Smith, R.A.; Bates, P.D.; Hayes, C. Evaluation of a coastal flood inundation model using hard and soft data. *Environ. Model. Softw.* **2012**, *30*, 35–46. [CrossRef]
31. Lewis, M.; Bates, P.; Horsburgh, K.; Neal, J.; Schumann, G. A storm surge inundation model of the northern Bay of Bengal using publicly available data. *Q. J. R. Meteorol. Soc.* **2013**, *139*, 358–369. [CrossRef]
32. Wadey, M.P.; Nicholls, R.J.; Haigh, I. Understanding a coastal flood event: The 10th March 2008 storm surge event in the Solent, UK. *Nat. Hazards* **2013**, *67*, 829–854. [CrossRef]
33. Coulthard, T.J.; Neal, J.C.; Bates, P.D.; Ramirez, J.; de Almeida, G.A.M.; Hancock, G.R. Integrating the LISFLOOD-FP 2D hydrodynamic model with the CAESAR model: Implications for modelling landscape evolution. *Earth Surf. Process. Landf.* **2013**, *38*, 1897–1906. [CrossRef]
34. Ramirez, J.; Lichter, M.; Coulthard, T.; Skinner, C. Hyper-resolution mapping of regional storm surge and tide flooding: Comparison of static and dynamic models. *Nat. Hazards* **2016**, *82*, 571–590. [CrossRef]
35. Pariartha, G.; Goonetilleke, A.; Egodawatta, P.; Mirfenderesk, H. The prediction of flood damage in coastal urban areas. In *IOP Conference Series: Earth and Environmental Science*; IOP Publishing: Bristol, UK, 2020; Volume 419, p. 012136.
36. Jibhakate, S.M.; Timbadiya, P.V.; Patel, P.L. Flood hazard assessment for the coastal urban floodplain using 1D/2D coupled hydrodynamic model. *Nat. Hazards* **2023**, *116*, 1557–1590. [CrossRef]
37. Pasquier, U.; He, Y.; Hooton, S.; Goulden, M.; Hiscock, K.M. An integrated 1D–2D hydraulic modelling approach to assess the sensitivity of a coastal region to compound flooding hazard under climate change. *Nat. Hazards* **2019**, *98*, 915–937. [CrossRef]
38. Pandey, S.; Rao, A.D.; Haldar, R. Modeling of Coastal Inundation in Response to a Tropical Cyclone Using a Coupled Hydraulic HEC-RAS and ADCIRC Model. *J. Geophys. Res. Ocean.* **2021**, *126*, e2020JC016810. [CrossRef]
39. Doong, D.-J.; Lo, W.; Vojinovic, Z.; Lee, W.-L.; Lee, S.-P. Development of a New Generation of Flood Inundation Maps—A Case Study of the Coastal City of Tainan, Taiwan. *Water* **2016**, *8*, 521. [CrossRef]
40. Riama, N.F.; Sari, R.F.; Sulistya, W.; Rahmayanti, H.; Pratama, K.R.; Pratama, B.E.; Suryo, A.W. Coastal inundation modeling and mapping for North Jakarta coast during a supermoon period. *Terr. Atmos. Ocean Sci.* **2021**, *32*, 375–390. [CrossRef]
41. Dasallas, L.; Lee, S. Topographical Analysis of the 2013 Typhoon Haiyan Storm Surge Flooding by Combining the JMA Storm Surge Model and the FLO-2D Flood Inundation Model. *Water* **2019**, *11*, 144. [CrossRef]

42. Nash, S.; Hartnett, M. Nested circulation modelling of inter-tidal zones: Details of a nesting approach incorporating moving boundaries. *Ocean. Dyn.* **2010**, *60*, 1479–1495. [CrossRef]
43. Olbert, A.I.; Comer, J.; Nash, S.; Hartnett, M. High-resolution multi-scale modelling of coastal flooding due to tides, storm surges and rivers inflows. A Cork City example. *Coast. Eng.* **2017**, *121*, 278–296. [CrossRef]
44. Yin, J.; Lin, N.; Yu, D. Coupled modeling of storm surge and coastal inundation: A case study in New York City during Hurricane Sandy. *Water Resour. Res.* **2016**, *52*, 8685–8699. [CrossRef]
45. Hu, R.; Fang, F.; Salinas, P.; Pain, C.C.; Domingo, N.S.; Mark, O. Numerical simulation of floods from multiple sources using an adaptive anisotropic unstructured mesh method. *Adv. Water Resour.* **2019**, *123*, 173–188. [CrossRef]
46. Brown, J.D.; Spencer, T.; Moeller, I. Modeling storm surge flooding of an urban area with particular reference to modeling uncertainties: A case study of Canvey Island, United Kingdom. *Water Resour. Res.* **2007**, *43*, W06402. [CrossRef]
47. Wang, H.V.; Loftis, J.D.; Liu, Z.; Forrest, D.; Zhang, J. The Storm Surge and Sub-Grid Inundation Modeling in New York City during Hurricane Sandy. *J. Mar. Sci. Eng.* **2014**, *2*, 226–246. [CrossRef]
48. Jelesnianski, C.; Chen, J.; Shaffer, W.; Gilad, A. SLOSH-A hurricane storm surge forecast model. In *IEEE OCEANS 1984*; IEEE: Toulouse, France, 1984; pp. 314–317.
49. Zhang, K.; Xiao, C.; Shen, J. Comparison of the CEST and SLOSH models for storm surge flooding. *J. Coast. Res.* **2008**, *24*, 489–499. [CrossRef]
50. Blumberg, A.F.; Georgas, N.; Yin, L.; Herrington, T.O.; Orton, P.M. Street-scale modeling of storm surge inundation along the New Jersey Hudson River waterfront. *J. Atmos. Ocean. Tech.* **2015**, *32*, 1486–1497. [CrossRef]
51. Georgas, N.; Blumberg, A.; Herrington, T.; Wakeman, T.; Saleh, F.; Runnels, D.; Jordi, A.; Ying, K.; Yin, L.; Ramaswamy, V.; et al. The stevens flood advisory system: Operational H3E flood forecasts for the greater New York/New Jersey Metropolitan Region. *Flood Risk Manag. Resp.* **2016**, *194*, 648–662.
52. Xie, L.; Pietrafesa, L.J.; Peng, M. Incorporation of a Mass-Conserving Inundation Scheme into a Three Dimensional Storm Surge Model. *J. Coast. Res.* **2004**, *204*, 1209–1223. [CrossRef]
53. Peng, M.; Xie, L.; Pietrafesa, L.J. Tropical cyclone induced asymmetry of sea level surge and fall and its presentation in a storm surge model with parametric wind fields. *Ocean. Model.* **2006**, *14*, 81–101. [CrossRef]
54. Dottori, F.; Todini, E. Developments of a flood inundation model based on the cellular automata approach: Testing different methods to improve model performance. *Phys. Chem. Earth* **2011**, *36*, 266–280. [CrossRef]
55. Zheng, Y.; Sun, H. An Integrated Approach for the Simulation Modeling and Risk Assessment of Coastal Flooding. *Water* **2020**, *12*, 2076. [CrossRef]
56. Hubbert, G.D.; McInnes, K.L. A storm surge inundation model for coastal planning and impact studies. *J. Coast. Res.* **1999**, 168–185.
57. McMillan, H.K.; Brasington, J. Reduced complexity strategies for modelling urban floodplain inundation. *Geomorphology* **2007**, *90*, 226–243. [CrossRef]
58. Hunter, N.M.; Bates, P.D.; Horritt, M.S.; Wilson, M.D. Simple spatially-distributed models for predicting flood inundation: A review. *Geomorphology* **2007**, *90*, 208–225. [CrossRef]
59. Fewtrell, T.; Bates, P.; de Wit, A.; Asselman, N.; Sayers, P. Comparison of varying complexity numerical models for the prediction of flood inundation in Greenwich, UK. In Proceedings of the FLOODrisk 2008, Oxford, UK, 30 September–2 October 2008.
60. Neal, J.; Villanueva, I.; Wright, N.; Willis, T.; Fewtrell, T.; Bates, P. How much physical complexity is needed to model flood inundation? *Hydrol. Process.* **2012**, *26*, 2264–2282. [CrossRef]
61. Tsakiris, G.; Bellos, V. A numerical model for two-dimensional flood routing in complex terrains. *Water Resour. Manag.* **2014**, *28*, 1277–1291. [CrossRef]
62. Bellos, V.; Tsakiris, G. Comparing various methods of building representation for 2D flood modelling in built-up areas. *Water Resour. Manag.* **2015**, *29*, 379–397. [CrossRef]
63. Zellou, B.; Rahali, H. Assessment of reduced-complexity landscape evolution model suitability to adequately simulate flood events in complex flow conditions. *Nat. Hazards* **2017**, *86*, 1–29. [CrossRef]
64. Afshari, S.; Tavakoly, A.A.; Rajib, M.A.; Zheng, X.; Follum, M.L.; Omranian, E.; Fekete, B.M. Comparison of new generation low-complexity flood inundation mapping tools with a hydrodynamic model. *J. Hydrol.* **2018**, *556*, 539–556. [CrossRef]
65. Hoch, J.M.; Eilander, D.; Ikeuchi, H.; Baart, F.; Winsemius, H.C. Evaluating the impact of model complexity on flood wave propagation and inundation extent with a hydrologic–hydrodynamic model coupling framework. *Nat. Hazards Earth Syst. Sci.* **2019**, *19*, 1723–1735. [CrossRef]
66. Favaretto, C.; Martinelli, L.; Ruol, P. A model of coastal flooding using linearized bottom friction and its application to a case study in Caorle, Venice, Italy. *Int. J. Offshore Polar Eng.* **2019**, *29*, 182–190. [CrossRef]
67. Dimitriadis, P.; Tegos, A.; Oikonomou, A.; Pagana, V.; Koukouvinos, A.; Mamassis, N.; Koutsoyiannis, D.; Efstratiadis, A. Comparative evaluation of 1D and quasi-2D hydraulic models based on benchmark and real-world applications for uncertainty assessment in flood mapping. *J. Hydrol.* **2016**, *534*, 478–492. [CrossRef]
68. Beven, K. Validation and equifinality. In *Computer Simulation Validation: Fundamental Concepts, Methodological Frameworks, and Philosophical Perspectives*; Beisbart, C., Saam, N.J., Eds.; Springer: Cham, Switzerland, 2019; pp. 791–809.
69. Her, Y.; Chaubey, I. Impact of the numbers of observations and calibration parameters on equifinality, model performance, and output and parameter uncertainty. *Hydrol. Process.* **2015**, *29*, 4220–4237. [CrossRef]

70. Bates, P.; Trigg, M.; Neal, J.; Dabrowa, A. *LISFLOOD-FP. User Manual*; School of Geographical Sciences, University of Bristol: Bristol, UK, 2013.
71. Hunter, N.M.; Horritt, M.S.; Bates, P.D.; Wilson, M.D.; Werner, M.G. An adaptive time step solution for raster-based storage cell modelling of floodplain inundation. *Adv. Water Res.* **2005**, *28*, 975–991. [CrossRef]
72. Neal, J.; Dunne, T.; Sampson, C.; Smith, A.; Bates, P. Optimisation of the two-dimensional hydraulic model LISFOOD-FP for CPU architecture. *Environ. Model. Softw.* **2018**, *107*, 148–157. [CrossRef]
73. Hall, J.W.; Tarantola, S.; Bates, P.D.; Horritt, M.S. Distributed sensitivity analysis of flood inundation model calibration. *J. Hydraul. Eng.* **2005**, *131*, 117–126. [CrossRef]
74. Werner, M.G.F.; Hunter, N.M.; Bates, P.D. Identifiability of distributed floodplain roughness values in flood extent estimation. *J. Hydrol.* **2005**, *314*, 139–157. [CrossRef]
75. Kulp, S.; Strauss, B.H. Global DEM errors underpredict coastal vulnerability to sea level rise and flooding. *Front. Earth Sci.* **2016**, *4*, 36. [CrossRef]
76. Xu, H. Modification of normalised difference water index (NDWI) to enhance open water features in remotely sensed imagery. *Int. J. Remote Sens.* **2006**, *27*, 3025–3033. [CrossRef]
77. Yang, X.; Zhao, S.; Qin, X.; Zhao, N.; Liang, L. Mapping of urban surface water bodies from Sentinel-2 MSI imagery at 10 m resolution via NDWI-based image sharpening. *Remote Sens.* **2017**, *9*, 596. [CrossRef]
78. Didier, D.; Baudry, J.; Bernatchez, P.; Dumont, D.; Sadegh, M.; Bismuth, E.; Bandet, M.; Dugas, S.; Sévigny, C. Multihazard simulation for coastal flood mapping: Bathtub versus numerical modelling in an open estuary, Eastern Canada. *J. Flood Risk Manag.* **2019**, *12*, e12505. [CrossRef]
79. Gallien, T.W.; Schubert, J.E.; Sanders, B.F. Predicting tidal flooding of urbanized embayments: A modeling framework and data requirements. *Coast. Eng.* **2011**, *58*, 567–577. [CrossRef]
80. Seenath, A.; Wilson, M.; Miller, K. Hydrodynamic versus GIS modelling for coastal flood vulnerability assessment: Which is better for guiding coastal management? *Ocean Coast. Manag.* **2016**, *120*, 99–109. [CrossRef]
81. Williams, L.L.; Lück-Vogel, M. Comparative assessment of the GIS based bathtub model and an enhanced bathtub model for coastal inundation. *J. Coast. Conserv.* **2020**, *24*, 1–15. [CrossRef]
82. Yunus, A.P.; Avtar, R.; Kraines, S.; Yamamuro, M.; Lindberg, F.; Grimmond, C.S.B. Uncertainties in tidally adjusted estimates of sea level rise flooding (bathtub model) for the Greater London. *Remote Sens.* **2016**, *8*, 366. [CrossRef]
83. Horritt, M.S.; Bates, P.D. Predicting floodplain inundation: Raster-based modelling versus the finite-element approach. *Hydrol. Process.* **2001**, *15*, 825–842. [CrossRef]
84. Horsburgh, K.; Horritt, M. The Bristol Channel floods of 1607–reconstruction and analysis. *Weather* **2006**, *61*, 272–277. [CrossRef]
85. Lewis, M.; Horsburgh, K.; Bates, P.; Smith, R. Quantifying the uncertainty in future coastal flood risk estimates for the UK. *J. Coast. Res.* **2011**, *27*, 870–881. [CrossRef]
86. Wadey, M.P.; Nicholls, R.J.; Hutton, C. Coastal Flooding in the Solent: An Integrated Analysis of Defences and Inundation. *Water* **2012**, *4*, 430–459. [CrossRef]
87. Quinn, N.; Lewis, M.; Wadey, M.P.; Haigh, I.D. Assessing the temporal variability in extreme storm-tide time series for coastal flood risk assessment. *J. Geophys. Res. Ocean.* **2014**, *119*, 4983–4998. [CrossRef]
88. Skinner, C.J.; Coulthard, T.J.; Parsons, D.R.; Ramirez, J.A.; Mullen, L.; Manson, S. Simulating tidal and storm surge hydraulics with a simple 2D inertia based model, in the Humber Estuary, UK. *Estuar. Coast. Shelf Sci.* **2015**, *155*, 126–136. [CrossRef]
89. Sadeghi, F.; Rubinato, M.; Goerke, M.; Hart, J. Assessing the Performance of LISFLOOD-FP and SWMM for a Small Watershed with Scarce Data Availability. *Water* **2022**, *14*, 748. [CrossRef]
90. Bradbrook, K.F.; Lane, S.N.; Waller, S.G.; Bates, P.D. Two-dimensional diffusion wave modelling of flood inundation using a simplified channel representation. *Int. J. River Basin Manag.* **2004**, *2*, 211–223. [CrossRef]
91. Hunter, N.M.; Horritt, M.S.; Bates, P.D.; Werner, M.G.F. Theoretical and practical limits to the use of storage cell codes for flood inundation modelling. In *Flood Risk Assessment*; Reeve, D., Ed.; Institute of Mathematics and its Applications: Southend-on-Sea, UK, 2004.
92. Wang, J.; Lei, X.; Cai, S.; Zhao, J. Location identification of river bathymetric error based on the forward and reverse flow routing. *Water Supply* **2022**, *22*, 5095–5110. [CrossRef]
93. Murdukhayeva, A.; August, P.; Bradley, M.; LaBash, C.; Shaw, N. Assessment of inundation risk from sea level rise and storm surge in northeastern coastal national parks. *J. Coast. Res.* **2013**, *29*, 1–16. [CrossRef]
94. Seenath, A. Modelling coastal flood vulnerability: Does spatially-distributed friction improve the prediction of flood extent? *Appl. Geogr.* **2015**, *64*, 97–107. [CrossRef]
95. Arcement, G.J.; Schneider, V.R. *Guide for Selecting Manning's Roughness Coefficients for Natural Channels and Flood Plains*; United States Government Printing Office: Denver, CO, USA, 1989.
96. Chow, V.T. *Open-Channel Hydraulics*; McGraw-Hill: New York, NY, USA, 1959.
97. Martinelli, L.; Zanuttigh, B.; Corbau, C. Assessment of coastal flooding hazard along the Emilia Romagna littoral, IT. *Coast. Eng.* **2010**, *57*, 1042–1058. [CrossRef]
98. Prime, T.; Brown, J.M. and Plater, A.J. Flood inundation uncertainty: The case of a 0.5% annual probability flood event. *Environ. Sci. Policy* **2016**, *59*, 1–9. [CrossRef]

99. Shen, Y.; Morsy, M.M.; Huxley, C.; Tahvildari, N.; Goodall, J.L. Flood risk assessment and increased resilience for coastal urban watersheds under the combined impact of storm tide and heavy rainfall. *J. Hydrol.* **2019**, *579*, 124159. [CrossRef]
100. Shustikova, I.; Domeneghetti, A.; Neal, J.C.; Bates, P.; Castellarin, A. Comparing 2D capabilities of HEC-RAS and LISFLOOD-FP on complex topography. *Hydrol. Sci. J.* **2019**, *64*, 1769–1782. [CrossRef]
101. Morsy, M.M.; Lerma, N.R.; Shen, Y.; Goodall, J.L.; Huxley, C.; Sadler, J.M.; Voce, D.; O'Neil, G.L.; Maghami, I.; Zahura, F.T. Impact of geospatial data enhancements for regional-scale 2D hydrodynamic flood modeling: Case study for the Coastal Plain of Virginia. *J. Hydrol. Eng.* **2021**, *26*, 05021002. [CrossRef]
102. Beven, K. Searching for the Holy Grail of Scientific Hydrology: Qt=H(S,R,ΔT)A as closure. *Hydrol. Earth Syst. Sci.* **2006**, *10*, 609–618. [CrossRef]
103. Copernicus Land Monitoring Service. Available online: https://land.copernicus.eu/pan-european/corine-land-cover (accessed on 19 March 2023).
104. Galiatsatou, P.; Makris, C.; Krestenitis, Y.; Prinos, P. Nonstationary Extreme Value Analysis of Nearshore Sea-State Parameters under the Effects of Climate Change: Application to the Greek Coastal Zone and Port Structures. *J. Mar. Sci. Eng.* **2021**, *9*, 817. [CrossRef]
105. Galiatsatou, P.; Makris, C.; Prinos, P.; Kokkinos, D. Nonstationary joint probability analysis of extreme marine variables to assess design water levels at the shoreline in a changing climate. *Nat. Hazards* **2019**, *98*, 1051–1089. [CrossRef]
106. De Vries, H.; Breton, M.; de Mulder, T.; Krestenitis, Y.; Proctor, R.; Ruddick, K.; Salomon, J.C.; Voorrips, A. A comparison of 2D storm surge models applied to three shallow European seas. *Environ. Softw.* **1995**, *10*, 23–42. [CrossRef]
107. Schwiderski, E.W. On charting global ocean tides. *Rev. Geophys.* **1980**, *18*, 243–268. [CrossRef]
108. Sakamoto, K.; Tsujino, H.; Nakano, H.; Hirabara, M.; Yamanaka, G. A practical scheme to introduce explicit tidal forcing into an OGCM. *Ocean Sci.* **2013**, *9*, 1089–1108. [CrossRef]
109. WaveForUs Operational Forecast System. Available online: http://wave4us.web.auth.gr/ (accessed on 19 March 2023).
110. METEO.GR Weather Forecast Node. Available online: https://www.meteo.gr/meteomaps/sea_level.cfm (accessed on 19 March 2023).
111. National Observatory of Athens. Available online: https://www.iersd.noa.gr/en/ (accessed on 19 March 2023).
112. Accu-Waves Operational Forecast System. Available online: https://accuwaves.eu/forecast/ (accessed on 19 March 2023).
113. Makris, C.; Baltikas, V.; Kontos, Y.; Androulidakis, Y.; Nagkoulis, N.; Kazakis, I.; Karambas, T.; Papadimitriou, A.; Metallinos, A.; Chondros, M.; et al. Integrated modeling of sea-state forecasts for safe navigation near and inside ports: The Accu-Waves platform. In Proceedings of the 31st ISOPE Conference, Rhodes, Greece, 20–25 June 2021; pp. 2307–2314.
114. CMS. Available online: https://data.marine.copernicus.eu/product/SEALEVEL_EUR_PHY_L4_MY_008_068/ (accessed on 19 March 2023).
115. UNESCO Intergovernmental Oceanographic Commission, IOC. Available online: https://www.ioc-sealevelmonitoring.org/ (accessed on 19 March 2023).
116. Landerer, F.W.; Volkov, D.L. The anatomy of recent large sea level fluctuations in the Mediterranean Sea. *Geophys. Res. Lett.* **2013**, *40*, 553–557. [CrossRef]
117. Bonaduce, A.; Pinardi, N.; Oddo, P.; Spada, G.; Larnicol, G. Sea-level variability in the Mediterranean Sea from altimetry and tide gauges. *Clim. Dyn.* **2016**, *47*, 2851–2866. [CrossRef]
118. Miglietta, M.M.; Rotunno, R. Development mechanisms for Mediterranean tropical-like cyclones (medicanes). *Q. J. R. Meteorol. Soc.* **2019**, *145*, 1444–1460. [CrossRef]
119. Dafis, S.; Claud, C.; Kotroni, V.; Lagouvardos, K.; Rysman, J.F. Insights into the convective evolution of Mediterranean tropical-like cyclones. *Q. J. R. Meteorol. Soc.* **2020**, *146*, 4147–4169. [CrossRef]
120. Lagouvardos, K.; Karagiannidis, A.; Dafis, S.; Kalimeris, A.; Kotroni, V. Ianos—A hurricane in the Mediterranean. *Bull. Am. Meteorol. Soc.* **2021**, *103*, E1621–E1636. [CrossRef]
121. Toomey, T.; Amores, A.; Marcos, M.; Orfila, A.; Romero, R. Coastal hazards of tropical-like cyclones over the Mediterranean Sea. *J. Geophys. Res. Oceans* **2022**, *127*, e2021JC017964. [CrossRef]
122. Europost.gr. Available online: https://europost.gr/kakokairia-mpallos-plimmyres-provlimata-se-oli-ti-chora-prognosi-kairoy/ (accessed on 19 March 2023).
123. Sinidisi.gr. Available online: https://sinidisi.gr/plimmires-vasiliki-lefkadas-den-anoixan-sxoleia/ (accessed on 19 March 2023).
124. MyLefkada YouTube. Available online: https://www.youtube.com/watch?v=bUeFHvBdaOY&ab_channel=MyLefkada (accessed on 19 March 2023).
125. NewsIt. Available online: https://www.newsit.gr/topikes-eidhseis/kakokairia-preveza-metraei-pliges-apo-plimmyres-kai-anemostrovilous/3419041/ (accessed on 19 March 2023).
126. NIKOS FATSIOS YouTube Channel about Igoumenitsa Floods. Available online: https://www.youtube.com/watch?v=lMyi5C05u1w&ab_channel=NIKOSFATSIOS (accessed on 19 March 2023).
127. Ethnos.gr. Available online: https://www.ethnos.gr/greece/article/124385/ (accessed on 19 March 2023).
128. Libre.gr. Available online: https://www.libre.gr/2020/09/18/ianos-voyliaxe-to-rio-deite-to-fainom/ (accessed on 19 March 2023).
129. CDSE. Available online: https://dataspace.copernicus.eu/browser/ (accessed on 19 March 2023).
130. Sentinel Hub. Available online: https://apps.sentinel-hub.com/eo-browser (accessed on 19 March 2023).
131. Gao, B.C. NDWI—A normalized difference water index for remote sensing of vegetation liquid water from space. *Remote Sens. Environ.* **1996**, *58*, 257–266. [CrossRef]

132. McFeeters, S.K. The use of the Normalized Difference Water Index (NDWI) in the delineation of open water features. *Int. J. Remote Sens.* **1996**, *17*, 1425–1432. [CrossRef]
133. Underwood, L.W.; Kalcic, M.T.; Fletcher, R.M. Resolution Enhancement of MODIS-derived Water Indices for Studying Persistent Flooding. In *AGU Fall Meeting Abstracts*; American Geophysical Union (AGU): San Francisco, CA, USA, 2012; p. OS11D–1689.
134. Kalcic, M.T.; Underwood, L.W.; Fletcher, R.M. A New Approach to Monitoring Coastal Marshes for Persistent Flooding. In *AGU Fall Meeting Abstracts*; American Geophysical Union (AGU): San Francisco, CA, USA, 2012; p. OS41C–1753.
135. Schmid, K.; Hadley, B.; Waters, K. Mapping and portraying inundation uncertainty of bathtub-type models. *J. Coast. Res.* **2014**, *30*, 548–561. [CrossRef]
136. Vousdoukas, M.I.; Voukouvalas, E.; Mentaschi, L.; Dottori, F.; Giardino, A.; Bouziotas, D.; Bianchi, A.; Salamon, P.; Feyen, L. Developments in large-scale coastal flood hazard mapping. *Nat. Hazards Earth Syst. Sci.* **2016**, *16*, 1841–1853. [CrossRef]
137. Karamouz, M.; Fereshtehpour, M. Modeling DEM errors in coastal flood inundation and damages: A spatial nonstationary approach. *Wat Resour. Res.* **2019**, *55*, 6606–6624. [CrossRef]
138. West, H.; Horswell, M.; Quinn, N. Exploring the sensitivity of coastal inundation modelling to DEM vertical error. *Int. J. Geogr. Inf. Sci.* **2018**, *32*, 1172–1193. [CrossRef]
139. Flowerdew, J.; Horsburgh, K.; Wilson, C.; Mylne, K. Development and evaluation of an ensemble forecasting system for coastal storm surges. *Q. J. R. Meteorol. Soc.* **2010**, *136*, 1444–1456. [CrossRef]
140. Doong, D.J.; Chuang, L.H.; Wu, L.C.; Fan, Y.M.; Kao, C.C.; Wang, J.H. Development of an operational coastal flooding early warning system. *Nat. Hazards Earth Syst. Sci.* **2012**, *12*, 379–390. [CrossRef]
141. Harley, M.D.; Valentini, A.; Armaroli, C.; Perini, L.; Calabrese, L.; Ciavola, P. Can an early-warning system help minimize the impacts of coastal storms? A case study of the 2012 Halloween storm, northern Italy. *Nat. Hazards Earth Syst. Sci.* **2016**, *16*, 209–222. [CrossRef]
142. Murty, P.L.N.; Padmanabham, J.; Kumar, T.S.; Kumar, N.K.; Chandra, V.R.; Shenoi, S.S.C.; Mohapatra, M. Real-time storm surge and inundation forecast for very severe cyclonic storm 'Hudhud'. *Ocean Eng.* **2017**, *131*, 25–35. [CrossRef]
143. Nikolić, Ž.; Srzić, V.; Lovrinović, I.; Perković, T.; Šolić, P.; Kekez, T. Coastal Flooding Assessment Induced by Barometric Pressure, Wind-Generated Waves and Tidal-Induced Oscillations: Kaštela Bay Real-Time Early Warning System Mobile Application. *Appl. Sci.* **2022**, *12*, 12776. [CrossRef]
144. Yu, D.; Lane, S.N. Urban fluvial flood modelling using a two-dimensional diffusion-wave treatment. Part 2: Development of a sub-grid-scale treatment. *Hydrol. Process.* **2006**, *20*, 1567–1583. [CrossRef]
145. Werner, M.G.F. A comparison of flood extent modelling approaches through constraining uncertainties on gauge data. *Hydrol. Earth Syst. Sci.* **2004**, *8*, 1141–1152. [CrossRef]

hydrology MDPI

Article

Water Level Forecasting in Tidal Rivers during Typhoon Periods through Ensemble Empirical Mode Decomposition

Yen-Chang Chen [1], Hui-Chung Yeh [2,*], Su-Pai Kao [3], Chiang Wei [4] and Pei-Yi Su [1]

1 Department of Civil Engineering, National Taipei University of Technology, Taipei 10617, Taiwan
2 Department of Land Resources, Chinese Culture University, Taipei 11114, Taiwan
3 TYLIN International Taiwan, Taipei 10657, Taiwan
4 Experimental Forest, National Taiwan University, Zhushan, Nantou 55750, Taiwan
* Correspondence: xhz@ulive.pccu.edu.tw

Abstract: In this study, a novel model that performs ensemble empirical mode decomposition (EEMD) and stepwise regression was developed to forecast the water level of a tidal river. Unlike more complex hydrological models, the main advantage of the proposed model is that the only required data are water level data. EEMD is used to decompose water level signals from a tidal river into several intrinsic mode functions (IMFs). These IMFs are then used to reconstruct the ocean and stream components that represent the tide and river flow, respectively. The forecasting model is obtained through stepwise regression on these components. The ocean component at a location 1 h ahead can be forecast using the observed ocean components at the downstream gauging stations, and the corresponding stream component can be forecast using the water stages at the upstream gauging stations. Summing these two forecasted components enables the forecasting of the water level at a location in the tidal river. The proposed model is conceptually simple and highly accurate. Water level data collected from gauging stations in the Tanshui River in Taiwan during typhoons were used to assess the feasibility of the proposed model. The water level forecasting model accurately and reliably predicted the water level at the Taipei Bridge gauging station.

Keywords: ensemble empirical mode decomposition (EEMD); flood period; tidal river; water level forecasting

Citation: Chen, Y.-C.; Yeh, H.-C.; Kao, S.-P.; Wei, C.; Su, P.-Y. Water Level Forecasting in Tidal Rivers during Typhoon Periods through Ensemble Empirical Mode Decomposition. *Hydrology* **2023**, *10*, 47. https://doi.org/10.3390/hydrology10020047

Academic Editors: Aristoteles Tegos, Alexandros Ziogas and Vasilis Bellos

Received: 22 December 2022
Revised: 8 February 2023
Accepted: 8 February 2023
Published: 10 February 2023

1. Introduction

An estuary is a transition zone with complex flow conditions in which a river enters the ocean. Complex factors contribute to the water level in tidal rivers; the water level is affected by not only the upstream river discharge but also ocean tides [1]. The water level in a tidal river changes because of the interaction between riverine and marine factors. Because of the rotation of the Earth and the varying strength of the gravitational pull from the Moon and Sun, the water level varies quasiperiodically every 12.25 h or twice every lunar day. [2]. Longer-period effects from storms and seasonal fluctuations influence salinity. Flooding from the upstream basin can alter the salinity profile and interrupt the tidal cycle [3]. A major climate factor affecting estuaries is wind; wind creates waves, which affect water circulation and the mixing of fresh and seawater [4]. Upon circulation and mixing, the 2% difference in the densities of fresh and seawater creates a pressure gradient in the horizontal direction that affects the water flow [5]. This density difference is largely caused by differences in temperature and salinity; however, salinity is by far the dominant factor affecting tidal river dynamics [6]. Considering all the aforementioned information, accounting for all physical processes in tidal rivers is challenging. These hydrological processes are complex, have mutual interactions, and are the driving forces [7] for other sedimentological, biological, and chemical processes. It is not easy to develop a model that can deal with all hydrological processes in tidal rivers. No simple conventional method can accurately forecast the discharge and water level in tidal rivers.

Because of the extremely unsteady flow conditions of tidal rivers or estuaries, forecasting their water levels is a difficult task. Theoretical and empirical approaches are commonly used to perform this task. The hydrological processes in a tidal river are unique, and the water level in a tidal river is continually changing because of the interactions of riverine and marine processes. The factors affecting water levels include the shape of the tidal river, astronomical tides, wind, salinity, temperature, sediment, flood, storm surge, and other factors that are too complex to model directly. Consequently, the hydrodynamic processes of tidal rivers are complex, nonstationary, and nonlinear [8]. Many of the concepts or principles identified by modeling other watercourses have been applied to forecast water levels in tidal rivers. The theoretical approach is based on continuity, momentum, and energy equations. However, a major disadvantage of theoretical methods is that the required parameters are usually difficult to determine from the observed data; in particular, the discharge is challenging to measure [9]. Although there are lots of open-source models available for free, programming and executing a newly developed model is time-consuming and costly. Some hydraulic models apply the mass conservation and momentum principle [10–12] to forecast water levels and current velocities during spring and neap tidal cycles. Hydrological routing, which is a simpler technique than that of hydraulic models, uses a continuity equation combined with a storage indication curve to forecast estuary water levels [13]. These hydraulic and hydrological models usually apply numerical methods to obtain results. Artificial neural networks (ANNs) have been widely used for data mining. An ANN is a black-box technique that can be used for water resource management and modeling hydrological processes [14–16]. An ANN can also be applied for forecasting tidal river water levels [8,17].

The variation in the water level of tidal rivers with time can be regarded as a signal. Some methods for signal processing analysis, such as the Fourier [18,19] and wavelet [20,21] transforms, are often used to analyze historical data for forecasting tidal river or estuary water levels. The Fourier transform can only be applied to linear and stationary processes, and wavelet transforms can only be applied to linear and nonstationary processes. However, the hydrological processes in tidal rivers are nonlinear and nonstationary. A novel method of handling nonstationary and nonlinear data is the Hilbert–Huang transform (HHT), which was proposed by Huang et al. [22,23]. The HHT is a method of decomposing an original signal into many intrinsic mode functions (IMFs) with a trend. The fundamental process of the HHT is the empirical mode decomposition (EMD) or ensemble EMD (EEMD) method, which involves breaking down a signal into various IMFs. Since their introduction, the EMD and EEMD methods have rapidly grown in popularity and have been effectively applied to estuaries [24,25], oceans [26,27], and other engineering fields, including water resources [28,29].

In this study, a conceptual model was developed for forecasting tidal river water levels during a flood period (Figure 1). The proposed model only requires water level data for prediction. EEMD is applied to decompose the water levels in tidal rivers into several IMFs. The IMFs decomposed through EEMD usually have a physically meaningful correspondence to physical data [30–33]. The water level in a tidal river is affected by many factors, such as tide, topography, friction, and river flow [24], in a complex manner. However, these data are difficult to obtain and thus cannot be used to develop a sophisticated model. By contrast, water level data can be easily collected. IMFs can be obtained through EEMD; however, because of a lack of data, the factors affecting IMFs cannot be determined. Therefore, the developed model was simplified by dividing IMFs into two groups: ocean and stream components. These components were used to establish regression methods for forecasting the contribution of each component to the water level. By adding the contributions from the two forecasted components, the water level in tidal rivers can be obtained. Finally, the water stages of the Tanshui River in Taiwan during typhoon periods were used as an example to demonstrate the calculation procedures and validate the reliability and accuracy of the proposed model.

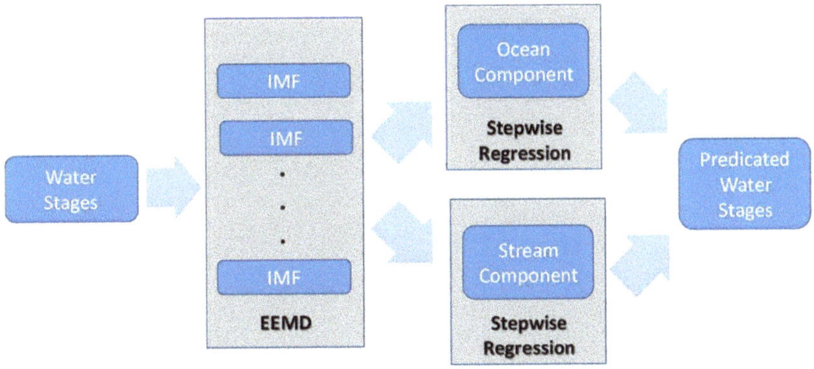

Figure 1. The approach used for forecasting water level in a tidal river.

2. EEMD and Stepwise Regression

2.1. EEMD Method

Huang et al. [23] proposed the EMD method, which is an intuitive and adaptive data analysis method. In EMD, basis functions are derived from the original signals. The aforementioned method directly resolves energies by using the intrinsic time scale of the original data, which are decomposed into several simple harmonic functions (i.e., IMFs) with different periodicity. An IMF is a simple oscillatory mode corresponding to a simple harmonic function and must satisfy the following two requirements. First, in the entire data set, the number of extrema and the number of zero-crossings must be equal or differ at most by 1. Second, at any point, the mean value of the envelopes defined by the local maximum and minimum is 0. Thus, EMD is used to decompose an original signal into multiple IMFs with different frequencies and a residual signal. These IMFs form a complete and nearly orthogonal basis for the original signal. An IMF can have variable amplitude and frequency along the time axis. The EMD method differs from wavelet and Fourier analysis in that the basis is not predetermined. Consequently, the characteristics of the original signal can be fully reflected. The EMD method is intuitive, direct, and self-adaptive.

The procedure of extracting an IMF is called sifting. Figure 2 presents an example of the sifting process for the time series of water level $X(t)$. This process involves the following steps:

A. The local maxima and minima in $X(t)$ are identified, as shown in Figure 2a.
B. Cubic spline is used on the local maximum and minimum values to generate two curves approximating the envelopes, namely, the upper and lower envelopes, as displayed in Figure 2b.
C. A mean curve is calculated from the two envelopes, as illustrated in Figure 2c. The mean is expressed as follows:

$$m(u) = \frac{E_u(t) + E_l(t)}{2} \quad (1)$$

where $m(t)$ is the mean, $E_u(t)$ is the upper envelope, and $E_l(t)$ is the lower envelope.

A variable $d(t)$ is defined as follows:

$$d(t) = X(t) - m(t) \quad (2)$$

where $d(t)$ is the difference between $X(t)$ and $m(t)$. If $d(t)$ does not meet the stopping criterion, $d(t)$ is set as the new $X(t)$ value, and the aforementioned steps are repeated to differentiate the extremes until $d(t)$ reaches the stopping criterion.

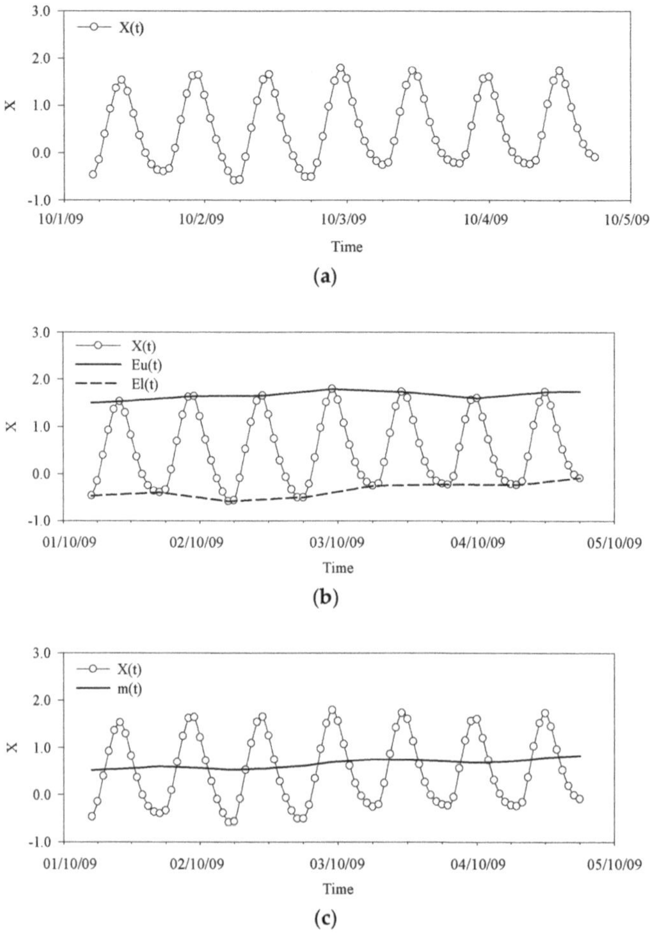

Figure 2. Definition of sifting: (**a**) Original data (water levels hydrograph); (**b**) Upper and lower envelopes; (**c**) Local mean value of the envelopes.

An excessive number of selection cycles can reduce the physical meaning of the IMF's instantaneous frequency and amplitude; thus, a stopping criterion must be set. The stopping criterion is based on the amplitude, energy, and phase. Common stopping criteria include the standard deviation, an *S*-number criterion [34], and an evaluation function [35]. In this study, the *S*-number criterion was used, where *S* is the maximum number of selection cycles. A selection cycle is terminated when the number of extreme values matches the number of zero-crossings.

The $d(t)$ value that meets the stopping criterion is set as an IMF, namely, $C_j(t)$, where j is a value from 1 to n. The residual $R_j(t)$ is the new $X_{j+1}(t)$ value, as expressed in the following equation:

$$R_j(t) = X_{j+1}(t) \tag{3}$$

EMD is then repeated to obtain additional IMFs. The final IMF n is recorded as $C_{j=n}(t)$. The term $X(t)$ represents the superposition of various IMF components ($C_j(t)$ and $R_n(t)$) and is expressed as follows:

$$X(t) = \sum_{j=1}^{n} C_j(t) + R_n(t) \tag{4}$$

In EMD, the problem of mode mixing occurs. Mode mixing is a problem in which an IMF produced through EMD decomposition contains components of different frequencies. Mode mixing is caused by intermittent signals and noise. In particular, mode mixing occurs because of unpredictable random noise contained in the original signal infiltrating the IMFs. This intermittent, irregular noise affects the determination of the upper and lower envelopes. Consequently, two signals of different time scales can be classified as one IMF, or signals of the same time scale might be separated into two IMFs. Mode mixing eliminates the physical significance of the IMF. To overcome this challenge, Wu and Huang [36] proposed the EEMD method, in which white noise is introduced to eliminate the effect of the original noise and obtain mode-consistent IMFs. EEMD is performed as follows. First, a white noise signal $w_i(t)$ is added to the original signal to form an ensemble. Second, the ensemble is subjected to EMD decomposition into several IMFs. Third, the first and second steps are repeated by adding white noise on each time scale.

Because white noise is stochastic and uniformly distributed on every component, its effect can be eliminated as its ensemble number increases; that is, if sufficient white noise addition cycles are performed, the obtained solution approaches the true answer, and the goal of eliminating noise and mode mixing can be achieved. According to statistical theory, the influence of the added noise and its relation to the ensemble number is expressed as follows:

$$\varepsilon_n = \frac{\varepsilon}{\sqrt{n}} \tag{5}$$

where n is the ensemble number, ε is the amplitude of the added white noise, and ε_n is the standard error. The noise-added signal based on the aforementioned relation is represented as follows:

$$X_E(t) = X(t) + \varepsilon \times noise(t) \tag{6}$$

The signal in Equation (6) is subjected to EMD decomposition. The IMFs at different frequencies are obtained from the ensemble average of each component.

Each IMF ($C_j(t)$) calculated through EEMD inherits the physical meaning of the original data. Therefore, EEMD is often applied in geographic research [36]. Tidal river water level is profoundly influenced by tides. If EEMD is used for analysis, water level can be decomposed into mutually independent IMFs with corresponding frequencies. Thus, the frequency of each IMF can be compared with the tidal frequency in the studied area. If an IMF has periodicity, it is likely to be related to tides. Therefore, IMFs generated from water level data can be classified into two groups: tidal functions and flood functions. By adding all tidal IMFs, the ocean component can be obtained; similarly, the stream component can be obtained by summing the remaining IMFs.

2.2. *Stepwise Regression Analysis*

Stepwise regression, which is a multiple linear regression technique, is an efficient method of selecting the most useful explanatory variables. This method is a modification of forward selection. The general idea behind stepwise regression is that at each stage of selection, all model variables are evaluated using the partial F-test based on a preselected critical value.

Initially, the candidate variables are identified. Stepwise regression with forward selection begins with no variables in the regression model. Let the set of all possible variables be $x_1, x_2, \ldots, x_m$. In stepwise regression, the model is initially fitted with only one variable. After fitting the variable x_i, the fit is checked using the critical F value. Models with two variables are then considered. The optimal regression model with variables x_i and x_j is selected using the F-test and is included in the model. This process continues until the F-test indicates that the inclusion of further functions is not useful, at which point a final model is obtained.

The key task in forecasting tidal river water levels involves constructing regression models for the ocean and stream components. The goal of the ocean- and stream-component regression models is to establish the relationship between the downstream and upstream water levels, respectively, and the forecasted values at the site of interest. By summing the forecasting values obtained from these two regression models, the water level of a tidal river can be predicted.

3. Study Area and Data Descriptions

In this study, water level data from the Tanshui River in Taiwan were used to evaluate the proposed model. As illustrated in Figure 3, the tributaries of the Tanshui River include the Keelung River, Hsin-Tien River, and Dahan River. The Tanshui River is formed by the merger of the Tanhan River and Xindian River. The largest tributary of the Tanshui River is the Keelung River. The Hsin-Tien River is approximately 21 km long and runs south to north through Taipei to the Taiwan Strait. The main stream of the Tanshui River has a length of 158.7 km and drains 2575 km^2 in north Taiwan. It originates from a 3529 m high mountain with an average gradient of 1:122. In Figure 3, the circle denotes the tidal area of the Tanshui River [37]. Tides in the Taiwan Strait primarily comprise the four tide components O1, K1, M2, and S2; the tide level data mostly comprise the principal lunar semidiurnal constituent. Semidiurnal tides are the most influential tides in the Tanshui River. The average tide level at the river mouth gauging station is 0.03 m, with the average tide range being 2.19 m, spring tide range being 2.89 m, and maximum tide range being 3 m because of the contraction of the channel cross section and wave propagation. The difference between the two tidal ranges each day is small, and the tidal range of diurnal tides is approximately 1/5th that of semidiurnal tides.

The Tanshui River flows past the Taipei metropolitan area, which is Taiwan's political and cultural center. Taipei, which is situated in a low-lying basin, is susceptible to flooding. A flood control system was constructed in Taipei beginning in 1970. This system includes dams, levees, pumping stations, floodways, and a warning system and is designed to withstand floods with a 200-year return period. Typically, no water flows in the Erchong Floodway on ordinary days. If extreme flooding occurs, the water from the Tahan River and Hsin-Tien River is redirected to the floodways and purged downstream in the Tanshui River. The flood warning system must accurately forecast water levels during flood periods. Therefore, gauging stations operated by the 10th River Management Office were established within the Tanshui River estuary region to collect water levels for flood routing; these stations include Tudigonbi, the Taipei Bridge on the Tanshui River, the Shinhai Bridge on the Tahan River, and the Chung Cheng Bridge on the Hsin-Tien River.

The narrowest cross section of the Tanshui River is located at the Taipei Bridge. Consequently, when flooding occurs, the velocity and water level at this spot increase considerably, which often results in serious damage. Therefore, forecasting the water level at the Taipei Bridge is an essential task for the flood warning system. In this study, EEMD was conducted to construct a water level forecasting model for flood warnings at the Taipei Bridge. The results of EEMD were used to assess the reliability and accuracy of the proposed model. Floods from the Tahan River and the Hsin-Tien River upstream of the Tanshui River and tides downstream of the Taipei Bridge affect the water level at the Taipei Bridge. Therefore, the stream component at the Taipei Bridge was forecast using data from the gauging station at the Shinhai Bridge on the Tahan River and the station at the Chung Cheng Bridge on the Hsin-Tien River, which is located upstream of the Taipei Bridge. The ocean component at the Taipei Bridge was forecast using data from the Tudigonbi station located downstream of the Taipei Bridge. Finally, by adding the forecasted stream and ocean components, the water level forecast at the Taipei Bridge was obtained.

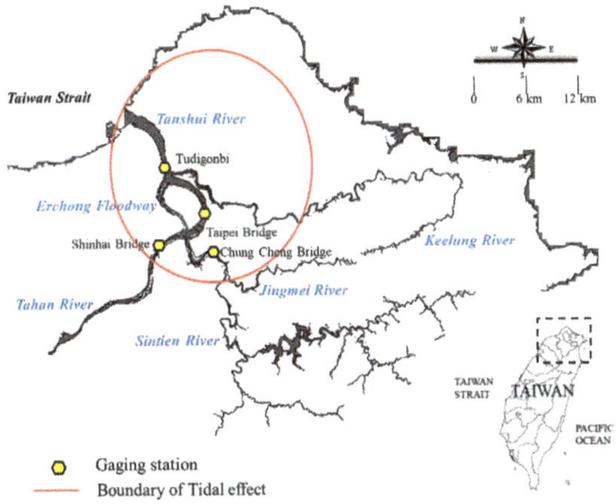

Figure 3. Study area and gauging stations.

The proposed model requires water level data for the Tudigonbi, Shinhai Bridge, and Chung Cheng Bridge stations to forecast the water level at the Taipei Bridge. The water level at each gauging station during typhoon periods differs considerably from that on ordinary days. Therefore, in this study, 15 typhoon or heavy storm events with complete water level data from 2004 to 2015 were used to establish a model for forecasting estuarine water levels. High-water data were selected as those from the period starting 1 day before the issue of a typhoon warning and ending the day after lifting the warning. A total of 10 out of the 15 events were further categorized for calibrating the proposed model; the remaining five events were used to verify the model. Table 1 lists the starting and ending times and the highest and lowest water levels at the Taipei Bridge for each typhoon event. Figure 4 presents the water level at each gauging station during Typhoon Soudelor; all gauging stations had an atypically high water level. The water level at the Chung Cheng Bridge and Shinhai Bridge, which are located at the boundary of the tidal area, increased sharply because of flooding. The water level at the Taipei Bridge also increased; however, this increase was smaller than those at the Chung Cheng Bridge and Shinhai Bridge. The only station close to the river mouth, namely, the Tudigonbi station, also exhibited a higher water level than usual; however, the difference was small. The periodic regularity of the water level disappeared for all gauging stations.

Table 1. Summary description of water levels at the Taipei Bridge during typhoons.

Phase	Typhoon	Duration	Water Level (m)	
			Max.	Min.
Calibration	Nock−Ten	22/10/2004–27/10/2004	1.88	−1.11
	Haitang	15/7/2005–21/7/2005	1.93	−1.34
	Matsa	2/8/2005–7/8/2005	3.62	−1.13
	Longwang	29/9/2005–4/10/2005	1.83	−1.12
	Fung−Wong	25/7/2008–30/7/2008	1.98	−1.13
	Fanapi	19/9/2010–21/9/2010	1.56	−0.91
	Saola	29/7/2012–4/8/2012	5.28	−1.09
	Soulik	10/7/2013–14/7/2013	2.88	−1.15
	Kong−Rey	26/8/2013–23/8/2013	1.68	−0.74
	Soudelor	5/8/2015–10/8/2015	5.17	−1.01
Verification	Mindulle	27/6/2004–4/7/2004	2.38	−1.58
	Kalmaegi	15/7/2008–19/7/2008	1.81	−1.12
	Sinlaku	10/9/2008–17/9/2008	3.45	−0.91
	Trami	19/8/2013–23/8/2013	2.85	−1.15
	Dujuan	26/9/2015–30/9/2015	3.78	−1.21

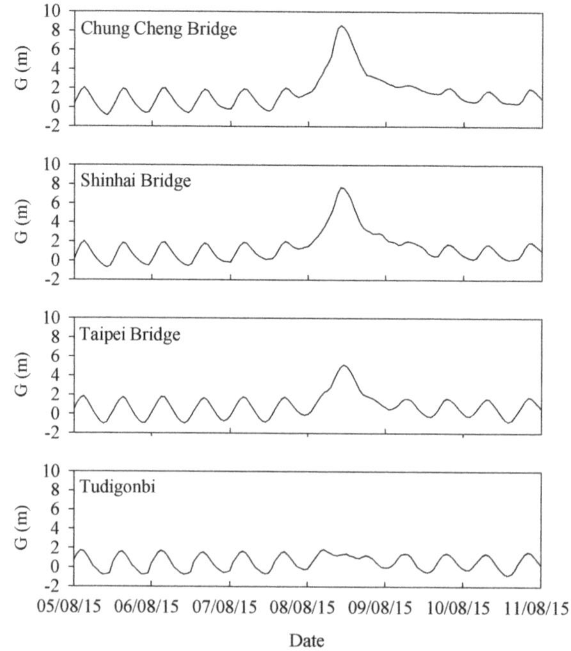

Figure 4. Water level hydrographs during Typhoon Soudelor.

4. Practical Applications

All hydrographs of the 15 events were connected, and EEMD was conducted to obtain the IMFs. The uppermost plot of Figure 5 presents the gauge height (*G*) at the Taipei Bridge, and the subsequent curves are represented in IMF1 to IMF8. IMF1, IMF2, IMF3, and IMF4 had periodicity; that is, these IMFs exhibited a pattern of cycles that repeat at intervals. Table 2 lists the periodicity for all IMFs at each gauging station. The frequencies (dividing the number of times an event occurs by the duration) of IMF1 and IMF2 for each station were approximately 0.0805 h^{-1}, which is similar to the M2 tidal component frequency presented in Table 3. This result suggests that IMF1 and IMF2 represent the influences of the semidiurnal tides. The periodicity of IMF3 for all stations was close to the principal solar or lunar diurnal constituent (P_1 and O_1 in Table 3), which indicated that diurnal tides contributed to the IMF3 component. IMF5, IMF6, IMF7, and IMF8 were clearly related to the tides. Therefore, IMF1–IMF4, which exhibited periodicity, were classified as ocean components, and the remaining IMFs were classified as stream components. Thus, the following equation is obtained:

$$OC = \mathrm{IMF1} + \mathrm{IMF2} + \mathrm{IMF3} + \mathrm{IMF4} \quad (7)$$

$$\mathrm{SC} = G - OC \quad (8)$$

where OC is the ocean component and SC is the stream component.

Table 2. Frequency (hr^{-1}) of the gauging stations.

Station	IMF1	IMF2	IMF3	IMF4	IMF5	IMF6	IMF7
Chung Cheng Bridge	0.0805	0.0805	0.0400	0.0218	0.0075	0.0049	0.0027
Shinhai Bridge	0.0805	0.0805	0.0415	0.0213	0.0098	0.0052	0.0025
Taipei Bridge	0.0805	0.0805	0.0388	0.0186	0.0100	0.0049	0.0025
Tudigonbi	0.0805	0.0805	0.0388	0.0174	0.0091	0.0049	0.0025

Table 3. The classical tidal type.

Type	Characterizes	Name	Cycle (Hour/Cycle)	Frequency (hr^{-1})
Semidiurnal tides	K_2	Lunisolar diurnal constituent	11.9672	0.0836
	S_2	Principal solar semidiurnal constituent	12.0000	0.0833
	M_2	Principal lunar semidiurnal constituent	12.4206	0.0805
	N_2	Large lunar elliptic semidiurnal constituent	12.6583	0.0775
	$2N_2$	Second large lunar elliptic semidiurnal constituent	12.9054	0.0790
Diurnal tides	K_1	Lunisolar diurnal constituent	23.9345	0.0418
	P_1	Principal solar diurnal constituent	24.0659	0.0416
	O_1	Principal lunar diurnal constituent	25.8193	0.0387
	Q_1	Large lunar elliptic constituent	26.8684	0.0372
Overtides	S_4	Solar quarter constituent	6.0000	0.1667
	M_4	Large lunar quarter constituent	6.2103	0.1610

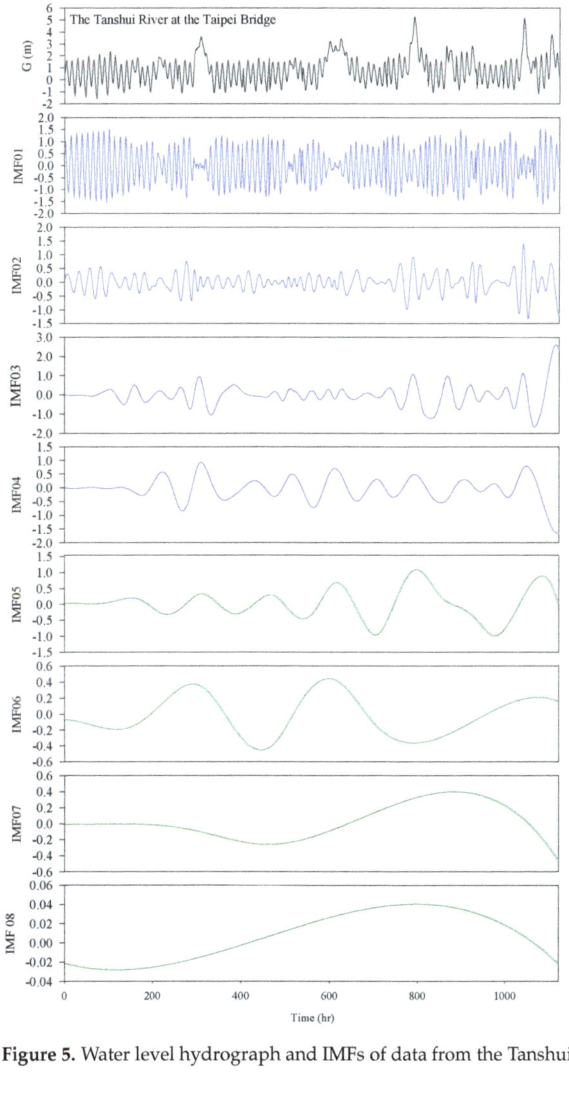

Figure 5. Water level hydrograph and IMFs of data from the Tanshui River at the Taipei Bridge.

Figure 6 presents the results of the EEMD decomposition of the water level at the Taipei Bridge into ocean and stream components. The results reveal how tides and upstream discharge affect the water level at the Taipei Bridge.

The lag time of the ocean components at the Taipei Bridge is related to the tides. Therefore, the regressors for forecasting the 1 h ahead ocean component (at $t + 1$) are the neighboring values of the ocean component at the Taipei Bridge and Tudigonbi for up to 3 h before the event (i.e., from $t - 2$ to t). A suitable linear regression model is given as follows:

$$OC_{T,\,t+1} = \beta_0 + \beta_1 OC_{T,t} + \beta_2 OC_{T,t-1} + \beta_3 OC_{T,t-2} + \beta_4 OC_{D,t} + \beta_5 OC_{D,t-1} + \beta_6 OC_{D,t-2} \quad (9)$$

where OC_T and OC_D indicate the forecasted ocean components at the Taipei Bridge and Tudigonbi, respectively; the subscripts $t - 2$, $t - 1$, t, and $t + 1$ indicate the time; and β_0, β_1, ..., β_6 are the regression coefficients. By fitting Equation (9) to the ocean component data of the calibration phase by using the stepwise regression method, the following equation is obtained:

$$OC_{T,\,t+1} = -0.004 + 1.791 OC_{T,t} - 0.456 OC_{T,t-1} + 0.213 OC_{T,t-2} - 1.052 OC_{D,t-1} \quad (10)$$

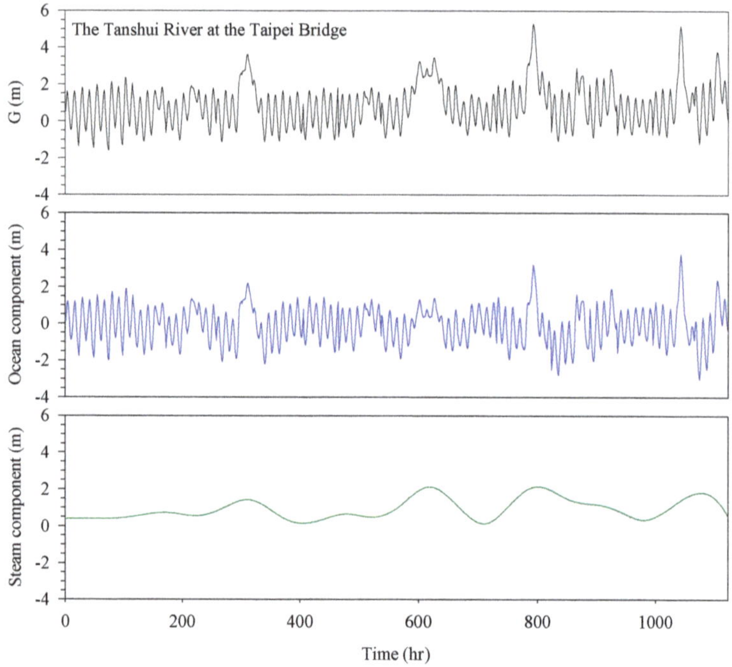

Figure 6. Water level, ocean component, and stream component hydrographs of the Tanshui River at the Taipei Bridge.

Figure 7 presents a comparison of the observed ocean component (OC_o) and forecasted ocean component (OC_p) and reveals that the water levels forecast through EEMD and stepwise regression are consistent with the observed water levels in the model calibration and verification processes. This figure also indicates that the proposed model can effectively reflect tidal dynamics.

Linear regression was conducted to forecast the stream component at the Taipei Bridge. The forecasted stream component at time $t + 1$ is a function of the stream components at the Chung Cheng Bridge and Shinhai Bridge at times t, $t - 1$, and $t - 2$. Stepwise regression was applied to produce the following stream component forecasting model:

$$SC_{T,\ t+1} = 0.042 + 1.333SC_{C,t} - 1.311SC_{C,t-1} + 0.503SC_{S,t} \tag{11}$$

where $SC_{T,t+1}$ is the forecasted stream component of the Taipei Bridge at time $t + 1$; $SC_{C,t-1}$ is the stream component of the Chung Cheng Bridge at time $t - 1$; and $SC_{C,t}$ and $SC_{S,t}$ are the stream components of the Chung Cheng Bridge and Shinhai Bridge at time t, respectively. Scatter plots of the observed and forecasted stream components in the calibration and verification phases are displayed in Figure 8. The terms SC_o and SC_p denote the observed and forecasted stream components, respectively. All the data points fall on or near the line of agreement between the observed and predicted results, which indicates the accuracy of the forecasted stream components.

Figures 9 and 10 present a comparison of the water levels forecast by the proposed model and the observed water levels in the calibration and validation phases. The forecasted water level is the sum of the forecasted ocean and stream components. The forecasted water levels of the proposed model are highly accurate. A comparison of the forecasted and observed water levels indicates that tidal amplitude, phase, and spring and neap tide modulations are accurately captured by the proposed model. Furthermore, the forecasted peaks are similar to the observed peaks. Therefore, the effect of floods on the water level in a tidal river can also be accurately forecast by the proposed EEMD model.

The quantitative metrics used for evaluating the accuracy of the proposed model were correlation coefficient (ρ) and root-mean-square error (RMSE), which are defined as follows:

$$\rho = \frac{\Sigma\,(G_p - \overline{G}_p)(G_o - \overline{G}_o)}{\sqrt{\Sigma\,(G_p - \overline{G}_p)^2\,\Sigma\,(G_o - \overline{G}_o)^2}} \tag{12}$$

$$RMSE = \sqrt{\frac{\Sigma\,(G_p - G_o)^2}{N}} \tag{13}$$

where G_p and G_o are the forecasted and observed water levels, respectively; $\overline{G}_p$ and $\overline{G}_o$ are the means of the forecasted and observed water levels, respectively; and N is the number of data sets. Table 4 lists the statistics corresponding to Figures 7–10. All correlation coefficients are close to unity. The RMSEs are between 0.10 and 0.17 m. These values are considerably smaller than the water level range. These statistical measures indicate that the proposed model is accurate, and its predictions are consistent with the observations; thus, this model can effectively forecast the water level in a tidal river.

Table 4. Summary of performance metrics carried out by comparing observations and forecasts.

	Phase	RMSE (m)	ρ
Ocean component	Calibration	0.10	0.986
	Verification	0.13	0.978
Stream component	Calibration	0.14	0.937
	Verification	0.12	0.969
Water level	Calibration	0.18	0.971
	Verification	0.17	0.976

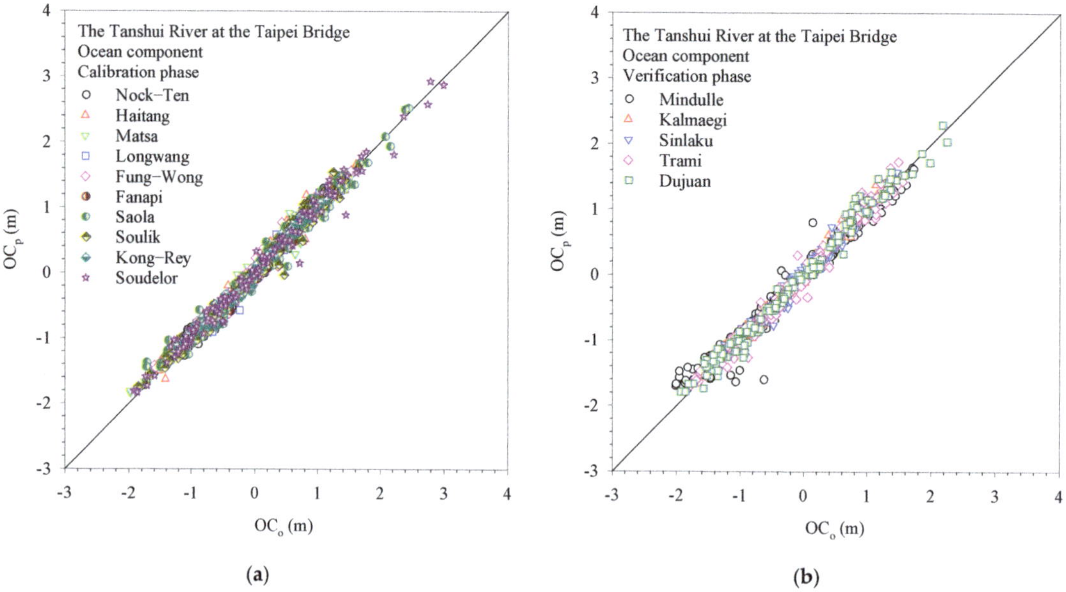

Figure 7. Accuracy of 1 h ahead ocean component forecasting during typhoons: (**a**) calibration phase; (**b**) Verification phase.

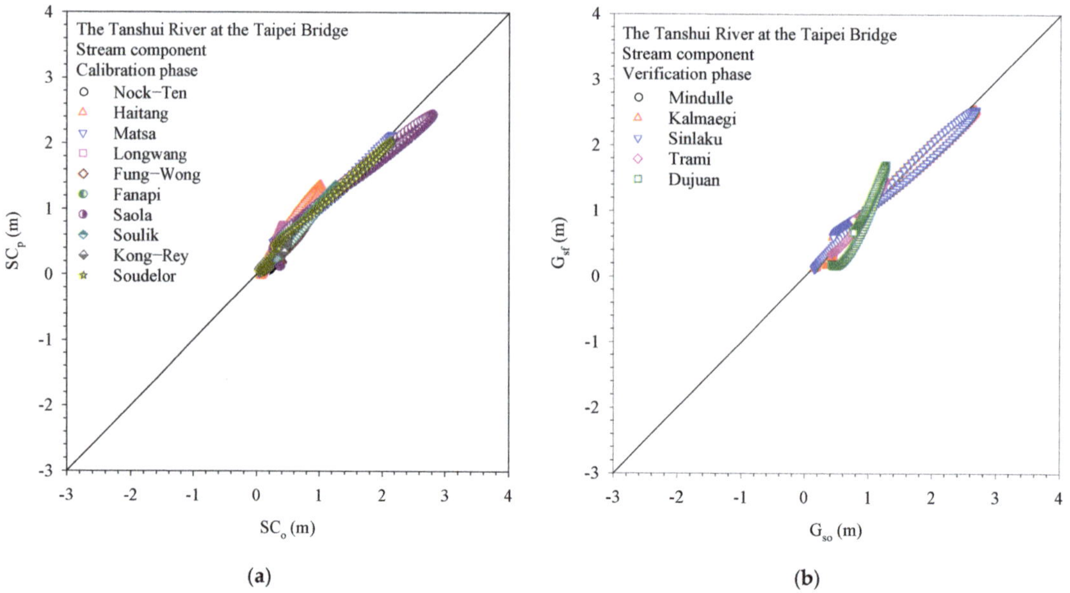

Figure 8. Accuracy of 1 h ahead stream component forecasting during typhoons: (**a**) Calibration phase; (**b**) Verification phase.

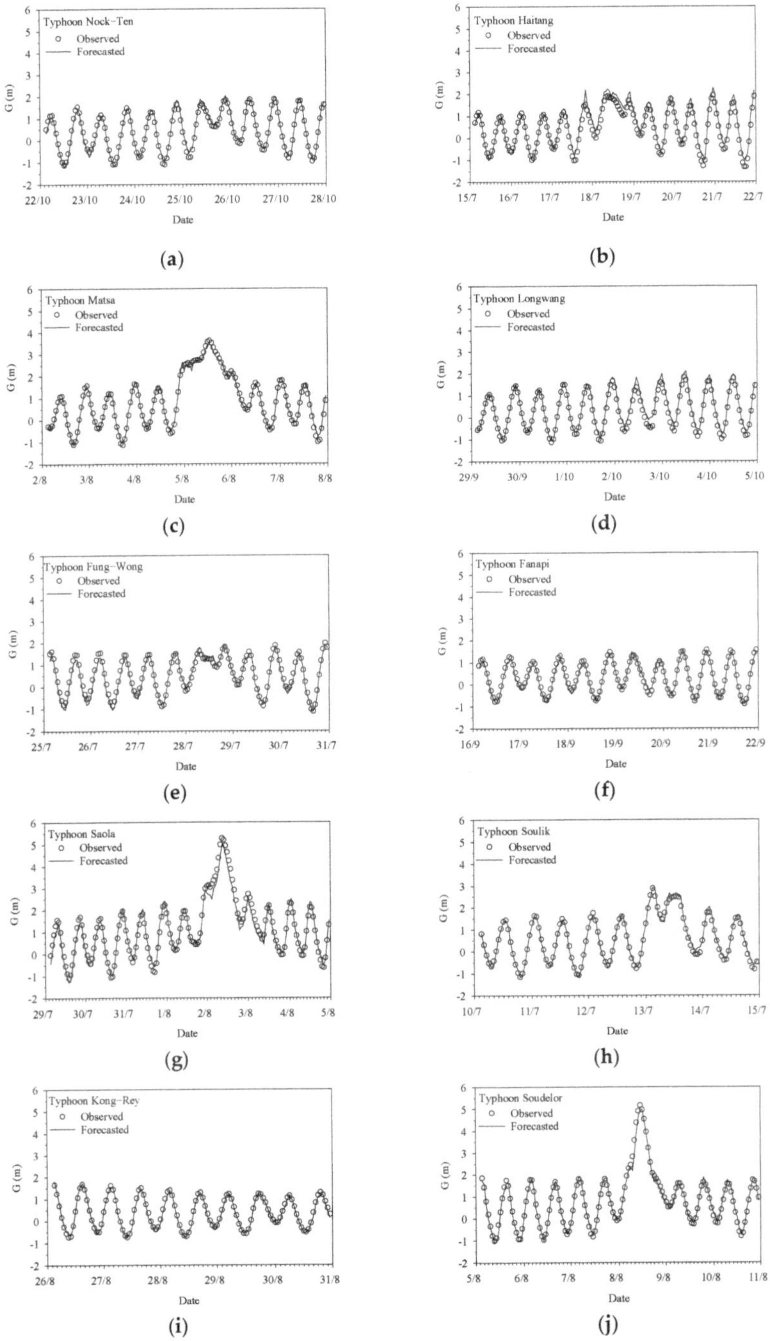

Figure 9. Comparison between 1 h ahead forecasted water levels and observed water levels during typhoons for calibration phase: (**a**) Typhoon Nock–Ten; (**b**) Typhoon Nock–Ten; (**c**) Typhoon Matsa; (**d**) Typhoon Longwang; (**e**) Typhoon Fung–Wang; (**f**) Typhoon Fannapi; (**g**) Typhoon Saola; (**h**) Typhoon Soulik; (**i**) Typhoon Kong–Rey; (**j**) Typhoon Soudelor.

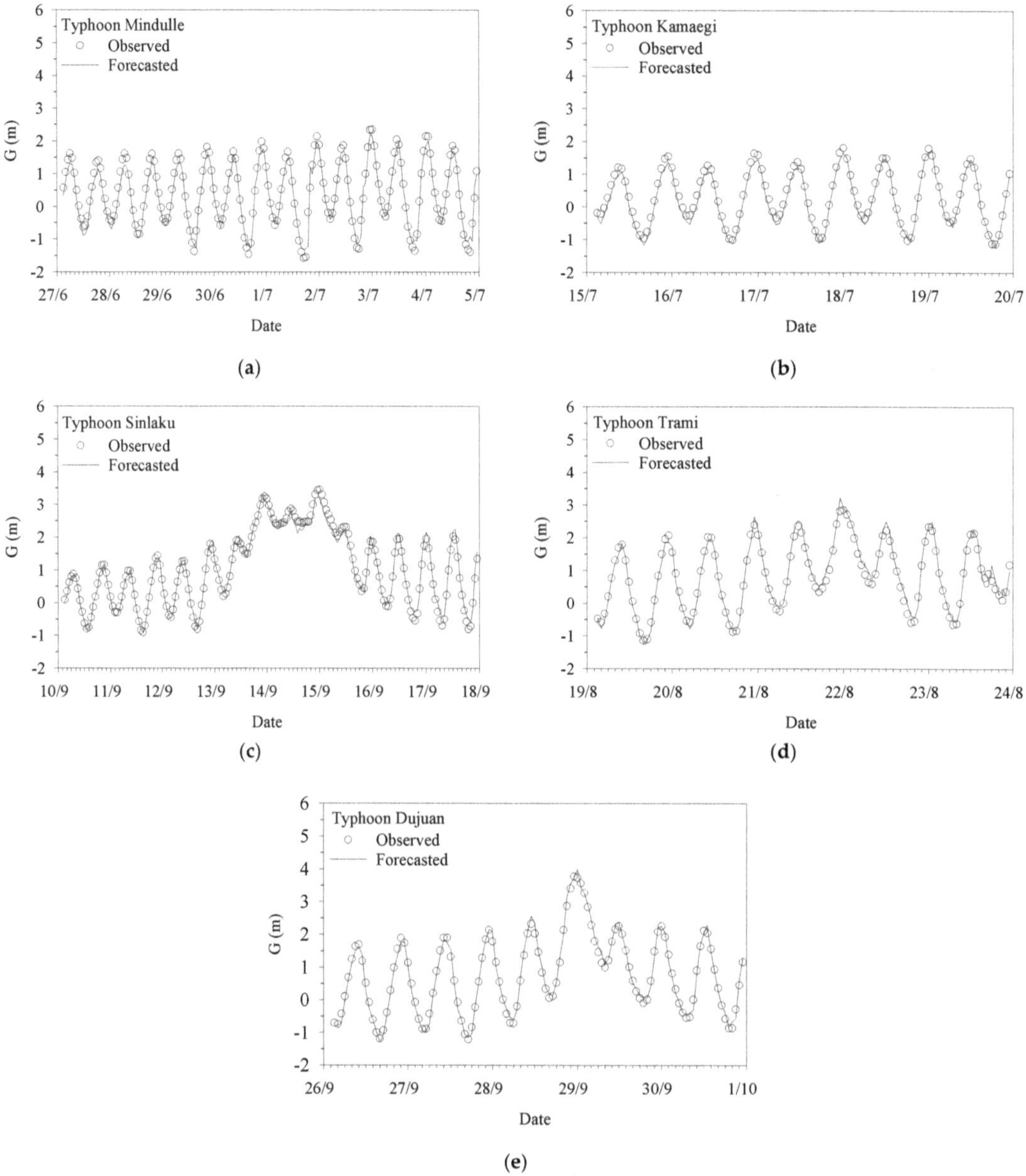

Figure 10. Accuracy of 1 h ahead water−stage forecasting during typhoons at verification phase: (**a**) Typhoon Mindulle l; (**b**) Typhoon Kamaegi; (**c**) Typhoon Sinlaku; (**d**) Typhoon Trami; (**e**) Typhoon Dujuan.

5. Summary and Conclusions

Numerous factors affect hydrological processes, and data collection in estuaries is challenging. Therefore, forecasting tidal river water levels is a difficult task. The proposed EEMD-based model is simpler than other hydrological and hydraulic models. EEMD does not require the numerous uncertain parameters used in other flooding simulation algorithms for forecasting water levels in tidal rivers, such as Manning's coefficient, channel bed elevation, energy slope, and cross-sectional area. The only input data required by the proposed model are water level data, which are comparatively easy to obtain. Moreover, the proposed simple model does not require complex theories or computations; only EEMD and stepwise regression are used. First, EEMD is used to decompose the water level into ocean and stream components as the regressors representing the two influential factors for the water level of tidal rivers: the tides and river flow. Estuarine water level forecasting can then be achieved by separately performing stepwise regression on the ocean and stream components at downstream and upstream locations, respectively, and summing the results for a target location.

A successful implementation of the proposed methodology was demonstrated in a case study of the Tanshui River, which is a tidal river. A water level forecasting model was constructed to forecast the 1 h ahead water level at the Taipei Bridge. The qualitative results, RMSEs, and correlation coefficients indicate that the developed model can achieve accurate water level forecasting during high-water-level periods in tidal rivers. Moreover, the clear physical meaning of each component reveals the simplicity and reliability of the proposed model.

The comparison of the proposed model and the other methods for forecasting water levels in tidal rivers, such as the Variational Mode Decomposition method, should be performed in the future. If additional data on tidal rivers can be obtained, water level components can be decomposed into other groups apart from only ocean and stream components, which can enable a more reliable and accurate model to be established for forecasting water levels in tidal rivers.

Author Contributions: Conceptualization, Y.-C.C.; Data collection, P.-Y.S., Formal analysis, S.-P.K.; Funding acquisition, Y.-C.C.; Investigation, S.-P.K. and P.-Y.S.; Methodology, Y.-C.C.; Supervision, H.-C.Y.; Validation, C.W.; Writing–original draft, H.-C.Y. All authors have read and agreed to the published version of the manuscript.

Funding: This article was based on work supported by the Ministry of Science and Technology, Taiwan (Grant no. MOST 111-2625-M-027-002-).

Data Availability Statement: The data presented in this study are available on request from the corresponding author or the Water Resources Agency, Taiwan.

Acknowledgments: The authors express their gratitude to the Ministry of Science and Technology, Taiwan (Grant no. MOST 111-2625-M-027-002-) for its support of this study.

Conflicts of Interest: The authors declare no conflict of interest.

References

1. Cameron, W.M.; Pritchard, D.W. Estuaries. In *The Sea*; Hill, N.N., Ed.; Harvard University Press: Cambridge, MA, US, 1963; Volume 2.
2. Nielsen, P. *Coastal and Estuarine Processes*; World Scientific: Singapore, 2009.
3. Lung, W.-S. *Water Quality Modeling for Wasteload Allocations and TMDLs*; Wiley: New York, NY, USA, 2001.
4. Ji, Z.-G. *Hydrodynamics and Water Quality–Modeling River, Lakes, and Estuaries*; Wiley: Hoboken, NJ, USA, 2008.
5. Hardisty, J. *Estuaries-Monitoring and Modeling the Physical System*; Blackwell: Malden, MA, USA, 2007.
6. Dronkers, J.; Leussen, W.V. *Physical Processes in Estuaries*; Springer: Heidelberg, Germany, 1988.
7. Dyer, K.R. *Estuaries—A Physical Introduction*; Wiley: New York, NY, USA, 2000.
8. Chang, F.-J.; Chen, Y.-C. Estuary water-stage forecasting by using radial basis function neural network. *J. Hydrol.* **2003**, *270*, 158–166. [CrossRef]
9. Chen, Y.-C.; Chiu, C.-L. An efficient method of discharge measurement in tidal streams. *J. Hydrol.* **2002**, *265*, 212–224. [CrossRef]
10. Hsu, M.-H.; Fu, J.-C.; Liu, W.-C. Flood routing with real-time stage correction method for flash flood forecasting in the Tanshui River, Taiwan. *J. Hydrol.* **2003**, *283*, 267–280. [CrossRef]
11. Bárcena, J.F.; García, A.; García, J.; Álvarez, C.; Revilla, J. Surface analysis of free surface and velocity to changes in river flow and tidal amplitude on a shallow mesotidal estuary: An application in Suances Estuary (Nothern Spain). . *Hydrol.* **2012**, *420–421*, 301–318. [CrossRef]
12. Saleh, F.; Ramaswamy, V.; Wang, Y.; Georgas, N.; Blumberg, A.; Pullen, J. A multi-scale ensemble-based framework for forecasting compound coastalriverine flooding: The Hackensack-Passaic watershed and Newark Bay. *Adv. Water Resour.* **2017**, *110*, 371–386. [CrossRef]

13. Meert, P.; Pereira, F.; Willems, P. Computationally efficient modelling of tidal rivers using conceptual reservoir-type models. *Enviorn. Modell. Softw.* **2016**, *77*, 19–31. [CrossRef]
14. Cuesta Cordoba, G.A.; Tuhovčák, L.; Tauš, M. Using artificial neural network models to assess water quality in water distribution networks. *Procedia Eng.* **2014**, *70*, 399–408. [CrossRef]
15. Shoaib, M.; Shamseldin, A.Y.; Melville, B.W. Comparative study of different wavelet based neural network models for rainfall–runoff modeling. *J. Hydrol.* **2014**, *515*, 47–58. [CrossRef]
16. Pham, B.T.; Le, L.M.; Le, T.T.; Bui, K.T.T.; Le, V.M.; Ly, H.B.; Prakash, I. Development of advanced artificial intelligence models for daily rainfall prediction. *Atmos. Res.* **2020**, *237*, 104845. [CrossRef]
17. Chen, W.-B.; Liu, W.-C.; Hsu, M.-H. Comparison of ANN approach with 2D and 3D hydrodynamic models for simulating estuary water stage. *Adv. Eng. Softw.* **2012**, *45*, 69–79. [CrossRef]
18. Zhang, X.-Q.; Bao, W.-M. Hydrodynamic simulation in tidal rivers using Fourier series. *J. Hydrol. Eng.-ASCE* **2013**, *18*, 1408–1415. [CrossRef]
19. Matte, P.; Secretan, Y.; Morin, J. Temporal and spatial variability of tidal-fluvial dynamics in theSt. Lawrence fluvial estuary: An application of nonstationary tidal harmonic analysis. *J. Geophys. Res.-Oceans* **2014**, *119*, 5724–5744. [CrossRef]
20. Jay, D.A.; Flinchem, E.P. Interaction of fluctuating river flow with a barotropic tide: A demonstration of wavelet tidal analysis methods. *J. Geophys. Res.* **1997**, *102*, 5705–5720. [CrossRef]
21. Wei, C.-C. Wavelet kernel support vector machines forecasting techniques: Case study on water-level predictions during typhoons. *Expert Syst. Appl.* **2012**, *39*, 5189–5199. [CrossRef]
22. Huang, N.E.; Long, S.R.; Sheng, Z. The mechanism for frequency downshift in nonlinear wave evolution. *Adv. Appl. Mech.* **1996**, *32*, 59–120. [CrossRef]
23. Huang, N.E.; Shen, Z.; Long, S.R.; Wu, M.C.; Shih, H.H.; Zheng, Q.; Yen, N.-C.; Tung, C.C.; Liu, H.H. The empirical mode decomposition and the Hilbert spectrum for nonlinear and non-stationary time series analysis. *Proc. Roy. Soc. Lond. A* **1998**, *454*, 903–995. [CrossRef]
24. Pan, H.; Guo, Z.; Wang, Y.; Lv, X. Application of the EMD method to river tides. *J. Atmos. Ocean. Technol.* **2018**, *35*, 809–819. [CrossRef]
25. Gan, M.; Pan, H.; Chen, Y.; Pan, S. Application of the Variational Mode Decomposition (VMD) method to river tides. *Estuar. Coast. Shelf Sci.* **2021**, *261*, 107570. [CrossRef]
26. Wu, L.-C.; Kao, C.C.; Hsu, T.-W.; Jao, K.-C.; Wang, Y.-F. Ensemble Empirical Mode Decomposition on storm surge separation from sea level data. *Coast. Eng. J.* **2011**, *53*, 223–243. [CrossRef]
27. Lee, H.S. Estimation of extreme sea levels along the Bangladesh coast due to storm surge and sea level rise using EEMD and EVA. *J. Geophys. Res. Oceans* **2013**, *18*, 4273–4285. [CrossRef]
28. Rezaie-Balf, M.; Kisi, O.; Chua, L.H.C. Application of ensemble empirical mode decomposition based on machine learning methodologies in forecasting monthly pan evaporation. *Hydrol. Res.* **2019**, *50*, 489–516. [CrossRef]
29. Roushangar, K.; Alizadeh, F. Scenario-based prediction of short-term river stage–discharge process using wavelet-EEMD-based relevance vector machine. *J. Hydroinform.* **2019**, *21*, 56–76. [CrossRef]
30. Flandrin, P.; Rilling, G.; Gonçalves, P. Empirical mode decomposition as a filter bank. *IEEE Signal Proc. Let.* **2003**, *11*, 112–114. [CrossRef]
31. Wu, Z.; Huang, N.E. A Study of the Characteristics of White Noise using the Empirical Mode Decomposition Method. *Proc. Roy. Soc. A* **2004**, *460*, 1597–1611. [CrossRef]
32. Huang, N.E. Introduction to the Hilbert-Huang transform and its related mathematical problems. In *Hilbert Huang Transform and Its Applications*; Huang, N.E., Shen, S.S., Eds.; World Scientific: Hackensack, NJ, USA, 2014. [CrossRef]
33. Chen, Y.-C.; Chen, R.-H.; Kao, S.-P.; Lin, M.-Y. Water stage component analysis in an estuary using the Hilbert Huang transform. *Terr. Atmos. Ocean. Sci.* **2018**, *29*, 215–229. [CrossRef]
34. Huang, N.E.; Wu, C.M.-L.; Long, S.R.; Shen, S.S.P.; Ou, W.; Gloersen, P.; Fan, K.L. A confidence limit for the Empirical Mode Decomposition and Hilbert spectral analysis. *Proc. R. Soc. A* **2003**, *459*, 2317–2345. [CrossRef]
35. Rilling, G.; Flandrin, P.; Goncalves, P. On empirical mode decomposition and its algorithms. In *2003 IEEE-EURASIP Workshop on Nonlinear Signal and Image Processing NSIP-03*; IEEE: Grado, Italy, 2003.
36. Huang, N.E.; Wu, Z. A review on Hilbert-Huang transform: Method and its applications to geophysical studies. *Rev. Geophys.* **2008**, *46*, RG2006. [CrossRef]
37. Chen, Y.-C.; Kao, S.-P.; Chiang, H.-W. Defining an estuary using the Hilbert-Huang transform. *Hydrolog. Sci. J.* **2013**, *58*, 841–853. [CrossRef]

hydrology

Article

Flood Exposure of Residential Areas and Infrastructure in Greece

Stefanos Stefanidis [1,*], Vasileios Alexandridis [2] and Theodora Theodoridou [3]

1 Laboratory of Mountainous Water Management and Control, School of Forestry and Natural Environment, Aristotle University of Thessaloniki, 54124 Thessaloniki, Greece
2 Independent Researcher, 54621 Thessaloniki, Greece
3 School of Spatial Planning and Development, Faculty of Engineering, Aristotle University of Thessaloniki, 54124 Thessaloniki, Greece
* Correspondence: ststefanid@gmail.com

Abstract: Worldwide, floods are the most common and widespread type of disaster during the 21st century. These phenomena have caused human fatalities, destruction of infrastructures and properties, and other significant impacts associated with human socioeconomic activities. In this study, the exposure of infrastructure (social, industrial and commercial, transportation) and residential areas to floods in Greek territory was considered. To accomplish the goal of the current study, freely available data from OpenStreetMap and Corine 2018 databases were collected and analyzed, as well as the flood extent zones derived under the implementation of the European Union's (EU) Floods Directive. The results will be useful for policy-making and prioritization of prone areas based not only on the extent of flood cover but also on the possible affected infrastructure types. Moreover, the aforementioned analysis could be the first step toward an integrated national-wide flood risk assessment.

Keywords: flood exposure; geospatial analysis; open-access data; infrastructure

Citation: Stefanidis, S.; Alexandridis, V.; Theodoridou, T. Flood Exposure of Residential Areas and Infrastructure in Greece. *Hydrology* **2022**, *9*, 145. https://doi.org/10.3390/hydrology9080145

Academic Editors: Aristoteles Tegos, Alexandros Ziogas and Vasilis Bellos

Received: 21 July 2022
Accepted: 12 August 2022
Published: 13 August 2022

1. Introduction

Floods are the most common type of natural disaster with devastating effects on local communities and infrastructure [1–4]. They can induce fatalities [5], major economic damage [6], and considerable effects on socioeconomic activities [7,8]. Thus, reliable flood risk assessment and resilience design of cities is a key priority for sustainable development. Despite the improvements in flood mitigation measures and technological advancements, floods continue to endanger human lives [9]. This is mainly due to the increasing human settlements and economic assets in floodplains, land-use change, and climate crisis [10,11].

The Sixth Assessment Report of the Intergovernmental Panel on Climate Change (IPCC) highlighted that extreme precipitation events will become more frequent in the near-future period over Europe [12]. Additionally, the natural water retention by land use is expected to decrease according to the forecasts of future urban land expansion [13]. Therefore, an increase in the likelihood and negative impacts of flood events is foreseen.

Floods are natural phenomena that cannot be prevented. Nevertheless, it is feasible and desirable to reduce their adverse outcomes, especially near residential areas and critical infrastructure. The costly floods that occurred at the beginning of the 21st century across Europe prompted the European Parliament to establish a Directive (2007/60/EC) on flood risk management. In the framework of this directive, the European Union (EU) Member States conducted flood risk management plans focused on the protection, prevention, and preparedness against flooding. Therefore, national-scale flood hazard maps were created, for different return period scenarios, by coupling hydrological and hydraulic modeling. Such maps provide crucial spatial information for flood risk assessment [14].

Several studies have been conducted on various aspects of floods. The majority of scholars look into post-flash flood analysis in terms of hydrological modeling and inundation mapping [15–19]. Nowadays, the use of Unmanned Aerial Vehicle (UAV) has been widely used as an alternative for post-flood surveys and data collection [20,21]. Moreover, the advantages of numerical weather prediction (NWP) models and rainfall radar were exploited, and flood forecasting and nowcasting approaches were developed [22–24]. Furthermore, numerous researchers have applied multi-criteria analysis (MCA) and machine learning (ML) techniques to provide flood susceptibility maps [25–28].

To the best of the author's knowledge, flood exposure analysis has garnered much global attention. However, flood exposure assessments of infrastructures are rare and focused on specific regions [29]. Large-scale approaches have been performed mainly in the United States [4,30], whereas, in Europe, the majority of the studies are focused on transportation networks. [3,31,32].

This study investigates the flood exposure of residential areas and infrastructure in Greece by combining open-access data with geospatial analysis. The proposed approach has the benefits of using easily accessible data, as well as simple and timeless GIS analysis for flood exposure assessment. Despite growing interest from academics and government agencies, this is the first quantitative nationwide assessment in the country. The outcomes provide insights for identifying areas where flood risk reduction should be prioritized. The methodology developed herein is easily transferable to other EU member states and can be scaled to a pan-European level.

2. Materials and Methods

2.1. Study Area

Greece is one of the EU's 27 member countries. It is located at the southern edge of the Balkan Peninsula (Southeast Europe), at the crossroads of Europe, Asia, and Africa, and shares borders with Albania to the northwest, Northern Macedonia, and Bulgaria to the north, and Turkey to the northeast. The Aegean Sea lies to the east of the mainland, the Ionian Sea to the west, and the Sea of Crete and the Mediterranean Sea to the south (Figure 1).

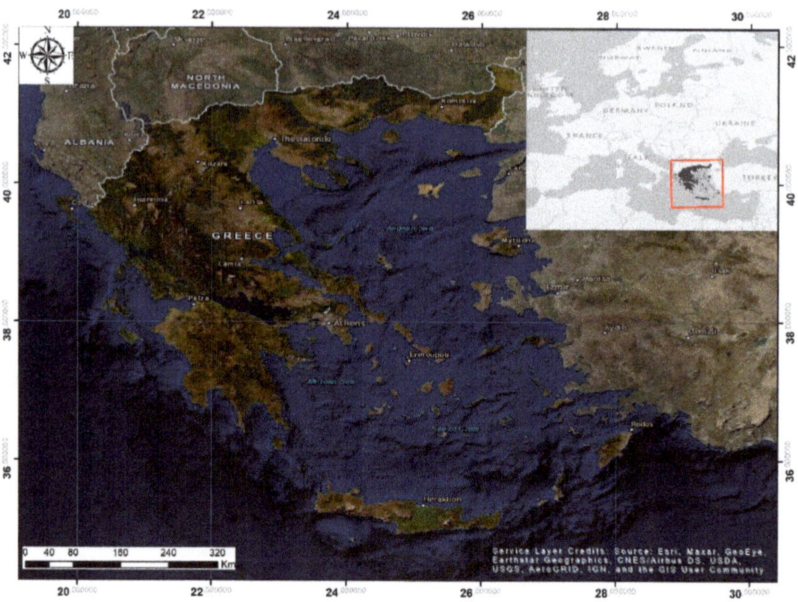

Figure 1. Location map of the study.

The country covers an area of approximately 132,000 km^2 and has a population of almost 10.7 million. It has a complex terrain, a highly diverse landscape, and the longest coastline in the Mediterranean (13,676 km), featuring numerous islands. According to the Köppen–Geiger climate classification, the climate is predominantly the temperate Mediterranean, with large areas of northern Greece classified as semi-arid and fewer regions, mostly at higher elevations, classified as humid continental [33]. However, due to the country's orography and climate type, precipitation over Greece presents great spatial and temporal variability. The precipitation pattern has significant seasonality, with the rainy season occurring in the fall, winter, and early spring and the dry season occurring throughout the summer months [34,35]. The Pindus Mountain range, which runs from northwest to southwest of the country, mainly affects the spatial variability of precipitation, and two distinct precipitation zones are determined. These are the wet zone to the west and the dry zone to the east [36]. Despite the fact that in the western part of Greece the highest amount of rainfall is recorded, most floods occur in the eastern part due to the proximity of urbanized areas to ephemeral torrential streams [37]. Also, the monthly distribution of flood events showed that November is the month with the richest flood records, followed by October [37].

2.2. Geospatial Analysis and Datasets

Flood exposure refers to valuable societal elements (such as people, infrastructure, etc.) located in floodplains [38]. The most common method is the spatial overlay between the flood hazard zones and assets. Spatial analysis of flood exposure presupposes the availability of geospatial data for assets and well-established flood hazard zones. This challenge is particularly addressed for national exposure analysis.

For the study's needs, various datasets were collected and processed. These datasets included residential areas, infrastructure, records of flood fatalities, and flood inundation maps. All the above datasets were organized in GIS thematic layers using the ArcGIS (v.10.7) software package. The outline of the methodology is presented in the following figure (Figure 2).

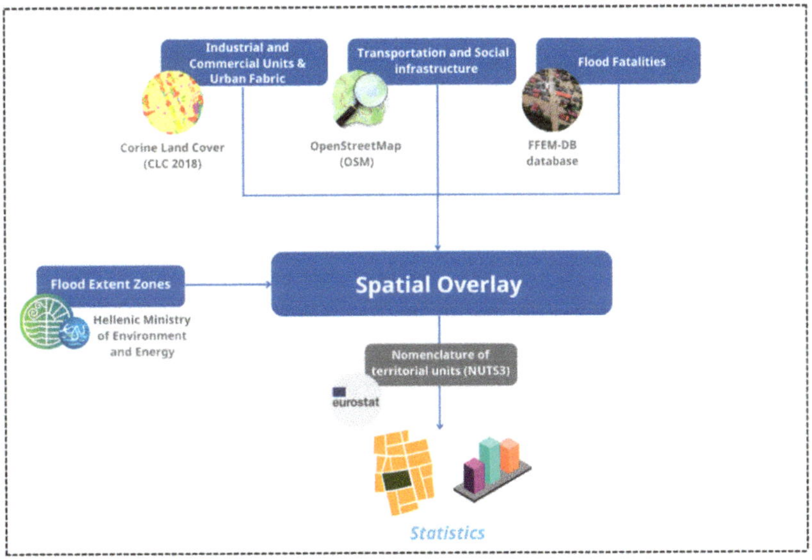

Figure 2. The overall workflow of the methodology.

Based on the Corine Land Cover (CLC 2018) dataset, the urban fabric (CLC codes: 1.1.1. & 1.1.2) and industrial and commercial units (CLC code: 1.2.1) were determined. The transportation infrastructure was extracted from the OpenStreetMap (OSM) dataset considering the major road types (motorway, trunk, primary and secondary roads) as well as the railway network. These features are nearly complete in OSM, since most European countries have more than 95% of their roads and railways mapped [39]. Additionally, OSM crowdsourced data is used to identify social infrastructure such as physical facilities and spaces where the community can access social services. These include health-care services, education and training, social housing programs, police, courts, and other systems for justice and public safety, as well as arts, cultural, and recreational facilities. To that end, the following vector data were exported and grouped: schools, universities, colleges, kindergartens, hospitals and clinics, nursing homes, community centers, sports centers, stadiums, campsites, archeological sites, monuments, art centers, theaters, museums, police and fire stations, court houses, airports and ports, and wastewater plants. Flood fatalities are analyzed by taking into account a recently developed dataset (FFEM-DB) for the Euro-Mediterranean region, covering the 1980–2020 period [40]. The flood hazard is represented by flood extent zones created as part of the implementation of the EU flood directive (2007/60/ EC) and the associate flood risk management plans. These maps are accessible through the Hellenic Ministry of Environment and Energy (Special Secretary for Water). The dataset includes three inundation depth maps corresponding to flood return periods of 50, 100, and 1000 years. In this analysis, the flood extent zones related to the probability of flood occurrence once 1 in 100 years were selected, as it is compatible with the national guidance on the design return period of flood defenses. Afterward, the Nomenclature of Territorial Units for Statistics Level 3 (NUTS 3) established by Eurostat was used for the comparative analysis of the results. A summary of the aforementioned datasets and their sources are presented in the following table (Table 1).

Table 1. Summary of the input datasets and sources used in this study.

Data	Dataset	Data Source	Data Accessibility	Format
Urban Fabric	Corine Land Cover (CLC 2018)	Copernicus Land Monitoring Service	https://land.copernicus.eu/pan-european/corine-land-cover/clc2018?tab=download (accessed on 10 October 2021)	vector
Industrial and Commercial Units				vector
Transportation Infrastructure	OpenStreetMap (OSM)	Geofabrik Download Server	https://download.geofabrik.de/europe/greece.html (accessed on 1 April 2022)	vector
Social Infrastructure				vector
Flood Fatalities Historical Records	Flood Fatalities of the Euro-Mediterranean region Database (FFEM-DB)	4TU Centre for Research Data	https://data.4tu.nl/articles/dataset/EUFF_2_0_European_Flood_Fatalities_database_/14754999/2 (accessed on 1 April 2022)	csv
Flood Extent Zones	Flood Risk Management Plans (2007/60/EC)	Hellenic Ministry of Environment and Energy (Special Secretary for Water)	http://floods.ypeka.gr:8080/geoserver/frmc2018100/wfs? (accessed on 15 November 2020)	vector
Nomenclature of Territorial Units for Statistics—level 3 (NUTS 3)	Eurostat	Geographic Information System of the Commission (GISCO)	https://ec.europa.eu/eurostat/web/gisco/geodata/reference-data/administrative-units-statistical-units/nuts (accessed on 1 April 2022)	vector

Analyzing flood exposure, the ratio of residential areas and infrastructure located in flood zones was estimated, considering the area of the urban fabric and industrial and commercial units, the length of transportation infrastructure, and the amount of social infrastructure.

3. Results and Discussion

The percentage coverage of flood extent zones per NUTS 3 provides an overview of the distribution of flood-prone areas over Greece, while historical records of flood fatalities give insights into areas where the surrounding environment may result in human losses during flood occurrences.

The highest coverage by flood extent zone is observed in Imathia (EL521) with a percentage equal to 24.3%, followed by Pella (EL524) and Florina (EL533) with percentages of 18.4% and 17.1%, all located in Northern Greece. Particularly high values (>10%) are also found in Karditsa and Trikala (EL611), Larissa (EL612), Kilkis (EL523), and Arta and

Preveza (EL541) (Figure 3). The results are justified by the fact that these areas are drained by large rivers and have correspondingly large floodplain areas.

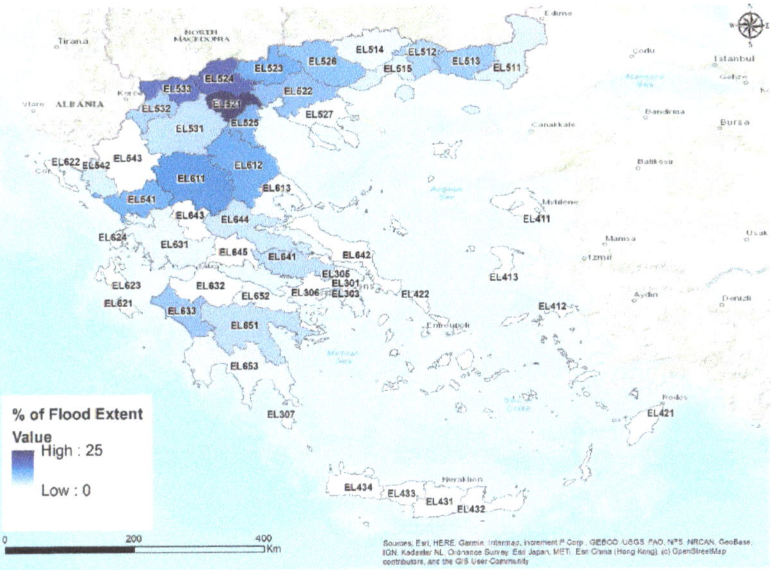

Figure 3. Spatial distribution of flood extent zones coverage per NUTS 3 in Greece.

On the contrary, the majority of flood fatalities were reported due to flash floods in ephemeral torrential streams [41]. Twenty-seven (27) deaths occurred in West Attika (EL306) mostly (21/27) as a consequence of the on 15 November 2017 (21/27) and twenty-one (21) deaths in Evia (EL642) as a result of two severe occurrences on 23 August 1990 (9/21) and 9 August 2020 (8/21). Furthermore, there were more than five deaths in the following areas: Cyclades Island (EL421), Argolida and Arkadia (EL651), Thessaloniki (EL522), Northern Athens (EL301), East Attica (EL305), and Corinthia (EL652). The distribution of findings shows that the deadliest floods occur in metropolitan centers and tourist areas (Figure 4). Economic development and population growth in these areas drive the expansion of built-up areas and human interventions within streambeds, intensifying flooding. Flood hazard assessment in such environments revealed that anthropogenic factors are the driving agents of flood genesis rather than natural factors [42]. Worth bearing in mind that most of these areas are typical wildland-urban interface (WUI) areas, as housing expands in and near forests [43]. Therefore, the probability of fire occurrence is higher. Despite the ecological disaster of a wildfire, flash floods follow due to the complete or partial loss of vegetation [44,45].

At a national level, the exposure ratio of residential areas and infrastructure located in flood zones are illustrated in the next figure (Figure 5) in ascending order. Only 5.5% of social infrastructures are located in flood zones at the lower end, compared to 12% of industrial and commercial units at the highest end. The ratio of urban fabric and transportation was found equal to 9.4% and 7.3%, respectively.

The spatial analyses show that the exposure ratios of the urban areas and infrastructures vary between NUTS 3. In general, northern and central Greece have the highest ratio in most of the examined categories, while particularly high values are also present in the Peloponnese (southern Greece).

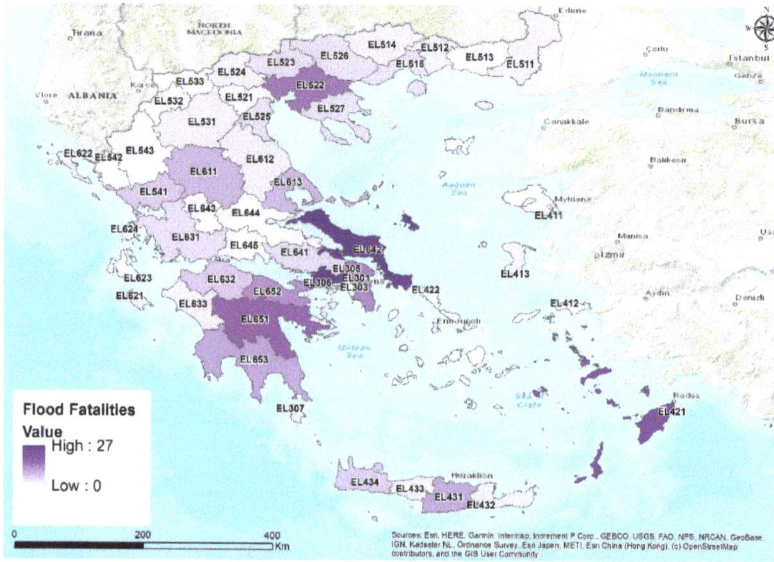

Figure 4. Spatial destitution of flood fatalities at the NUTS 3 level over Greece.

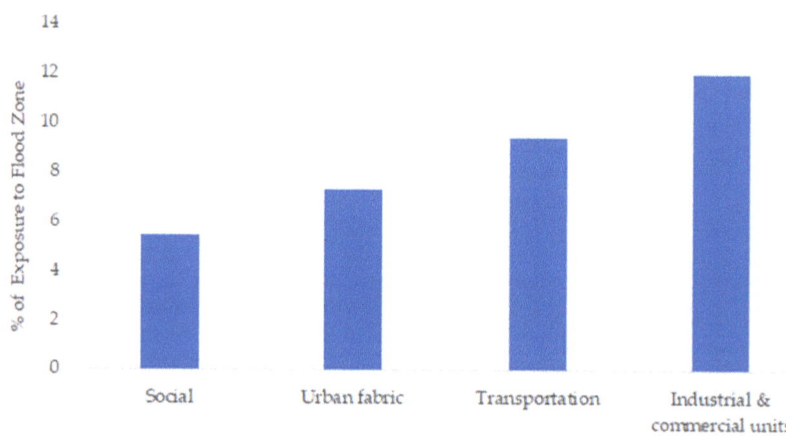

Figure 5. The ratios of residential areas and infrastructure located in flood zones in Greece.

The areas of an industrial and commercial unit are occupied by manufacturing, commerce, financial operations, and services. The existence of this infrastructure in floodplains affects various sectors of the economy, with cascading effects on the local community. As a result, methodologies for estimating commercial damage in flood risk assessments and developing probabilistic models suitable for pan-European applications using openly available data have been developed [46]. The flood exposure analysis of these areas revealed that the higher exposure ratio (37.6%) was found in Karditsa and Trikala (EL611), followed by 34.3% in Pella (EL524) and 33.6% in Argolida and Arkadia (EL651). Also, the two most populated metropolitan areas in Greece, the Central Athens sector (EL303) and Thessaloniki (EL522), have a large proportion of industrial and commercial units located in flood zones (29.2% and 28.5%, respectively). The spatial distribution and the analytical graphical representation of the results can be seen in Figures 6 and 7 respectively.

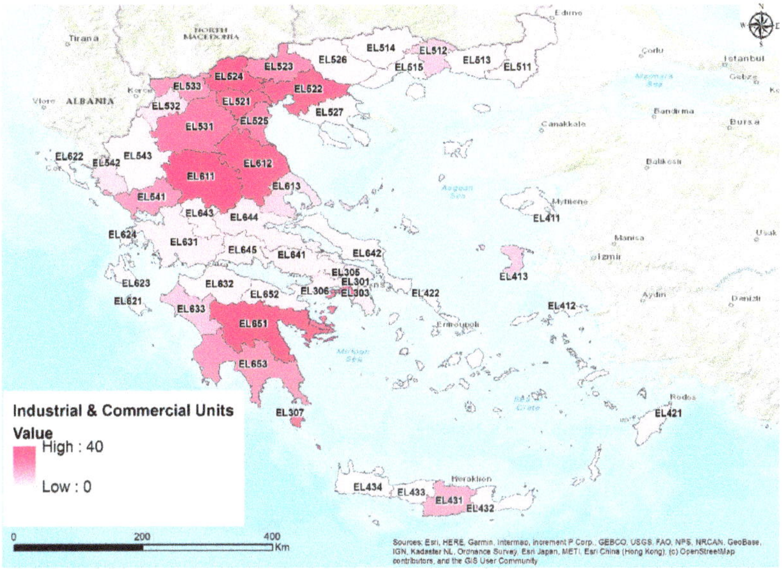

Figure 6. Spatial distribution of the ratio of industrial and commercial units in flood zones per NUTS 3.

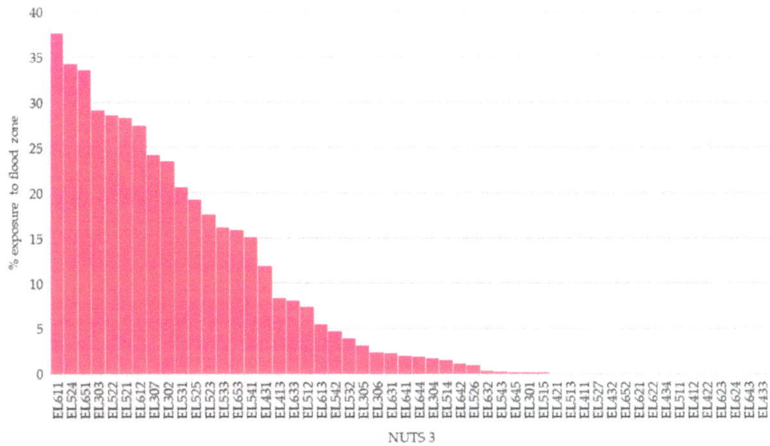

Figure 7. Graphical representation of the ratio of industrial and commercial units in flood zones per NUTS 3 in descending order.

Another crucial element, regarding flood risk, is the transportation infrastructure. The direct effects include material damage to infrastructure, disturbances in the traffic management systems, difficulties in evacuation and rescue operations, and last but not least, fatalities. Indirect effects may include passenger and cargo delay costs [47]. The accessibility of the road network during flood events is fundamental for evacuations and avoiding casualties [48]. Vehicle-related incidents account for an important part of flood fatalities both internationally [49,50] and in Greece [51]. It has also been acknowledged that individuals ignore warning signs or even drive into flooded waterways [52]. To that end, flood risk assessment of the transportation infrastructure is a necessity and integrated approaches have been applied [3]. Recently, national scale studies examined the resilience assessment of transport assets in a multi-hazard environment [53,54]. Our analysis emerged that 45.3% of transportation network length is located in the flood extent zone in Imathia

(EL521) and 43.0% in Pella, followed by Peiraeus Nisoi (EL307) (37.4%) and Thessaloniki (EL522) (23.4%). Rather high percentages (>20%) were also found in Argolida and Arkadia (EL651), Karditsa and Trikala (EL611), and Florina (EL533). The spatial distribution of the ratio transportation infrastructure located in the floodplain can be seen in Figure 8 and the graphical analysis of the results in descending order in Figure 9.

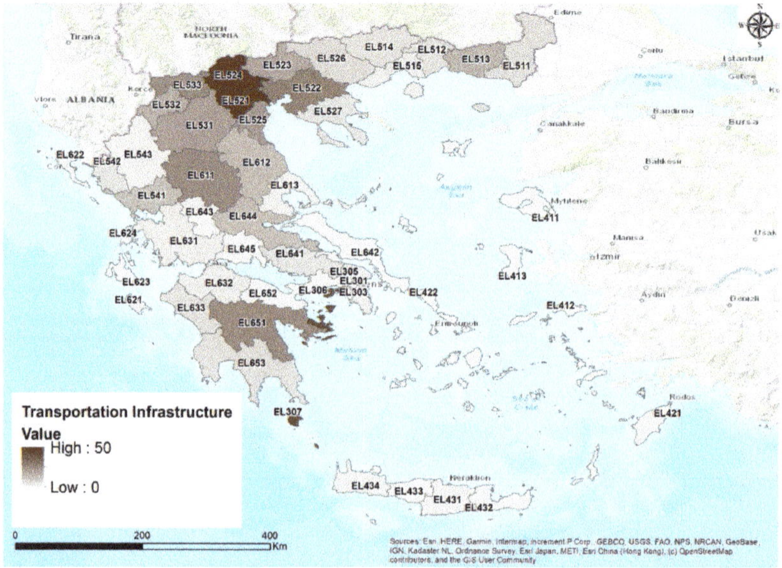

Figure 8. Spatial distribution of the ratio of transportation infrastructure in flood zones per NUTS 3.

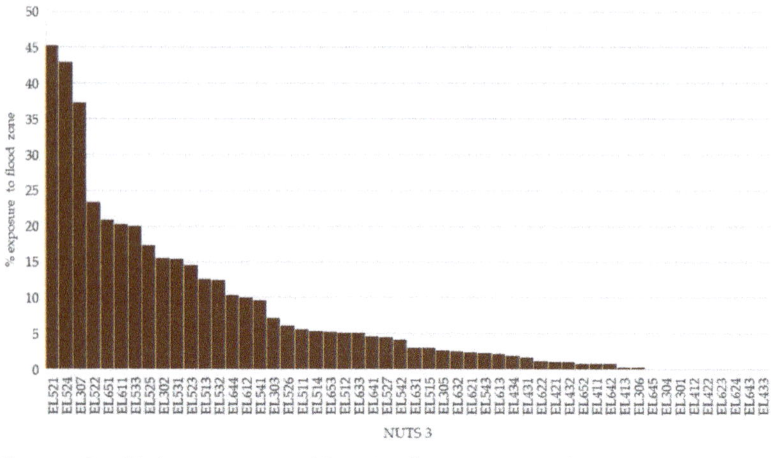

Figure 9. Graphical representation of the ratio of transportation infrastructure in flood zones per NUTS 3 in descending order.

The identification of residential areas located in floodplain zones is very important as it is directly related to economic damage to individuals' properties and is more likely to have adverse effects on local communities. Moreover, it can affect real estate values and be a tool in the housing market [55]. Currently, most homeowners are uninsured against flood damage, while the obligation for flood insurance is enforced when a purchase is completed through the establishment of a new bank loan. Insurance against floods should be a requirement for houses nearby ephemeral streams or rivers. The ratio and the spatial distribution of the urban fabric in flood zones could be the first step for the determination of the insurance fees [56]. The end-user, insurance companies, in this case, could use these data as services (DaaS). Regarding the Greek territory, the highest ratio of the urban fabric in flood zones (36.7%) was found in Imathia (EL521), followed by Florina (EL533) and Pella (EL524) with ratios equal to 35.8% and 31.4% respectively. Noteworthy that these were the regions with the largest flood extent zones. Also, high ratios, approximately 20.0% were recorded in Karditsa and Trikala (EL611) and Argolida and Arkadia (EL651) (Figures 10 and 11).

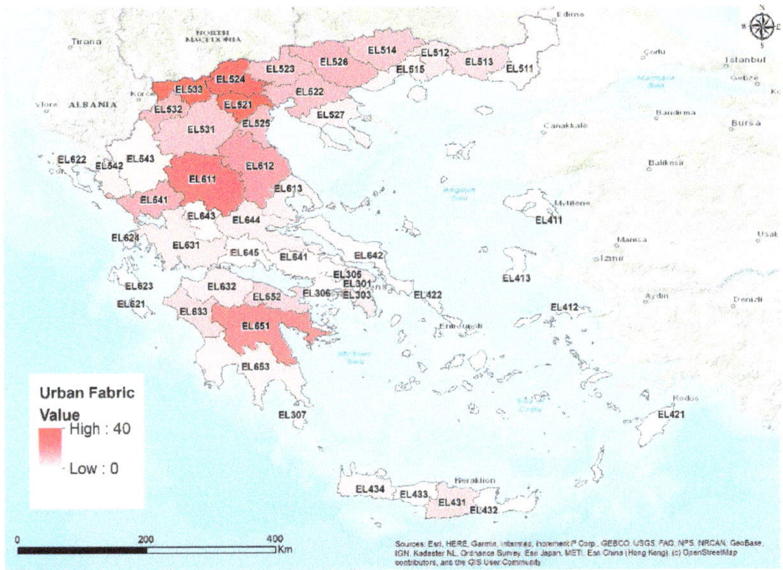

Figure 10. Spatial distribution of the ratio of the urban fabric in flood zones per NUTS 3.

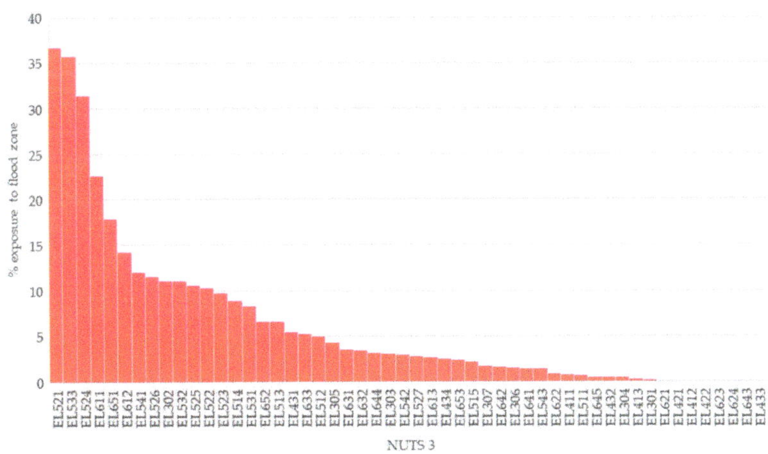

Figure 11. Graphical representation of the ratio of the urban fabric in flood zones per NUTS 3 in descending order.

Social infrastructures are related to national well-being and security. Due to their significance, reducing flood risk to these infrastructures has raised the concern of the scientific community [30]. The exposure of social infrastructure to flood endangers vulnerable groups of the population. In such places, the evacuation and rescue are more complex. Moreover, the damage to certain social infrastructure during flood events makes the coordination and operational function of local authorities more difficult. The geospatial analysis emerged that the highest ratio of social infrastructure in flood zones appeared in Larisa (EL612) (61.8%) followed by Pieria (EL525) (52.8%) and Argolida and Arkadia (EL651) (43.6%). Notably, seven other NUTS 3 units, namely Arta and Preveza (EL541), Pella (EL524), Kilkis (EL523) Laconia and Messenia (EL653), Magnisia (EL613), Florina (EL533) and Karditsa and Trikala (EL611), have more than 20% of their social infrastructure in floodplains. The spatial and graphical representation of the results are given in the following figures (Figures 12 and 13).

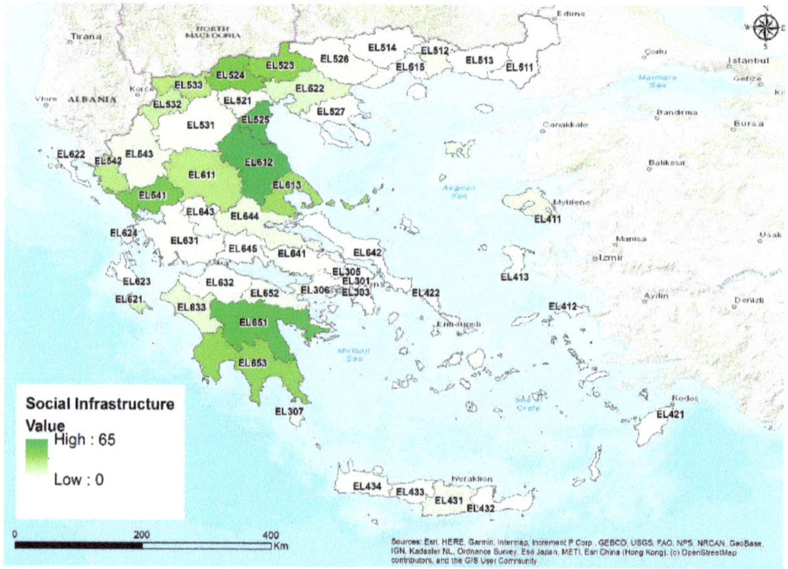

Figure 12. Spatial distribution of the ratio of social infrastructure in flood zones per NUTS 3.

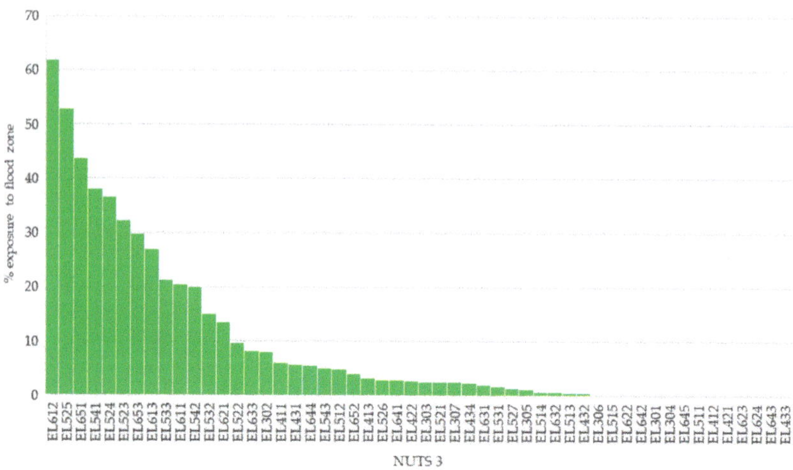

Figure 13. Graphical representation of the ratio of social infrastructure in flood zones per NUTS 3 in descending order.

Summarizing the results, it was found that Karditsa and Trikala (EL611), as well as Pella (EL525), had more than a 20% flood exposure ratio for all the examined types of infrastructures and urban fabric.

The analysis highlights critical infrastructure exposure to floods and identifies the areas with the highest ratios in the Greek territory. This research can be the first step toward an integrated physical and social vulnerability assessment [57]. Furthermore, it provides useful insights to stakeholders and policymakers for spatial planning and scheduling of flood prevention projects. Besides the classical structural measures, natural-based solutions must be considered, such as the management of forest ecosystems not only for wood production but also to enhance their protective role. Therefore, the protection of forests from abiotic and biotic disturbances in prone areas should be a priority to avoid vegetation damage in the mountainous watersheds, which subsequently increases flooding in the lowland areas. The findings of such studies should not be restricted to the scientific community but should be communicated to the general public in order to raise awareness about human interventions in streambeds and the protection of the environment as a flood prevention measure.

The spatial overlay of assets and infrastructure with floodplains is particularly important as it has cascading effects on local communities. These results could be a toolkit for local authorities, which are in charge of operational functions, obligations, and civil protection tasks for the protection of life, property, and the local economy. The knowledge of elements at risk facilitates procedures in prevention, preparedness, and response as well as enhances resilience at a local scale.

This knowledge sets the way for the introduction of nature-based solutions as local mitigation efforts move forward. The term "Nature-Based Solutions" (NBS) refers to a recent approach shift for flood risk management (FRM) towards solutions that employ elements, procedures, and management techniques that arise from nature to enhance water retention and reduce flooding [58]. They benefit low-level floods in smaller, more often flooded watersheds and help communities become more resilient to the effects of climate change, such as flooding. They also slow the passage of rain through the terrain into streams and rivers, preventing coastal flooding from tidal seas. Using nature-based solutions offer other benefits in addition to reducing flooding. For instance, they can reduce soil erosion in rivers and streams, increase species diversity in rivers and streams, and help fight global warming by storing carbon. Although nature-based solutions can lower the danger of flooding, they are not a component of traditional risk management [59]. More people must embrace nature-based solutions as the go-to infrastructure for combating climate change. These solutions should be viewed as important infrastructure to reduce climate change and safeguard our communities in order to build resilience to its effects.

Our approach is efficient on a national scale, although some limitations exist. The flood extent zones used in this study are derived from the Hellenic Flood Risk Management Plans (FRMP) conducted in the frame of the 2007/60/EC directive implementation. According to the project technical specifications, hydraulic modeling was not performed in streams with small watersheds (10 km^2), and floodplain areas of less than 25 km^2 were not further investigated unless significant historical flood records were reported. To that end, some streams were excluded from the analysis and are not considered herein. A detailed mapping of flood extent zones has to be conducted at a local scale and will be the basis for a holistic flood exposure analysis. A target of future research could be the expansion of the analysis to a pan-European scale and also evaluate the effect of flood exposure on land prices.

4. Conclusions

This study introduces the first nationwide spatial assessment of flood exposure in residential areas and infrastructures in Greece. Spatial analysis and open access data were coupled to illustrate the variation of flood exposure at the national and NUTS 3 levels. Specifically, the ratio of the urban fabric, transportation, social, industrial, and commercial infrastructures in 100-year flood zones was evaluated as well as the spatial pattern of the exposure. These categories were selected due to their devastating effects on local communities.

The flood exposure ratio of the aforementioned assets and facilities ranges from 5.5% to 12% at a national level. Nevertheless, some NUTS 3 level regions show particularly high ratios in certain categories. The results indicate that northern and central Greece generally have a high flood exposure ratio. Moreover, the outputs of this study detect places where further actions should be prioritized to evaluate and reduce flood risk.

The developed methodology could act as a roadmap for integrated flood risk assessment. The spatial results can be easily overlaid with other spatial data for further analysis, while the methodology is highly transferable as it is based on open-access geospatial data.

Author Contributions: Conceptualization, S.S.; methodology, S.S.; software, S.S.; formal analysis, S.S.; investigation, S.S.; data curation, S.S.; writing—original draft preparation, S.S.; writing—review and editing, S.S., V.A. and T.T. visualization, S.S., V.A. and T.T. All authors have read and agreed to the published version of the manuscript.

Funding: This research received no external funding.

Data Availability Statement: Not applicable.

Conflicts of Interest: The authors declare no conflict of interest.

References

1. Centre for Research on the Epidemiology of Disasters (CRED). CRED Crunch 66-Disasters Year in Review 2021. Available online: https://cred.be/sites/default/files/CREDCrunch66N.pdf (accessed on 10 May 2022).
2. Papilloud, T.; Röthlisberger, V.; Loreti, S.; Keiler, M. Flood exposure analysis of road infrastructure—Comparison of different methods at national level. *Int. J. Disaster Risk Reduct.* **2020**, *47*, 101548. [CrossRef]
3. Van Ginkel, K.C.; Dottori, F.; Alfieri, L.; Feyen, L.; Koks, E.E. Flood risk assessment of the European road network. *Nat. Hazards Earth Syst. Sci.* **2021**, *21*, 1011–1027. [CrossRef]
4. Porter, J.R.; Shu, E.; Amodeo, M.; Hsieh, H.; Chu, Z.; Freeman, N. Community Flood Impacts and Infrastructure: Examining National Flood Impacts Using a High Precision Assessment Tool in the United States. *Water* **2021**, *13*, 3125. [CrossRef]
5. Petrucci, O.; Aceto, L.; Bianchi, C.; Bigot, V.; Brazdil, R.; Pereira, S.; Kahraman, A.; Kikiç, O.; Kotroni, V.; Llasat, M.C.; et al. Flood Fatalities in Europe, 1980–2018: Variability, Features, and Lessons to Learn. *Water* **2019**, *11*, 1682. [CrossRef]
6. Merz, B.; Kreibich, H.; Schwarze, R.; Thieken, A. Review article "Assessment of economic flood damage". *Nat. Hazards Earth Syst. Sci.* **2010**, *10*, 1697–1724. [CrossRef]
7. Allaire, M. Socio-economic impacts of flooding: A review of the empirical literature. *Water Secur.* **2018**, *3*, 18–26. [CrossRef]
8. Giannaros, C.; Kotroni, V.; Lagouvardos, K.; Oikonomou, C.; Haralambous, H.; Papagiannaki, K. Hydrometeorological and socio-economic impact assessment of stream flooding in southeast mediterranean: The case of Rafina catchment (Attica, Greece). *Water* **2020**, *12*, 2426. [CrossRef]
9. Cornwall, W. Europe's deadly floods leave scientists stunned. *Science* **2021**, *373*, 372–373. [CrossRef] [PubMed]
10. Sassi, M.; Nicotina, L.; Pall, P.; Stone, D.; Hilberts, A.; Wehner, M.; Jewson, S. Impact of climate change on European winter and summer flood losses. *Adv. Water Resour.* **2019**, *129*, 165–177. [CrossRef]
11. Faccini, F.; Luino, F.; Paliaga, G.; Roccati, A.; Turconi, L. Flash Flood Events along the West Mediterranean Coasts: Inundations of Urbanized Areas Conditioned by Anthropic Impacts. *Land* **2021**, *10*, 620. [CrossRef]
12. IPCC. Climate Change 2021: The physical science basis. In *Contribution of Working Group, I to the Sixth Assessment Report of the Intergovernmental Panel on Climate Change*; Cambridge University Press: Cambridge, UK, 2021.
13. Myronidis, D.; Ioannou, K. Forecasting the urban expansion effects on the design storm hydrograph and sediment yield using artificial neural networks. *Water* **2018**, *11*, 31. [CrossRef]
14. Zhang, K.; Shalehy, M.H.; Ezaz, G.T.; Chakraborty, A.; Mohib, K.M.; Liu, L. An integrated flood risk assessment approach based on coupled hydrological-hydraulic modeling and bottom-up hazard vulnerability analysis. *Environ. Model. Softw.* **2022**, *148*, 105279. [CrossRef]
15. Myronidis, D.; Stathis, D.; Sapountzis, M. Post-Evaluation of flood hazards induced by former artificial interventions along a coastal Mediterranean settlement. *J. Hydrol. Eng.* **2016**, *21*, 05016022. [CrossRef]
16. Hdeib, R.; Abdallah, C.; Colin, F.; Brocca, L.; Moussa, R. Constraining coupled hydrological-hydraulic flood model by past storm events and post-event measurements in data-sparse regions. *J. Hydrol.* **2018**, *565*, 160–176. [CrossRef]
17. Senatore, A.; Davolio, S.; Furnari, L.; Mendicino, G. Reconstructing flood events in Mediterranean coastal areas using different reanalyses and high-resolution meteorological models. *J. Hydrometeorol.* **2020**, *21*, 1865–1887. [CrossRef]
18. Petrović, A.M.; Novković, I.; Kostadinov, S. Hydrological analysis of the September 2014 torrential floods of the Danube tributaries in the Eastern Serbia. *Nat. Hazards* **2021**, *108*, 1373–1387. [CrossRef]

19. Tegos, A.; Ziogas, A.; Bellos, V.; Tzimas, A. Forensic Hydrology: A Complete Reconstruction of an Extreme Flood Event in Data-Scarce Area. *Hydrology* **2022**, *9*, 93. [CrossRef]
20. Andreadakis, E.; Diakakis, M.; Vassilakis, E.; Deligiannakis, G.; Antoniadis, A.; Andriopoulos, P.; Spyrou, N.I.; Nikolopoulos, E.I. Unmanned aerial systems-aided post-flood peak discharge estimation in ephemeral streams. *Remote Sens.* **2020**, *12*, 4183. [CrossRef]
21. Vélez-Nicolás, M.; García-López, S.; Barbero, L.; Ruiz-Ortiz, V.; Sánchez-Bellón, Á. Applications of Unmanned Aerial Systems (UASs) in hydrology: A review. *Remote Sens.* **2021**, *13*, 1359. [CrossRef]
22. Furnari, L.; Mendicino, G.; Senatore, A. Hydrometeorological ensemble forecast of a highly localized convective event in the Mediterranean. *Water* **2020**, *12*, 1545. [CrossRef]
23. Spyrou, C.; Varlas, G.; Pappa, A.; Mentzafou, A.; Katsafados, P.; Papadopoulos, A.; Anagnostou, M.N.; Kalogiros, J. Implementation of a nowcasting hydrometeorological system for studying flash flood events: The case of Mandra, Greece. *Remote Sens.* **2020**, *12*, 2784. [CrossRef]
24. Bournas, A.; Baltas, E. Investigation of the Gridded Flash Flood Guidance in a Peri-Urban Basin in Greater Athens area, Greece. *J. Hydrol.* **2022**, *610*, 127820. [CrossRef]
25. Nachappa, T.G.; Piralilou, S.T.; Gholamnia, K.; Ghorbanzadeh, O.; Rahmati, O.; Blaschke, T. Flood susceptibility mapping with machine learning, multi-criteria decision analysis and ensemble using Dempster Shafer Theory. *J. Hydrol.* **2020**, *590*, 125275. [CrossRef]
26. Costache, R.; Pham, Q.B.; Sharifi, E.; Linh, N.T.T.; Abba, S.I.; Vojtek, M.; Vojteková, J.; Nhi, P.T.T.; Khoi, D.N. Flash-flood susceptibility assessment using multi-criteria decision making and machine learning supported by remote sensing and GIS techniques. *Remote Sens.* **2020**, *12*, 106. [CrossRef]
27. Pourghasemi, H.R.; Kariminejad, N.; Amiri, M.; Edalat, M.; Zarafshar, M.; Blaschke, T.; Cerda, A. Assessing and mapping multi-hazard risk susceptibility using a machine learning technique. *Sci. Rep.* **2020**, *10*, 3203. [CrossRef] [PubMed]
28. Abedi, R.; Costache, R.; Shafizadeh-Moghadam, H.; Pham, Q.B. Flash-flood susceptibility mapping based on XGBoost, random forest and boosted regression trees. *Geocarto Int.* **2021**, 1–18. [CrossRef]
29. Pant, R.; Thacker, S.; Hall, J.W.; Alderson, D.; Barr, S. Critical infrastructure impact assessment due to flood exposure. *J. Flood Risk Manag.* **2018**, *11*, 22–33. [CrossRef]
30. Qiang, Y. Flood exposure of critical infrastructures in the United States. *Int. J. Disaster Risk Reduct.* **2019**, *39*, 101240. [CrossRef]
31. Argyroudis, S.A.; Mitoulis, S.A.; Winter, M.G.; Kaynia, A.M. Fragility of transport assets exposed to multiple hazards: State-of-the-art review toward infrastructural resilience. *Reliab. Eng. Syst. Saf.* **2019**, *191*, 106567. [CrossRef]
32. Papilloud, T.; Keiler, M. Vulnerability patterns of road network to extreme floods based on accessibility measures. *Transp. Res. Part D Transp. Environ.* **2021**, *100*, 103045. [CrossRef]
33. Beck, H.E.; Zimmermann, N.E.; McVicar, T.R.; Vergopolan, N.; Berg, A.; Wood, E.F. Present and future Köppen-Geiger climate classification maps at 1-km resolution. *Sci. Data* **2018**, *5*, 180214. [CrossRef] [PubMed]
34. Livada, I.; Charalambous, G.; Assimakopoulos, M.N. Spatial and temporal study of precipitation characteristics over Greece. *Theor. Appl. Climatol.* **2008**, *93*, 45–55. [CrossRef]
35. Markonis, Y.; Batelis, S.C.; Dimakos, Y.; Moschou, E.; Koutsoyiannis, D. Temporal and spatial variability of rainfall over Greece. *Theor. Appl. Climatol.* **2017**, *130*, 217–232. [CrossRef]
36. Nastos, P.T.; Politi, N.; Kapsomenakis, J. Spatial and temporal variability of the Aridity Index in Greece. *Atmos. Res.* **2013**, *119*, 140–152. [CrossRef]
37. Diakakis, M.; Mavroulis, S.; Deligiannakis, G. Floods in Greece, a statistical and spatial approach. *Nat. Hazards* **2012**, *62*, 485–500. [CrossRef]
38. Koks, E.E.; Jongman, B.; Husby, T.G.; Botzen, W.J. Combining hazard, exposure and social vulnerability to provide lessons for flood risk management. *Environ. Sci. Policy* **2015**, *47*, 42–52. [CrossRef]
39. Barrington-Leigh, C.; Millard-Ball, A. The world's user-generated road map is more than 80% complete. *PLoS ONE* **2017**, *12*, e0180698. [CrossRef]
40. Papagiannaki, K.; Petrucci, O.; Diakakis, M.; Kotroni, V.; Aceto, L.; Bianchi, C.; Brázdil, R.; Gelabert, M.G.; Inbar, M.; Kahraman, A.; et al. Developing a large-scale dataset of flood fatalities for territories in the Euro-Mediterranean region, FFEM-DB. *Sci. Data* **2022**, *9*, 166. [CrossRef]
41. Diakakis, M. Characteristics of Infrastructure and Surrounding Geo-Environmental Circumstances Involved in Fatal Incidents Caused by Flash Flooding: Evidence from Greece. *Water* **2022**, *14*, 746. [CrossRef]
42. Stefanidis, S.; Stathis, D. Assessment of flood hazard based on natural and anthropogenic factors using analytic hierarchy process (AHP). *Nat. Hazards* **2013**, *68*, 569–585. [CrossRef]
43. Mitsopoulos, I.; Mallinis, G.; Dimitrakopoulos, A.; Xanthopoulos, G.; Eftychidis, G.; Goldammer, J.G. Vulnerability of peri-urban and residential areas to landscape fires in Greece: Evidence by wildland-urban interface data. *Data Brief* **2020**, *31*, 106025. [CrossRef] [PubMed]
44. Mitsopoulos, I.; Eftychidis, G.; Papathanasiou, C.; Makropoulos, C.; Mimikou, M. Assessing post fire flood risk potential in a typical Mediterranean Wildland-Urban Interface of Greece. In Proceedings of the International Conference on Changing Cities II Spatial, Design, Landscape & Socio-Economic Dimensions, Porto Heli, Peloponnese, Greece, 22–26 June 2015; pp. 1127–1135.

45. Karkani, A.; Evelpidou, N.; Tzouxanioti, M.; Petropoulos, A.; Santangelo, N.; Maroukian, H.; Spyrou, E.; Lakidi, L. Flash Flood Susceptibility Evaluation in Human-Affected Areas Using Geomorphological Methods—The Case of 9 August 2020, Euboea, Greece. A GIS-Based Approach. *GeoHazards* **2021**, *2*, 366–382. [CrossRef]
46. Wu, Z.; Lv, H.; Meng, Y.; Guan, X.; Zang, Y. The determination of flood damage curve in areas lacking disaster data based on the optimization principle of variation coefficient and beta distribution. *Sci. Total Environ.* **2021**, *750*, 142277. [CrossRef]
47. Pregnolato, M. Bridge safety is not for granted–A novel approach to bridge management. *Eng. Struct.* **2019**, *196*, 109193. [CrossRef]
48. Sohn, J. Evaluating the significance of highway network links under the flood damage: An accessibility approach. *Transp. Res. Part A Policy Pract.* **2006**, *40*, 491–506. [CrossRef]
49. Ahmed, M.A.; Haynes, K.; Taylor, M. Vehicle-related flood fatalities in Australia, 2001–2017. *J. Flood Risk Manag.* **2020**, *13*, e12616. [CrossRef]
50. Petrucci, O. Factors leading to the occurrence of flood fatalities: A systematic review of research papers published between 2010 and 2020. *Nat. Hazards Earth Syst. Sci.* **2022**, *22*, 71–83. [CrossRef]
51. Diakakis, M.; Deligiannakis, G. Vehicle-related flood fatalities in Greece. *Environ. Hazards* **2013**, *12*, 278–290. [CrossRef]
52. Hamilton, K.; Peden, A.E.; Pearson, M.; Hagger, M.S. Stop there's water on the road! Identifying key beliefs guiding people's willingness to drive through flooded waterways. *Saf. Sci.* **2016**, *89*, 308–314. [CrossRef]
53. Argyroudis, S.A.; Mitoulis, S.A.; Hofer, L.; Zanini, M.A.; Tubaldi, E.; Frangopol, D.M. Resilience assessment framework for critical infrastructure in a multi-hazard environment: Case study on transport assets. *Sci. Total Environ.* **2020**, *714*, 136854. [CrossRef]
54. Karatzetzou, A.; Stefanidis, S.; Stefanidou, S.; Tsinidis, G.; Pitilakis, D. Unified hazard models for risk assessment of transportation networks in a multi-hazard environment. *Int. J. Disaster Risk Reduct.* **2022**, *75*, 102960. [CrossRef]
55. Kousky, C.; Kunreuther, H.; LaCour-Little, M.; Wachter, S. Flood risk and the US housing market. *J. Hous. Res.* **2020**, *29* (Suppl. S1), S3–S24. [CrossRef]
56. Luke, A.; Sanders, B.F.; Goodrich, K.A.; Feldman, D.L.; Boudreau, D.; Eguiarte, A.; Serrano, K.; Reyes, A.; Schubert, J.E.; AghaKouchak, A.; et al. Going beyond the flood insurance rate map: Insights from flood hazard map co-production. *Nat. Hazards Earth Syst. Sci.* **2018**, *18*, 1097–1120. [CrossRef]
57. Rehman, S.; Sahana, M.; Hong, H.; Sajjad, H.; Ahmed, B.B. A systematic review on approaches and methods used for flood vulnerability assessment: Framework for future research. *Nat. Hazards* **2019**, *96*, 975–998. [CrossRef]
58. Raška, P.; Bezak, N.; Ferreira, C.S.S.; Kalantari, Z.; Banasik, K.; Bertola, M.; Bourke, M.; Cerdà, A.; Davids, P.; de Brito, M.M.; et al. Identifying barriers for nature-based solutions in flood risk management: An interdisciplinary overview using expert community approach. *J. Environ. Manag.* **2022**, *310*, 114725. [CrossRef]
59. Lallemant, D.; Hamel, P.; Balbi, M.; Lim, T.N.; Schmitt, R.; Win, S. Nature-based solutions for flood risk reduction: A probabilistic modeling framework. *One Earth* **2021**, *4*, 1310–1321. [CrossRef]

hydrology

MDPI

Article

Wetland Vulnerability Metrics as a Rapid Indicator in Identifying Nature-Based Solutions to Mitigate Coastal Flooding

Narcisa Gabriela Pricope [1,*] and Greer Shivers [1,2]

1 Department of Earth and Ocean Sciences, University of North Carolina Wilmington, 601 S College Rd., Wilmington, NC 28403, USA
2 Wilmington Urban Area Metropolitan Planning Organization (WMPO), City of Wilmington, 305 Chestnut Street, Wilmington, NC 28401, USA
* Correspondence: pricopen@uncw.edu; Tel.: +1-910-962-3499

Abstract: Flood mitigation in low-gradient, tidally-influenced, and rapidly urbanizing coastal locations remains a priority across a range of stakeholders and communities. Wetland ecosystems act as a natural flood buffer for coastal storms and sea level rise (SLR) while simultaneously providing invaluable benefits to urban dwellers. Assessing the vulnerability of wetlands to flood exposure under different SLR scenarios and vegetation responses to climatic variability over time allows for management actions, such as nature-based solutions, to be implemented to preserve wetland ecosystems and the services they provide. Nature-based solutions (NBSs) are a type of green infrastructure that can contribute to flood mitigation through the management and restoration of the ecosystems that provide socio-environmental benefits. However, identifying the flood mitigation potential provided by wetlands and the suitability for NBS implementation depends on the ecological condition and environmental exposure. We propose that wetland vulnerability assessments can be used as a rapid method to quantify changes in ecosystem dynamics and flood exposure and to prioritize potential locations of NBSs implementation. We quantified exposure risk using 100- and 500-year special flood hazard areas, 1–10 ft of sea level rise scenarios, and high-tide flooding and sensitivity using timeseries analyses of Landsat 8-derived multispectral indices as proxies for wetland conditions at subwatershed scales. We posit that wetland areas that are both highly vulnerable to recurrent flooding and degrading over time would make good candidate locations for NBS prioritization, especially when they co-occur on or adjacently to government-owned parcels. In collaboration with local governmental agencies responsible for flood mitigation in the coastal sub-watersheds of the City of New Bern and New Hanover County, North Carolina, we conducted field verification campaigns and leveraged local expert knowledge to identify optimal NBS priority areas. Our results identified several government-owned parcels containing highly vulnerable wetland areas that can be ranked and prioritized for potential NBS implementation. Depending on the biophysical characteristics of the area, NBS candidate wetland types include brackish and freshwater marshes and riverine swamp forests, even though the predominant wetland types by area are managed loblolly pinelands. This study underscores the critical importance of conserving or restoring marshes and swamp forests and provides a transferable framework for conducting scale-invariant assessments of coastal wetland condition and flood exposure as a rapid method of identifying potential priority areas for nature-based solutions to mitigate coastal flooding.

Keywords: wetlands; nature-based solutions; flood mitigation; coastal flooding; tidal watersheds

Citation: Pricope, N.G.; Shivers, G. Wetland Vulnerability Metrics as a Rapid Indicator in Identifying Nature-Based Solutions to Mitigate Coastal Flooding. *Hydrology* **2022**, *9*, 218. https://doi.org/10.3390/hydrology9120218

Academic Editors: Aristoteles Tegos, Alexandros Ziogas and Vasilis Bellos

Received: 27 October 2022
Accepted: 29 November 2022
Published: 2 December 2022

1. Introduction

Wetlands are essential ecosystems that provide value and services to society such as flood mitigation, pollutant sequestration, and valuable natural habitat areas [1–3]. Wetland ecosystems are widely responsible for the purification and infiltration of excess stormwater in rapidly urbanizing communities, resulting in decreases in permeable surface cover. A wetland that is 1 acre in size can store approximately one million gallons of water when in functional condition [4]. Wetland water storage capacity along with the slowed velocity of floodwaters moving through wetlands can lower flood amplitudes and reduce the potential for destruction caused by a flooding event [4]. Watersheds located in the temperate climate zone require at least 3–7% wetland land cover to provide adequate flood mitigation and suitable water quality to surrounding built and natural communities [2]. In 2011, the global monetary value of coastal wetland ecosystem services was estimated to be worth $20.4 trillion per year or 43.1% of global ecosystem service value per year despite accounting for roughly 15% of all natural wetlands [5]. While coastal wetlands may make up a small portion of globally distributed wetland types, the value and services that they provide society are far too great to be lost.

Natural ecosystems such as wetlands are subject to natural and anthropogenic pressures, including climate change, that increase vulnerability to natural hazards and result in decreased ecosystem resilience [6,7]. Vulnerability refers to the degree that a system is at risk to or unable to cope with impacts brought forth by climate change or other environmental stressors and is composed of exposure, sensitivity and adaptive capacity [7,8]. Resilience refers to a system's ability to rebound following a disturbance, such as a natural hazard, or cope with changes fueled by an altering global climate [9]. When occurring in tandem, natural and anthropogenic pressures on wetlands can lead to poor management practices, conversion to other land uses, or wholesale ecosystem destruction [10–12]. The marginal value of wetlands and the services they provide increase as populations increase, to a point, and then degrades to the point that ecosystem services including flood and pollution mitigation abilities are diminished or lost [2]. Decreases in wetland areas and their ecosystem service capacities increase risk to surrounding areas due to a decreased ability for wetlands to capture, store, and slow down inundation following a flood event. Identifying wetland areas that are experiencing decreases in vegetative health metrics and increases in flood risk is beneficial in determining possible management interventions to sustain ecosystem services in rapidly urbanizing coastal communities [10,11,13]. Geospatial technologies allow for the collection, inventory, and analysis of wetland vulnerability related to natural hazard and ecosystem-based data by entities seeking to assess a wetland's ability to perform ecosystem services [1,10,13]. Coastal communities frequently affected by natural hazards, such as hurricanes or storms, are at risk from extensive flooding and related infrastructure damage with long-term detrimental effects on the natural and built environments, as well as human livelihoods and life losses. The USA Atlantic coastal region experiences multiple types of flooding due to the presence of the Atlantic Ocean, rivers, and natural hazards that can frequently lead to compound flood risks and, as such, compound flooding needs that must be considered when devising management solutions to mitigate flood risk to coastal communities.

Typically, flood mitigation is managed by gray infrastructure solutions and more recently green infrastructure solutions. Gray infrastructure practices refer to traditional approaches to water management and natural hazards mitigation through human-engineered solutions, such as digging drainage ditches, creating concrete stormwater systems, and utilizing hardened structures like seawalls to protect coastlines [14,15]. Green infrastructure practices refer to hybridized infrastructure systems that improve societal and ecosystem resiliency to natural hazards concurrently by relying on natural ecosystems to address flooding. Stormwater wetlands, rain gardens, and permeable pavement are examples of the types of green infrastructure solutions being increasingly implemented in urban areas [14]. A subset of green infrastructure, nature-based solutions (NBSs), are management actions that require the use of ecosystems and their services to address societal issues, such as climate change impacts or flood mitigation and abatement [16]. Wetlands as a nature-based solution are highly effective at providing areas for water regulation while also providing recreation opportunities and thus present an optimal solution to flood mitigation in urban areas [17]. Prioritizing wetland ecosystems for restoration to aid in natural hazard reduction requires knowledge of what locations reduce exposure to flooding from natural hazards and future sea levels and protect vulnerable communities and infrastructure [18,19]. Necessary restoration steps can be taken to decrease the severity of flood damage by preserving landscapes proven to mitigate flood waters and implementing NBSs that promote healthy wetlands by identifying degraded wetlands based on site-specific metrics. Yet, for the introduction of NBSs to be a successful approach at mitigating flood inundation in at-risk wetland ecosystems, the process must also be cost effective in order to garner support from the implementing parties [20,21]. We posit that by assessing the vulnerability level of wetlands to increased flooding and as a function of long-term vegetative health metrics, rapid assessments of potential NBS suitability can be undertaken.

Loss of wetland cover and vegetation can be accelerated by increased frequencies of inundation events fueled by natural hazards, sea level rises, or saltwater intrusion into freshwater systems [22]. Even small fractions of wetland loss caused by sea level rise require management interventions to reestablish and prevent further degradation and, in this context, NBSs have proven to be a cost-effective way to leverage degrading wetland areas to enhance coastal resilience [22]. However, conducting wetland vulnerability assessments in highly populated coastal communities with fragmented wetland cover remains a challenge, particularly when high-resolution site-specific data may not be available and the complexity of a localized geospatial assessment may go beyond the skill sets of local jurisdictions. To determine the applicability of NBSs on a localized scale, the ability to perform a vulnerability assessment with localized data must be addressed first so that determining suitability for NBSs is not overwhelming to accomplish. While the IPCC has laid the groundwork for definitions and an approach to undertake ecosystem vulnerability assessments [8], it would be shortsighted to imply that a single approach is the only available method to conduct a vulnerability assessment [23].

Coastal wetlands can be difficult to monitor with traditional methods that are largely comprised of collecting in situ data due to resource needs and the reduced accessibility of wetland landscapes [24]. Utilizing remote sensing and geospatial technologies to monitor wetlands can help overcome some burdens of access and resource needs due to the presence of publicly available data and the ability to collect data in areas that are not easily accessible with the use of uncrewed aerial vehicles (UAVs), crewed aerial vehicles, and satellite technology [24]. While remote sensing and GIS can help alleviate some of the hardships of collecting wetland data, oftentimes the spatial and temporal resolution of the imagery are inappropriate for the scales necessary to assess localized, fragmented wetland status. Yet, existing satellite and aerial vehicle-collected data can provide medium to high resolution geospatial data at temporal resolutions high enough to use for accurate analysis [24].

We rely solely on freely and publicly available geospatial data to demonstrate a rapid approach to identifying vulnerable wetland sites and determine wetland locations where NBS implementation can be used to support flood mitigation efforts across a gradient of watershed development in a coastal area. We analyzed wetland condition and wetland restoration areas in New Hanover and the City of New Bern, NC by: (1) computing flood exposure levels to coastal flood inundation, (2) undertaking a time series analysis of Landsat 8 multispectral data to determine trends in vegetation health metrics from 2014–2021 (wetland sensitivity), (3) creating a combined wetland vulnerability assessment composed of both exposure and sensitivity results, and (4) analyzing and prioritizing suitable parcels for NBS implementation centered around highly vulnerable wetland areas.

2. Materials and Methods

2.1. Study Areas

The climate of North Carolina (NC) is humid and subtropical, marked by cold winters and warm summers that bring forth large amounts of precipitation, with total annual precipitation amounts varying from 34.8 inches (2007) to 68.4 inches (2018), and moderate to high vulnerability to tropical storms. Coastal NC experiences a hurricane level storm roughly once every 3 years because of its location along the coast of the Atlantic

Ocean in addition to smaller storms with damaging impacts stemming from winds, storm surge flooding, and heavy rainfall; rising sea levels have also contributed to increases in tidal flooding events that damage infrastructure by overwhelming the transportation and stormwater networks [25]. The low-lying topography of the coastal part of the state not only compounds flood vulnerability but also means that the region is characterized by expansive wetlands, such as managed loblolly pineland/upland forested wetlands (24.7% of all wetlands) or bottomland hardwood/riverine swamp forest (22.3%), composed primarily by flood-resistant species such as cypress, black gum, and red maple trees [26].

This research was designed in consultation with local government representatives in two NC municipalities at different stages of resilience and flood mitigation planning: the City of New Bern Development Services, where resilience measures to reduce flood impacts are well underway, and the New Hanover County (NHC) Office of Recovery and Resilience, where coordinated resilience planning is still in the early stages. We selected two HUC-12 subwatersheds in each of the two respective communities to conduct a wetland vulnerability analysis, engage repeatedly with relevant stakeholders, and create an NBS prioritization scheme. In New Hanover County, with input from NHC representatives, we selected the Smith Creek and Masonboro Island–Mason Inlet subwatersheds (Figure 1), which are both heavily populated, tidally-influenced urban watersheds characterized primarily by freshwater and salt/brackish marshes respectively that lacked any vulnerability assessment on exposure and sensitivity to natural hazards (Figures S1 and S2). For New Bern, we selected the City of New Bern–Trent River and City of New Bern–Neuse River (Figure 1), both urban, semi-tidally influenced watersheds characterized by the presence of primarily freshwater marsh and bottomland hardwood/riverine swamp forest mixed with managed pinelands located in Craven County (Figures S3 and S4). Both watersheds historically contain problematic areas for stormwater inundation and, hence, the City of New Bern, the NC Ecosystems Enhancement Program, and the NC Clean Water Management Trust Fund are currently working to create functional wetlands that can sequester floodwater inundation through NBSs in the Neuse River.

2.2. Data Sources

This research integrates existing publicly available geospatial data to create a wetland vulnerability assessment and determine exposure and sensitivity based on flood inundation scenarios and spectral indices that are used as proxies for vegetative health metrics. All data used in this study is summarized in Table 1 by source, data format, and location. The data is made available and described sequentially below in Section 2.3.

All data were projected to the North American Datum (NAD) 1983 StatePlane North Carolina Federal Information Processing Standards (FIPS) 3200 Feet and we used the most current North Carolina Department of Environmental Quality (NCDEQ) Wetlands GIS dataset as a location mask for this study since we focused on wetland areas specifically. Exposure and sensitivity results are shown only for locations deemed to be a wetland by this dataset. The sea level rise and Federal Emergency Management Agency (FEMA) special flood hazard areas in vector format were resampled to a common 30 m spatial resolution upon rasterization and inclusion into the final model.

2.3. Wetland Vulnerability Analysis

2.3.1. Flood Exposure

The data utilized to determine the exposure risk included 100- and 500-year FEMA flood zones, high-tide flooding, and sea level rise datasets provided by the National Oceanic and Atmospheric Administration (NOAA) (Table 1). The various exposure datasets provide potential flood inundation data that will present the overall flood risk of an area by examining multiple types and projections of flood hazards, not a single inundation circumstance. Utilizing FEMA flood zones to determine flood risk has the main drawback of not always accurately predicting every area that may experience flooding, but it is a nationally recognized dataset that influences government responses to flood events. While flooding can certainly occur outside of the typical 100- and 500-yr special flood hazard areas, locations in flood zones face a higher likelihood of recurrent flooding at the parcel levels. The NOAA high-tide flooding is the only feature already in raster format with a spatial resolution of 8.94 ft (2.72 m) for New Hanover County and 8.88 ft (2.70 m) for the City of New Bern. The 2 special flood hazard areas and 10 sea level rise scenarios, 1 through 10 ft, were rasterized to the same spatial resolution and coordinate system as the high-tide flooding layer for optimal overlay of raster cells and accuracy. Lastly, the ArcGIS Pro Cell Statistic tool was used to create a sum overlay between the 13 exposure datasets (Figure 2). The final summed dataset was then categorized into three grouped categories using a quantile classification scheme, with values 1 through 4 grouped and reclassified into a new low exposure class (1), where (1) indicates low flood exposure, values 5–8 were reclassified as (2) and indicated medium exposure to flood inundation, and values 9–12 grouped and reclassified into a high flood exposure class (3) [23]. The final exposure results were extracted to only the areas of wetlands determined by the NCDEQ wetland and wetland restoration areas datasets.

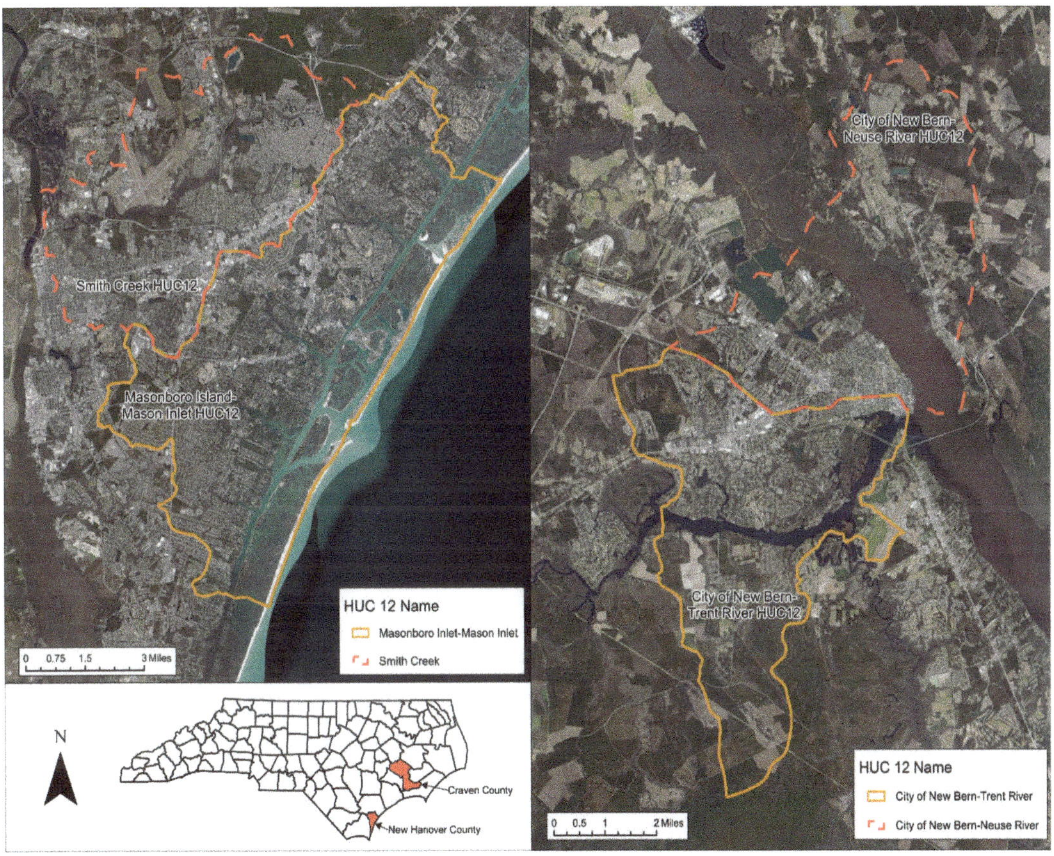

Figure 1. New Hanover County, NC study areas consisting of the Masonboro Island–Mason Inlet and Smith Creek subwatersheds.

Table 1. Wetland vulnerability assessment datasets and sources.

Dataset	Data Source	Data Format (Spatial Resolution)	Link to Dataset and Dataset Technical Documentation
High-tide flooding	NOAA	Raster (~2.7 m)	https://coast.noaa.gov/slrdata/ (accessed on 1 January 2022)
Sea level rise (1–10 ft)	NOAA	Vector	https://coast.noaa.gov/slrdata/ (accessed on 1 January 2022)
FEMA special flood hazard areas (100- and 500-yr)	NC FRIS	Vector	https://fris.nc.gov/fris/Home.aspx?ST=NC (accessed on 1 January 2022)
Landsat 8 multispectral imagery derived indices (NDVI, NDMI, and SAVI); 12 time steps 2014–2021	USGS	Raster (30 m)	https://earthexplorer.usgs.gov/ (accessed on 1 January 2022)
Wetlands and wetland restoration areas	NCDEQ	Vector	https://deq.nc.gov/about/divisions/coastal-management/coastal-management-gis-data/download-coastal-wetlands-spatial (accessed on 1 January 2022)
Parcels	NC One Map	Vector	https://www.nconemap.gov/pages/parcels (accessed on 1 January 2022)

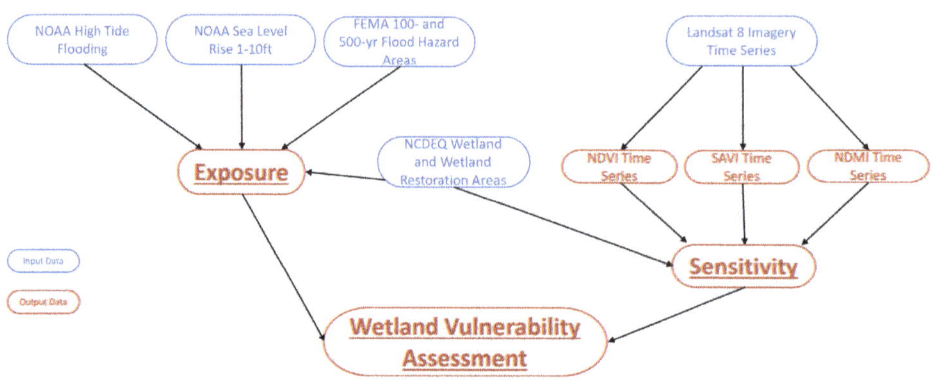

Figure 2. Summary of workflow steps to compute a wetland vulnerability assessment.

2.3.2. Wetland Vegetation Sensitivity

Wetland sensitivity was determined by identifying trends in vegetation indices derived from moderate-resolution Landsat 8 multispectral data for every year with data available since 2014, with less than 10% cloud cover at peak vegetative productivity (June to September). Based on drought data published by the US Drought Monitor produced by the National Oceanic and Atmospheric Association, the United States Department of Agriculture, and the University of Nebraska–Lincoln, no conditions above abnormally dry were reported for the dates when the Landsat 8 data was collected aside from one instance of slight moderate drought around the 07/27/2019 collection date [27]. No hurricanes occurred directly before or on the date that any of the LS imagery was collected. We computed yearly spectral indices to identify the chlorophyll concentrations present in the vegetation and vegetation moisture content for wetland areas within our study sites. By including the results of commonly utilized spectral indices for each study site over multiple years into a change detection analysis, changes in vegetative productivity and the directionality changes in productivity are identifiable and can be integrated into future management practices aiming to protect wetland ecosystems and their services. We posit that wetlands that are experiencing a decrease in productivity or condition can be readily identified and possible management can be introduced to restore the vegetation in a given area if there is a significant trend over time. We used three indices: the Normalized Difference Vegetation Index (NDVI), a widely used index to determine vegetation chlorophyll concentrations as a measure of productivity and a plant's ability to photosynthesize [28]. Secondly, we computed the Soil-Adjusted Vegetation Index (SAVI) because of its ability to measure chlorophyll present in live biomass while also mitigating the impacts of soil reflectivity [29]. Integration of SAVI in this coastal area is important because of the common presence of bare soil found in fluctuating tidal environments. Thirdly, we computed the Normalized Difference Moisture Index (NDMI) as a proxy for moisture in soils and vegetation and when monitoring vegetation disturbances [30,31]. Finally, we combined the three vegetation indices in the LandTrendr tool in ArcGIS Pro to compute changes over the time steps of Landsat data with a snapping date of 6/30 for the 12 time steps of imagery collected between 2014 and 2021 (Table S1). The LandTrendr output was a change analysis dataset that contains model coefficients that were then input into the Run Generate Trend Raster tool used to perform a Mann–Kendall significance test (Figure S5). We then utilized the z-score band to break the data into confidence intervals, creating three categorized values with 1 being increasing trends in vegetation index metrics, 2 being no trend, and 3 being decreasing trends.

2.3.3. Combined Wetland Vulnerability Assessment

Exposure and sensitivity results were combined and evaluated to determine where areas of high flood exposure and decreasing trends in vegetative condition overlap using the Weighted Sum tool whereby exposure was given a 55% weight while the sensitivity rasters will have a combined weight of 45% between the three indices (Figure 2). By having the exposure data at a slightly higher weight, areas of physical vulnerability to flooding will hold a slight dominance over decreases in vegetation health, which was deemed helpful when attempting to locate areas to implement flood-mitigating nature-based solutions [32,33]. Sekovski et al. (2020) determined the weights for their study by calling on experts in the environmental science field who were familiar with the localized processes of their study area to determine what extent each of their variables were contributing to coastal vulnerability. We then classified the results of the weighted sum analysis into low, medium, and high wetland vulnerability to inundation. Lastly, the highest risk areas were extracted to determine potential NBSs suitability relative to existing government-owned parcels under the assumption that areas characterized by a statistically significant decrease in vegetative health metrics between 2014–2021 and high exposure to flood risks are more susceptible to wetland loss.

2.3.4. Site Suitability for Nature-Based Solutions

To determine suitable sites for nature-based solutions based on the combined wetland vulnerability analysis, we used state parcel data created by NC One Map to find parcels that are already in government ownership, preferably unused, and located close to areas deemed highly vulnerable following analysis. The types of nature-based solutions being recommended for implementation following this study would be those that improve wetland ecosystem services and extent, such as bioretention, wetland restoration or the creation of new stormwater wetlands. The type of wetland nature-based solution able to be implemented is dependent on the size of a parcel deemed suitable. We conducted several site visits to locations in both New Hanover and Craven Counties during the month of July 2021 that consisted of geotagged photos and field inspections with relevant managing stakeholders to identify areas of the highest recurrent flood risk as well as expert-proposed candidates for NBS implementation following the presentation of our modeling results. The site visits not only provided critical on-the-ground information on existing risk and vulnerability to the built infrastructure but helped further solidify our working relationships with the managing stakeholders and potential decision makers in adopting and implementing NBSs at the selected sites.

3. Results

As anticipated, in both New Hanover County and the City of New Bern, all four subwatersheds examined show the highest flood exposure risk for wetlands located on the coast and near the major hydrologic features in the region. In the Masonboro Island–Mason Inlet subwatershed, the majority of the wetlands on the barrier islands and directly surrounding the tidal creek hydrologic features are exposed to flood inundation (Figure 3). The main portions of the tidal creeks and barrier islands marking the coastline show high flood exposure risk with 20.3% of the entire subwatershed at some risk to flood inundation (Table 2). All the tidal creeks located within the Masonboro Island–Mason Inlet subwatershed are also highly populated areas that are historically at risk from repeated flooding during relatively small storm events or even sunny day or high-tide flooding. The Smith Creek subwatershed experiences the highest flood risk along wetlands located close to the Cape Fear River (Figure 3) and experiences compounding flood risks, such as a combination of tropical storms, riverine flooding, and sea level rise exacerbated by high-tide flooding. While the inland portion of Smith Creek shows a lower flood exposure overall, the presence of cleared, drained, and cutover wetland types in the area could contribute to decreased flood sequestration due to lack of functional wetlands able to perform ecosystem services (see Figure S2). In total, 7.4% of the Smith Creek subwatershed experiences some level of flood exposure (Table 2) as does the downtown area of the city of Wilmington, and the communities of Kings Grant, Forest Hills, and Murrayville.

In the City of New Bern–Neuse River and the City of New Bern–Trent River subwatersheds, the highest flood exposure characterizes wetland areas along the Neuse River and mouth of the Trent River (Figure 3). In the City of New Bern–Neuse River subwatershed, 14.7% is exposed to recurrent flooding, with 5.1% of the entire subwatershed falling into the high exposure level (Table 2). The wetland areas with highest vulnerability to flooding are those found directly at the mouth of the Trent River where it branches off from the larger Neuse River. This subwatershed contains the historic downtown New Bern along with a vast majority of the city's incorporated areas and citizens. Overall, 5.7% of the City of New Bern–Trent River subwatershed is exposed to tidal and SLR-induced floods, with 2.9% of the entire subwatershed falling into the high exposure category (Table 2). High flood exposure makes up the largest area of flood exposure in the City of New Bern–Trent River subwatershed followed by low flood exposure composing 1.9% of the entire subwatershed. The Neuse River Bridge dissects through wetlands that show to have a high flood exposure on the western side of the structure and a mixture of low to medium flood exposure on the eastern side. Having a major transportation structure located within areas with high exposure to flooding can lead to infrastructure damage or failure over time as sea level rising scenarios and natural hazards causing flood inundation to continue to amplify [34]. These analyses do provide a cursory and standardized approach to visualizing areas of the highest risk and exposure to compound inundation and thus represent useful planning tools in the process of NBS suitability assessment.

Our next step involved conducting a 12-year vegetation change analysis aimed at quantifying areas of vegetation condition decline, stability, or improvements for our four subwatersheds focusing solely on wetlands as defined by the NCDEQ. Even though we calculated trends (and their relative statistical significance) based on SAVI (Figure S6) and NDMI (Figure S7) indices and all three vegetation indices directionality are included in our final combined vulnerability metric, below we only show and discuss in detail the results of the NDVI trend analysis. In the Masonboro Island/Inlet area, more than 24% of the wetlands included in our analysis show a decreasing vegetation trend based on the time series of NDVI extracted from Landsat imagery (Table 3 and Figure 4). This area was highlighted as having a decreased vegetation condition that is contiguous along the barrier island and the main stems of the major tidal creek inlets, and characterizes a much larger proportion of the study area compared to all of the other three study locations. For the other locations, the majority of wetlands analyzed exhibit no significant trend in vegetation response, while only marginal increases in vegetation response are recorded, with the highest proportion occurring in the Smith Creek watershed region. For our Craven County study locations, we see a similar pattern of highest vegetation response declines in the tidally influenced zones, although the Neuse River is located a distance away from the coast (Figure 4).

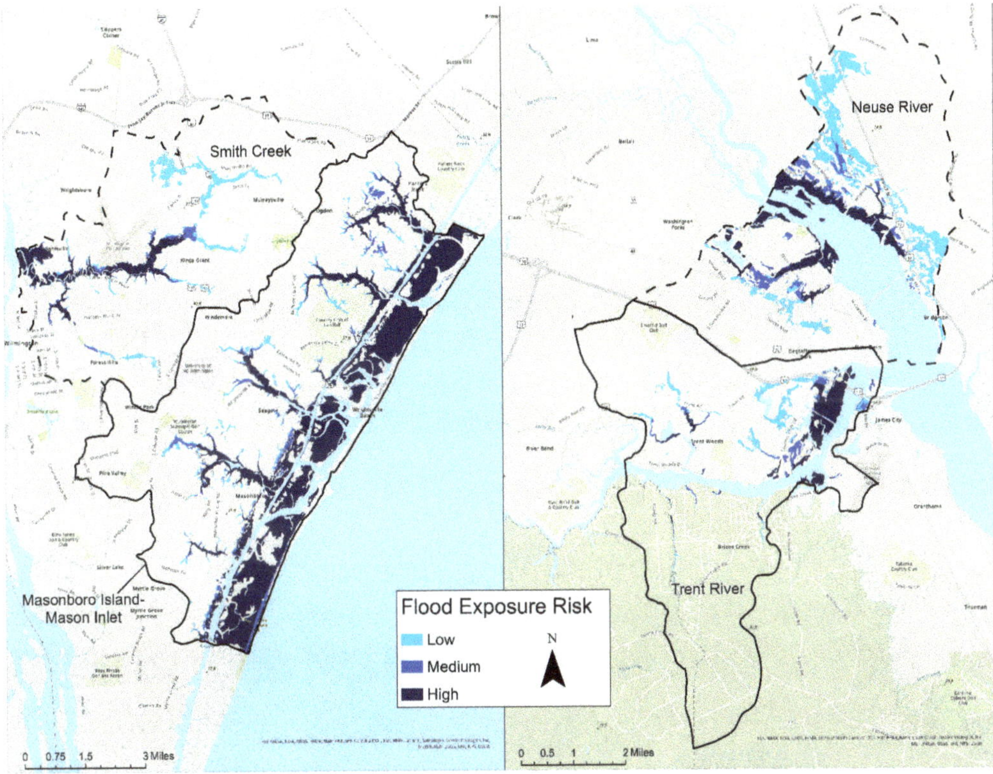

Figure 3. Flood exposure results of the Masonboro Island–Mason Inlet and Smith Creek subwatersheds in New Hanover County, NC (**left map**) and City of New Bern–Neuse River and the City of New Bern–Trent River subwatersheds in New Hanover County, NC (**right map**).

Table 2. Flood exposure risk for the four study areas as total area (acres) and a percentage of the HUC affected.

Masonboro Island–Mason Inlet			**Smith Creek**	
Flood exposure level	Area impacted	% of HUC-12 impacted	Areas impacted	% of HUC-12 impacted
Low	214.0	1.6%	262.5	3.1%
Medium	209.3	1.6%	68.2	0.8%
High	2265.5	17.1%	300.6	3.5%
City of New Bern–Neuse River			**City of New Bern–Trent River**	
Flood exposure level	Areaimpacted	% of HUC-12 impacted	Areaimpacted	% of HUC-12 impacted
Low	428.5	7.5%	108.6	1.9%
Medium	118.0	2.1%	55.0	0.9%
High	293.7	5.1%	170.6	2.9%

Table 3. Summary of the NDVI-based time series analysis z-score results for our four study areas in acres and as a percentage of the HUC-12 impacted.

Masonboro Island–Mason Inlet			**Smith Creek**	
Time series z-score trend	Area impacted	% of HUC-12 impacted	Area impacted	% of HUC-12 impacted
Increasing	192.3	1.4%	156.4	1.8%
No Trend	2225.4	16.8%	2323.9	27.2%
Decreasing	3226.1	24.3%	1254.5	14.7%
City of New Bern–Neuse River			**City of New Bern–Trent River**	
Time series z-score trend	Area impacted	% of HUC-12 impacted	Area impacted	% of HUC-12 impacted
Increasing	94.8	1.6%	39.7	0.7%
No Trend	1268.0	22.0%	1224.1	20.9%
Decreasing	732.4	12.7%	628.2	10.7%

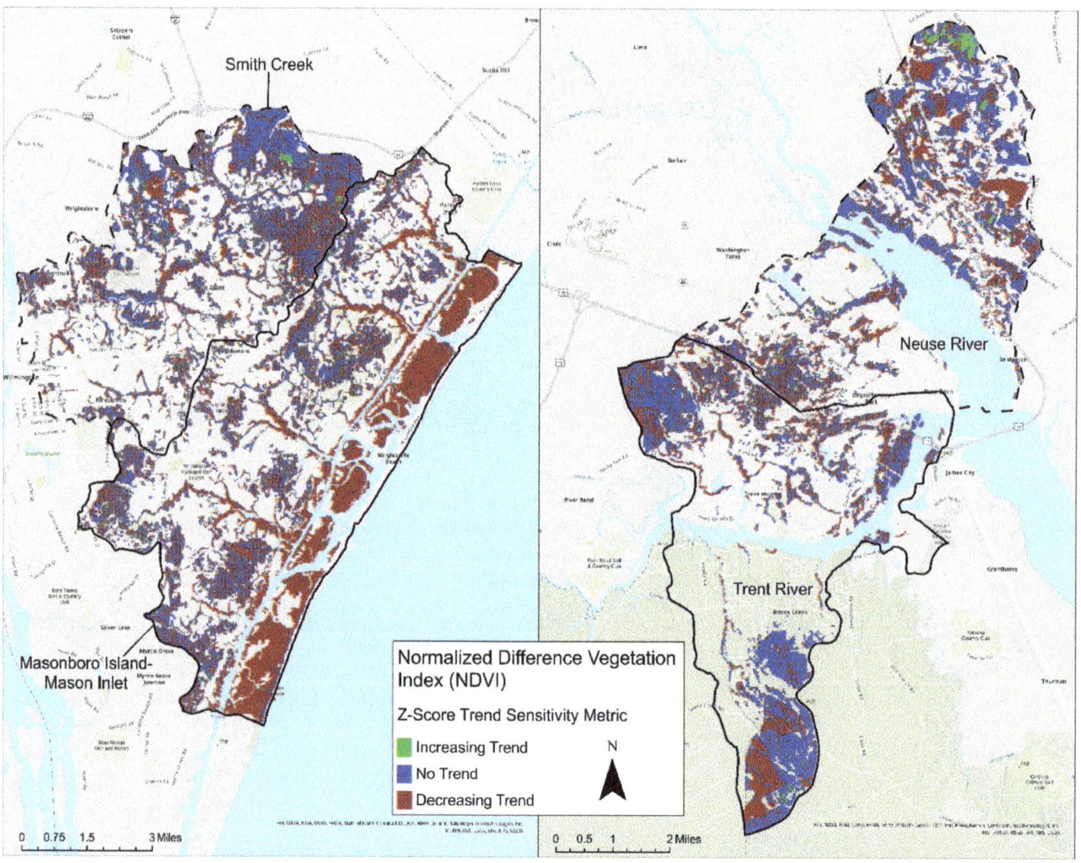

Figure 4. NDVI time series wetland sensitivity results for the Masonboro Island–Mason Inlet and Smith Creek subwatersheds in New Hanover County, NC.

Finally, we combined the flood exposure and vegetation sensitivity metrics into a single metric called wetland vulnerability, reclassified from its numeric values into three categories: low, medium, and high vulnerability to inundation as a function of wetland condition (Table 4). In New Hanover County, the majority of wetland areas in the Masonboro Island–Mason Inlet subwatershed located on the Atlantic Coast are ranked as highly vulnerable areas (Figure 5). The portions of the five smaller hydrologic features within the Masonboro Island–Mason Inlet subwatershed located directly on the coast are also ranked as highly vulnerable when combining exposure to inundation and the high rate of vegetation decline resulting from the sensitivity analysis. The Smith Creek subwatershed shows high vulnerability immediately adjacent to the Cape Fear River but the ranking declines towards the medium ranking as the hydrologic feature moves inland. Overall, as was the case with the vegetation analysis, the Masonboro Island–Mason Inlet shows many highly vulnerable areas with roughly 11.7% of the subwatershed's total area deemed highly vulnerable to the combined effects of inundation and declining vegetation conditions (Table 4).

Table 4. Combined wetland vulnerability assessment metrics for the four study areas in acres and as a percentage of the HUC-12 impacted.

Masonboro Island–Mason Inlet			**Smith Creek**	
Wetland Vulnerability Rating	Areaimpacted	% of HUC-12 impacted	Area impacted	% of HUC-12 impacted
Low	196.8	1.5%	245.9	2.9%
Medium	743.8	5.6%	273.9	3.2%
High	1555.5	11.7%	70.2	0.8%
City of New Bern–Neuse River			**City of New Bern–Trent River**	
Wetland Vulnerability Rating	Area impacted	% of HUC-12 impacted	Area impacted	% of HUC-12 impacted
Low	381.0	6.6%	94.0	1.6%
Medium	325.9	5.7%	158.4	2.7%
High	40.9	0.7%	38.6	0.7%

In Craven County, a relatively smaller proportion of the study areas were ranked as highly vulnerable to the combined effects of inundation and vegetation declines, with the highest concentration of highly vulnerable areas in the northeastern portion of the City of New Bern–Trent River subwatershed, where there is existing transportation infrastructure and two hydrologic features, the Neuse and Trent rivers, along with smaller hydrologic features that branch off from the Trent River (Figure 5). The wetland areas in the City of New Bern–Neuse River subwatershed that were determined to be at high vulnerability are those directly bordering the Neuse River and the smaller hydrologic systems along the river's coastline, primarily classified as medium or low overall vulnerability (Table 4).

The final step consisted of stratifying our results based on parcel ownership in order to identify potential sites for nature-based solution implementation, relative to data collected in the field during our site visits (Figures 6 and 7). The optimal type of parcel for potential NBS implementation is already in government ownership located close to wetland areas identified as high vulnerability or containing high vulnerability areas, even if not spatially contiguous. Interestingly, the locations considered by county managers to be highly problematic in the New Hanover Pages Creek watershed, for instance, show medium overall vulnerability in our model ranking and are not co-located with government-owned parcels, which partially explains the relative lack of success mitigating for repeated flooding occurring in those locations (Figure 6).

In Craven County on the other hand, where resilience planning efforts are currently well underway, our vulnerability ranking overlapped closely with areas considered highly at risk by City of New Bern Development Services officials who are implementing NBSs in the Stanley White Recreation Area and the Jack Smith Creek locations (Figure 7).

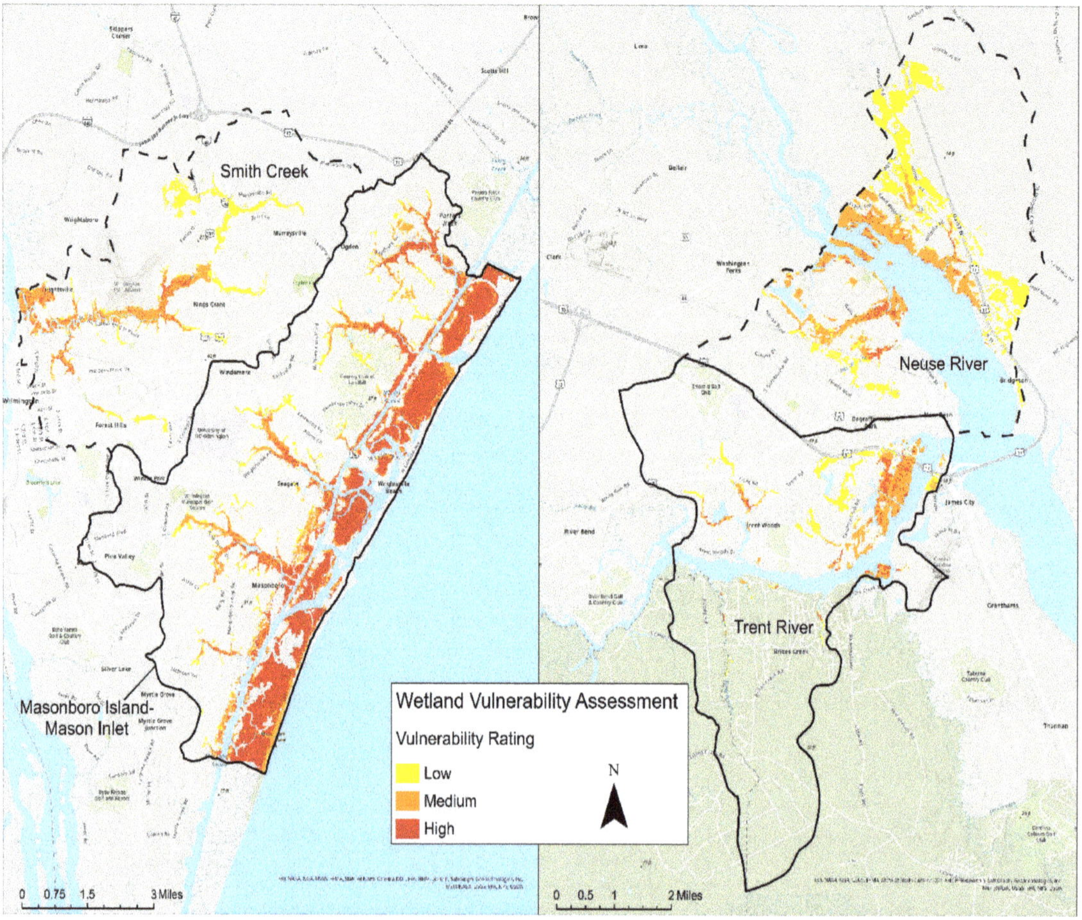

Figure 5. Wetland vulnerability assessment of the Masonboro Island–Mason Inlet and Smith Creek subwatersheds, NC showing categorical vulnerability rankings.

Spatially, it is important to identify specific locations and parcels that contain high proportions of wetlands ranked as highly vulnerable to inundation and present potential opportunities for NBS implementation (Table 5). In some instances, parcels larger than 65 hectares are deemed almost 70% vulnerable and, when those parcels are also contiguous and cover large spatial extents that protect significant residentials areas (Figure 8B), they present huge opportunities for brackish or saltwater marsh restoration that can benefit both ecosystems and humans. In New Hanover County, the Smith Creek subwatershed contains 25 government-owned parcels in total, while the Masonboro Island–Mason Inlet subwatershed contains 83 government-owned parcels identified as highly vulnerable, which are all nearly exclusively freshwater marsh ecosystems. Introducing a wetland-centered, nature-based solution project within any of these parcels has the potential to help mitigate flood inundation from the downtown Wilmington and Hightsville areas that contain historic structures and densely populated areas (Table 5 and Figure 3).

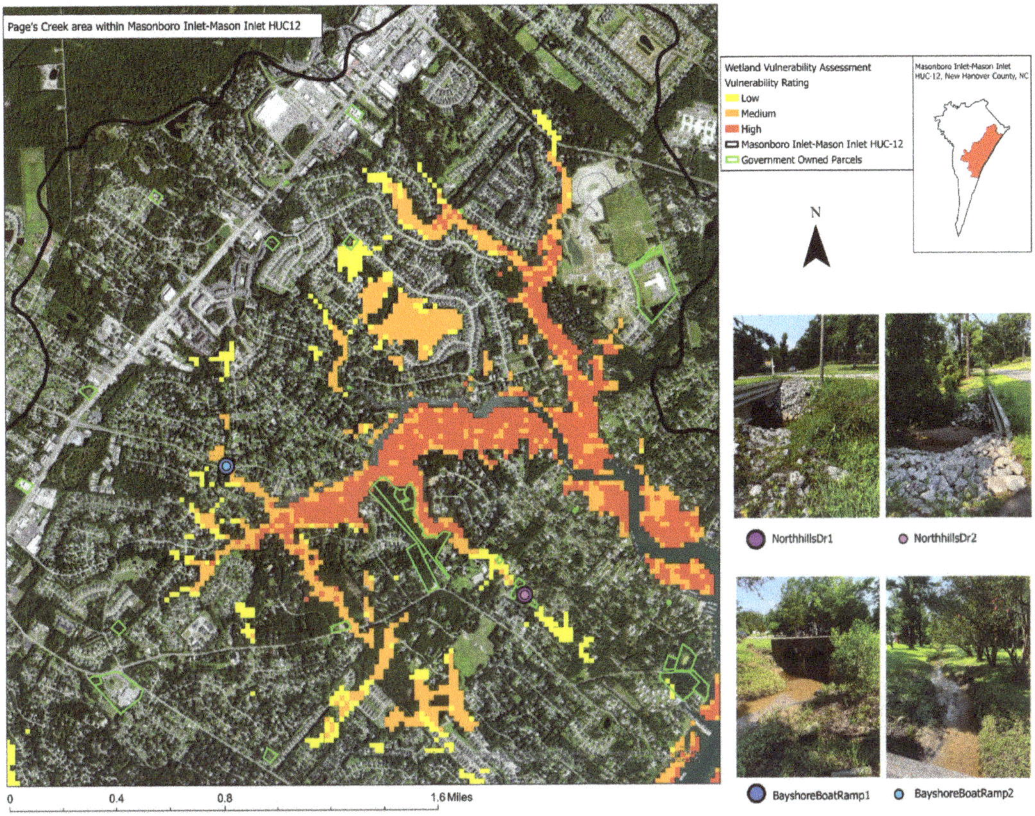

Figure 6. Wetland vulnerability assessment results, government-owned parcels, and locations identified in the field during site visits in the Page's Creek area within the Masonboro Island–Mason Inlet subwatershed, NC overlaid on NAIP imagery.

In New Bern, the City of New Bern–Neuse River subwatershed contains six government-owned parcels and the City of New Bern–Trent River subwatershed contains two government-owned parcels exhibiting high vulnerability to inundation and vegetation declining in condition, primarily located in the Neuse River, making them difficult to access and not the best area for implementation to protect societal infrastructure from flood inundation. The two government-owned parcels in the City of New Bern–Trent River subwatershed are considered highly vulnerable areas that are exclusively freshwater marshes (Table 5). In the City of New Bern–Trent River, 10.2% of the Lawson Creek Park parcel showed the presence of highly vulnerable areas while only 2.53% of the Clermont parcel showed high vulnerability (both characterized by freshwater marshes and riverine swamp forests in about equal proportion), making Lawson Creek Park the most optimal government-owned location for NBS implementation subwatershed (Figure 9C). Utilizing a park as a location to implement a NBS is a fantastic opportunity because not only is the parcel already government owned, but it is also in a location that is visited frequently by the public that could be utilized as an environmental educational tool. While not exhaustive, these results show opportunities afforded when using freely available and public data for the quick identification of potential NBSs locations as a function of exposure to risk and ecosystem sensitivity that can be used to both enhance ecosystem functions and benefit the built and infrastructural system.

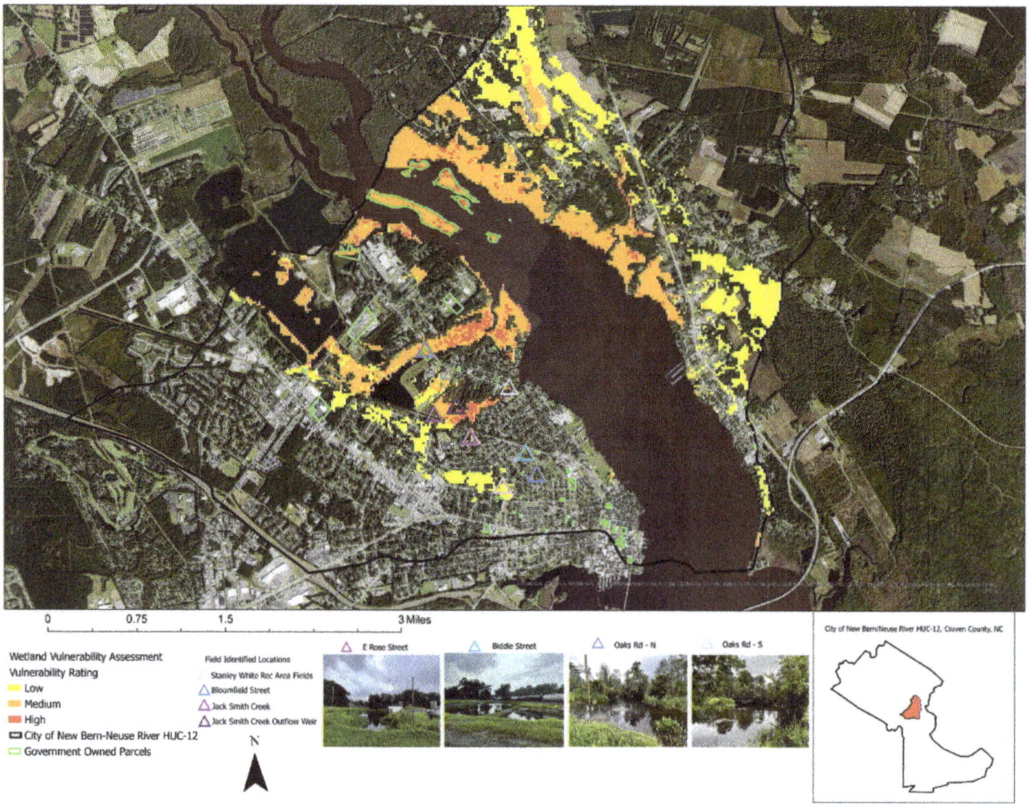

Figure 7. Wetland vulnerability assessment results, government-owned parcels, and locations identified in the field as potential NBS sites in the Trent River subwatershed in New Bern, NC overlaid on NAIP imagery.

Table 5. Highest vulnerability parcels in the Smith Creek subwatershed, Masonboro Island–Mason Inlet, Neuse subwatershed, and the Trent River watershed.

Parcel Name	Total Parcel Area (ha)	Total Area Vulnerable (ha)	% of Parcel Vulnerable
	Smith Creek		
Marne Dr Tract	20.11	5.36	26.68
Lot 19A Princess Place Park	0.26	0.08	33.84
PT Love	5.51	2.06	37.46
	Masonboro Island–Mason Inlet		
Landfall Natural & Scenic Preservation Area 1	67.23	45.01	66.94
Channel Acres	1.92	1.13	59.03
Landfall Natural & Scenic Preservation Area 2	17.94	8.86	49.39
	Neuse River		
Goose Island	1.76	0.50	28.7
Bluff island	7.04	1.69	24.1
Trim Cypress Island	2.43	0.44	18.4
Oaks Rd	1.11	0.14	13.4
	Trent River		
Lawson Creek Park	56.51	5.78	10.2
Clermont	5.91	0.14	2.53

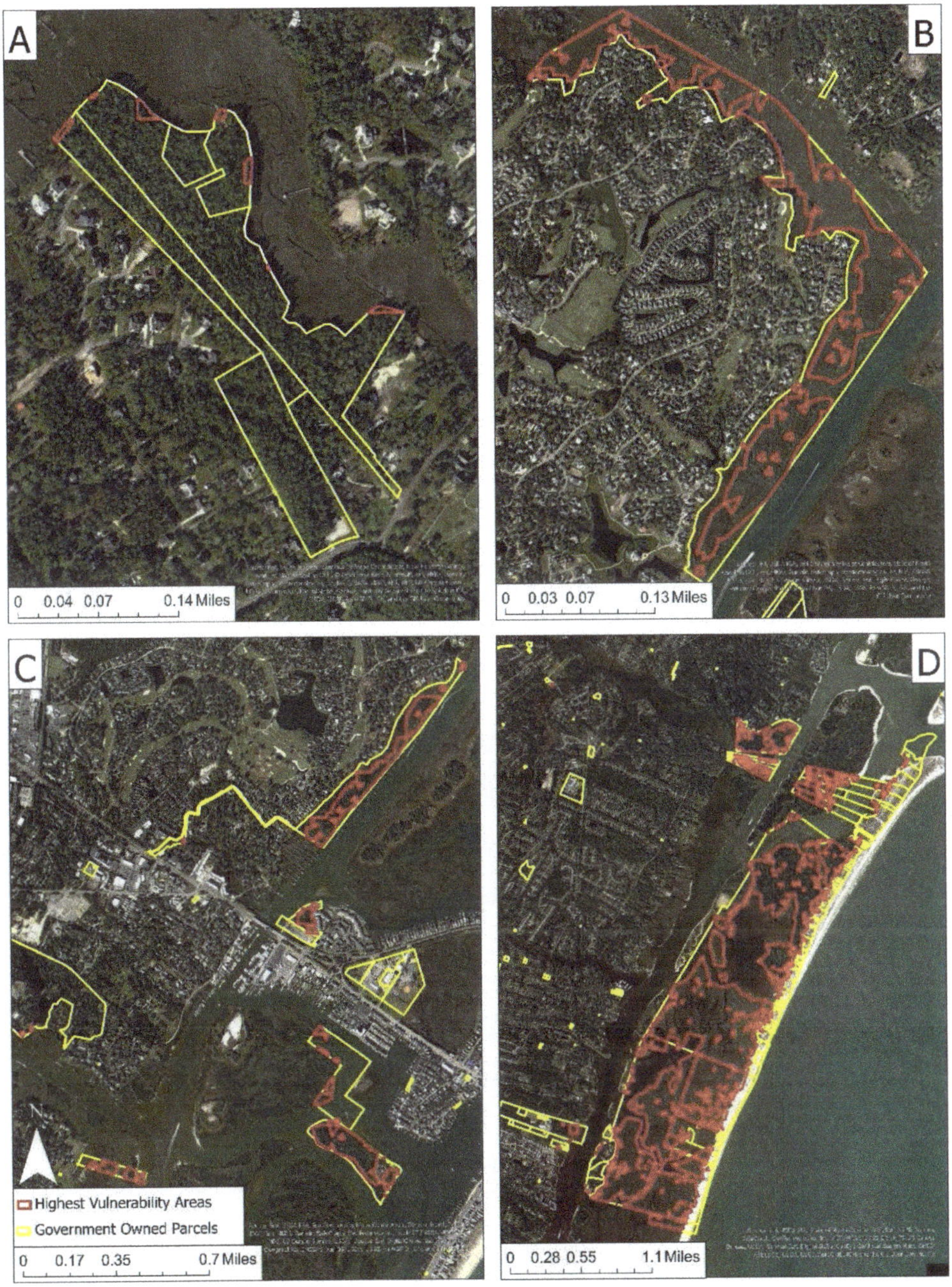

Figure 8. Overlap between the highest vulnerability areas and government-owned parcels of the Masonboro Island–Mason Inlet superimposed on NAIP imagery; (**A**) shows the Page's Creek area, (**B**) shows the Howe's Creek area, (**C**) shows the areas around Wrightsville Beach before the main island, and (**D**) shows Masonboro Island.

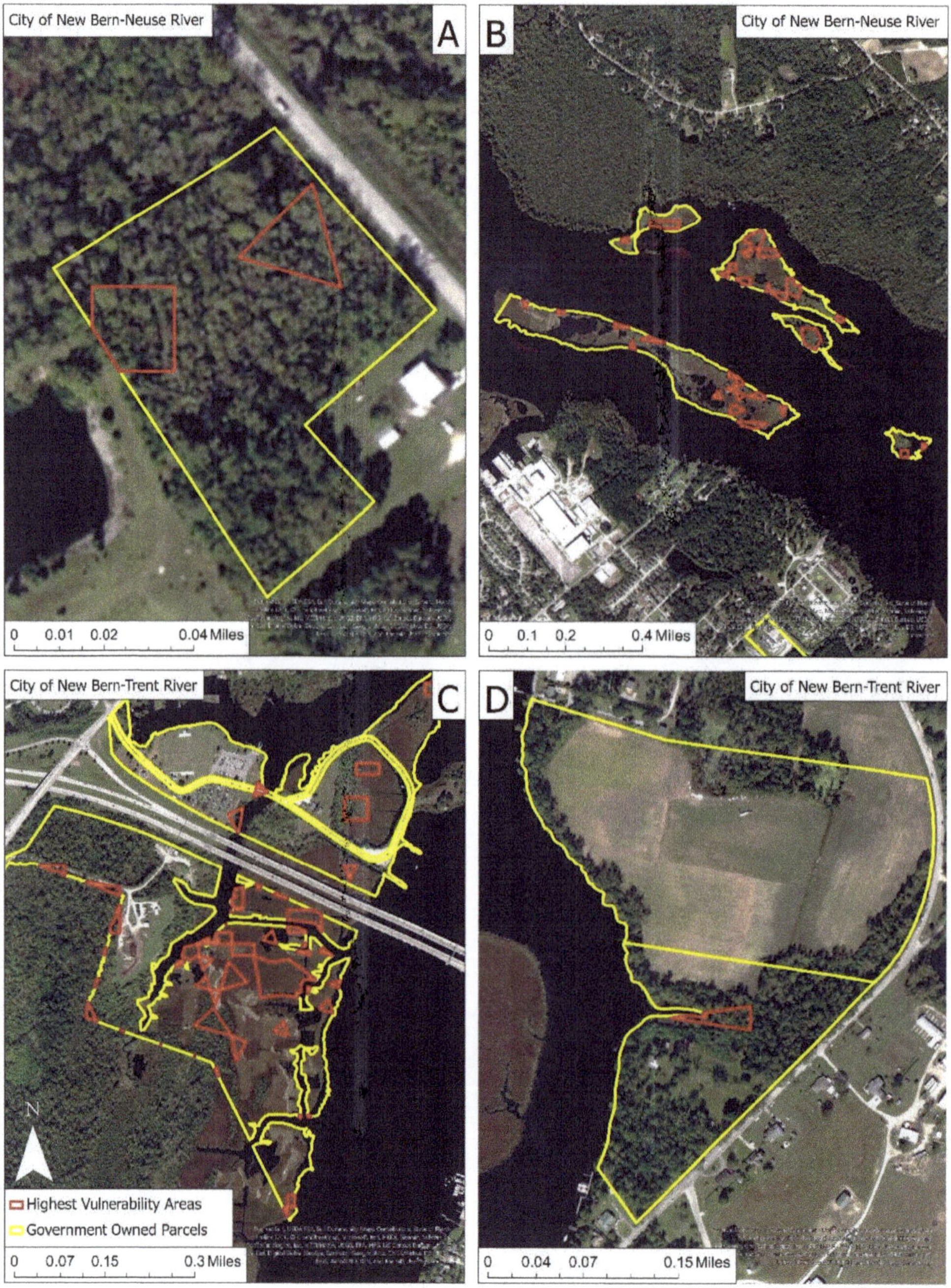

Figure 9. Overlap between the highest vulnerability areas and government-owned parcels in both New Bern study areas superimposed on NAIP imagery; (**A**) shows the Oaks Rd parcel, (**B**) shows the vulnerable islands in the Neuse River, (**C**) shows the Lawson Creek Park parcel, and (**D**) shows the Clermont parcel.

4. Discussion

The Masonboro Island–Mason Inlet subwatershed contains the highest area of wetlands and wetland restoration areas (20.3% of the entire subwatershed), which are vulnerable to flood inundation based on the sea level rise, high-tide flooding, and FEMA 100-yr and 500-yr flood zone data collected for the flood exposure analysis. It also contains the largest area experiencing a decreasing trend in wetland vegetation conditions (22.5% of the area showing a decreasing trend in NDMI values and a 23.4% decreasing trends in both NDVI and SAVI over the last 12 years was considered). While all four subwatershed examined here are coastal, the Masonboro Island–Mason Inlet area is the only one that directly borders the Atlantic Ocean, resulting in a higher inundation exposure due to the lack of a buffer region between the ocean and the wetland landscapes located closely to urbanized areas. The City of New Bern–Trent River subwatershed has the lowest area vulnerable to high flood exposure (2.9% of the total area) and decreases in wetland vegetation health metrics of the study areas were examined (10.7% for NDVI and SAVI and 11.8% for NDMI metrics per total area). Wetlands directly along the coast may experience much more rapid deterioration of vegetation with increasing sea levels, tidal inundation, and more frequent instances of flood events. We advance the notion that NBS implementation would be beneficial in wetland areas located near major hydrologic features that have a high exposure level to flood inundation scenarios and are experiencing decreasing health metrics indicative of healthy and productive vegetation, such as chlorophyll and moisture concentrations. Leveraging existing wetland areas as a nature-based solution is a natural tool to mitigate climate change impacts in coastal urban areas where flood inundation is common due to sea level rise, tidal flooding, and natural hazards bringing forth heavy precipitation and storm surges [17]. The introduction of NBSs to conserve and restore wetlands, such as brackish and freshwater marshes and riverine swamp forests, in urban areas simultaneously preserves biodiversity and the ecological balance of existing waterbodies by providing water sequestration and purification services [14]. Finally, the introduction of NBSs further provides vegetated areas with aesthetic and recreational values that allow citizens to connect with nature and enhance their mental wellbeing [14]. Overall, while NBSs in the coastal environment (either on the land part of a coast or in nearshore waters) are very efficient at mitigating the effects of extreme fast-evolving processes that lead to episodic coastal flooding (waves and surges), they are less efficient in decreasing the inundation effects of very slow long-term processes, such as mean SLR induced by climate change [17,18]. A notable exception is provided by marsh ecosystems that have the ability to collect sand and sediments and raise their surface elevation over time as SLR progresses [18].

As global climate change and sea level rises along the world's developed coastlines intensify, coastal zones will continue to experience higher vulnerability to flood inundation and ecosystem degradation [22]. While localized scale actions cannot mitigate these effects, implementing NBSs to protect and restore wetland ecosystems and associated ecosystem services can contribute to mitigating flood inundation and the impacts of climate extremes in urbanized coastal communities [16,17]. Especially in low-lying coastal areas, increasing compounding flood risks associated with storm surge and heavy precipitation are likely to continue to co-occur [34]. Areas that are typically only submerged during high tides will likely experience increases in time spent submerged with the sea level rise, which can lead to complete submergence over time and declines in vegetation conditions as species are unable to cope with changing environmental conditions or migrate quickly enough [34]. The impacts of rising water levels from sea level rise and natural hazard events will lead to structures located along the bodies of water to experience submergence and inundation, which will in turn impact societal infrastructure such as stormwater systems, buildings, and roadways [34]. Yet many municipalities or small urban centers do not have the capacity to run complex models to simulate their exposure to flood risks and prioritize potential interventions in areas that experience recurrent inundation and that may even be bought out (New Hanover County Office of Recovery and Resilience, personal communication, 2021).

Although proposed as a rapid assessment method with high transferability and replicability, this work has several important limitations and areas of improvement, primarily in computing the wetland sensitivity metric as presented here. Currently, North Carolina does not have any wetland-species-specific classification data available for geospatial analyses. While it would be difficult to create a statewide or subwatershed-level dataset identifying individual wetland vegetation species, it would certainly increase the accuracy of geospatial assessments of wetland vegetation conditions without relying on the collection of field observations. The NCDEQ wetland data utilized for this study does identify wetland types but does not go into depth regarding a full species breakdown for every location given the statewide scope of the dataset. The NCDEQ data does provide an idea of what types of species are commonly present within each wetland type to give those utilizing the data an idea of what species are more than likely present in each wetland community. Determining specific wetland vegetation species can be necessary for a remote sensing-based wetland assessment due to variations in spectral properties between species, along with precipitation variations that control species spectral responses [35,36]. Certain species of wetland vegetation—such as *Juncus roemerianus*, commonly known as black needlebrush—can have levels of chlorophyll and moisture slightly lower than the typical healthy vegetation that can result in false declining condition trends when utilizing remotely sensed data and spectral indices alone [35]. Species like cordgrass and juncus have a much sparser canopy composition relative to other wetland vegetation like phragmites, which leads to higher reflectance values in the red wavelength range and lower values in the near infrared than other wetland species [35]. Higher values in the red wavelength region infer a lower chlorophyll concentration which could impact the results of spectral indices if not accurately accounted for when undertaking a geospatial assessment of wetlands using remotely sensed data [35]. Ground truthing and taking site observations of vegetation communities are helpful steps to confirm species present in a study area so the ecosystem dynamics and structure of wetland areas are accurately described. Although some species of wetland vegetation may present lower spectral profiles than typical healthy vegetation due to structure and composition, the goal of this analysis was to determine the overall trends in vegetation conditions based on spectral signatures present in moderate-resolution satellite data and not to observe the values of the indices alone. For a wetland vegetation patch to be determined as increasing or decreasing in terms of its spectrally based indices (whether NDVI, SAVI or NDMI), there would have to be a shift in values over time that is atypical and maintains temporal consistency. If a species has consistently lower values in the metrics calculated here, the lower values alone would not cause an area to present either decreasing or increasing trends when considered over time. Secondly, to increase the accuracy of future vulnerability assessments, higher resolution multispectral data can be utilized, such as Sentinel-2 or in situ collected unoccupied aerial systems (UAS) imagery. We chose to utilize Landsat 8 data because it is widely accepted and available in the geospatial community along with the Analyze Changes using LandTrendr specifically created by Esri for the ArcGIS Pro software to be used with Landsat data. Drawbacks of utilizing a method that requires obtaining in situ UAS data is the difficulty of collecting data on a large scale, such as at the HUC-12 level, in developed areas and the resource intensity in the way of equipment, time, and workforce [24]. Increasing publicly available wetland inventory data that elaborates on types of species found in an area would be beneficial for geospatial analysis when retrieval of small scale, field-based data is not possible. With advancements in technology leading to high spatial and temporal resolution imagery data becoming widely available, remote sensing and GIS could be employed more frequently to monitor wetland areas, especially in areas that are difficult to traverse on the ground.

To further strengthen the results of this study, more in-depth evaluations of various wetland types along the Atlantic coast would be a great indicator to determine if subwatersheds on the coast are overall experiencing greater impacts from increased flood intensity and wetland vegetation degradation than their riverine counterparts. Locations for potential NBSs do not have to be implemented on government-owned parcels alone. Utilizing locations such as parks and public schools located in or near wetland areas provides good alternatives given educational opportunities for schoolchildren and the public. The state of North Carolina has a buyout system for property due to past inundation or high flood risks that were bought out from private citizens largely funded by the Federal Emergency Management Agency (FEMA) [37]. These lots would be another type of location that could be utilized for the introduction of NBSs. FEMA buyout lots present many advantages for NBS implementation for flood mitigation because they are already owned by a governmental entity, are in a known area of high risk to flood inundation, and are excluded from future re-developments [37]. Community outreach is also an avenue that could be explored to determine if any local landowners are willing and able to implement some form of nature-based solutions on private land to help combat future flood inundation.

Finally, while parcels optimal for NBSs can be identified through the results of a wetland vulnerability assessment, there are multiple factors to consider prior to implementation. Local conditions should be considered as failing to do so can cause negative impacts because mismatches between a solution and the socio-spatial context may lead to the nature-based solution envisioned being no longer fit to address both environmental and social needs [38]. The salvageability of the wetland vegetation, the connectivity of a potential implementation location to hydrologic features, and the size of a parcel identified for implementation can influence the effectiveness of an NBS. Wetland restoration efforts, such as nature-based solutions, are likely to fail if the sources of degradation continue to influence the area [39]. Due to the possibility of restoration failing because of upstream stressors that cause degradation, it is necessary to identify the causes of degradation to an area prior to implementation and eliminate or decrease said stressors [39]. Non-wetland areas near wetland areas experiencing flood risks and decreases in vegetation health metrics should also be considered for nature-based solution implementation, especially in developed and developing urban areas, to reduce flood exposure and devastating socio-economic effects [40–42].

5. Conclusions

This study presents a simple and replicable workflow for identifying potential sites for NBS implementation predicated on the utilization of freely and publicly available geospatial datasets so that interested decision-making parties can easily implement it in natural areas such as brackish and freshwater marshes and riverine swamp forests. In identifying potentially suitable NBS implementation sites that leverage existing wetland locations, irrespective of location near the coast or relatively inland, we rely on the assumption that wetland areas at high flood risk and that exhibit decreasing vegetation condition trends present an ideal opportunity to address both wetland conservation or restoration as well as the impacts of flood inundation in urban areas. Especially when co-occurring with large and underutilized or unutilized land parcels under government ownership, the double advantage of preserving or restoring wetland ecosystem functions (depending on the vegetation condition status highlighted by the multi-year trend analyses) and providing inundation mitigation for the built environment makes nature-based solutions an important consideration. Looking beyond wetland areas and government-owned parcels may present other implementation areas that could assist in further restoring wetlands in close proximity to those areas and helping mitigate flood risk in developed and urbanized coastal communities at risk from recurrent inundation. Finally, this study underscores the critical importance of conserving or restoring brackish and freshwater marshes and swamp forests even though, proportionally and depending on location, they represent a minority of wetland types present in the highly populated Atlantic Coastal Plain region.

Supplementary Materials: The following are available online at https://www.mdpi.com/article/10.3390/hydrology9120218/s1. Figure S1. Wetland types of the Masonboro Island-Mason Inlet subwatershed (32,814.41 acres) per NCDEQ wetland data. Figure S2. Wetland types of the Smith Creek subwatershed (21,136.41 acres) per NCDEQ wetland data. Figure S3. Wetland types of the City of New Bern-Neuse River subwatershed (14,458.74 acres) per NCDEQ wetland data. Figure S4. Wetland types of the City of New Bern-Trent River subwatershed (14,210.71 acres) per NCDEQ wetland data. Table S1 Dates of Landsat 8 data utilized for the sensitivity time series analysis. Figure S5. Detailed wetland vulnerability assessment workflow showing intermediary computational steps for ArcGIS Pro implementation. Figure S6. SAVI time series wetland sensitivity results of the Masonboro Island-Mason Inlet and Smith Creek subwatersheds in New Hanover County, NC. Figure S7. NDMI time series wetland sensitivity results of the Masonboro Island-Mason Inlet and Smith Creek subwatersheds in New Hanover County, NC.

Author Contributions: Funding was acquired by N.G.P.; conceptualization by N.G.P. Formal data analysis by G.S. and N.G.P. Original draft, edits, and final manuscript N.G.P. and G.S. All authors have read and agreed to the published version of the manuscript.

Funding: This research was funded by the North Carolina Sea Grant Karl Havens Regional Resilience Initiative (2020–2022) awarded to PI Narcisa Pricope at the University of North Carolina Wilmington.

Data Availability Statement: All publicly available datasets used in this research are available at the locations indicated in Table 1 or upon request in their processed version from the corresponding author.

Acknowledgments: We would like to thank the North Carolina Sea Grant first and foremostly for providing us with funding and support over the last two years to begin exploring this area of research. Secondly, we would like to thank our collaborators on the North Carolina Sea Grant Karl Havens Regional Resilience Initiative for being incredible colleagues. Finally, we are truly thankful to Alice Wilson from the City of New Bern and her colleagues, as well as the New Hanover County Office of Recovery and Resilience for their support during the project and various field visits.

Conflicts of Interest: The authors declare no conflict of interest. The funders had no role in the design of the study; in the collection, analyses, or interpretation of data; in the writing of the manuscript; or in the decision to publish the results.

References

1. Garg, J.K. Wetland Assessment, Monitoring and Management in India Using Geospatial Techniques. *J. Environ. Manag.* **2015**, *148*, 112–123. [CrossRef]
2. Mitsch, W.J.; Gosselink, J.G. The Value of Wetlands: Importance of Scale and Landscape Setting. *Ecol. Econ.* **2000**, *35*, 25–33. [CrossRef]
3. Sivakumar, R.; Ghosh, S. Wetland Spatial Dynamics and Mitigation Study: An Integrated Remote Sensing and GIS Approach. *Nat. Hazards* **2015**, *80*, 975–995. [CrossRef]
4. Environmental Protection Agency. *Wetlands: Protecting Life and Property from Flooding*; EPA: Washington, DC, USA, 2006.
5. Davidson, N.C.; van Dam, A.A.; Finlayson, C.M.; McInnes, R.J. Worth of Wetlands: Revised Global Monetary Values of Coastal and Inland Wetland Ecosystem Services. *Mar. Freshw. Res.* **2019**, *70*, 1189–1194. [CrossRef]
6. Anjos, L.J.S.; de Toledo, P.M. Measuring Resilience and Assessing Vulnerability of Terrestrial Ecosystems to Climate Change in South America. *PLoS ONE* **2018**, *13*, e0194654. [CrossRef]
7. Xu, K.; Wang, X.; Jiang, C.; Sun, O.J. Assessing the Vulnerability of Ecosystems to Climate Change Based on Climate Exposure, Vegetation Stability and Productivity. *For. Ecosyst.* **2020**, *7*, 23. [CrossRef]
8. Sharma, J.; Ravindranath, N.H. Applying IPCC 2014 Framework for Hazard-Specific Vulnerability Assessment under Climate Change. *Environ. Res. Commun.* **2019**, *1*, 051004. [CrossRef]
9. Turner, B.L.; Kasperson, R.E.; Matson, P.A.; McCarthy, J.J.; Corell, R.W.; Christensen, L.; Eckley, N.; Kasperson, J.X.; Luers, A.; Martello, M.L.; et al. A Framework for Vulnerability Analysis in Sustainability Science. *Proc. Natl. Acad. Sci. USA* **2003**, *100*, 8074–8079. [CrossRef]
10. Halls, J.N.; Magolan, J.L. A Methodology to Assess Land Use Development, Flooding, and Wetland Change as Indicators of Coastal Vulnerability. *Remote Sens.* **2019**, *11*, 2260. [CrossRef]
11. McCauley, L.A.; Jenkins, D.G.; Quintana-Ascencio, P.F. Isolated Wetland Loss and Degradation over Two Decades in an Increasingly Urbanized Landscape. *Wetlands* **2012**, *33*, 117–127. [CrossRef]
12. Rodríguez, J.F.; Saco, P.M.; Sandi, S.; Saintilan, N.; Riccardi, G. Potential Increase in Coastal Wetland Vulnerability to Sea-Level Rise Suggested by Considering Hydrodynamic Attenuation Effects. *Nat. Commun.* **2017**, *8*, 16094. [CrossRef] [PubMed]

13. Orimoloye, I.R.; Mazinyo, S.P.; Kalumba, A.M.; Nel, W.; Adigun, A.I.; Ololade, O.O. Wetland Shift Monitoring Using Remote Sensing and GIS Techniques: Landscape Dynamics and Its Implications on Isimangaliso Wetland Park, South Africa. *Earth Sci. Inform.* **2019**, *12*, 553–563. [CrossRef]
14. Stefanakis, A.I. The Role of Constructed Wetlands as Green Infrastructure for Sustainable Urban Water Management. *Sustainability* **2019**, *11*, 6981. [CrossRef]
15. O'Donnell, E.C.; Thorne, C.R.; Yeakley, J.A.; Chan, F.K.S. Sustainable Flood Risk and Stormwater Management in Blue-Green Cities; an Interdisciplinary Case Study in Portland, Oregon. *JAWRA J. Am. Water Resour. Assoc.* **2020**, *56*, 757–775. [CrossRef]
16. Cohen-Shacham, E.; Walters, G.; Janzen, C.; Maginnis, S. (Eds.) *Nature-Based Solutions to Address Global Societal Challenges*; International Union for Conservation of Nature (IUCN): Gland, Switzerland, 2016; ISBN 9782831718125.
17. Haase, D. Urban Wetlands and Riparian Forests as a Nature-Based Solution for Climate Change Adaptation in Cities and Their Surroundings. In *Nature-Based Solutions to Climate Change Adaptation in Urban Areas*; Theory and Practice of Urban Sustainability Transitions; Springer: Cham, Switzerland, 2017; pp. 111–121. [CrossRef]
18. Arkema, K.K.; Guannel, G.; Verutes, G.; Wood, S.A.; Guerry, A.; Ruckelshaus, M.; Kareiva, P.; Lacayo, M.; Silver, J.M. Coastal Habitats Shield People and Property from Sea-Level Rise and Storms. *Nat. Clim. Chang.* **2013**, *3*, 913–918. [CrossRef]
19. Gearey, M. Re-Naturing Cities: Reducing Flood Risk through Nature-Based Solutions. *Geography* **2018**, *103*, 105–109. [CrossRef]
20. Taillardat, P.; Thompson, B.S.; Garneau, M.; Trottier, K.; Friess, D.A. Climate Change Mitigation Potential of Wetlands and the Cost-Effectiveness of Their Restoration. *Interface Focus* **2020**, *10*, 20190129. [CrossRef]
21. Seddon, N.; Chausson, A.; Berry, P.; Girardin, C.A.J.; Smith, A.; Turner, B. Understanding the Value and Limits of Nature-Based Solutions to Climate Change and Other Global Challenges. *Philos. Trans. R. Soc. B Biol. Sci.* **2020**, *375*, 20190120. [CrossRef]
22. Liu, Z.; Fagherazzi, S.; Cui, B. Success of Coastal Wetlands Restoration Is Driven by Sediment Availability. *Commun. Earth Environ.* **2021**, *2*, 44. [CrossRef]
23. Szlafsztein, C.; Sterr, H. Coastal Zone Management Tool: A GIS-Based Vulnerability Assessment to Natural Hazards. In *Mangrove Dynamics and Management in North Brazil*; Ecological Studies; Springer: Berlin/Heidelberg, Germany, 2010; Volume 211, pp. 365–385. [CrossRef]
24. Morgan, G.R.; Hodgson, M.E.; Wang, C.; Schill, S.R. Unmanned Aerial Remote Sensing of Coastal Vegetation: A Review. *Ann. GIS* **2022**, *28*, 385–399. [CrossRef]
25. Kunkel, K.E. State Climate Summaries for the United States 2022. NOAA Technical Report NESDIS 150. Available online: https://statesummaries.ncics.org/chapter/nc/ (accessed on 20 November 2021).
26. Download Coastal Wetlands Spatial Data | NC DEQ. Available online: https://deq.nc.gov/about/divisions/coastal-management/coastal-management-gis-data/download-coastal-wetlands-spatial (accessed on 1 June 2021).
27. United States Department of Agriculture; University of Nebraska-Lincoln. National Oceanic and Atmospheric Association Time Series U.S. Drought Monitor. Available online: https://droughtmonitor.unl.edu/DmData/TimeSeries.aspx (accessed on 20 October 2021).
28. Schultz, M.; Clevers, J.G.P.W.; Carter, S.; Verbesselt, J.; Avitabile, V.; Quang, H.V.; Herold, M. Performance of Vegetation Indices from Landsat Time Series in Deforestation Monitoring. *Int. J. Appl. Earth Obs. Geoinf.* **2016**, *52*, 318–327. [CrossRef]
29. Candiago, S.; Remondino, F.; De Giglio, M.; Dubbini, M.; Gattelli, M. Evaluating Multispectral Images and Vegetation Indices for Precision Farming Applications from UAV Images. *Remote Sens.* **2015**, *7*, 4026–4047. [CrossRef]
30. Harmse, C.J.; Gerber, H.; van Niekerk, A. Evaluating Several Vegetation Indices Derived from Sentinel-2 Imagery for Quantifying Localized Overgrazing in a Semi-Arid Region of South Africa. *Remote Sens.* **2022**, *14*, 1720. [CrossRef]
31. Ochtyra, A.; Marcinkowska-Ochtyra, A.; Raczko, E. Threshold- and Trend-Based Vegetation Change Monitoring Algorithm Based on the Inter-Annual Multi-Temporal Normalized Difference Moisture Index Series: A Case Study of the Tatra Mountains. *Remote Sens. Environ.* **2020**, *249*, 112026. [CrossRef]
32. Sekovski, I.; Del Río, L.; Armaroli, C. Development of a Coastal Vulnerability Index Using Analytical Hierarchy Process and Application to Ravenna Province (Italy). *Ocean Coast. Manag.* **2020**, *183*, 104982. [CrossRef]
33. Singh, G.; Pandey, A. Flash Flood Vulnerability Assessment and Zonation through an Integrated Approach in the Upper Ganga Basin of the Northwest Himalayan Region in Uttarakhand. *Int. J. Disaster Risk Reduct.* **2021**, *66*, 102573. [CrossRef]
34. Azevedo de Almeida, B.; Mostafavi, A. Resilience of Infrastructure Systems to Sea-Level Rise in Coastal Areas: Impacts, Adaptation Measures, and Implementation Challenges. *Sustainability* **2016**, *8*, 1115. [CrossRef]
35. Zomer, R.J.; Trabucco, A.; Ustin, S.L. Building Spectral Libraries for Wetlands Land Cover Classification and Hyperspectral Remote Sensing. *J. Environ. Manag.* **2009**, *90*, 2170–2177. [CrossRef]
36. PRISM Climate Group. PRISM Climate Data; Oregon State University. Available online: https://prism.oregonstate.edu/ (accessed on 15 September 2021).
37. ReBuild NC Strategic Buyout Program Frequently Asked Questions | ReBuild NC. Available online: https://www.rebuild.nc.gov/strategic-buyout-program-frequently-asked-questions (accessed on 23 January 2022).
38. Albert, C.; Brillinger, M.; Guerrero, P.; Gottwald, S.; Henze, J.; Schmidt, S.; Ott, E.; Schröter, B. Planning Nature-Based Solutions: Principles, Steps, and Insights. *Ambio* **2021**, *50*, 1446–1461. [CrossRef]
39. US EPA. OW Principles of Wetland Restoration | US EPA. Available online: https://www.epa.gov/wetlands/principles-wetland-restoration (accessed on 29 September 2021).

40. Stefanidis, S.; Alexandridis, V.; Theodoridou, T. Flood Exposure of Residential Areas and Infrastructure in Greece. *Hydrology* **2022**, *9*, 145. [CrossRef]
41. Yi, Q. Flood exposure of critical infrastructures in the United States. *Int. J. Disaster Risk Reduct.* **2019**, *39*, 101240. [CrossRef]
42. Pricope, N.G.; Hidalgo, C.; Pippin, J.S.; Evans, S. Shifting landscapes of risk: Quantifying pluvial flood vulnerability beyond the regulated floodplain. *J. Environ. Manag.* **2022**, *304*, 114221. [CrossRef] [PubMed]

hydrology

MDPI

Article

Assessing the Impact of the Urban Landscape on Extreme Rainfall Characteristics Triggering Flood Hazards

Yakob Umer [1,*], Victor Jetten [1], Janneke Ettema [1] and Gert-Jan Steeneveld [2]

[1] Faculty of Geo-Information Science and Earth Observation (ITC), University of Twente, P.O. Box 217, 7500 AE Enschede, The Netherlands
[2] Meteorology and Air Quality Section, Wageningen University, P.O. Box 47, 6700 AA Wageningen, The Netherlands
* Correspondence: y.m.umer@utwente.nl

Abstract: This study configures the Weather Research and Forecasting (WRF) model with the updated urban fraction for optimal rainfall simulation over Kampala, Uganda. The urban parameter values associated with urban fractions are adjusted based on literature reviews. An extreme rainfall event that triggered a flood hazard in Kampala on 25 June 2012 is used for the model simulation. Observed rainfall from two gauging stations and satellite rainfall from Climate Hazards Group InfraRed Precipitation with Station data (CHIRPS) are used for model validation. We compared the simulation using the default urban fraction with the updated urban fraction focusing on extreme rainfall amount and spatial-temporal rainfall distribution. Results indicate that the simulated rainfall is overestimated compared to CHIRPS and underestimated when comparing gridcell values with gauging station records. However, the simulation with updated urban fraction shows relatively better results with a lower absolute relative error score than when using default simulation. Our findings indicated that the WRF model configuration with default urban fraction produces rainfall amount and its spatial distribution outside the city boundary. In contrast, the updated urban fraction has peak rainfall events within the urban catchment boundary, indicating that a proper Numerical Weather Prediction rainfall simulation must consider the urban morphological impact. The satellite-derived urban fraction represents a more realistic urban extent and intensity than the default urban fraction and, thus, produces more realistic rainfall characteristics over the city. The use of explicit urban fractions will be crucial for assessing the effects of spatial differences in the urban morphology within an urban fraction, which is vital for understanding the role of urban green areas on the local climate.

Keywords: extreme rainfall; default urban fraction; Kampala; urban parameter; updated urban fraction; WRF model

Citation: Umer, Y.; Jetten, V.; Ettema, J.; Steeneveld, G.-J. Assessing the Impact of the Urban Landscape on Extreme Rainfall Characteristics Triggering Flood Hazards. *Hydrology* **2023**, *10*, 15. https://doi.org/10.3390/hydrology10010015

Academic Editors: Aristoteles Tegos, Alexandros Ziogas and Vasilis Bellos

Received: 5 November 2022
Revised: 24 December 2022
Accepted: 26 December 2022
Published: 6 January 2023

1. Introduction

Numerical weather prediction (NWP) models, such as Weather Research and Forecasting (WRF), are nowadays used for flood hazard modeling and forecasting in urban areas [1,2]. However, simulating the spatial–temporal rainfall characteristics and structures that trigger flood hazards is challenging and complex to predict. The mechanisms affecting rainfall are affected by many factors: the quality of initial and boundary conditions, domain set, and parametrization schemes in model [3]. The urban landscape characteristics, such as urban fraction and urban parameters, are essential factors in the urban parameterization schemes that affect the simulated extreme rainfall triggering floods in the urban areas. Changes in the urban landscape alter the near-surface radiation and energy budgets, momentum, and water vapor in urban areas, which affect the initiation and intensification of convective processes over a city [4]. Moreover, with an urban expansion, a larger thermal contrast between the urban areas and the water body can result in stronger low-level circulation [5]. Consequently, meteorological conditions alter; thus, it determines the formation of convection storms and intensive rainfall over urban areas [6–8]. In the numerical weather prediction models, the impact of such urban surface change can be handled through land-surface modeling implemented in parameterization schemes.

The WRF model is widely used to examine and assess the impact of the urban landscape on hydrometeorological processes leading to changes in high-intensity rainfall events. In the WRF model, these processes are addressed using urban parameterization schemes, for instance, the Single-Layer Urban Canopy Model (SLUCM) [9]. A variety of studies have been carried out using the WRF model to improve its skills in assessing the impact of the urban landscape on meteorological fields leading to extreme rainfall events [10–14].

However, the default urban fraction and the corresponding parameter in the WRF model incorrectly represent the true extent and values of the urban surfaces for individual cities. For instance, the default urban fraction in the WRF model, which is provided by Moderate-resolution Imaging Spectroradiometer (MODIS) observational data, cannot fully represent the correct extent and position of the urban area. Urban parameter values are also site-specific and incorrectly represent a city's extent and position; hence, it needs to be updated, as suggested by [7,15–17]. Therefore, the correct representation of the city's urban fraction is required to optimally simulate the high-intensity rainfall distribution over the urban area. This study uses the detailed and high-resolution urban fraction generated using Landsat image instead of the default MODIS urban fraction. This Landsat image is highly detailed and able to capture detailed urban features, such as wetlands and individual urban fractions at 30 m resolution, which proved to be applicable for hydrological process modeling leading to flooding in Kampala. Here, we updated the default urban fraction in the WRF model primarily to represent the correct position and extent of the city and, secondly, to use a consistent urban fraction for integrated flood hazard modeling and the WRF model.

Several other urban fractions have been developed for different atmospheric modeling purposes. For example, local climate zoning (LCZ) is designed to study the thermal characteristics of urban areas [8,18], and in-homogeneous urban canopy parameters (UCP) are developed for air quality modeling [6]. In this study, the urban fraction derived from the Landsat image of 2016, initially developed for urban land-use planning and flood management [19], is used in the WRF model. Therefore, this study is the first order to study the role of the new urban fraction on rainfall and can improve the simulated rainfall. In comparison, the next study can consider the detailed urban morphology and also compare the current procedure with the existing procedure for a detailed study of hydro-meteorological processes in urban areas. Assessing the effect of urbanization factors on extreme rainfall is important to improve our understanding of how urban growth and expansion affect localized meteorological and hydrological processes. In this research, we ask how the proposed updated urban fraction for the WRF model is expected to improve the accuracy of extreme rainfall simulation required for flood management, particularly in data-scarce areas.

The aim of this paper is to configure the WRF model optimally with urban fraction specifically developed for the city of Kampala, Uganda, and to evaluate the impact of adjusting urban fraction and parameters on the simulated rainfall event. The use of the WRF model to study the deep convection over Kampala requires a special configuration, which requires the proper position and extent of the city for better consideration of the spatial contrast between the city and Lake Victoria. This study is a pioneer in using an explicit and alternative satellite-derived urban fraction in the WRF model and evaluating its application in deep convection triggering the localized flood. The study provides a detailed analysis of (1) the WRF model's configuration with the adjusted urban fraction in comparison with the default urban fraction; (2) the impact of the updated urban landscape, which include both the updated urban fraction and adjusted urban parameters on the simulated rainfall.

2. Materials and Methods

This section presents the selected rainfall event and WRF model configuration, followed by the methodology used in this study and model verification. The methodology followed in this study begins with the WRF model in Section 2.2, which introduces the model's setting and configuration and the choice of the physical parameterizations used. Section 2.3 presents the method to incorporate the updated urban fraction in the model, followed by Section 2.4, which introduces different model simulations carried out in this study. In Section 2.5, we present a strategy to verify and analyze the simulated rainfall results.

2.1. Study Area and Selected Event

This study was conducted in Kampala, the capital city of Uganda (polygon in dark lines, Figure 1, Right), as a case study to test the method. The city is an ideal location to test the method because it is one of the exemplary sub-Saharan African city's experiencing tremendous urban expansion over the last three decades that contributed to flooding. The city is positioned on the shore of Lake Victoria and has an area of about 290 km^2. Combined with urban expansion, high-intensity rainfall events from tropical weather conditions, soil infiltration properties, and lack of proper drainage systems are the main triggering mechanisms for flooding [20].

The 25 June 2012 rainfall event that caused a localized flood event in Kampala was selected for this study. For this event, two types of rainfall observations are present; rain gauge measurements and satellites. On 25 June 2012, two rain gauge stations were in operation in Kampala city: Automatic Weather Station (AWS) at the Makerere University campus, recording at 10 min intervals, and Kampala Central station at 24 h intervals. The 24 h rainfall data of Kampala central station were collected from the Global Summary of the Day (GSOD) dataset provided by the National Climatic Data Center (NCDC). At Makerere University, a daily total of 66.2 mm was recorded, and Kampala Central station recorded 60 mm, which is a typical 2-year return period event [21].

In addition, satellite-estimated rainfall from Climate Hazards Group InfraRed Precipitation with Station data (CHIRPS) [22] was retrieved for model evaluation. CHIRPS is considered one of the best rainfall products for decision-making in East Africa [23,24]. The CHIRPS rainfall data has 0.05 degree (~5.5 km) spatial and daily temporal resolutions. For the WRF model evaluation, the CHIRPS rainfall data is rescaled using linear interpolation to the innermost domain of WRF spacing, which is 1 km $\times$ 1 km.

The selected rainfall event occurred in the transition between the two main rainy seasons. Its weather systems are mesoscale and local scale systems, mostly convection systems associated with the interaction of the urban areas with lake circulation and the surrounding mountains [25,26]. The rainfall is often very localized and is characterized by high-intensity rainfall events as it is associated with highly variable weather systems; hence, available rain gauges are not sufficient to capture the spatial variability of these events.

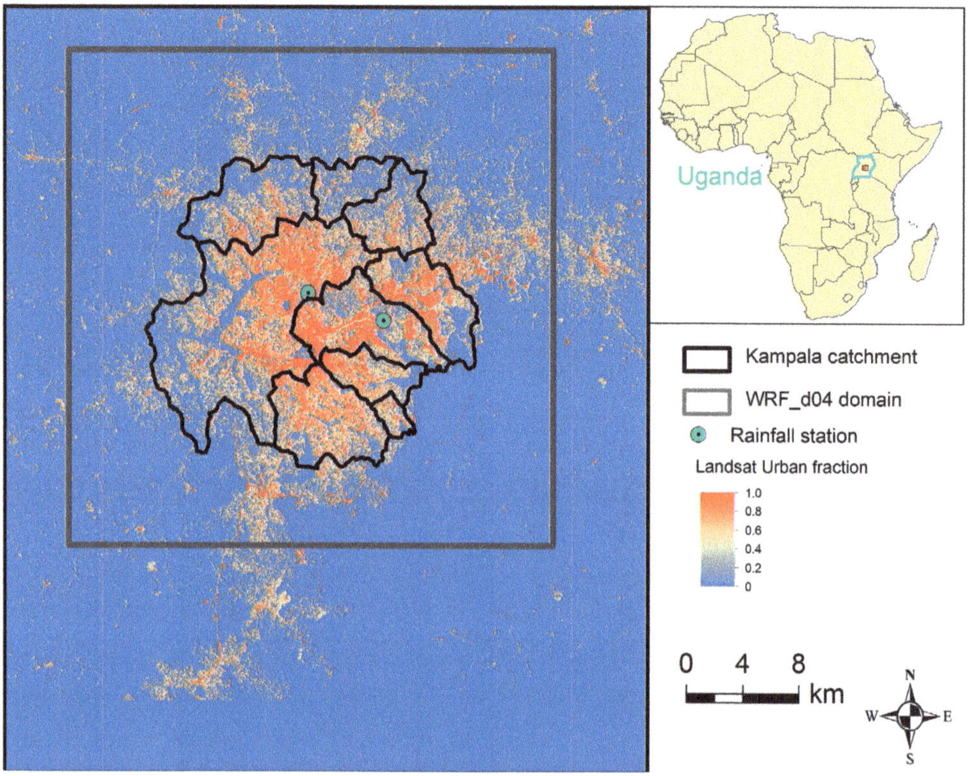

Figure 1. Study area: Kampala catchment boundary represented by dark line polygon; Gray line rectangle indicates WRF d04 domain boundary and map of urban fraction in Kampala as derived from the Landsat image [27].

2.2. The WRF Model Setting and Configuration

This study uses the WRF-ARW version 4 [28] with a two-way nested domain configuration. The WRF model setup consists of four domains centered on Kampala. The four domains are a 27 km outer fixed domain (d01) and three fixed nest domains of 9 km (d02), 3 km (d03), and 1 km (d04) grid spacing, and all domains had 31 × 31 grid points as shown in Figure 2 and Table 1, and conformed to the most recommended ratio of 1:3 by [29]. Each model domain used the Mercator projection system with 38 vertical levels and a pressure top of 50 hPa. As shown in Figure 2, Kampala is central in all four domains. Under Figure 2, we further show land-use categories per gridcell and the default urban representation in the innermost domain d04. The number of gridcells that will be changed to urban when we used the updated urban fraction is later discussed in the results part (see Figure 3).

Rainfall simulation using mesoscale NWP models, such as WRF, requires a proper selection of physics parameterization schemes. These parameterization schemes include microphysics, Planetary boundary layer (PBL), cumulus, radiations, urban canopy, surface layer, and land surface schemes. Based on the sensitivity assessment described in [3], we selected Morrison microphysics, Grell Freitas cumulus parametrization, and ACM2 PBL parameterization combinations for the 25 June 2012 event as the main rainfall-controlling physics in the area. All parameterization schemes in the WRF model are applied for all domains, while the urban canopy parameterization is only applied for the 1 km domain following the procedure suggested by the WRF model manual. Initial and boundary conditions are retrieved from the ERA5 global Reanalysis Model, with a resolution of 30 km [30]. Following [26], the static lake surface temperature of Lake Victoria was set to 24 °C. The model simulation covers three days, from 24 June at 00:00 UTC to 26 June 2012 at 24:00 UTC, to allow spin-up of the atmospheric processes.

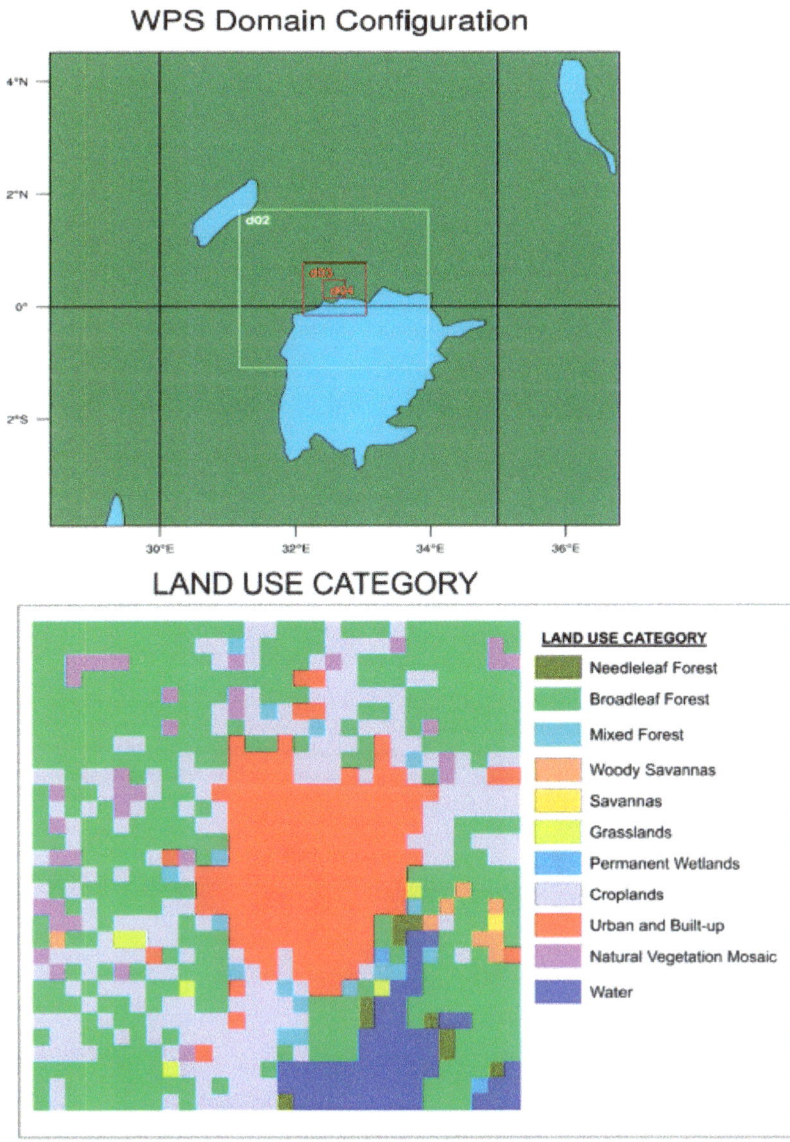

Figure 2. Upper: The Weather Research and Forecasting (WRF) model configuration using four domains (d01, d02, d03, and d04); Bottom: Land-use categories and the default urban fraction representation (RED POLYGON) in the innermost domain d04 of WRF.

Table 1. Weather Research and Forecasting model settings used in the current study.

Model	WRF v 4.1			
Characteristics	**Domain 1 (d01)**	**Domain 2 (d02)**	**Domain 3 (d03)**	**Domain 4 (d04)**
Horizontal grid spacing	27 km	9 km	3 km	1 km
Horizontal Dimensions	31 × 31 × 31	31 × 31 × 31	31 × 31 × 31	31 × 31 × 31
Time step 60 s	adaptive time step	adaptive time step	adaptive time step	adaptive time step
Initial-boundary conditions	ERA-5 (30 km)	simulation of domain 1	simulation of domain 2	simulation of domain 3
Model run period	0000 UTC 24 June–1800 UTC 26 June 2012			

The urban canopy model is one of the optional parameterization schemes implemented in the WRF model to account for urbanization (Urban fraction) and associated parameters for the meteorological processes through the energy partitioning modeling system. The available urban canopy schemes in WRF are the multi-layer urban canopy model (MUCM) [31] and the SLUCM [9]. The MUCMs incorporate building effect parametrization (BEP) and a building energy model (BEM) [32], which are used to deal with sources and sinks of heat. The SLUCM neglects the variation in building height and density in the model grids and uses only a simplified street canyon (i.e., walls, roof, and roads) geometry to represent urban surfaces. A study indicates that the MUCMs better simulate the extreme rainfall amount and its spatial distribution compared to when using SLUCM [4]. However, MUCM requires detailed building data and parameters, which are not easy to be acquired based on the literature reviews or remotely sensed information; thus, it is challenging to apply in a data-scarce area, such as Kampala.

In this study, the SLUCM scheme [33] was used to accommodate the urban surface's effects on simulated rainfall within the WRF model. The SLUCM scheme employs a common single-layer street canon representation of urban areas with its numerical framework well-elaborated [34]. The scheme is simple mainly because it uninvolved the effect of building parameterization (e.g., variation in building height and building density) as in the case of the Multi-Layer urban canopy model (MUCM) [31]. The SLUCM in the WRF model is coupled to the NoahMP land-surface model through a parameter called "two-dimensional urban fraction (FRC_URB2D)". The NoahMP land surface model handles the non-urban fraction (vegetation cover) of the grid, while the SLUCM handles the urban fraction part. The detailed physics options and parametrization used in the SLUCM are found in [35–37].

The SLUCM requires an urban fraction (urban map) and urban parameters linked to the urban fraction for model simulation. As the default urban fraction acquired from the MODIS with all urban extent assigned to a single urban value does not represent the true extent and position of a city, we updated the urban fraction based on the satellite-derived urban fraction of Kampala.

2.3. Adjusted Urban Fraction

By default, the WRF model uses the land-use categories based on Moderate-resolution Imaging Spectroradiometer (MODIS) observations [35]. With the WRF version 4 release, the MODIS land-use data is updated and available at a resolution of 30 s with 20 land-use categories [36]. This dataset contains the land-cover classification of the international Geosphere-Biosphere program and is modified for the Noah land-surface model [37]. Within this land-use classification, the default urban fraction (base map in the WRF model) is represented by the homogeneous urban fraction with all cell values assigned to 0.9 (HIR) (Figure 1). The default urban parameters dataset that is linked with this default urban fraction is also provided as static data, as shown in Table 2 (second column). In this study, the urban land-use fraction developed by [19] that is used for urban planning and integrated urban flood modeling in Kampala was used. Simultaneously, the urban parameters linked to the urban fraction were adjusted through a literature review [15,38,39].

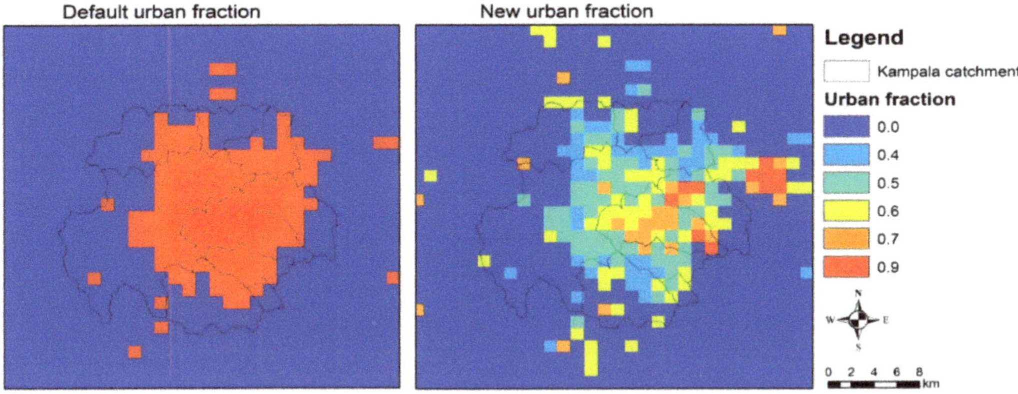

Figure 3. The default and updated urban fraction representation used in the model simulations. All pixels in the inner domain of WRF represent 1 km. The default urban fraction (**left**) is from Noah LSM based on MODIS observation [35], whereas the updated urban fraction is derived from a Landsat image developed using the cellular automata model [19].

For the WRF modeling, we used the built-up fraction of the Landsat image (Figure 1) to replace the default urban fraction in the WRF model's preprocessing following a similar procedure [40]. The updated urban fraction is derived based on the 30 m resolution Landsat image 2016 [19]. This adjusted urban fraction is generated using a supervised classification by sorting the satellite image pixels into three major urban land cover categories: Built-up, including buildings and pavements; non-built, and bare soil. These three urban land cover classes are developed as an array of cells, each with an associated fraction of land cover (for built-up, vegetation, and bare soil), and finally, add up to 1, see [19] for details. Figure 1 shows that the urban fraction value is close to 1 in the high-intensity urban areas (i.e., areas around the city center), while in the suburban areas, the urban fraction value approaches zero. Here, the higher the intensity of built-up areas (urban fraction 1), the lower the vegetation cover and vice versa. The new urban fraction exists at a higher spatial resolution (i.e., 30 m) than the WRF innermost domain cell size, which is 1 km. To match the WRF cell size, the new urban fraction cell size is rescaled, and the adjusted urban land-use fraction is inserted into WRF following the input data format and processes [41,42].

2.4. Model Simulation Strategy

Three simulations are performed to distil the impact of changing urban fractions and adjusted urban parameters used in the SLUCM. The first simulation (hereafter DUF_DUP) uses the default urban fraction with the default urban parameters (Table 2, second column) as a benchmark. The second simulation (hereafter DUF_AUP) uses the default urban fraction (Figure 3) with adjusted urban parameters (Table 2, third column), where values were adjusted based on literature [7,20], as shown in Table 2 (third column). The third simulation (hereafter SUF_AUP) is with an updated land-use fraction based on the Landsat 2016 image, with the adjusted urban parameters. For the SUF_AUP simulation, we have replaced the default homogeneous urban fraction with a heterogeneous urban fraction to define Kampala's more realistic urban representation. As already mentioned in the introduction, the default urban parameter values in WRF, when no information is provided, do not represent the urban surfaces of any city. Therefore, it is recommended not to use these values as is. Hence, the possible fourth simulation using the updated urban fraction with the default urban parameter values is not considered here.

Table 2. Default and adjusted urban parameter values that are assigned to the urban fraction.

	Default Urban Parameter Value	Adjusted Urban Parameter Value
Roof height (m)	7.5	15
Road width (m)	9.8	10
Roof width (m)	9.4	20
Standard deviation of roof height (m)	3	1.5
Albedo (−)		
Roof	0.2	0.1
Wall	0.2	0.1
Road	0.2	0.15
Emissivity (−)		
Roof	0.9	0.85
Wall	0.9	0.9
Road	0.95	0.95
Conductivity of materials (Cal cm^{-1} s^{-1} $°C^{-1}$)		
Roof	0.67	0.4
Wall	0.67	1
Road	0.404	0.8
Heat capacity of materials (Cal cm^{-3} $°C^{-1}$)		
Roof	1.00×10^6	1.20×10^6
Wall	1.00×10^6	1.20×10^6
Road	1.40×10^6	1.50×10^6
Total thickness of material layers (m)		
Roof	0.05	0.50
Wall	0.05	0.30
Road	0.25	1.00

2.5. Model Verification

To evaluate the simulated rainfall in the innermost domain d04, we used the relative error (RE) index of [43]. Model performance simulating the event is evaluated using observed rainfall data from two gauging stations and CHIRPS data. The comparison with the two gauging stations is carried out with respect to the gridcell daily rainfall amount at the station locations. The comparison with the CHIRPS was carried out as daily accumulated rainfall distribution over the Kampala catchment and the catchment area-averaged amount as a relative error. The catchment area is the area covering the greater Kampala, represented by a polygon indicated using a black line in Figure 1.

The RE index (Equation (1)) in percentages computes the simulated accumulated 24 h rainfall, S, with respect to observed rainfall at the station location, O. In the case of comparing WRF with CHIRPS data, S and O are the average values of all grids inside the innermost domain of WRF, while in case of 2 stations, the simulated WRF values, S, of gridcell are taken which is located at the rain gauge station, O.

$$RE = \frac{S - O}{O} \times 100 \quad (1)$$

Each simulation (DUF_DUP, DUF_AUP, and SUF_AUP) will result in three RE measures; one for each gauging station and one for the area-averaged compared with CHIRPS. To measure the overall magnitude of error for each simulation, the average relative error (ARE) of the three evaluation locations is calculated based on the three absolute RE (i.e., RE at two gauging stations and area-averaged). The impact of adjusted model settings on simulated extreme rainfall is also evaluated in the form of spatial distribution for objective analysis in two main aspects: maximum rainfall amount and its spatial distribution in the catchment and time evolution. The event's time evolution over two hours from 11:00 to 12:50 UTC is presented, similar to when the 25 June 2012 observed rainfall event occurred.

3. Results

This section presents the impact of the updated urban fraction in the WRF model, including urban parameters on the simulated rainfall in terms of maximum accumulated 24 h rainfall amount and its spatial distribution and the time evolution of peak rainfall amount distribution over two hours. Three simulations are intercompared as well as validated against the observed rainfall from the CHIRPS and two rain gauge stations. The evaluation focuses on the high-intensity rainfall event of 25 June 2012 that triggered the Kampala flood hazard.

The following section describes the representation of the satellite-derived urban fraction in the WRF model and its comparison with the default urban fraction. The impact of adjusted urban parameters on the simulated rainfall and the comparisons are presented in Sections 3.2 and 3.3, followed by a discussion and conclusion in Sections 4 and 5, respectively.

3.1. Updated Urban Fraction Representation

The new urban fraction for Kampala is different from the default urban fraction in two main aspects: the fraction of urbanization and the spatial extent of the city.

The urban fraction parameter in WRF defines the percentage of the gridcell covered by impervious urban surfaces, while the remaining fraction is treated as a pervious, vegetated surface. Figure 3 shows the default urban fraction of 0.9, implying that the city is represented by a homogeneous high-intensity residential urban fraction (pixels value 0.9 as described by red color). Based on the Landsat image classification, the updated urban fraction cells have an average value of 0.64, representing a lower-intensity urban residential category. In the right-sided map, the city center is partly characterized by orange color because uninhabitable wetlands (blue areas in Figure 3) are located next to high-intensity pixels in the LandSat image. The highest urban fraction is (0.9) found on the city's eastern outskirts, where an all-terrain is suitable for constructing a building.

Another critical aspect of the new urban fraction map is its spatial extent. As shown in Figure 3, the new urban fraction covers a broader area of about 50 pixels compared to the default urban fraction. Croplands initially represented about 40 pixels, Broadleaf Forest represented 7 pixels, and the rest with Natural Vegetation mosaics (see Figure 1). The changes in croplands into the urban fraction are mainly located in the city's eastern and southern parts. In contrast, the change of Broadleaf Forest to the urban fraction is located in the Northern part of the city.

3.2. Model Validation

The ability of the WRF model to properly simulate the event is evaluated through a comparison with the gridcell daily rainfall amount in the d04 domain with the two gauging stations. Table 3 summarizes the comparison of three WRF simulated gridcell-total accumulated rainfall with the observation at the station locations AWS and GSOD and area-averaged rainfall with that of CHIRPS. All three WRF simulations underestimated rainfall compared to the observations at rain gauging locations, as indicated by RE's large negative values. In contrast, compared to the CHIRPS area-averaged rainfall amount over the innermost WRF domain, all simulations are overestimated, but the SUF_AUP simulation relatively performs better with RE = 13% for SUF-AUP vs. 50% for DUF_DUP. Comparing the three simulations using the ARE, the SUF_AUP simulation performs better with a relatively lower absolute error value of 53%.

Table 3. Comparison of WRF rainfall with the stations (AWS and GSOD) and Area-averaged regridded CHIRPS rainfall for DUF_DUP, DUF_AUP, and SUF_AUP simulations for 25 June 2012 rainfall events in Kampala, Uganda. The areal rainfall amount is the average of all grids in the innermost domain of WRF.

Simulations	Gridcell-Rainfall AWS (mm)			Gridcell-Rainfall GSOD (mm)			Area-Averaged Rainfall (mm)			ARE
	AWS	WRF	RE (%)	GSOD	WRF	RE (%)	CHIRPS	WRF	RE (%)	ARE (%)
DUF_DUP	66	25	−62	60	13	−78	16	24	+50	63
DUF_AUP	66	16	−75	60	13	−78	16	23	+44	66
SUF_AUP	66	21	−68	60	13	−47	16	18	+13	53

The spatial distribution of the total 24 h rainfall amount from CHIRPS [22] and 3 WRF simulations are shown in Figure 4. Based on the CHIRPS rainfall, the maximum rainfall accumulations are located to the southeast of the Kampala city catchment along Lake Victoria's coastline, with a peak accumulation of 43 mm at location *X*. It is worth noting that the CHIRPS 24 h rainfall amount at the gauging stations is 30 mm, which is about a half less than the amount observed at gauging stations. The insets in Figure 4b–d show the different plots of the simulation with CHIRPS. The results of three WRF simulations, both in terms of maximum accumulated rainfall and its spatial distribution, show not in good agreement with that of CHIRPS rainfall. The difference in accumulated rainfall between the CHIRPS and the DUF_DUP and DUF_AUP simulations is about 17 and 22 mm (dark yellow color in the insets in the bottom-right corner in Figure 4b,c), respectively, while in the other locations, the difference is negative (light yellow color). The result indicates that the model simulations captured the spot with less daily rainfall than CHIRPS. In the SUF_AUP simulation, the difference in accumulated rainfall between the CHIRPS and the SUF_AUP simulation is 40 mm at the spotted location (red color in the insets in the bottom-right corner in Figure 4d), which indicates that the location of simulated maximum daily rainfall displaced place compared to CHIRPS. The maximum negative difference in the accumulated rainfall above −60 mm is located in the city center in the case of SUF_AUP (dark Blue color), which indicates that the peak simulated rainfall is displaced compared to CHIRPS in all cases.

3.3. *Impact on 24 h Rainfall Amount*

This section presents the impact of the urban landscape on 24 h rainfall amount and the inter-comparison between the three simulations (the DUF_DUP, DUF_AUP, and SUF_AUP) and benchmarked with the CHIRPS observation.

In the DUF_DUP simulation, the maximum rainfall accumulation (80 mm) is located in the southwest part of the Kampala catchment, as indicated by *X* in Figure 4b. Heavy rainfall amount greater than the observed rainfall (i.e., 60 mm) extends from Lake Victoria in the south/southeast to the northwest part of the Kampala catchment. In the DUF_AUP simulation, the accumulated rainfall's spatial distribution follows a similar distribution pattern as DUF_DUP, except that the peak accumulation is higher (89 mm) at location *X* Figure 4c. Moreover, the location of the cluster of peak accumulation moves further to the northwest of the city (as indicated by *Y* in Figure 4c). In the simulation with the updated urban fraction and its parameters (SUF_AUP), the spatial rainfall pattern changed. The heavy rainfall is concentrated at the center of the city with a peak accumulation of 82 mm, as shown in location *X* in Figure 4d. The heavy rainfall distribution indicates the cluster of peak rainfall at three different locations (two along the coastline of Lake Victoria and one in the city center).

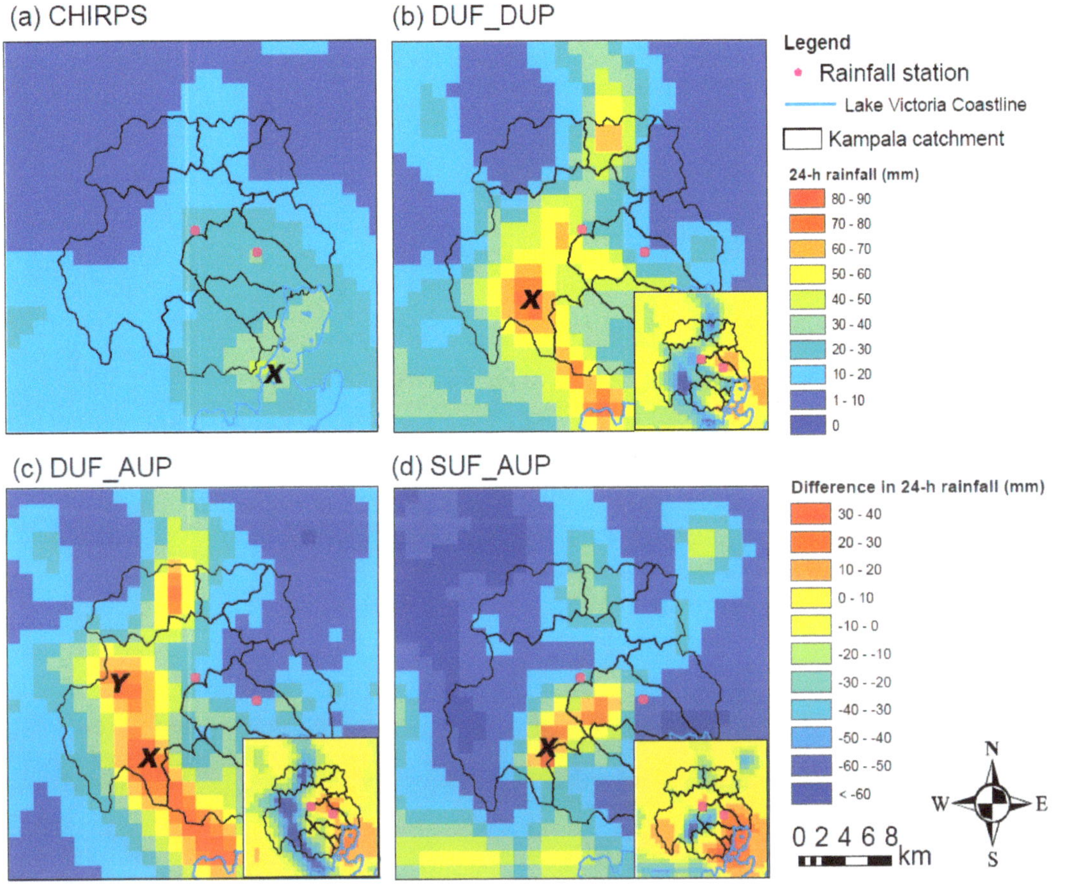

Figure 4. 24 h accumulated rainfall for (**a**) CHIRPS, (**b**) DUF_DUP, (**c**) DUF_AUP, and (**d**) SUF_AUP simulations. The insets in the bottom-right corner in (**b–d**) are the difference in the 24 h accumulated rainfall in the DUF_DUP, DUF_AUP, and SUF_AUP simulations from the CHIRPS observation, respectively.

3.4. Impact on 2 h Rainfall Amount

The WRF model simulations are also examined to understand the precipitation event's time evolution over the catchment by giving special attention to the timing of the observed event. Based on the Automatic Weather Station data, we know that the 25 June 2012 rainfall event lasted for two hours, from 11:10 UTC to 12:50 UTC, and we use this duration as a reference for the time evolution analysis. Figure 5 shows the cumulative rainfall curves for the observation and three WRF simulations at the AWS location. For all simulations, the event's start is very close to the observation (i.e., about +/− 30 min), which is an outstanding result given that rainfall is highly erratic. However, all simulations do not adequately capture the end duration and time to the peak. Compared to observations, the modeled storms start a half-hour earlier for SUF_AUP and a half-hour later for DUF_DUP and DUF_AUP simulations. The time to peak (the time at the steepest slope attain) is about an hour after the observation for both DUF_DUP and DUF_AUP simulations. In the SUF_AUP simulation, the time to peak coincides well with the observed event but with a lower rain rate per minute. The result is the cumulative rainfall for the duration equivalent to the observation at the AWS location. However, due to the spatial and temporal variability of the simulated rainfall event, the analysis for a longer duration and also the analysis at different gridcell can result in a different outcome. For example, the 2 h accumulated rainfall of about 60–70 mm is simulated by all three simulations but at different locations than the AWS (see Figure 6).

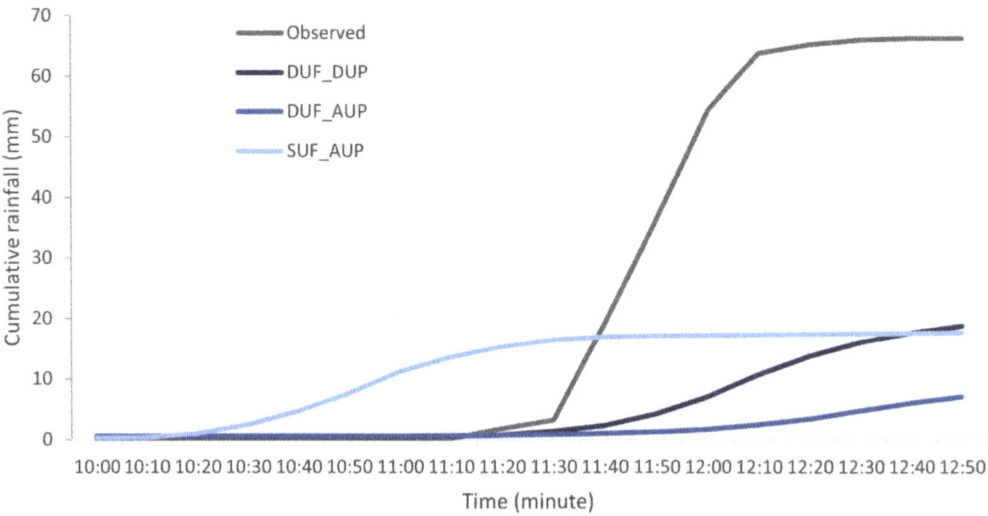

Figure 5. Cumulative rainfall curves for observation and three WRF simulations at the AWS location. Gridcell-rainfall curves for the DUF_DUP, DUF_AUP, and SUF_AUP simulations are shown in three-hour time windows from 10:00 to 12:50, equivalent to the observation at the AWS location. The 2 h rainfall analysis focuses on the duration between 11:00 to 12:50, where the maximum peak intensity was captured by AWS observation.

The spatial distribution of the 2 h rainfall over the catchment is examined in Figure 6. In the same period of two hours, from 11:00 to 12:50, in the DUF_DUP simulation, the cluster of maximum rainfall accumulation (61 mm) is located to the southeast of the Kampala catchment area (i.e., on the edge of the catchment boundary) and extended further to the northwest of the catchment boundary. In the DUF_AUP simulation, the pattern of rainfall distribution over the catchment is similar to that of the DUF_DUP, but the maximum rainfall accumulation (72 mm) is located in the northwest of the catchment area. In the simulation in which an updated urban fraction is used (SUF_AUP simulation), the rainfall pattern is different, with a single maximum rainfall accumulation (75 mm) located in the city center. Moreover, with an updated urban fraction, the total volume of rainfall over the urban catchment is less than when using the default urban fraction. The results suggest that in addition to updated urban parameters, the change in the urban fraction (i.e., changes the intensity of urban fraction and urban extent) alters the amount, structure, and propagation of high-intensity rainfall over the city. Moreover, compared to the 24 h rainfall analysis, the pattern and location of the simulated maximum rainfall clusters for 2 h are similar to that of 24 h. However, the amount is less in the case of a 2 h duration. For instance, in the case of a 2 h duration, the maximum rainfall accumulation is reduced by 17 mm for DUF_AUP and 7 mm for SUF_AUP simulation. The reduction in rainfall amount for a 2 h duration is mainly due to temporal variability of the simulated rainfall, which is linked to the more prolonged instability in the atmosphere.

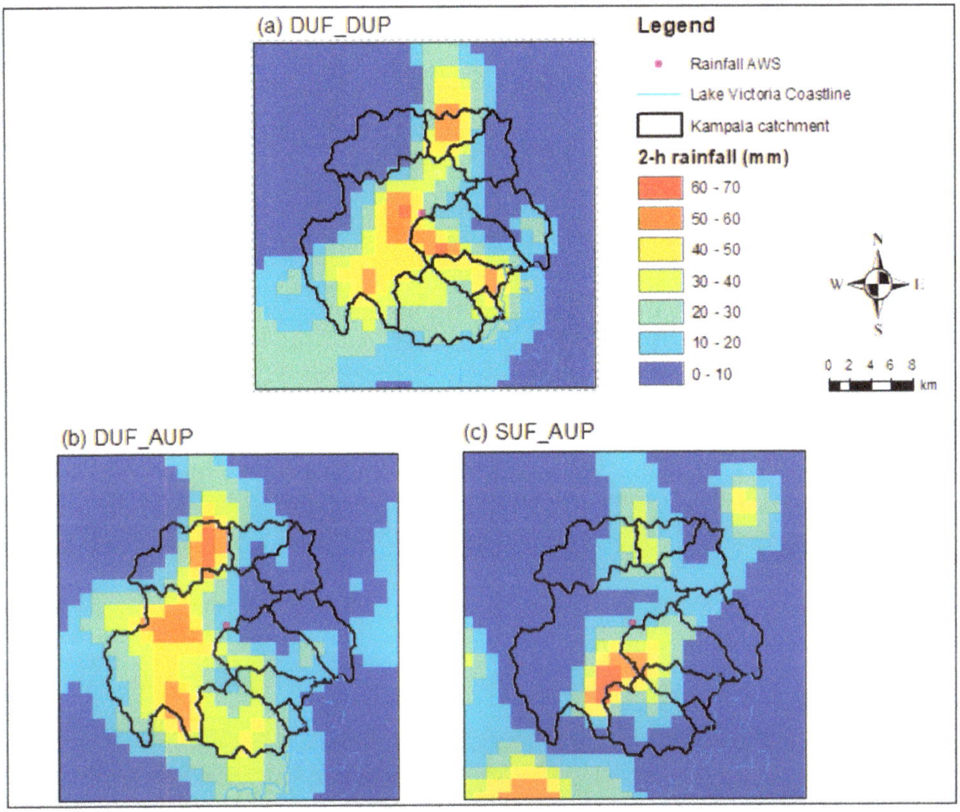

Figure 6. 2 h accumulated rainfall distribution for model simulations for the period of 11:00–12:50 UTC on 25 June 2012 (the same period as the observed rainfall using AWS in Kampala): (**a**) DUF_DUP, (**b**) DUF_AUP, and (**c**) SUF_AUP simulations.

A comparison of DUF_AUP and SUF_AUP simulations with the DUF_DUP simulation over the catchment represents the impact of urban setting changes (Figure 7). The figure presents the subtraction of the DUF_DUP simulation from the two experiments, as the DUF_DUP is the reference against which we compare. The red color in the figure shows where the simulation with the experiments produces more rainfall, while the blue color gives the places the simulation with the default urban landscape produces more rainfall. The average difference in accumulated rainfall between the DUF_AUP and the DUF_DUP simulations is 22 mm (Figure 7a). The negative difference of less than −30 mm is located at the Kampala catchment's north, central, and southeast (dark blue color). In the SUF_AUP simulation, the positive accumulated rainfall difference with the DUF_DUP simulation is, on average, +14 mm, particularly in the city center and the northern part, outside the Kampala catchment boundary (Figure 7b). The negative differences of less than −30 mm are simulated at several locations, mainly in the city center, north and southeast of the catchment (dark blue color).

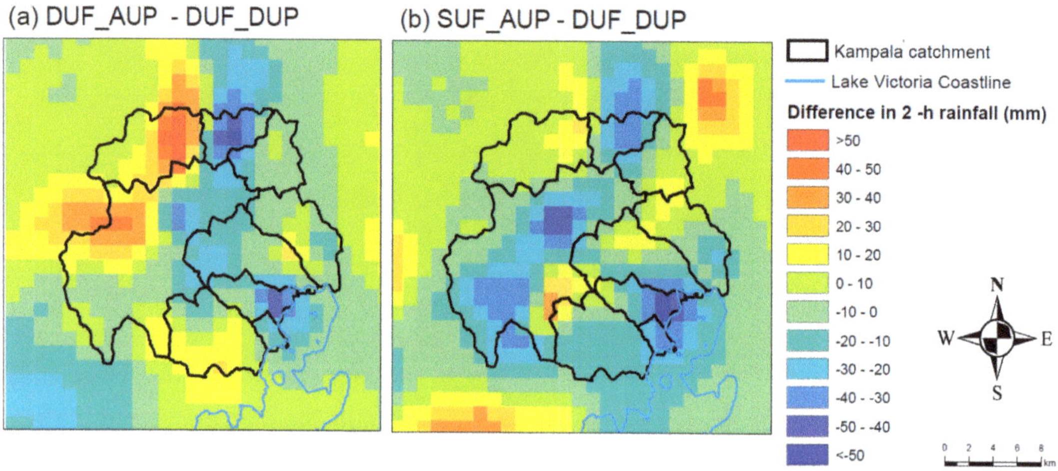

Figure 7. 2 h accumulated rainfall difference for the period of 11:00–12:50 UTC on 25 June 2012 (the same period as the observed rainfall using AWS in Kampala): subtractions of DUF_DUP simulation from (**a**) DUF_AUP and (**b**) SUF_AUP simulations.

The result also shows the inter-difference in the pattern and structure of the simulated rainfall between the DUF_AUP and SUF_AUP simulations. As shown in the figure, except in the southeast over Kampala city (red) and the furthest north of the city (blue), where the differences are the same, the patterns differ a lot elsewhere.

4. Discussion

Updating urban fractions in the WRF model can have a scientific significance in urban hydrometeorological modeling particularly to properly address the impact of urban surface heterogeneity on meteorological variables. In this research, we assessed and evaluated the impact of the updated urban fraction and urban parameters on the simulated high-intensity rainfall that can be used for proper flood hazard modeling in the urbanized and data-scarce area of Kampala, Uganda. The results indicate that the spatial distribution of the simulated event is extremely changed with the updated urban fraction. The results can be further improved by considering the effects of the explicit spatial differences in the urban morphology within the urban fraction, which is essential for a detailed analysis of hydro-meteorological processes and their impact on urban climate. This can be through comparison with the previous landcover and urban fraction development approach commonly used the World Urban Database and Access Portal Tools (WUDAPT) or Local Climate Zones (LCZ) in the WRF model. In particular, the LCZ map that the WRF model often uses was already developed for Kampala to study the variations in urban temperature and their links to health issues in the urban area [18]. Furthermore, the quality of the modeling can be improved by refining the WRF innermost domain resolution as close as that of the satellite pixel sizes to capture the actual information provided by the satellite image.

The model simulation accuracy evaluation results with the relative error (RE) index showed that the model reliably detects a cluster of maximum peak events over the city when using the updated urban fraction. When comparing the overall magnitude of error for each simulation, the SUF_AUP simulation performs better with a lower ARE score of 53%. The better performance when using an updated urban fraction is mainly due to the correct simulation of the spatial distribution of extreme grid-value rainfall events compared to when using the default urban fraction. The intercomparison between the simulations indicated that the presence of the urban landscape alters both the pattern and propagation of high-intensity rainfall over the city, mainly due to the modification of moisture transport, heat, and wind fields by the urban landscape, as indicated by [4,5].

However, there was an overall lower performance in terms of spatial distribution that was observed for all simulations, which indicates that the simulated events are spatially dislocated compared to observation. The main reason for the spatial discrepancies of the simulated events is due to a lack of sufficient observation in the urban catchment of the city for verification. Thus, two gauging stations cannot capture the simulated events across the Kampala catchment. The CHIRPS, which is the grided observed data, also indicated that the actual daily maximum rainfall in the southeastern part of the city as opposed to the WRF simulation. Overcoming the data-scarcity problem requires the installation of sufficient instruments in the catchment.

Our analysis showed that the updated urban landscape affected the location and pattern and the simulated rainfall amount over the city in two ways:

(1) Impact on location and pattern of the simulated event: The location and pattern of the simulated rainfall event are primarily affected by the updated urban parameters. Notably, unlike the DUF_DUP simulation, in the DUF_AUP simulation, the clusters of extreme rainfall concentration are increased and stretched from lake Victoria (south) to the northwest outskirt of the city. Adjusting the urban parameters (Table 2), where most values changed in favor of heat absorption by buildings during the daytime, is the main factor causing the area's simulated rainfall displacement, which is in agreement with similar studies, for example, [7,20]. Moreover, the location and pattern of the simulated event are also affected by the updated urban fraction, as indicated in the two-hour rainfall analysis (Figure 6). Unlike the simulations that use the default urban fraction, the urbanization intensity is low in the case of the SUF_AUP simulation, which results in low drag resistance, leading to the cluster of peak events occurring in the city center. Moreover, compared to the default MODIS-based Noah urban fraction [35], the SUF_AUP simulation uses the more realistic extent of the city and fraction, resulting in a more realistic rainfall pattern compared to when using the default urban fraction, particularly in capturing the location of the event triggers the flood event. Consequently, the simulated extreme rainfall was moving south–north over the urban area. In contrast, the results of DUF_DUP and DUF_AUP simulations indicate that the cluster of extreme rainfall moves southwest, then northwest direction while decreases in rainfall amount in the urban area. Similar studies also demonstrated that the urban surface, through its surface resistance and drag force, plays a vital role in hindering the movement and speed of rainfall systems from moving toward urban areas [44,45].

(2) Impact on the amount of simulated peak event: updating urban parameters and urban fraction affected the amount of the simulated peak event. With the DUF_AUP simulation, peak rainfall event increases compared to the DUF_DUP simulation, particularly at a 24 h time scale, which is expected as the adjusted urban parameters enhance instability in the boundary layer. The presence of a high-urban fraction may act as a barrier that might split the convective cells over the city, and the high moisture in the atmosphere, as indicated by [44], may lead to an increase in the simulated rainfall. However, with SUF_AUP simulation that uses a low-urban fraction, rainfall amount is decreased compared to when using the default urban fraction (Figure 6), which is due to the reduced sensible heat flux that might lead to hindered instability in the atmosphere. The result further indicated that for the simulation with an updated urban fraction, although the considered urban extent (i.e., the area covered by urban grids) is more compared to the default urban fraction (Figure 3), the cluster of simulated heavy rainfall is fewer. This result implies that urban land surface heterogeneities are essential in affecting the mechanisms leading to the amount and spatial distribution of high-intensity rainfall events.

In the end, it is essential to outline some limitations to using the current procedure in the WRF model. One of the limitations is the stretched urban grid values created due to the resampling of the Landsat urban fraction from 30 m resolution to the model's 1 km resolution. This shrinks the original grid values that represent high-intensity resident areas into grid values that resample more of low-intensity urban areas. Future studies further refining the WRF downscaling into high resolution, for instance, 250 m, will reduce the limitation introduced by rescaling grids. Another issue worth mentioning is that this study applied the current procedure in a single city using a single rainfall event. Future studies that will consider a greater number of rainfall events in the different urban areas can further strengthen the simulation results.

5. Conclusions

The mesoscale WRF model's standard representation of the urban areas is often not representative of specific cities. Efforts have been made to incorporate the satellite-driven urban fraction into the WRF model for the proper simulation of extreme rainfall over Kampala, Uganda. The main approach is to utilize the high-resolution urban fraction derived based on Landsat into WRF model configured at 1 km and compare the results with the default urban fraction. Satellite urban fraction was initially developed for urban flood modeling and then following the procedure in the WRF manual we used that for WRF modeling, and urban parameters linked to the urban fractions were adjusted based on the literature review. The 25 June 2012 rainfall event that caused a localized flood event in Kampala was selected for this study. Three simulations were compared: first, with the default settings for urban fraction and its parameters (DUF_DUP); second, the simulation adjusting only the urban parameters (DUF_AUP); and finally, the simulation implementing the new urban fraction in combination with adjusted urban parameters (SUF_AUP). The peak rainfall and its spatial distribution over the Kampala catchment were evaluated using observed data from two gauging stations and the CHIRPS satellite precipitation dataset.

The results indicated benefits in several aspects of using the updated urban fraction over the default urban fraction. Mainly, the updated urban fraction represents the correct position and extent of the city leading to changes in storm structure, evolution and intensity). The results of this research indicate that the updated urban fraction in the WRF model based on the Landsat image is valuable information for the proper simulation of a high-intensity convective storm.

Compared to the observation, the spatial distribution and timing of convective storms are well captured by the WRF model when using the updated urban fraction. However, with all simulations, the WRF model overestimates rainfall compared to the CHIRPS and underestimates compared to gridcell values at gauging stations. The discrepancies between the model simulations and the CHIRPS observations are the known limitation of CHIRPS in capturing the maximum rainfall amount. Additionally, due to the absence of a dense urban gauging station network, there is no proper spatio-temporal record of the rainfall event over the city. Based on the available observations, the SUF_AUP simulation with a more realistic urban fraction and adjusted urban parameters shows relatively better performance with the lowest ARE score compared to the other two simulations.

To assess the impact of the updated urban landscape on the simulated rainfall, we analyzed rainfall distribution and amount for 24 h to understand the impact on the simulated daily rainfall and 2 h for which flood hazards occurred. Our results showed that the WRF model configuration with default urban fraction produces more peak rainfall amounts over the city with its spatial distribution covering wider areas. In contrast, the updated urban fraction has less cluster of peak rainfall events with less spatial distribution coverage within the urban catchment boundary. The results indicated benefits in several aspects of using the updated urban fraction over the default urban fraction. Mainly, the updated urban fraction represents the correct position and extent of the city that produces peak rainfall amount close to the observation. Moreover, the updated urban fraction represents the correct urban intensity that leads to less effect on the simulated rainfall. This study demonstrated that the explicit use of the satellite-derived urban fraction for NWP modeling is advantageous over the standard urban classification, mainly in two aspects: First, it represents the correct extent and position of the urban area, and second, it is possible to produce a future prediction of urbanization and then used in the NWP model for future impact assessment and local climate study. Thus, the study contributes to the emerging understanding of the usability of the high-resolution urban fractions from the remote sensing image in the NWP model to properly account for the impact of urban heterogeneity on extreme rainfall events. Moreover, the proper updating of land-use/land cover information in the NWP model contributes to improving model forecasting ability, particularly for the localized early-warning system.

Author Contributions: Conceptualization, Y.U.; methodology, Y.U. and J.E.; software, Y.U., J.E., and G.-J.S.; validation, Y.U.; formal analysis, Y.U.; investigation, Y.U.; resources, J.E. and V.J.; data curation, Y.U.; writing—original draft preparation, Y.U.; writing—review and editing, Y.U., G.-J.S., and J.E.; visualization, Y.U.; supervision, J.E. and V.J.; project administration, V.J.; funding acquisition, J.E. and V.J. All authors have read and agreed to the published version of the manuscript.

Funding: This research was conducted under the regular research program at the University of Twente.

Data Availability Statement: Data will be made available on request.

Acknowledgments: We would like to express our thanks to the University of Twente for funding this research. The authors would also like to thank Gemechu Fanta and Reinder Ronda for their support on managing and adjusting the urban fraction and vegetation cover fraction in the WRF model.

Conflicts of Interest: The authors declare no conflict of interest.

References

1. Wang, H.; Hu, Y.; Guo, Y.; Wu, Z.; Yan, D. Urban flood forecasting based on the coupling of numerical weather model and stormwater model: A case study of Zhengzhou city. *J. Hydrol. Reg. Stud.* **2022**, *39*, 100985. [CrossRef]
2. Umer, Y.; Jetten, V.; Ettema, J.; Lombardo, L. Application of the WRF model rainfall product for the localized flood hazard modeling in a data-scarce environment. *Nat. Hazards* **2022**, *111*, 1813–1844. [CrossRef]
3. Umer, Y.; Ettema, J.; Jetten, V.; Steeneveld, G.J.; Ronda, R. Evaluation of the WRF Model to Simulate a High-Intensity Rainfall Event over Kampala, Uganda. *Water* **2021**, *13*, 873. [CrossRef]
4. Paul, S.; Ghosh, S.; Mathew, M.; Devanand, A.; Karmakar, S.; Niyogi, D. Increased spatial variability and intensification of extreme monsoon rainfall due to urbanization. *Sci. Rep.* **2018**, *8*, 3918. [CrossRef] [PubMed]
5. Ryu, Y.-H.; Smith, J.A.; Bou-Zeid, E.; Baeck, M.L. The influence of land surface heterogeneities on heavy convective rainfall in the Baltimore–Washington metropolitan area. *Mon. Weather Rev.* **2016**, *144*, 553–573. [CrossRef]
6. Dai, J.; Wang, X.; Dai, W.; Chang, M. The impact of inhomogeneous urban canopy parameters on meteorological conditions and implication for air quality in the Pearl River Delta region. *Urban Clim.* **2019**, *29*, 100494. [CrossRef]
7. Wouters, H.; Demuzere, M.; Blahak, U.; Fortuniak, K.; Maiheu, B.; Camps, J.; Tielemans, D.; Van Lipzig, N. The efficient urban canopy dependency parametrization (SURY) v1. 0 for atmospheric modelling: Description and application with the COSMO-CLM model for a Belgian summer. *Geosci. Model Dev.* **2016**, *9*, 3027–3054. [CrossRef]
8. Stewart, I.D.; Oke, T.R. Local climate zones for urban temperature studies. *Bull. Am. Meteorol. Soc.* **2012**, *93*, 1879–1900. [CrossRef]
9. Chen, F.; Kusaka, H.; Bornstein, R.; Ching, J.; Grimmond, C.; Grossman-Clarke, S.; Loridan, T.; Manning, K.W.; Martilli, A.; Miao, S. The integrated WRF/urban modelling system: Development, evaluation, and applications to urban environmental problems. *Int. J. Climatol.* **2011**, *31*, 273–288. [CrossRef]
10. Pathirana, A.; Denekew, H.B.; Veerbeek, W.; Zevenbergen, C.; Banda, A.T. Impact of urban growth-driven landuse change on microclimate and extreme precipitation—A sensitivity study. *Atmos. Res.* **2014**, *138*, 59–72. [CrossRef]
11. Sati, A.P.; Mohan, M. Impact of urban sprawls on thunderstorm episodes: Assessment using WRF model over central-national capital region of India. *Urban Clim.* **2021**, *37*, 100869. [CrossRef]
12. Shepherd, J.M.; Burian, S.J. Detection of Urban-Induced Rainfall Anomalies in a Major Coastal City. *Earth Interact.* **2003**, *7*, 1–17. [CrossRef]
13. Salamanca, F.; Martilli, A.; Tewari, M.; Chen, F. A study of the urban boundary layer using different urban parameterizations and high-resolution urban canopy parameters with WRF. *J. Appl. Meteorol. Climatol.* **2011**, *50*, 1107–1128. [CrossRef]
14. Yang, L.; Smith, J.; Niyogi, D. Urban Impacts on Extreme Monsoon Rainfall and Flooding in Complex Terrain. *Geophys. Res. Lett.* **2019**, *46*, 5918–5927. [CrossRef]
15. Brousse, O.; Martilli, A.; Foley, M.; Mills, G.; Bechtel, B. WUDAPT, an efficient land use producing data tool for mesoscale models? Integration of urban LCZ in WRF over Madrid. *Urban Clim.* **2016**, *17*, 116–134. [CrossRef]
16. Alexander, P.J.; Bechtel, B.; Chow, W.T.L.; Fealy, R.; Mills, G. Linking urban climate classification with an urban energy and water budget model: Multi-site and multi-seasonal evaluation. *Urban Clim.* **2016**, *17*, 196–215. [CrossRef]
17. Hu, C.; Fung, K.Y.; Tam, C.-Y.; Wang, Z. Urbanization Impacts on Pearl River Delta Extreme Rainfall Sensitivity to Land Cover Change Versus Anthropogenic Heat. *Earth Space Sci.* **2021**, *8*, e2020EA001536. [CrossRef]

18. Brousse, O.; Georganos, S.; Demuzere, M.; Vanhuysse, S.; Wouters, H.; Wolff, E.; Linard, C.; Nicole, P.-M.; Dujardin, S. Using local climate zones in Sub-Saharan Africa to tackle urban health issues. *Urban Clim.* **2019**, *27*, 227–242. [CrossRef]
19. Perez Molina, E. Spatial Planning, Growth, and Flooding: Contrasting Urban Processes in Kigali and Kampala. Ph.D. Thesis, University of Twente, Enschede, The Netherlands, 2019.
20. Umer, Y. Advances in Localized Flood Hazard Modelling in Urbanized and Data-Scare Areas. Ph.D. Thesis, University of Twente, Enschede, The Netherlands, 2022.
21. Sliuzas, R.; Flacke, J.; Jetten, V. Modelling urbanization and flooding in Kampala, Uganda. In Proceedings of the Network-Association of European Researchers on Urbanisation in the South (N-AERUS) XIV, Enschede, The Netherlands, 12–14 September 2013.
22. Funk, C.; Peterson, P.; Landsfeld, M.; Pedreros, D.; Verdin, J.; Shukla, S.; Husak, G.; Rowland, J.; Harrison, L.; Hoell, A. The climate hazards infrared precipitation with stations—A new environmental record for monitoring extremes. *Sci. Data* **2015**, *2*, 150066. [CrossRef]
23. Diem, J.E.; Konecky, B.L.; Salerno, J.; Hartter, J. Is equatorial Africa getting wetter or drier? Insights from an evaluation of long-term, satellite-based rainfall estimates for western Uganda. *Int. J. Climatol.* **2019**, *39*, 334–3347. [CrossRef]
24. Cattani, E.; Merino, A.; Guijarro, J.; Levizzani, V. East Africa rainfall trends and variability 1983–2015 using three long-term satellite products. *Remote Sens.* **2018**, *10*, 931. [CrossRef]
25. Anyah, R.O. *Modeling the Variability of the Climate System over Lake Victoria Basin*; North Carolina State University: Raleigh, NC, USA, 2005.
26. Sun, X.; Xie, L.; Semazzi, F.; Liu, B. Effect of lake surface temperature on the spatial distribution and intensity of the precipitation over the Lake Victoria basin. *Mon. Weather Rev.* **2015**, *143*, 1179–1192. [CrossRef]
27. Pérez-Molina, E.; Sliuzas, R.; Flacke, J.; Jetten, V. Developing a cellular automata model of urban growth to inform spatial policy for flood mitigation: A case study in Kampala, Uganda. *Comput. Environ. Urban Syst.* **2017**, *65*, 53–65. [CrossRef]
28. Powers, J.G.; Klemp, J.B.; Skamarock, W.C.; Davis, C.A.; Dudhia, J.; Gill, D.O.; Coen, J.L.; Gochis, D.J.; Ahmadov, R.; Peckham, S.E. The weather research and forecasting model: Overview, system efforts, and future directions. *Bull. Am. Meteorol. Soc.* **2017**, *98*, 1717–1737. [CrossRef]
29. Liu, J.; Bray, M.; Han, D. Sensitivity of the Weather Research and Forecasting (WRF) model to downscaling ratios and storm types in rainfall simulation. *Hydrol. Process.* **2012**, *26*, 3012–3031. [CrossRef]
30. Hersbach, H.; Bell, B.; Berrisford, P.; Hirahara, S.; Horányi, A.; Muñoz-Sabater, J.; Nicolas, J.; Peubey, C.; Radu, R.; Schepers, D. The ERA5 global reanalysis. *Q. J. R. Meteorol. Soc.* **2020**, *146*, 1999–2049. [CrossRef]
31. Martilli, A.; Clappier, A.; Rotach, M.W. An urban surface exchange parameterisation for mesoscale models. *Bound.-Layer Meteorol.* **2002**, *104*, 261–304. [CrossRef]
32. Salamanca, F.; Krpo, A.; Martilli, A.; Clappier, A. A new building energy model coupled with an urban canopy parameterization for urban climate simulations—Part I. Formulation, verification, and sensitivity analysis of the model. *Theor. Appl. Climatol.* **2010**, *99*, 331. [CrossRef]
33. Kusaka, H.; Kondo, H.; Kikegawa, Y.; Kimura, F. A simple single-layer urban canopy model for atmospheric models: Comparison with multi-layer and slab models. *Bound.-Layer Meteorol.* **2001**, *101*, 329–358. [CrossRef]
34. Song, J.; Wang, Z.-H. Interfacing the urban land–atmosphere system through coupled urban canopy and atmospheric models. *Bound.-Layer Meteorol.* **2015**, *154*, 427–448. [CrossRef]
35. Ran, L.; Gilliam, R.; Binkowski, F.S.; Xiu, A.; Pleim, J.; Band, L. Sensitivity of the Weather Research and Forecast/Community Multiscale Air Quality modeling system to MODIS LAI, FPAR, and albedo. *J. Geophys. Res. Atmos.* **2015**, *120*, 8491–8511. [CrossRef]
36. Wang, W.; Bruyere, C.; Duda, M.; Dudhia, J.; Gill, D.; Kavulich, M.; Keene, K.; Lin, H.-C.; Michalakes, J.; Rizvi, S. *User's Guides for the Advanced Research WRF (ARW) Modeling System*; Version 3; National Center For Atmospheric Research Mesoscale and Microscale: Boulder, CO, USA, 2018.
37. Gilliam, R.C.; Pleim, J.E. Performance assessment of new land surface and planetary boundary layer physics in the WRF-ARW. *J. Appl. Meteorol. Climatol.* **2010**, *49*, 760–774. [CrossRef]
38. Cai, M.; Ren, C.; Xu, Y.; Lau, K.K.-L.; Wang, R. Investigating the relationship between local climate zone and land surface temperature using an improved WUDAPT methodology—A case study of Yangtze River Delta, China. *Urban Clim.* **2018**, *24*, 485–502. [CrossRef]
39. Oliveros, J.M.; Vallar, E.A.; Galvez, M.C.D. Investigating the Effect of Urbanization on Weather Using the Weather Research and Forecasting (WRF) Model: A Case of Metro Manila, Philippines. *Environments* **2019**, *6*, 10. [CrossRef]
40. Skamarock, W.C.; Klemp, J.B.; Dudhia, J.; Gill, D.O.; Barker, D.M.; Wang, W.; Powers, J.G. *A Description of the Advanced Research WRF Version 2*; National Center for Atmospheric Research Boulder Co Mesoscale and Microscale Meteorology Division: Boulder, CO, USA, 2005.
41. Wang, W.; Barker, D.; Bray, J.; Bruyere, C.; Duda, M.; Dudhia, J.; Gill, D.; Michalakes, J. *User's Guide for Advanced Research WRF (ARW) Modeling System Version 3*; Mesoscale and Microscale Meteorology Division–National Center for Atmospheric Research (MMM-NCAR): Boulder, CO, USA, 2007.
42. Guide, W.U. Chapter 3: WRF Preprocessing System (WPS). Land Use and Soil Categories in the Static Data. 2009. Available online: http://www.2.mmm.ucar.edu/wrf/users/docs/user_guide_V3/users_guide_chap3.htm (accessed on 1 March 2014).

43. Tian, J.; Liu, J.; Yan, D.; Li, C.; Yu, F. Numerical rainfall simulation with different spatial and temporal evenness by using a WRF multiphysics ensemble. *Nat. Hazards Earth Syst. Sci.* **2017**, *17*, 563–579. [CrossRef]
44. Miao, S.; Chen, F.; Li, Q.; Fan, S. Impacts of urban processes and urbanization on summer precipitation: A case study of heavy rainfall in Beijing on 1 August 2006. *J. Appl. Meteorol. Climatol.* **2011**, *50*, 806–825. [CrossRef]
45. Zhang, Y.; Miao, S.; Dai, Y.; Bornstein, R. Numerical simulation of urban land surface effects on summer convective rainfall under different UHI intensity in Beijing. *J. Geophys. Res. Atmos.* **2017**, *122*, 7851–7868. [CrossRef]

hydrology

MDPI

Article

Predicting Urban Flooding Due to Extreme Precipitation Using a Long Short-Term Memory Neural Network

Raphaël A. H. Kilsdonk , Anouk Bomers * and Kathelijne M. Wijnberg

Water Engineering and Management Department, University of Twente, 7500 AE Enschede, The Netherlands; rahkilsdonk@gmail.com (R.A.H.K.); k.m.wijnberg@utwente.nl (K.M.W.)

* Correspondence: a.bomers@utwente.nl; Tel.: +31-5-3489-1062

Abstract: Extreme precipitation events can lead to the exceedance of the sewer capacity in urban areas. To mitigate the effects of urban flooding, a model is required that is capable of predicting flood timing and volumes based on precipitation forecasts while computational times are significantly low. In this study, a long short-term memory (LSTM) neural network is set up to predict flood time series at 230 manhole locations present in the sewer system. For the first time, an LSTM is applied to such a large sewer system while a wide variety of synthetic precipitation events in terms of precipitation intensities and patterns are also captured in the training procedure. Even though the LSTM was trained using synthetic precipitation events, it was found that the LSTM also predicts the flood timing and flood volumes of the large number of manholes accurately for historic precipitation events. The LSTM was able to reduce forecasting times to the order of milliseconds, showing the applicability of using the trained LSTM as an early flood-warning system in urban areas.

Keywords: machine learning; sewer model; LSTM neural network; urban sewer flooding

Citation: Kilsdonk, R.A.H.; Bomers, A.; Wijnberg, K.M. Predicting Urban Flooding Due to Extreme Precipitation Using a Long Short-Term Memory Neural Network. *Hydrology* **2022**, *9*, 105. https://doi.org/10.3390/hydrology9060105

Academic Editors: Aristoteles Tegos, Alexandros Ziogas and Vasilis Bellos

Received: 6 May 2022
Accepted: 6 June 2022
Published: 10 June 2022

Publisher's Note: MDPI stays neutral with regard to jurisdictional claims in published maps and institutional affiliations.

1. Introduction

Extreme precipitation events, of both short and long duration, can cause inundations locally or downstream of a catchment due to raising river water levels [1]. This research focuses on local flooding due to extreme precipitation events and more specifically on urban flooding due to the exceedance of the sewer capacity. Pluvial urban flooding can occur quite suddenly, and therefor, early flood warning systems with a short run time are desired such that proper flood mitigation measures can be taken in time. Urban flooding differs from flooding in other areas because of the large amount of impervious surface area negating infiltration and increasing the load on sewer systems. Flooding in an urban environment is caused by short extreme precipitation events where infiltration is negligible. It is expected that flood probabilities will increase in the future due to an increase in impervious surface area, causing more runoff to the sewer system. In addition, due to climate change, it is expected that rainfall intensities will increase locally, resulting in higher runoff volumes [2,3].

Numerical models are generally used to investigate the effects of extreme precipitation events on inundation extents and to design sewer systems accordingly. These physics-based models are computationally expensive. Since precipitation forecasts are generally highly uncertain, especially for extreme local events, a probabilistic approach is required to simulate all potential flood scenarios. Consequently, detailed physics-based models cannot be used as a flood early warning systems. However, a fast prediction of the inundated areas during extreme events ensures that flood mitigation measures can be taken on time. For this reason, other approaches for the faster computation of flood predictions have been studied in recent years (e.g., [4,5]). A commonly applied method to reduce computational load is surrogate modelling, representing a second-level abstraction from the original system. Response surface surrogate models, such as machine learning (ML) algorithms, are data-driven models trained based on the input–output relations of a physically based model or field measurements. As a result, ML algorithms do not capture any physical components of the original system. They are, once trained, extremely fast in predicting the output based on a given input [6] and can do so on a continuous basis. For this reason, ML algorithms have frequently been applied for water resources applications [6,7]. More specifically, many studies have already shown the applicability of ML algorithms to predict (historic) stream flow conditions, weather conditions, water quality and dike breaches accurately (e.g., [8–13]). However, the use of ML algorithms for sewer applications is still limited, but they have great possibilities in predicting sewer overflows based on precipitation forecasts.

Recent examples of ML algorithms for sewer system applications are presented by [14,15]. Rjeily et al. [14] developed a data-driven modelling approach to predict water depth variations within the most critical manholes in an urban drainage system. This early flood warning system was trained using measurements of 10 storm events simulated with a hydraulic model. Measured rainfall intensities and modelled water depth variations in five manholes were used as the input and target output data, respectively. Zang et al. [15] studied the accuracy of multiple ML algorithms to predict sewer overflow of a combined sewer system into open water bodies causing heavy pollution. In total, 26 rainfall events resulting in sewer overflow were used to train the various ML algorithms. Although both studies showed the potential of using ML algorithms as an early warning system for sewer applications, these studies only used a few historic events to train the algorithms, while using more samples can ensure better model performance since it is more likely that the global minimum of the error function is found [16]. Therefore, it is questionable if the trained the ML algorithms are able to generalise the system behaviour. Furthermore, because of expected climate change, more extreme precipitation events may occur than observed so far, but these events are not considered in the training data sets if historic events are considered. Therefore, a synthetic data set with a wide variety of rainfall events in terms of both rainfall intensities and rainfall patterns will be used in this study. Additionally, the studies conducted so far only predicted sewer overflow at a few predefined output locations while an overview of the entire sewer system is required to make fair flood mitigation measures during extreme events. For this reason, the objective of this research is to set up an ML algorithm that predicts flood volume time series for all manholes present in a specific urban area, trained on a wide variety of rainfall events. Only then will the developed ML algorithm have the potential to be used as an early flood warning system by decision makers.

The methodology of this research is shown in Figure 1. First, the case study and the numerical sewer model used to create the training data are described (Section 2). A synthetic precipitation data set is constructed since no sufficient historic rainfall events resulting in flood inundations exist and to enable the inclusion of a wider variety of precipitation events than observed so far (Section 3). These synthetic rainfall events are used as input of the numerical sewer model. An ML algorithm is constructed which is able to predict flood volume time series for all manholes in the area as the target output, given a precipitation time series as input (Section 4). The constructed ML algorithm is validated to determine the final performance of the algorithm (Section 5.1). Furthermore, the algorithm is tested based on radar rainfall measurements of a few historic extreme precipitation events (Section 5.2). This paper ends with a discussion (Section 6) and the main conclusions (Section 7).

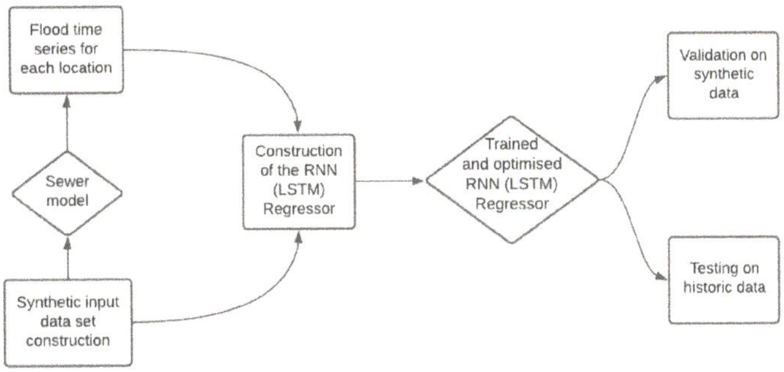

Figure 1. Flow chart of the steps taken in the present research to set up an LSTM that is able to predict inundation volumes at manhole locations.

2. Case Study and the Numerical Sewer Model

The residential area of Hooglanderveen in the city of Amersfoort, the Netherlands, is chosen as a case study since frequent pluvial flooding occurs in this region. Although the region of Hooglanderveen is chosen as a case study, the proposed methods in this study are applicable to any residential area with a similar sewer system and topographical features.

Hooglanderveen is located in the northeast of Amersfoort (see Figure 2) and has a surface area of approximately 1.75 km^2.

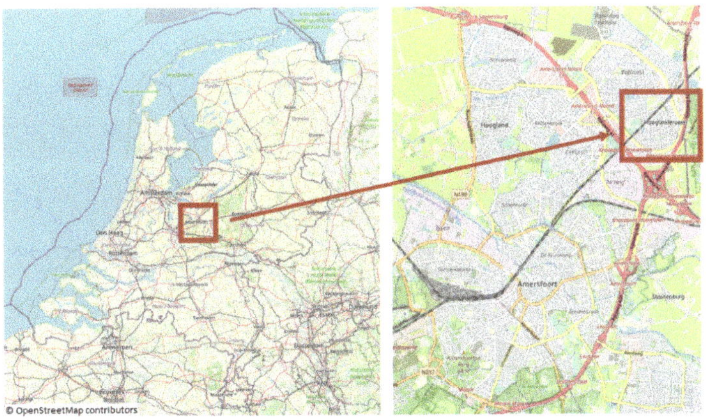

Figure 2. Location of the study area of Hooglanderveen in Amersfoort, The Netherlands.

Especially in the northwestern region of Hooglanderveen, frequent pluvial flooding is experienced, where surface levels are relatively low. The combined sewer system present in Hooglanderveen is a type of gravity sewer and has 230 manholes, 4 pumps, and 3 overflows (Figure 3). These are all connected with sewer pipes (Figure 3). The sewer system transports both precipitation runoff and domestic sewage to a sewage treatment plant and can be divided into two components: (1) the major sewer system, consisting of streets, inlets, ditches, and surface water channels, and (2) the minor sewer system, composed of interconnected pipes, manholes, and pumps [1]. The major system can be characterised as the surface system, whereas the minor system represents the subsurface system. Flooding occurs whenever and wherever the discharge capacity of the inlet into the minor system is exceeded. This can have several causes. First, flooding can occur when precipitation intensity exceeds the discharge capacity of the inlet. Water cannot enter the minor system and remains at the surface level. Second, the discharge capacity may be lower between some sewer pipes due to, e.g., clogging or smaller pipe diameters causing water to flow back onto the streets through the inlets or manholes. Third, the combined gravity-driven sewer system has a larger discharge capacity than the pump at the end of the system. Therefore, a storage is designed in the minor system to accommodate this difference in capacity. This storage is equivalent to approximately 7–9 mm of precipitation in the Netherlands [17]. When the storage capacity is exceeded and more water enters the system, storm water will exit via the overflows. If the capacity of the overflows is exceeded, storm water will flood the streets.

In this study, an ML algorithm is set up to predict flooding in Hooglanderveen in real-time precipitation forecasts. An ML algorithm is generally trained using field measurements based on historical events or outcomes of model simulations. Since insufficient measurements are available of historic precipitation events resulting in flooding in the study area, a numerical sewer model will be used to generate the training data. The numerical sewer model is a validated model built with the software Infoworks ICM. The sewer model represents a one-dimensional (1D) model of the minor system and uses the shallow water equations to solve the 1D flow. Only the surface area of the major system, without considering topographic gradients, is included in the model. Based on these areas, the shortest flow paths to the nearest inlet is determined to compute the inflow from the major system into the minor system. Henonin et al. [18] further details the modelling approach of such a 1D sewer model. The sewer model was calibrated using measurements and is used by local ministries for flood risk evaluation.

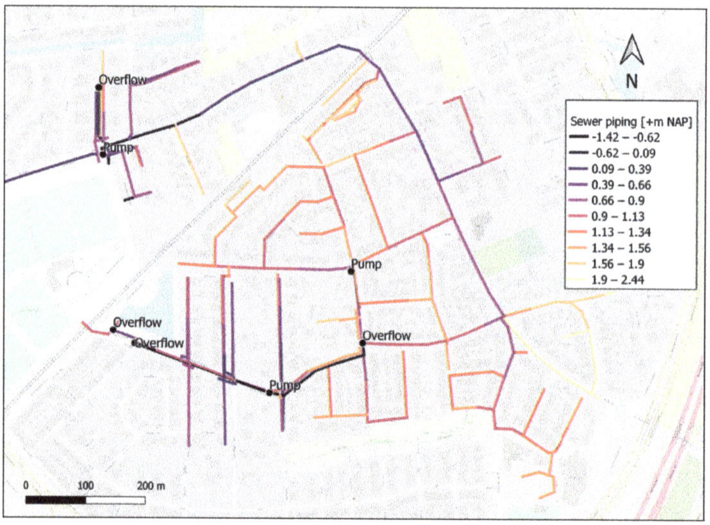

Figure 3. Locations of important structures in the studied area and the level of sewer piping.

The sewer system has a slope from the southeastern to northwestern part of the study area. Since it is a gravity-based sewer system, the general direction of the sewer flow follows this slope. The model has as input a spatially uniform precipitation event and provides as output flood volumes at each manhole in the area. Note that the output is a flood volume and not a flood level, as topographic gradients of the surface level and the flow along these topographic gradients are not included in the model.

3. Training and Testing Data

3.1. The Synthetic Precipitation Events

The sewer model computes flood volumes based on an input precipitation event. In this study, synthetic events are considered to enable the inclusion of a wide variety of precipitation events. These synthetic precipitation events are based on design events to test the sewer systems using numerical models in the Netherlands [19]. Spatially uniform precipitation events are considered because of the relatively small size of the studied area. For the construction of the synthetic precipitation training data set, statistics of the following three precipitation characteristics are used [19]: precipitation duration, precipitation intensity, and precipitation pattern. Combinations between the three characteristics are made to generate unique precipitation events.

Due to the inherent early warning system that is proposed in the present research, we focus on short-term, high-intensity flood events. For this, [19] recommends a precipitation duration of 4, 8, or 12 h. The minimum and maximum precipitation intensities corresponding to a return period of 2 to 1000 years for a duration of 4 and 12 h are 28 mm and 139 mm, respectively (Figure 4 shows the intensity curves for a return period of 2 to 1000 years). To generate the training data set, the precipitation intensities are divided into six values with a minimum and maximum of 30 mm and 105 mm, respectively. The minimum value is taken as the rounded minimum value given by the precipitation curves (Figure 4). The maximum value is set to a lower value than provided by the precipitation curves since increasing the intensity to a value larger than 105 mm did not result in any differences in model output in terms of flood complexity since the number of flooded manholes remained constant. Only the flood volumes increased linearly.

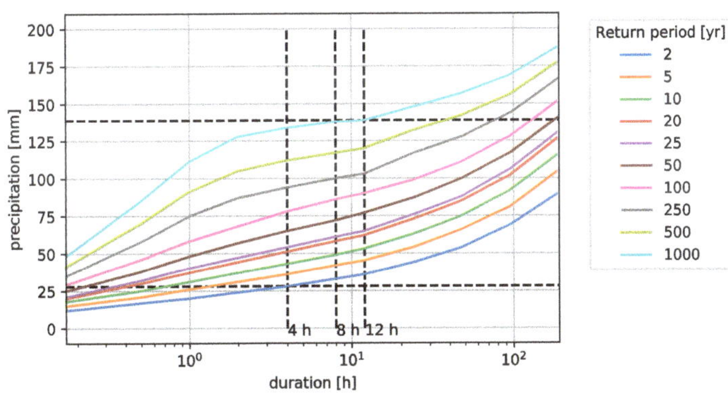

Figure 4. Precipitation intensity curves, the dashed black lines indicate maximum and minimum for the 4, 8, and 12 h durations.

In addition to the precipitation duration and intensity, seven distinct precipitation patterns for short-term events are considered in the Dutch water policy [19]. These patterns consist of a fraction of the total precipitation per hour. The seven precipitation patterns can be described as follows (Figure 5):

- Uniform: General uniform shape with minor changes in precipitation intensity during the event;
- One peak—12.5%: Pattern with one peak that has 12.5% of the total intensity in the peak;
- One peak—37.5%: Pattern with one peak that has 37.5% of the total intensity in the peak;
- One peak—62.5%: Pattern with one peak that has 62.5% of the total intensity in the peak;
- One peak—87.5%: Pattern with one peak that has 87.5% of the total intensity in the peak;
- Two peaks—short distance: Pattern with two peaks that has a small temporal distance between the two peaks;
- Two peaks—large distance: Pattern with two peaks that has a large temporal distance between the two peaks;

With six precipitation intensity values, seven precipitation patterns, and three precipitation durations, the total amount of unique precipitation events is 126. The majority of papers reviewed by [16] use a minimum data set size of 100 samples to train the ML algorithms, indicating that the size of the data set should be sufficiently large to train the ML algorithm properly. All possible values of each precipitation feature are shown in Table 1.

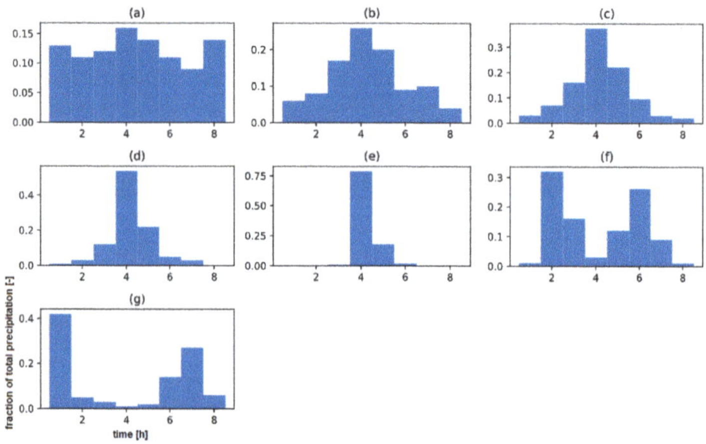

Figure 5. Seven precipitation patterns for a duration of 8 h, with (**a**) Uniform; (**b**) 1 peak—12.5%; (**c**) 1 peak—37.5%; (**d**) 1 peak—62.5%; (**e**) 1 peak—87.5%; (**f**) 2 peaks—short; and (**g**) 2 peaks—long.

Table 1. All possible values for each precipitation event feature.

$P_{intensities}$	$P_{patterns}$	$P_{duration}$
30 mm	Uniform	4 h
40 mm	1 peak—12.5%	8 h
60 mm	1 peak—37.5%	12 h
75 mm	1 peak—62.5%	
90 mm	1 peak—87.5%	
105 mm	2 peaks—short	
	2 peaks—long	

3.2. Interpolation of Precipitation Patterns

The precipitation patterns provided by [19] have a time step of one hour, while the time step of the sewer model is set to one minute to ensure accurate model results. For this reason, the precipitation patterns are linearly interpolated to create realistic precipitation events. Furthermore, to facilitate the operationally of a flood early warning system, the input time series is made to mimic a conventional precipitation forecast. Based on expert opinion, it was found that for short-term precipitation forecasts, a time step of 5 min is generally used. Therefore, the input time series will be a cascading precipitation pattern with a time step of 1 min, which changes its value after every 5 min (Figure 6). Due to this interpolation method, the total precipitation is, at maximum, 2% lower than the value as defined.

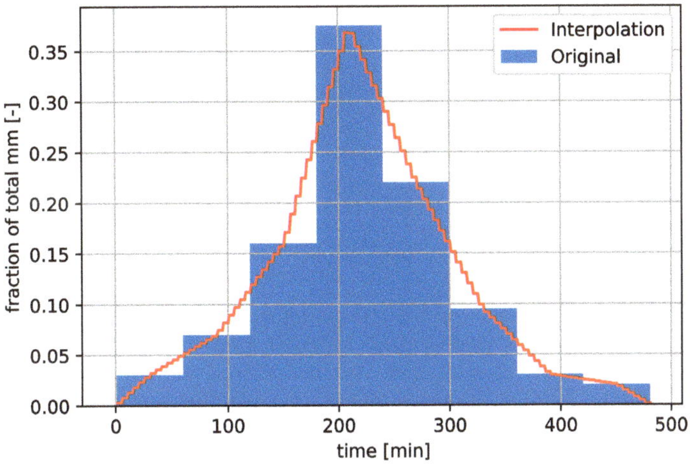

Figure 6. Example interpolation of an eight hour precipitation pattern with a peak of 37.5% of the total precipitation (precipitation pattern as given in Figure 5c).

3.3. Historic Data

The synthetic data set is used to train, validate, and test the LSTM. However, this raises the question whether the LSTM, trained on synthetic data, is capable of reproducing the results of the sewer model on real-world precipitation data. To evaluate this, radar precipitation data from historic extreme precipitation events were obtained. A list of three reported flood events in Hooglanderveen was provided by the municipality of Amersfoort, and related precipitation time series were obtained from precipitation radar data provided by Hydrologic (Figure 7) and used as input for the sewer model. The time series start one day prior to the date that a flood was reported, as there can be a delay between flooding and reporting. All events show large peaks in precipitation up to 106 mm/h. This precipitation peak is higher than the value used in the synthetic data set, having a maximum precipitation of 88 mm/h.

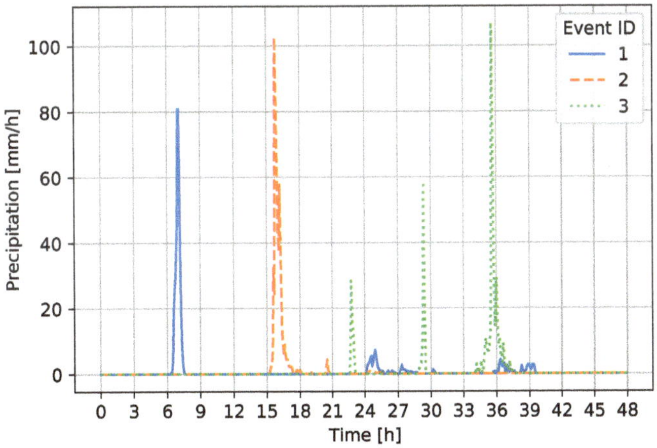

Figure 7. Precipitation time series for historic flood events in Hooglanderveen. All time series start one day prior to the reported flooding, as there can be a delay in reporting. This can be seen with precipitation events 1 and 2.

4. Construction of the Long Short-Term Memory (LSTM) Neural Network

In this study, the LSTM neural network proposed by [20] is used to predict flood volumes for the 230 manholes in the sewer system of Hooglanderveen. Although many neural network structures exist, LSTMs have shown to be most successful and are generally applied to predict time series [21]. More specifically, LSTM has become the focus of deep learning because of their powerful learning capacity in comparison to other recurrent neural network (RNN) approaches [21]. To explain the concept of an LSTM, we first briefly explain artificial neural networks (ANN) and recurrent neural networks (RNN).

4.1. The Concept of Neural Networks

An ANN is a network of interconnected neurons that translate an input to an output using weights and transfer functions. A flowchart of a simple ANN is shown in Figure 8. Here, the inputs (x_i) are multiplied by their weights (w_i), with the result being summed and used as input for the transfer function of the neuron. The result of the transfer function is then used as input for the output function. This output function is a linear function for regression. The output function gives the output (y). The difference between the predicted value and the observed value is then used to change the weights of the ANN. This can be performed using various techniques, with the most common approach being back-propagation with stochastic gradient descent [22]. The transfer function of a neuron can be a linear function, sigmoid function, or any other function. When the ANN is expanded to use more inputs and neurons, all inputs are connected to every neuron with individual weights. One can add as many neurons, inputs. and outputs as desired and can also vary the amount of layers of neurons. The parameters not trained by the neural network, such as the choice of the number of neurons and the type of transfer functions, are called hyper-parameters.

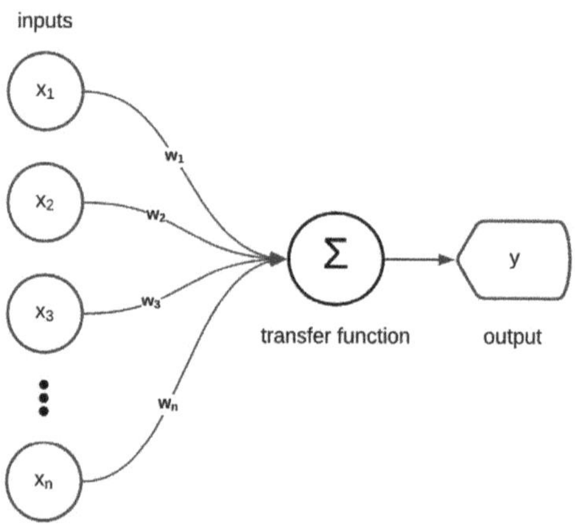

Figure 8. An illustration of a simple ANN. Here, we have multiple inputs (x_i), connected to the neuron with weights (w_i). This output of the neuron is passed to the output (y) via a linear function.

A recurrent neural network is a type of artificial neural network (ANN) that uses the output of previous time steps (y_{t-1}) as input for the current time step (y_t). Therefore, the RNN is better equipped to predict time series than traditional ANNs [23]. However [24] have shown that a simple RNN can barely store information for longer than 10 time steps. Therefore, other approaches to an RNN have been studied, with one of the most commonly applied being the LSTM proposed by [20]. More specifically, [15] compared the accuracy of various neural network approaches in predicting sewer overflows. Even though the LSTM had a relatively slower learning curve, the results of this type of neural network were most promising for multi-step-ahead predictions [15]. This is because an LSTM has an added cell state that is updated using transfer functions at each time step. This cell state is also used to predict the output of each time step, making it possible to store information for a longer period.

4.2. The LSTM Set-Up

The sewer model input is a spatially uniform precipitation intensity time series, and the output is a flood volume for each time step at each manhole in the studied area. The LSTM is a 'one-to-one' recurrent neural network. This means that for each timestep of the input, an output is calculated. The timesteps for the precipitation input time series, sewer model output, and LSTM predictions are thus all equal to 5 min. Furthermore, the LSTM set-up is similar to that of the sewer model with 1 input, 1 hidden layer, and 230 outputs (1 for each manhole). The number of neurons in the hidden layer and the learning rate are determined using hyper-parameter optimisation. The LSTM is constructed using Keras [25]. Keras is a high-level library used for machine learning applications. Keras runs on Tensorflow [26], which is an open source machine learning software released by Google in 2015.

The synthetic input–output data set, created with the sewer model, is split into training, testing, and validation data sets. The training data are used to find an optimal set of connection weights, the test data are used to choose the best network configuration (i.e., the hyperparameters: in this study, the number of neurons and the learning rate), and the validation set is only used to evaluate the LSTM's final performance in terms of generalization ability [27].

The data set is divided according to the average of studies studied by [16]. They found that 60%, 18%, and 22% of the total data were used for training, testing, and validation, respectively. In the present study, a split of 60%, 20%, and 20% is used. The input precipitation time series are normalised to a $[0,1]$ range.

For the determination of the hyper-parameters of the LSTM, Bayesian hyper-parameter optimisation is used. Due to the long training times for each configuration of the LSTM (60 min+), grid search or random search hyper-parameter optimisation was not feasible. The hyper-parameters determined were the number of neurons of the LSTM layer and the learning rate. The sequential model built with Keras is comprised of two layers. The first layer is the LSTM layer, in which the transfer functions were set to the standard functions. The second layer is a Dense layer. This layer is a standard ANN layer of neurons with a linear activation function. The layer consists of 230 units, which coincides with the amount of target outputs in the model. The sequential model is compiled using the MAE loss function for training.

4.3. The Performance Indicators

The performance indicators used to assess the predictive capability of the trained LSTM are Nash-Sutcliffe efficiency (NSE) and coefficient of determination R^2. The MAE is used to train and test the LSTM, and the NSE and R^2 are used to assess the predictive ability of the LSTM on the validation data set.

The calculation of the MAE is shown in Equation (1). A value of 0 shows a perfect fit between the observed and predicted values:

$$MAE = \frac{1}{n}\sum_{i=1}^{n}|y_i - \hat{y}_i|, \tag{1}$$

in which y_i is the i-th predicted value, and $\hat{y}_i$ is the i-th observed value.

The NSE is commonly used as a predictive measure of hydrological models. For some precipitation events, manholes in the north of the area had NSE values approaching negative infinity. No flooding occurred at these manholes and the (negative) flood volumes in the sewer model results. However, the LSTM still predicted relatively high fluctuations. The scale of these fluctuations were small, causing no wrong predictions in flooding. These fluctuations around the mean did result in the NSE values approaching negative infinity. Therefore, the bounded version of the NSE, proposed by [28] and called C2M (see Equation (2)), is applied instead. NSE values are now bounded to the interval $[-1, 1]$, providing a more usable mean NSE value of all manholes in the area:

$$C2M = \left(1 - \frac{\sum_{i=1}^{n}(y_i - \hat{y}_i)^2}{\sum_{i=1}^{n}(y_i - y_\mu)^2}\right) / \left(1 + \frac{\sum_{i=1}^{n}(y_i - \hat{y}_i)^2}{\sum_{i=1}^{n}(y_i - y_\mu)^2}\right), \quad (2)$$

in which y_μ is the mean of the predicted values, and $\hat{y}_\mu$ is the mean of the observed values.

The last performance indicator used is the R^2. The R^2 measures the correlation between the observed and predicted values. The Equation for R^2 is shown in Equation (3):

$$R^2 = \frac{\left(\sum_{i=1}^{n}(y_i - y_\mu)(\hat{y}_i - \hat{y}_\mu)\right)^2}{\sum_{i=1}^{n}(y_i - y_\mu)^2 \sum_{i=1}^{n}(\hat{y}_i - \hat{y}_\mu)^2} \quad (3)$$

5. Results

5.1. LSTM Validation Based on Synthetic Precipitation Events

After Bayesian optimisation, the LSTM has 636 neurons to predict the flood volumes at the 230 manholes accurately and a learning rate of 0.01. The total run time of the LSTM on the 25 precipitation events present in the validation data set was 1.89 s. During this validation, the LSTM was capable of predicting if a manhole will flood with an accuracy 99.60% (with a threshold value of 1 m^3). Only in 0.26% of the precipitation events was a flood predicted by the LSTM, while no flooding occurred during the sewer model simulation (LSTM prediction > 1 m^3 and sewer model prediction < 1 m^3 in Figure 9). Only in 0.14% of the precipitation events was the opposite applied, meaning that the LSTM did not predict a flood while flooding occurred according to the sewer model (LSTM prediction < 1 m^3 and sewer model prediction > 1 m^3 in Figure 9). This high accuracy, in combination with the extremely low computation time, shows the potential of using an LSTM as an early flood-warning system.

Furthermore, the flood volumes were predicted with high accuracy by the trained LSTM. An average R^2 of 0.99 and an average NSE of 0.87 for all manholes was found (Table 2). However, only 38% of the manholes in the studied area experienced flooding on the validation data set. The manholes that did not flood show a relatively low goodness-of-fit. In these cases, the sewer model predicted mostly an almost constant negative flood volume that varied slightly over time. A negative flood volume predicted by the sewer model means that the water level is below the surface level and thus no flooding occurs. For these situations, the LSTM predicts larger negative flood volume fluctuations since the LSTM is sensitive to any change in the input parameters: even a small change in the precipitation results in a different predicted flood volume. However, these volume fluctuations predicted by the LSTM were still below 0.1 m^3 and not relevant for flood forecasting purposes.

Table 2. The hyper-parameter and evaluation values of the LSTM sequential model after Bayesian optimisation.

Performance Indicator	Value
NSE (all manholes)	0.87
NSE (flooding manholes)	0.92
R^2	0.99

Since the manholes that do not flood are not interested from an early flood warning perspective, we only focus on the results of the flooded manholes. Figure 9 shows the predicted flood volumes of the LSTM and sewer model for each time step of the 25 precipitation events present in the validation data. It shows that the LSTM predictions closely resemble the sewer model output since most data points follow the linear 1:1 line. However, the LSTM tends to slightly underpredict the flood volumes, and especially the peak, compared to the sewer model output. On average, the peak values are underpredicted by 8.5%

by the LSTM. This behaviour is a well-known problem with neural networks since they are prone to systematically underpredict flood series for extreme events [13]. If accurate prediction of the peak values is of high importance, LSTM performance can be increased by, for example, postprocessing the flood volume predictions by applying an unscented Kalman filter [29].

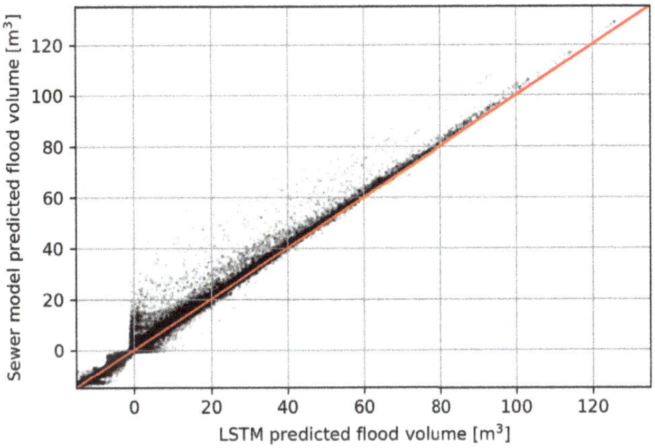

Figure 9. Scatter plot of the predicted and actual flood volumes for the LSTM regressor evaluated on the synthetic validation data set ($R^2 = 0.99$). Negative flood volume are plotted until -15 m^3, no more false negative or false positive values are observed past this value.

A map with the NSE values for the flooded manholes is shown in Figure 10. The NSE values vary between 0.39 and 0.99, with an average value of 0.92. Higher NSE values are generally found in the centre and northwest of the study area, where the most severe flooding occurs. The LSTM predictions were less accurate in the southeastern region of the study area, where the manholes only experience minor flooding because of the relatively high surface levels.

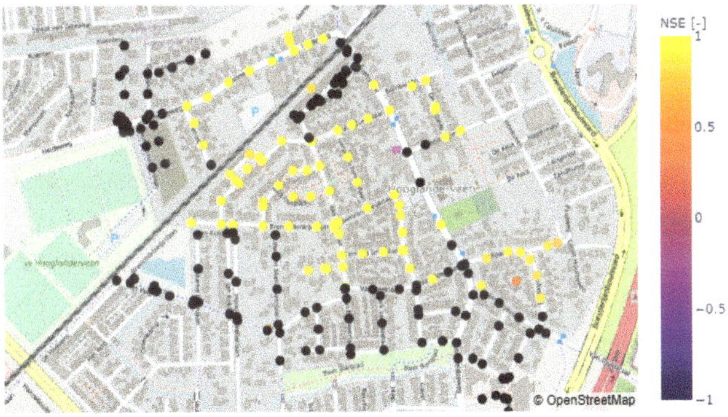

Figure 10. The NSE values for each manhole in the case study area that experienced flooding from the validation data set (NSE = 0.92). The NSE values were calculated with the predicted flood volume time series by the LSTM network and sewer model. The NSE is calculated for each time series and a mean is taken for each manhole. Dark grey manholes indicate locations where no flooding occurs.

Figure 11 shows the predicted flood volumes both by the LSTM and sewer model for a manhole located in the centre of the study area, where extreme flooding occurs at most manholes. This manhole has an average NSE of 0.95. A lag is generally present between the peak of the precipitation event and the moment that flooding of the manholes starts to occur. The LSTM is able to predict this lag with high accuracy when compared to the sewer model output. Furthermore, the LSTM is capable of predicting the general shape of the flood volume hydrograph accurately, both in terms of the timing that flooding starts to occur as well as the timing of the peak flood volume. However, again, the slight tendency of the LSTM to underpredict the peak flood volumes is visible.

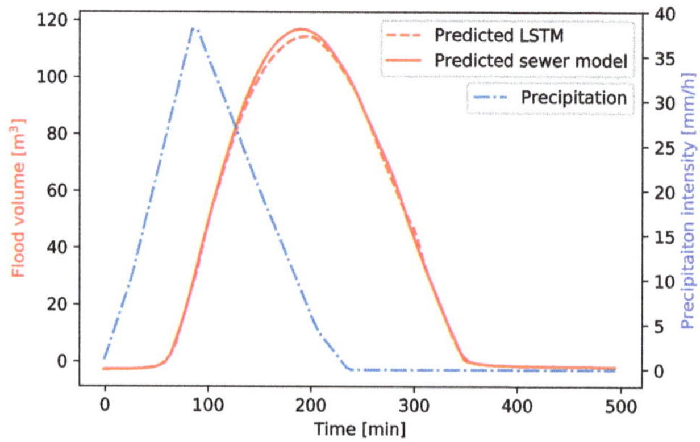

Figure 11. Flood volume time series, for the LSTM network validated on synthetic data, at a manhole in the centre of the area (NSE = 0.95).

The predicted flood volumes by the sewer model and LSTM for a manhole located in the southeastern part of the study area are shown in Figure 12. Here, the LSTM has an average NSE of 0.39. Again, the shape of the flood hydrograph is predicted accurately, even when a two-peaks event is considered. However, the underprediction of the peak value is larger in this region of the study area. It seems that the LSTM has more difficulties in accurately predicting flood volumes in cases of relatively sharp flood volume hydrographs, with large differences between the flood volumes in two consecutive time steps. The accuracy of the LSTM predictions can therefore be improved by reducing the time step of the training data set such that the change in flood volume within two consecutive time steps is reduced.

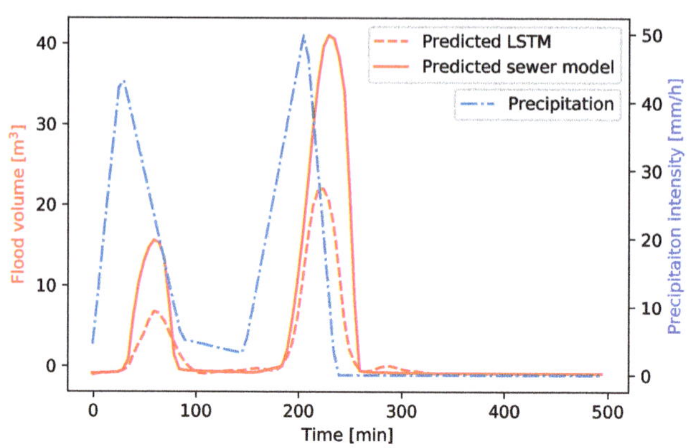

Figure 12. Flood volume time series, for the LSTM network validated on synthetic data, at a manhole in the southeast of the area (NSE = 0.39).

5.2. LSTM Evaluation Based on Historic Precipitation Events

To further test the LSTM, three historic precipitation events that caused flooding in the area were identified. These historic precipitation events were simulated both by the sewer model and LSTM network to predict corresponding flood volumes. Again, the performance of the LSTM model is compared against the sewer model predictions since this model is used to train the LSTM. For this reason, the LSTM performance is at maximum as good as the sewer model, and comparing LSTM predictions with field measurements does not give a proper indication of the LSTM performance.

Also on the historic data set, the LSTM shows the high potential to be used as an early flood warning system. In 94.4% of the precipitation events, the LSTM predicted correctly if flooding occurred at one of the manholes (with a threshold of 1 m^3). Only in 4.6% of the precipitation events was a flood predicted by the LSTM, while no flooding occurred during the sewer model simulation. Only in 1.0% of the precipitation events did the LSTM not predict a flood while flooding occurred. This shows that the number of false positive and false negative flood predictions has not increased compared to the validation using the synthetic data set. Therefore, the ability of the LSTM to predict if a flooding occurs even holds for scenarios deviating from those used during the training procedure.

Figure 13 shows the predicted flood volumes by the LSTM and sewer model for each time step of the three historic precipitation events. This figure also shows that the LSTM is able to predict if flooding occurs accurately. However, the tendency to underpredict flood volumes is again present and is even more severe compared to the validation results based on the synthetic data set. On average, the peak flood volumes are underpredicted by 34.3%.

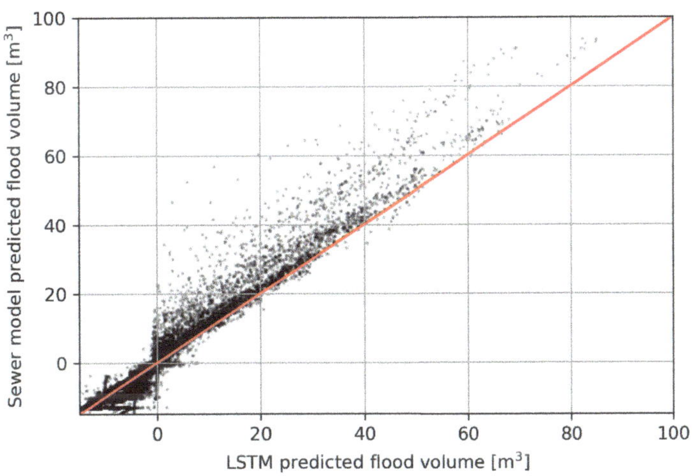

Figure 13. Scatter plot of the predicted and actual flood volumes for the LSTM regressor evaluated on the historic precipitation data set ($R^2 = 0.99$). Negative flood volume are plotted until -15 m^3, no more false negative or false positive values are observed past this value.

During the validation based on the synthetic data set (Section 5.1), we found that the average NSE increases if only the manholes that experience flooding are considered. When we test the LSTM performance on historic precipitation events, we find an average NSE of 0.57 if only the flooded manholes are considered, while an average NSE of 0.61 is found for all manholes (Table 3). This is probably caused by the low LSTM performance for the manholes in the southeastern region (Figure 14), where the flood volume time series show complex behaviours.

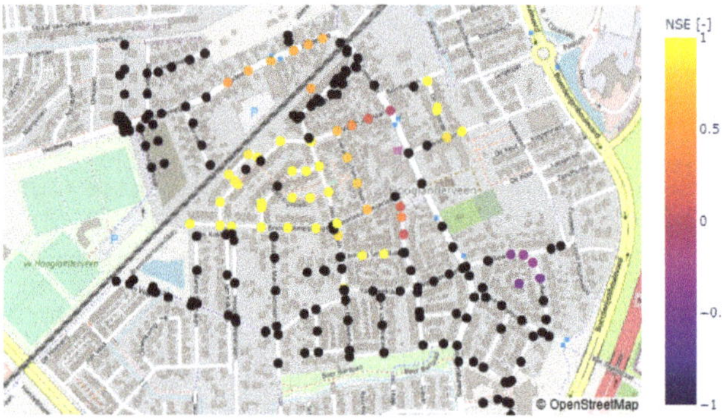

Figure 14. NSE values for each manhole in the case study area that experiences flooding from the historic data set (mean NSE = 0.57). NSE values have been calculated with the predicted flood volume time series by the LSTM network and sewer model. The NSE is calculated for each time series, and a mean is taken for the manholes. Dark grey manholes indicate locations where no flooding occurs.

Table 3. Performance evaluation for the LSTM tested on historic data.

Performance Indicator	Value
NSE (all manholes)	0.61
NSE (flooding manholes)	0.57
R^2	0.99

Figures 15 and 16 show the predicted flood volumes by the LSTM and sewer model for a manhole in the centre (NSE = 0.96) and southeast (NSE = −0.50) of the study area, respectively. The hydrograph shape, in terms of the timing that flooding starts to occur and the timing of the peak value, are predicted with high accuracy for the manhole located in the centre of the study area. This shows that the LSTM performance does not significantly change compared to the validation results on the synthetic data set for the region, where the most frequent and severe flooding occurs. On the other hand, the predictive ability in the southeastern region has decreased (Figure 16). Especially, the peak flood volume is underpredicted significantly. However, again, the timing that flooding starts to occur and the timing of the peak value are captured accurately by the LSTM. This shows that, despite the fact that the total flood volumes are underpredicted, the LSTM still has potential to be used as an early flood warning system in these regions.

The lower LSTM performance on the historic data set, compared to the synthetic data set, is probably caused by the fact that the historic precipitation peaks are confined in a smaller time span, compared to the synthetic training data set. Also in the synthetic training data set, we already found that the the LSTM's performance decreases for the manholes where the flooding occurred in a relatively small time span (Figure 12). Furthermore, the lower performance of the LSTM on historic rainfall events can be explained by the small fluctuations and/or noise in the precipitation data. This shows that, in general, the LSTM performs best when large and smooth precipitation intensities are given as input, resulting in large flood volume time series and matching the precipitation patterns from the synthetic training data set.

To increase the predictive ability of the LSTM, two adjustments are proposed: First, the time step used in this study was 5 min. Due to the sudden nature of extreme precipitation events, this relatively long time step results in a large increase in the flood volumes in two consecutive time steps. Therefore, we recommend reducing this time step, which will only increase the computation time of the sewer model used to generate the training data and barely that of the LSTM. Second, the precipitation statistics were given in patterns with a time step of 1 h. In this study, this pattern was linearly interpolated. By adjusting this interpolation approach, the sharp hydrographs observed in the historic data can be recreated in the synthetic data set, ensuring that more events with confined peaks are included in the training data set.

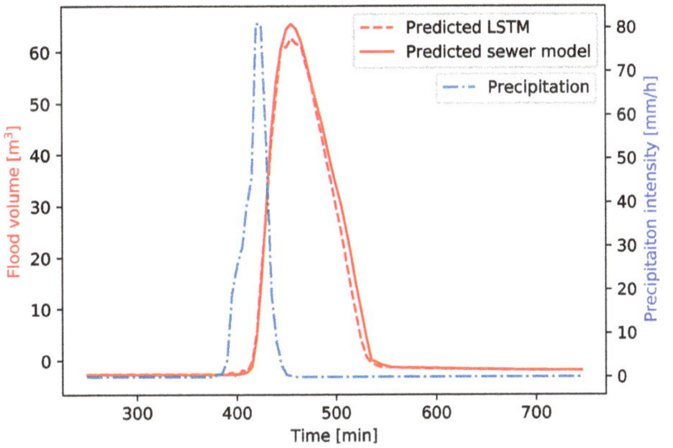

Figure 15. Flood volume time series, for the LSTM network validated on historic data, at a manhole in the centre of the area (NSE = 0.96).

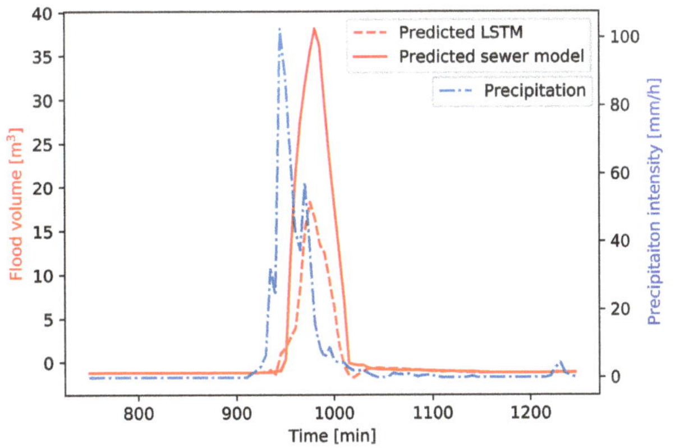

Figure 16. Flood volume time series, for the LSTM network validated on historic data, at a manhole in the southeast of the area (NSE = −0.50).

6. Discussion

Many studies use historic data to train neural networks (e.g., [5,8,10,15]). However, in this study, input–output relations of a numerical sewer model were used to train the LSTM network. Furthermore, synthetic precipitations events were used to create the training data set, adding two additional levels of abstraction from reality (e.g., [13,30,31]). Making use of synthetic precipitation events ensures that a wide range of precipitation characteristics, in terms of precipitation pattern, intensity, and duration, can be included systematically. Section 5.2 showed that, even though the LSTM was trained on synthetic precipitation events, it still accurately predicts which manholes will flood. This indicates that the LSTM is able to respond to precipitation events not present in the training data accurately due to the wide variety of events included in the training data set. This even applies for precipitation events having higher rainfall intensities than present in the training data.

It must be noted that the developed LSTM only predicts flood volumes at maximum as accurate as the sewer model used to train the LSTM. This means that errors present in the sewer model are inherently also present in the LSTM. Additionally, the LSTM is only capable of predicting reliable outputs for the conditions it was trained for. For two historic flood events, not presented in this paper, we found that flooding was observed by inhabitants of Hooglanderveen while the sewer model, and consequently the LSTM, did not predict any flooding. During these events, the measured precipitation intensities were relatively low and would most likely not lead to any flooding in the area under normal circumstances. Therefore, it might be that the inflow of some manholes was blocked by leaves during the precipitation event, causing the inundation of the streets. The sewer model was not designed to model these rare events and hence the LSTM is also not able to include these processes in the predictions.

The computational costs of the LSTM are extremely low, with forecasting times in the order of milliseconds for a single event. Due to the inherent variability in extreme flood events, and the need for ensemble forecasting, many simulations are required. The LSTM can be applied successfully for this purpose, providing a probability of flood volumes instead of a deterministic forecast. This can be helpful for decision makers in their assessment of possible damages caused by the extreme precipitation event.

Regarding the set-up of the LSTM, it was decided to develop a single LSTM network for the entire Hooglanderveen sewer system. This has as advantage that flood volumes at all manholes are computed based on a single input precipitation event. However, setting up an LSTM network for the entire system increases the complexity of the network significantly, compared to having a separate LSTM for each manhole. Consequently, the training time is also significantly higher. Kratzert et al. [8] analysed the effect of setting up a single LSTM to predict rainfall runoff for multiple catchments compared to using multiple regional LSTMs each trained for a single catchment. They found that using a single LSTM network to predict the runoff for multiple catchments results in slightly more accurate predictions, especially in cases with a strong correlation in the predicted output at the various catchments. Furthermore, they suggest that using a single LSTM for an entire network reduces the risk of overfitting compared to setting up an LSTM network for each desired output location [8]. For these reasons, setting up a single LSTM network to predict all manholes in a sewer system is recommended despite the long training times involved.

7. Conclusions

The objective of this research was to construct an LSTM neural network that can predict location-based flooding due to extreme precipitation in an urban environment. For the first time, such an LSTM was developed for a large sewer system covering many manholes. Because insufficient measured data of extreme precipitation events were available, a numerical sewer model was used to generate the training data covering a wide variety of synthetic precipitation events in terms of precipitation intensities and patterns. The LSTM was set up for the whole area of Hooglanderveen in Amersfoort containing 230 manholes. The trained LSTM, having 636 neurons, predicted the flood volume time-series of all flooded manholes with high accuracy, resulting in an average NSE of 0.92. Furthermore, the temporal aspects of the flood wave, in terms of the duration of the flooding, as well as the timing of the peak flood volume, were accurately predicted by the LSTM. Especially the locations with frequent and severe flooding are predicted with high accuracy. Therefore, we conclude that the behaviour of the existing numerical sewer model and its characteristics were successfully reproduced by the LSTM.

Testing of the LSTM on observed historic data shows that the LSTM can also accurately predict the temporal aspects of the flooding for historic precipitation events. Using a large variety of synthetic precipitation events in the training data set ensured that the trained LSTM was able to generalise, even though the historic precipitation patterns differ from the synthetic data since the historic precipitation events are confined to a relatively short interval with high-intensity precipitation. However, it was found that the LSTM tends to underpredict flood volumes, especially for the relatively sharp flood volume hydrographs, with large differences between the flood volumes in two consecutive time steps. In this study, a relatively large time step of five minutes was used to train the LSTM. Therefore, the accuracy of the LSTM predictions can easily be improved by reducing this time step such that the change in flood volume within two consecutive time steps is reduced.

The computational costs of forecasting a single event is exceptionally low, reducing the forecasting time to the order of milliseconds, making the LSTM highly functional as an early flood warning system. Furthermore, this extremely low computational cost makes it possible to compute ensemble forecasts of pluvial flooding, using stochastic precipitation forecasts instead of a single deterministic time series.

Author Contributions: Conceptualization, R.A.H.K., A.B. and K.M.W.; methodology, R.A.H.K. and A.B.; software, R.A.H.K.; validation, R.A.H.K.; data curation, R.A.H.K.; writing—original draft preparation, R.A.H.K. and A.B.; writing—review and editing, R.A.H.K. and A.B.; visualization, R.A.H.K.; supervision, A.B. and K.M.W. All authors have read and agreed to the published version of the manuscript.

Funding: This project has received funding from the European Union's Horizon 2020 Research and Innovation Programme under grant agreement no. 820751.

Institutional Review Board Statement: Not applicable.

Informed Consent Statement: Not applicable.

Data Availability Statement: Restrictions apply to the availability of the input data. The results of the synthetic validation data can be viewed on the following website: https://hooglanderveen-riolering-opti.herokuapp.com/, accessed on 5 May 2022. The results of the historic data test can be viewed on the following website: https://hooglanderveen-riolering-hist.herokuapp.com/, accessed on 5 May 2022.

Acknowledgments: The authors would like to thank Hydrologic for their guidance and expert advice during the research. The authors would also like to thank the Municipality of Amersfoort for providing the data related to the observed historical precipitation events. Furthermore, the authors would like to thank Arcadis for providing the input–output data of the sewer model used to train the LSTM network in this study.

Conflicts of Interest: The author declares no conflict of interest.

Abbreviations

The following abbreviations are used in this manuscript:

MDPI	Multidisciplinary Digital Publishing Institute
DOAJ	Directory of Open Access Journals
LSTM	Long term-short term neural network
ML	Machine learning
ANN	Artificial neural networks
RRN	Recurrent neural networks

References

1. Szöllösi-Nagy, A.; Zevenbergen, C. *Urban Flood Management*; CRC Press: Boca Raton, FL, USA, 2004.
2. May, W. Potential future changes in the characteristics of daily precipitation in Europe simulated by the HIRHAM regional climate model. *Clim. Dyn.* **2008**, *30*, 581–603. [CrossRef]
3. Min, S.K.; Zhang, X.; Zwiers, F.W.; Hegerl, G.C. Human contribution to more-intense precipitation extremes. *Nature* **2011**, *470*, 378–381. [CrossRef] [PubMed]
4. Ayazpour, Z.; Bakhshipour, A.E.; Dittmer, U. Combined Sewer Flow Prediction Using Hybrid Wavelet Artificial Neural Network Model. In *Proceedings of the New Trends in Urban Drainage Modelling*; Mannina, G., Ed.; Springer International Publishing: Cham, Switzerland, 2019; pp. 693–698.
5. Mounce, S.R.; Shepherd, W.; Sailor, G.; Shucksmith, J.; Saul, A.J. Predicting combined sewer overflows chamber depth using artificial neural networks with rainfall radar data. *Water Sci. Technol.* **2014**, *69*, 1326–1333. [CrossRef]
6. Razavi, S.; Tolson, B.A.; Burn, D.H. Review of surrogate modeling in water resources. *Water Resour. Res.* **2012**, *48*, 1–32. [CrossRef]
7. Sit, M.; Demiray, B.Z.; Xiang, Z.; Ewing, G.J.; Sermet, Y.; Demir, I. A comprehensive review of deep learning applications in hydrology and water resources. *Water Sci. Technol.* **2020**, *82*, 2635–2670. [CrossRef] [PubMed]

8. Kratzert, F.; Klotz, D.; Brenner, C.; Schulz, K.; Herrnegger, M. Rainfall–runoff modelling using Long Short-Term Memory (LSTM) networks. *Hydrol. Earth Syst. Sci.* **2018**, *22*, 6005–6022. [CrossRef]
9. Bomers, A.; van der Meulen, B.; Schielen, R.M.J.; Hulscher, S.J.M.H. Historic flood reconstruction with the use of an Artificial Neural Network. *Water Resour. Res.* **2019**, *55*, 9673–9688. [CrossRef]
10. Liu, P.; Wang, J.; Sangaiah, A.K.; Xie, Y.; Yin, X. Analysis and Prediction of Water Quality Using LSTM Deep Neural Networks in IoT Environment. *Sustainability* **2019**, *11*, 2058. [CrossRef]
11. Poornima, S.; Pushpalatha, M. Prediction of Rainfall Using Intensified LSTM Based Recurrent Neural Network with Weighted Linear Units. *Atmosphere* **2019**, *10*, 668. [CrossRef]
12. Zou, R.; Lung, W.S.; Wu, J. An adaptive neural network embedded genetic algorithm approach for inverse water quality modeling. *Water Resour. Res.* **2007**, *43*, 8. [CrossRef]
13. Bomers, A. Predicting Outflow Hydrographs of Potential Dike Breaches in a Bifurcating River System Using NARX Neural Networks. *Hydrology* **2021**, *8*, 87. [CrossRef]
14. Rjeily, Y.A.; Abbas, O.; Sadek, M.; Shahrour, I.; Chehader, F.H. Flood forecasting within urban drainage systems using NARX neural network. *Water Sci. Technol.* **2017**, *76*, 2401–2412. [CrossRef] [PubMed]
15. Zhang, D.; Lindholm, G.; Ratnaweera, H. Use long short-term memory to enhance Internet of Things for combined sewer overflow monitoring. *J. Hydrol.* **2018**, *556*, 409–418. [CrossRef]
16. Rajaee, T.; Ebrahimi, H.; Nourani, V. A review of the artificial intelligence methods in groundwater level modeling. *J. Hydrol.* **2019**, *572*, 336–351. [CrossRef]
17. Rioned, S. Kennisbank. Available online: https://www.riool.net/kennisbank (accessed on 27 November 2020).
18. Henonin, J.; Russo, B.; Mark, O.; Gourbesville, P. Real-time urban flood forecasting and modelling—A state of the art. *J. Hydroinform.* **2013**, *15*, 717–736. [CrossRef]
19. Beersma, J.; Hakvoort, H.; Jilderda, R.; Overeem, A.; Versteeg, R. *Neerslagstatistiek en Reeksen Voor Het Waterbeheer*; Stichting Toegepast Onderzoek Waterbeheer: Amersfoort, The Netherlands, 2019.
20. Hochreiter, S.; Schmidhuber, J. Long Short-Term Memory. *Neural Comput.* **1997**, *9*, 1735–1780. [CrossRef]
21. Yu, Y.; Si, X.; Hu, C.; Zhang, J. A Review of Recurrent Neural Networks: LSTM Cells and Network Architectures. *Neural Comput.* **2019**, *31*, 1235–1270._a_01199. [CrossRef]
22. Goodfellow, I.; Bengio, Y.; Courville, A.; Bengio, Y. *Deep Learning*; MIT Press: Cambridge, MA, USA, 2016; Volume 1.
23. Elman, J.L. Finding Structure in Time. *Cogn. Sci.* **1990**, *14*, 179–211._1. [CrossRef]
24. Bengio, Y.; Simard, P.; Frasconi, P. Learning long-term dependencies with gradient descent is difficult. *IEEE Trans. Neural Netw.* **1994**, *5*, 157–166. [CrossRef]
25. Chollet, F. Keras. 2015. Available online: https://github.com/fchollet/keras (accessed on 15 November 2020).
26. Abadi, M.; Agarwal, A.; Barham, P.; Brevdo, E.; Chen, Z.; Citro, C.; Corrado, G.S.; Davis, A.; Dean, J.; Devin, M.; et al. TensorFlow: Large-Scale Machine Learning on Heterogeneous Systems. 2015. Available online: tensorflow.org (accessed on 15 November 2020).
27. Bowden, G.J.; Maier, H.R.; Dandy, G.C. Optimal division of data for neural network models in water resources applications. *Water Resour. Res.* **2002**, *38*, 2-1–2-11. [CrossRef]
28. Mathevet, T.; Michel, C.; Andréassian, V.; Perrin, C. A bounded version of the Nash-Sutcliffe criterion for better model assessment on large sets of basins. *IAHS Publ.* **2006**, *307*, 211–219.
29. Zhou, Y.; Guo, S.; Xu, C.Y.; Chang, F.J.; Yin, J. Improving the Reliability of Probabilistic Multi-Step-Ahead Flood Forecasting by Fusing Unscented Kalman Filter with Recurrent Neural Network. *Water* **2020**, *12*, 578. [CrossRef]
30. Kabir, S.; Patidar, S.; Xia, X.; Liang, Q.; Neal, J.; Pender, G. A deep convolutional neural network model for rapid prediction of fluvial flood inundation. *J. Hydrol.* **2020**, *590*, 125481. [CrossRef]
31. Lin, Q.; Leandro, J.; Wu, W.; Bhola, P.; Disse, M. Prediction of Maximum Flood Inundation Extents With Resilient Backpropagation Neural Network: Case Study of Kulmbach. *Front. Earth Sci.* **2020**, *8*, 332. [CrossRef]

hydrology

Article

Differentiated Spatial-Temporal Flood Vulnerability and Risk Assessment in Lowland Plains in Eastern Uganda

Godwin Erima [1,*], Isa Kabenge [2], Antony Gidudu [3], Yazidhi Bamutaze [4] and Anthony Egeru [1]

1 Department of Environmental Management, Makerere University, Kampala P.O. Box 7062, Uganda
2 Department of Agricultural and Bio-Systems Engineering, Makerere University, Kampala P.O. Box 7062, Uganda
3 Department of Geomatics and Land Management, Makerere University, Kampala P.O. Box 7062, Uganda
4 Department of Geography, Geo-Informatics and Climatic Sciences, Makerere University, Kampala P.O. Box 7062, Uganda
* Correspondence: godwin.erima@mak.ac.ug; Tel.: +256-772-929181

Abstract: This study was conducted to map flood inundation areas along the Manafwa River, Eastern Uganda using HECRAS integrated with the SWAT model. The study mainly sought to evaluate the predictive capacity of SWAT by comparisons with streamflow observations and to derive, using HECRAS, the flood inundation maps. Changes in Land-use/cover showed by decrease in forest areas and wetlands, and conversions into farmlands and built-up areas from 1995 to 2017 have resulted in increased annual surface runoff, sediment yield, and water yield. Flood frequency analysis for 100-, 50-, 10-, and 5-year return periods estimated peak flows of 794, 738, 638, and 510 m^3/s, respectively, and total inundated areas of 129, 111, 101, and 94 km^2, respectively. Hazard classification of flood extent indicated that built-up areas and commercial farmlands are highly vulnerable, subsistence farmlands are moderately to highly vulnerable, and bushland, grassland, tropical high forest, woodland, and wetland areas are very low to moderately vulnerable to flooding. Results demonstrated the usefulness of combined modeling systems in predicting the extent of flood inundation, and the developed flood risk maps will enable the policy makers to mainstream flood hazard assessment in the planning and development process for mitigating flood hazards.

Keywords: Eastern Uganda; flood plains; flood hazard maps; HEC-RAS; return period; SWAT

Citation: Erima, G.; Kabenge, I.; Gidudu, A.; Bamutaze, Y.; Egeru, A. Differentiated Spatial-Temporal Flood Vulnerability and Risk Assessment in Lowland Plains in Eastern Uganda. *Hydrology* **2022**, *9*, 201. https://doi.org/10.3390/hydrology9110201

Academic Editors: Aristoteles Tegos, Alexandros Ziogas and Vasilis Bellos

Received: 10 October 2022
Accepted: 1 November 2022
Published: 9 November 2022

1. Introduction

In recent years, variability in natural disasters has increased due to the changes in global climate, land use/cover, and socio-economic development [1]. Statistics show that 318 natural disasters affected 122 countries worldwide in the year 2017 alone, the impacts of which resulted in 9503 deaths, 96 million people affected, and USD 314 billion as economic damages, and floods accounted for 38.3% of these disasters and 35% of deaths, affecting 59.6% of people's livelihoods and 6.2% of economic damages [2]. Uganda, like other low-income countries, is vulnerable to extreme weather events such as droughts and floods [3]. In Eastern Uganda, the low-lying areas of Butaleja district are vulnerable to flooding [4], and more recently, in December 2019, floods led to four deaths, and over 2000 people were displaced [5].

Flood inundation mapping plays an important role in designing sustainable plans, protecting human properties and lives, and mitigating disaster risks [6]. It is also a crucial step in developing flood hazard maps and conducting proper flood assessments [7]. Flood inundation mapping usually requires repeated observations of the flooded area and inundation extents through remote sensing images [8] or ground observations [9]. Obtaining representative meteorological data for watershed-scale hydrological modeling can be difficult and time-consuming [10]. The difficulty in collecting data can be attributed to the following reasons: (i) lack of reliable equipment, (ii) absence of a good archiving system and software to store and process the data, and (iii) lack of funds to organize data collection campaigns [11]. It is also worth mentioning that once the data have been captured and archived, accessing them is quite costly [12]. Weather stations based on the ground do not always adequately represent the weather occurring over a watershed because they can have gaps in their data series and can be far from the watershed of interest or recent data are not available [13]. For data-scarce areas, hydrological and hydrodynamic models as such play a critical role in flood simulations and risk assessment [14].

Hydrologic models rely on the parameterization of watershed properties and rainfall patterns and depths to produce a flood hydrograph of discharge at discrete time steps [15]. These models have become widely used in flood forecasting, stream flow prediction, and quantifying effects of climate change and land use impacts or other spatially distributed properties. However, their limited routing methods do have some drawbacks in simulating flows in large watersheds. Examples of hydrologic modeling tools include the wflow, Hydrologic Engineering Center–Hydrologic Modeling System (HEC–HMS), the Hydrologic Simulation Program–FORTRAN (HSPF), Soil and Water Assessment Tool (SWAT), and MIKE-SHE [15]. On the other hand, hydrodynamic modeling tools are based on the solutions to St. Venant equations to calculate open channel flow. The most commonly used of these models are either one-dimensional or two-dimensional. Widely used hydrodynamic modeling tools include FLO-2d, Lisflood-FP (1D and 2D models), Water Quality Analysis Simulation Program (WASP), CE-QUALW2, Environmental Fluid Dynamics Code (EFDC), EPDRIV1, Hydrologic Engineering Center River Analysis System (HECRAS), MIKE11 (1-D model), MIKE21 (2-D model), and SOBEK [15]. The comparison between models has been a significant issue of debate in the scientific fraternity [16,17]. The resulting differences are attributed mainly to the quality of topographic and input data [18] and less to the complexity of the phenomenon itself [19]. Several studies have compared the performance of 1D and 2D hydraulic models for river flood simulations [16,20] and have concluded that all models have proven sufficiently accurate, but they still have discovered that flood inundation modeling involves several sources of uncertainty such as (1) input data (boundary and initial condition data, digital elevation models and channel bathymetry, hydraulic structures, roughness parameterization), (2) model structure (1D, 2D, quasi 2D, 1D/2D), and (3) internal model parameters. Furthermore, they emphasize the fact that no matter the quality of the input data, provided the user does not properly fit the data into the appropriate geometrical description of the model, the final results of the simulation will be of considerably lower accuracy [16].

Combining hydrodynamic models with hydrological models often compliments and overcomes the shortcomings of either type of modeling approach [21]. In the current study, the hydrologic modeling tool, namely the Soil and Water Assessment Tool (SWAT), is used to derive flow hydrographs at designated locations, which were then fed into the hydrodynamic modeling tool, namely the Hydrologic Engineering Center's River Analysis System (HECRAS) for flood prediction. The SWAT and HECRAS programs were adopted in this study because they are freely available, user-friendly, peer-reviewed, are continuously improved and developed. The SWAT modeling system is a long-term, continuous model simulation of the watershed developed by the United States Department of Agricultural (USDA) [22]. SWAT has proven to perform well in streamflow and base-flow simulations around the world and in complex catchments with extreme events [23] since it allows the interconnections of different physical processes [24]. Additionally, the model is recognized as suitable for investigating long-term impacts, particularly in watersheds without conventional gauges [25]. HECRAS is one of the most commonly used modeling systems to analyze channel flow and floodplain delineation [26]. HECRAS uses geometric data representation as well as geometric and hydraulic computation routines for a network of natural and constructed channels of the river. HECRAS has the ability to make the calculations of water surface profiles for steady and gradually varied flow as well as for subcritical, super critical, and mixed flow regimes. HECRAS is also capable of doing modeling for sediment transport, which is notoriously difficult. The HECGeoRAS is a GIS extension with a set of procedures, tools, and utilities for the preparation of river geometry GIS data to import into HECRAS, and it is used to generate the final inundation map [27].

In recent decades, many researchers have performed flood hazard mapping in various parts of the world, as reported elsewhere [26]. Generally, basin-scale flood hazard mapping is performed worldwide [26]; however, limited research exists in the literature for Uganda river basins [28]. The main objective of this study was to analyze the inundation area along the Manafwa River network and to assess the flood hazard in the Manafwa catchment. The specific objectives pertaining to this study were to: (1) assess the land-use/cover changes, (2) evaluate the impact of LULC on the hydrologic characteristics, (3) evaluate the predictive capacity of the SWAT modeling system by comparisons with streamflow observations, and (4) derive using HECRAS the flood hazard maps. In this study, we aim to address one scientific question: (1) how suitable is the coupled hydrology-inundation model for producing probability maps of flood plain areas for mapping vulnerability and risk areas in a data-scarce area? Integrated modeling is the focus of this study because by using it to simulate the rainfall depths at different probabilities, complete flood hazard maps are obtained. Additionally, it can be used for other purposes in the design and analysis of flood mitigation measures, as well as flood forecasting and warning systems. The novelty of the present study is to combine the physically based distributed hydrologic model SWAT with the hydraulic model HEC-RAS for flood prediction in Eastern Uganda, which has not been conducted before for tropical catchments and for small watersheds. The study area is an important hydrological region in Uganda, very populous with extensive areas of rice cultivation, and no similar studies (to the authors' knowledge) have been conducted in the past on the Manafwa Catchment. The Office of the Prime Minister (OPM) in Uganda will be in a position to strengthen the catchment planning process, and this will be a platform for further studies to be carried out on other catchments in the country.

2. Materials and Methods

2.1. Study Area

The Manafwa catchment covers a total area of 502 km^2 in the Mt Elgon region, located in the eastern region of Uganda (Figure 1). The catchment is characterized by high relief in the East, with altitudes ranging from 1041 to 4301 m above sea level, and its main stream drains from Mt Elgon to Lake Kyoga in downstream. The annual mean temperature is 23 °C, and the mean annual rainfall is 1500 mm. The annual rainfall follows a bimodal pattern, marked by the dry season covering the period of June–August (JJA) and December–February (DJF); and the rainy season occurs during the months of March to May (MAM) and short rains in September–November (SON). The geology in the Mt. Elgon region comprises mainly Pre-Cambrian and Cainozoic rock formations, including volcanics, granites, and sediments. The predominant soil type is Vertisols, regionally known as "black cotton soils". Generally, the soils in the highlands are clays, while those in the midlands and the lowlands are clay loams or sandy. Land-use/cover changes in the catchment are characterized by the conversion from natural forest to other land-use/cover types, especially crop lands and grazing, due to the high population growth rate of 3.5% increasing demand for arable lands for crop production. The catchment is also characterized by low-income generating activities and weak infrastructural and service facilities.

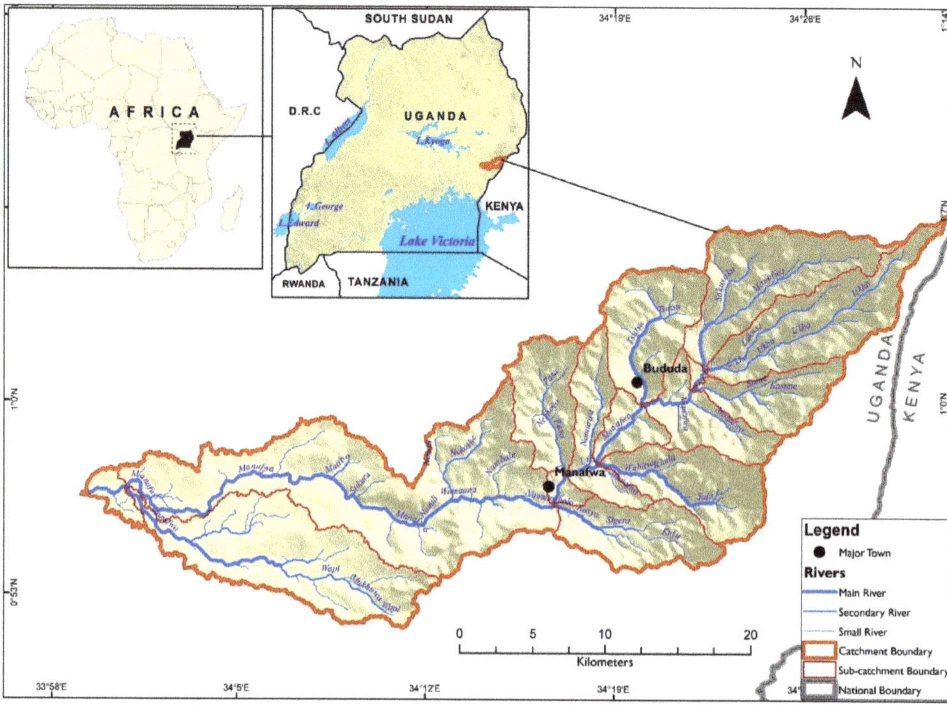

Figure 1. Map of the study area.

2.2. *Hydrological Modelling*

2.2.1. Model Input Data

In this study, the SWAT model [22] was used to simulate discharge data at the required station in the catchment for the chosen time period. A 30-m spatial resolution digital elevation model (DEM) from the Shuttle Radar Topography Mission (SRTM) downloaded from https://earthexplorer.usgs.gov/ (accessed on 10 July 2018) was used to derive the topographic information used for drainage pattern definition. A soil map was obtained from FAO, Scale- 1:50,000 (2000); Land cover maps of 1995, 2008, and 2017 (Figure 2a–c) with a spatial resolution of 30 × 30 m were obtained from National Forestry Authority (NFA), which is the mandated institution required to frequently monitor land use/cover changes in Uganda; Relative humidity, wind speed, solar radiation, and the minimum and maximum air temperatures were obtained from the Climate Forecast System Reanalysis (CFSR), which was designed based on the forecast system of the National Centers for Atmospheric Prediction (NCEP) from 1981 to 2013 https://globalweather.tamu.edu (accessed on 10 July 2018). The rain gauge network of the area is very sparse, and as such, the precipitation data were downloaded from CHIRPS for the 1981–2013 period. Daily discharge data were acquired from the Directorate of Water Resources Management, MWE for the period of 1981–2013 obtained from the Manafwa river gauge (station ID 82212)

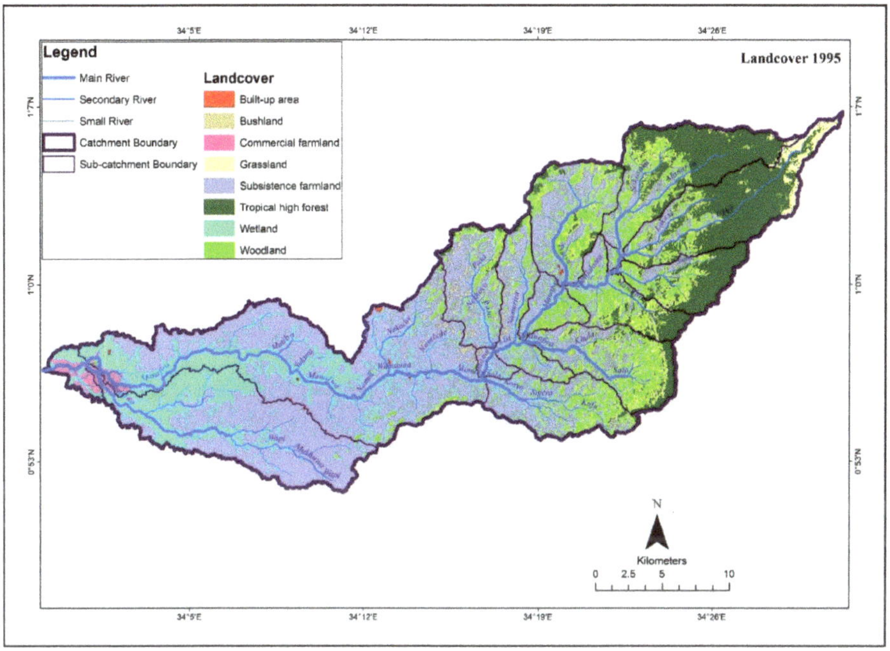

(**a**)

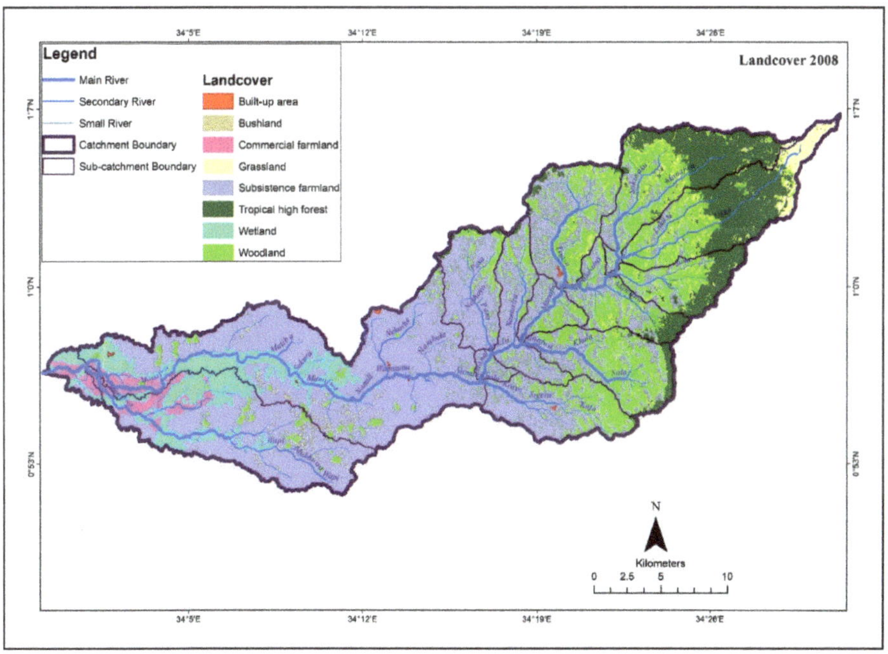

(**b**)

Figure 2. *Cont.*

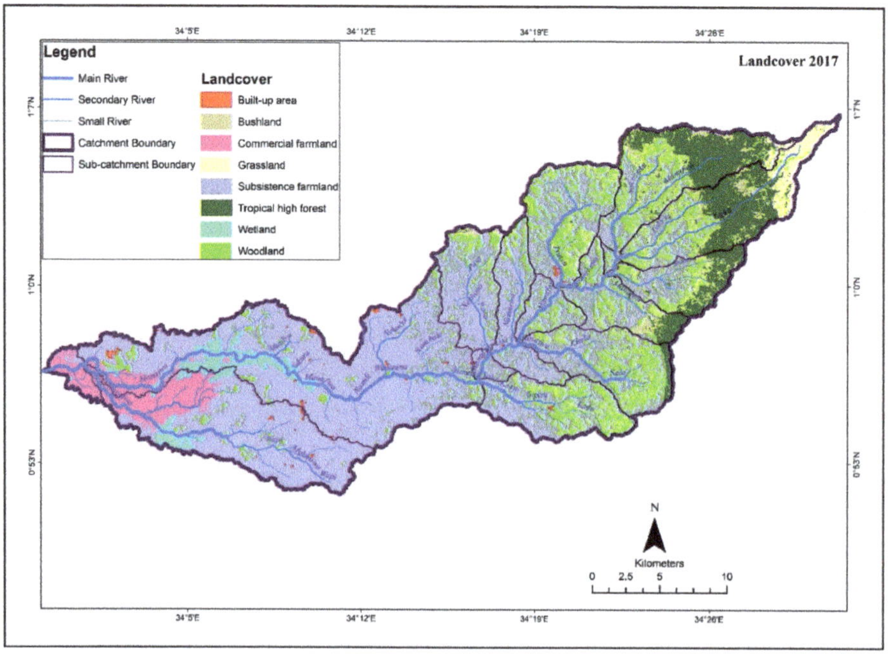

(c)

Figure 2. (**a**) Land use land cover Map of 1995. (**b**) Land use land cover Map of 2008. (**c**) Land use land cover Map of 2017.

2.2.2. Model Set-Up and Calibration

The initial model setup was carried out with the Arc SWAT 2012 (revision 664) using the DEM. A total of 19 sub-catchments were defined with 73 Hydrologic Response Units (HRUs) from a unique combination of land use/cover, soil type, and slope at thresholds over the sub-catchment area of 10% for all categories. Surface runoff and infiltration were computed using the Soil Conservation Services (SCS) curve number method. Evapotranspiration was calculated based on the Penman Monteith method using the obtained climate data (mean daily temperature, solar radiation, and wind speed). The lateral flow was calculated using a kinematic storage model described in [22]. After the initial setup, the model was calibrated and validated at a daily resolution using the Sequential Uncertainty Fitting (SUFI-2) algorithm in the SWAT Calibration and Uncertainty Program (SWAT-CUP, version 5.1.6.2) [22], following the procedures of [23]. The SUFI-2 program was applied for parameter optimization, and Latin Hypercube sampling iteratively discarded the worst simulations by rejecting the 2.5th and 97.5th percentile of the cumulative distribution. Thus, the best 95% of simulations generated a parameter range (95% prediction uncertainty, 95PPU) rather than a single final parameterization. The uncertainty band (95PPU) was used to account for the modeling uncertainty [22]. Calibration of the streamflow was performed from the year 2000 to 2010, and the validation was performed from the year 2011 to 2013.

2.2.3. Model Performance Evaluation

In this study, the model performance during calibration and validation was evaluated based on three quantitative statistics: specifically, the coefficient of determination (R^2) using Equation (1), the Nash-Sutcliffe efficiency (NSE) using Equation (2), and the percent bias (PBIAS) using Equation (3). The coefficient of determination (R^2) ranges between 0 and 1.0, with high values indicating less error variance. The NSE, which was used as the objective function, ranges between $-\infty$ and 1.0. An NSE of 1.0 indicates a perfect fit between the simulated and observed data [29]. The optimal value of PBIAS is 0%, with positive and negative values indicating model underestimation and overestimation bias, respectively. The model performance was considered to be satisfactory if NSE $\geq$ 0.50, $R^2 \geq 0.50$, and PBIAS $\leq \pm 25\%$ [29].

$$R^2 = \frac{\left[\sum_{i=1}^{n}\left(O_i - \grave{O}\right)\left(P_i - \acute{P}\right)\right]^2}{\sum_{i=1}^{n}\left(O_i - \grave{O}\right)^2 \sum_{i=1}^{n}\left(P_i - \acute{p}\right)^2} \tag{1}$$

$$NSE = 1 - \frac{\sum_{i=1}^{n}\left(O_i - P_i\right)^2}{\sum_{i=1}^{n}\left(O_i - \grave{O}\right)^2} \tag{2}$$

$$PBIAS = 100 * \frac{\sum_{i=1}^{n}\left(O_i - P_i\right)}{\sum_{i=1}^{n} O_i} \tag{3}$$

where O_i and P_i are the measured and simulated data, respectively, $\grave{O}$ and $\acute{P}$ are the means of measured and simulated data, and n is the number of observations. The modeling uncertainty was quantified as the P- and R-factor [22]. The P-factor measures the ability of the model to bracket the observed data with the 95PPU. The P-factor is between 0 and 1, where 1 means a 100% bracketing of the observed data. The R factor represents the width of the 95PPU, ranges from 0 to 8, and should be below 1, implying a small uncertainty band [22].

2.3. Hydraulic Modelling Using HECRAS

The hydraulic model used for our study is based on Hydraulic Engineering Center's River Analysis System (HEC-RAS), version 5.0.3 [30]. This model was designed to perform 1D steady flow as well as 2D unsteady flow simulations for a river flow analysis and sediment transport and water temperature/quality modeling. The model uses geometric data representation as well as geometric and hydraulic computation routines for a network of natural and constructed channels of the river. The model required to discharge, DEM as a boundary condition, and Manning's roughness coefficient derived from LULC for calibration. The Model was discretized into an equal number of grid cells of 30 m $\times$ 30 m, i.e., equal geometry to maintain spatial uniformity. HECRAS modeling within the Manafwa floodplain followed three steps (Figure 3).

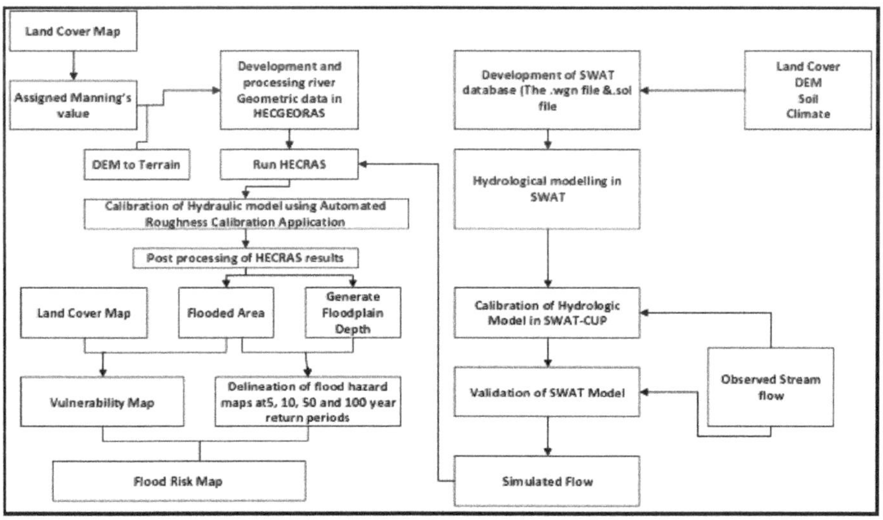

Figure 3. Schematic of Data and Models for flood prediction and analysis.

Step one—This preprocessing stage involved the manual digitization of thematic vector layers (e.g., river network, stream centerline, river banks, flow paths, cross sections) in ArcGIS 10.5. software based on the STRM DEM with 30 m spatial resolution and generation of the attribute table for each of them. The DEM was used as an input in the RAS mapper of HECRAS 2D to develop a Digital Terrain Model (DTM). The DEM data were added to the New Terrain Layer dialogue of the RAS mapper in HECRAS 2D. A new terrain layer was created with terrain files that were used for evaluation. This information was saved in the terrain folder in GeoTiff format. In addition to GeoTiff, two more files were created in. hdf and. vrt formats. The file. hdf was created in the RAS mapper, which contained information on the raster data. The. vrt file helped visualize and display multiple data. For visualizing flood plain, the model geometry was coupled to DTM, and the DTM acted as a basis to create a 2D mesh using the polygon shapefile. The DEM and DTM were used for computing the water surface elevations to visualize floodplain geometry and flood risk analysis. The importance of DEM's accuracy has been highlighted by several authors, especially in two-dimensional hydraulic–hydrodynamic modeling applications [20].

Step two (Figure 3)—This processing stage involved the import of the required parameters (e.g., Manning roughness coefficient, hydrological data) into HECRAS software to run the 2D flood simulation. Thereby, the Manning roughness coefficient (n) was calculated based on land use/cover classes in combination with typical roughness coefficient tables for each cross-section, stream centerline, and river bank intersections with values (built-up area: n = 0.3; farmland: n = 0.025; bushland: n = 0.035; tropical high forest: n = 0.1; woodland: n = 0.06; wetland: n = 0.04 [31]. Steady flow analysis was used instead of unsteady flow analysis because, in the second case, the HECRAS software needs a hydrograph, which we could not obtain from the local authorities. Thereby, to overcome this limitation, we used the flow rate for the gauging station.

Step three—This post-processing stage involved exporting the HECRAS results to the software and generating the flood patterns with the different recurrence intervals. The validation of the results was performed by comparing the real discharge recorded at the gauging station with the computed discharge hydrographs. A detailed description of HECRAS is provided by [32,33].

2.4. Flood Hazard Analysis

To assess flood hazard, the DEM was converted into Triangulated Irregular Network (TIN) format, and TIN showed that the Elevation of the study area ranged from 1070 to 4260 m (Figure 4). After that, the river cross-sections, stream centerline, stream bank lines, flow lines, and other river geometry information were extracted from the TIN for the HECGeoRAS model. The geometric data of the Manafwa River basin are shown in Figure 5. At the same time, the Manning roughness coefficient (n) was calculated based on land use/cover classes in combination with typical roughness coefficient tables for the study area [30,31]. After the RAS geometry data preparation, the HEC-GeoRAS model was used to generate the RAS GIS import file (final river geometry file) that was used as input for HECRAS.

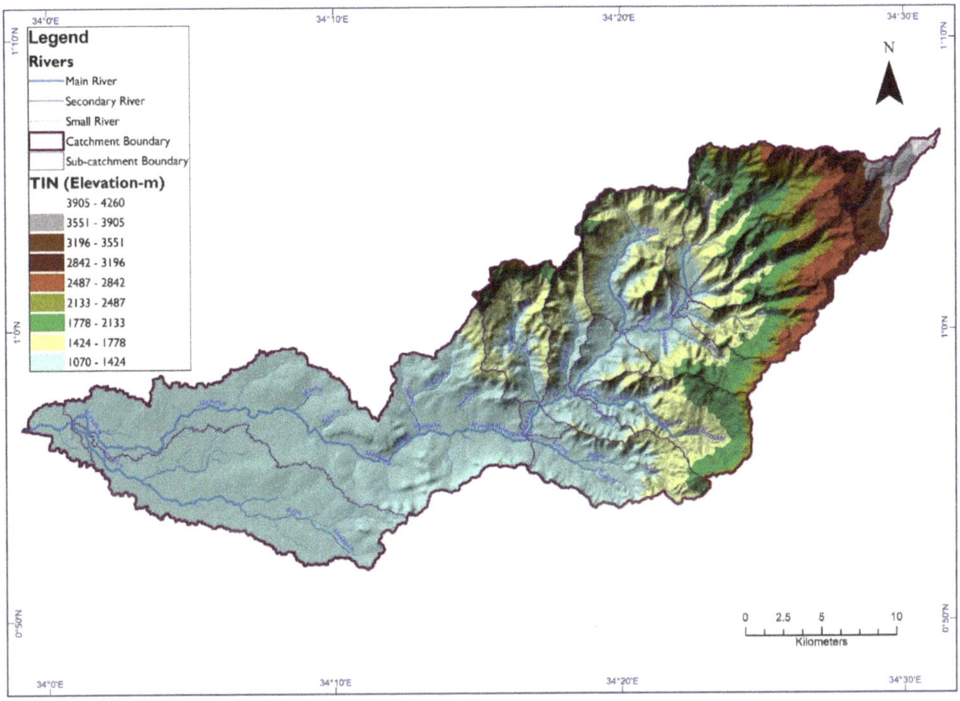

Figure 4. TIN of the Manafwa River Basin.

Figure 5. Schematic view of the geometry of study river.

The outputs of HECGeoRAS preprocessing provided GIS to RAS import files; thereafter, two-dimensional hydrodynamic models were created in HECRAS 5.0.3 for the flood frequency analysis of 5, 10, 50, and 100 years return periods. The Manning's 'n' value, flow data, and boundary conditions were inputted in the imported GIS2RAS file, and the

HECRAS results were obtained. To perform HECRAS analysis, River discharge data were used as the upstream boundary condition, while normal depth was used as the downstream boundary condition. These boundary conditions require the input of the Energy Grade Line (EGL) slope at the downstream boundary. The flow data obtained from SWAT and geometry obtained from the DEM created were also inputted. The cross-sections were created in HEC-GeoRAS. Thereafter, the water surface profiles were obtained, and the sufficiency of cross-section coverage was checked. River cross-sections were used at 100 m distance each, and 500 cross-sections were generated for the analysis. The 1D model was connected to the floodplain, and a 2D computational mesh was created at 100 × 100 m grid size. Although the cell size is rather large, considerable hydraulic details are still retained within a cell using the 2D Geometric Preprocessor. The algorithm preprocesses cells and cell faces to develop detailed hydraulic property tables (elevation versus wetted perimeter, elevation versus area, roughness, etc.) based on the underlying terrain (5 × 5 m in this case). As such, HEC-RAS can produce detailed results (for example, a cell can be partially wet), which is an advantage over other models that use a single elevation for each cell [30]. The outputs were exported to GIS from HECRAS, and water surface TINs were created in the Arc GIS environment. Thereafter, flood plain extent and depth grids were obtained, and flood hazard maps for 5-, 10-, 50-, and 100-year return periods were prepared using Arc GIS.

A flood hazard assessment was undertaken based on the flood water depth indicated by the prepared flood map of the Manafwa watershed. For this, hazard levels were ranked in terms of water depth, and these levels were determined by reclassifying the flood grid water depth bounding cells. Five hazard levels were categorized based on water depth such as very low (<0.5 m), low (0.5–1 m), moderate (1–1.5 m), and high (1.5–2 m) and very high (>2 m) area bounded by each level calculated by modification of the scale used in the MLIT methodology [34] and flood hazard maps were prepared.

2.5. Flood Vulnerability Analysis

The first step in vulnerability analysis was to identify the elements at risk in the study area. In this study, elements at risk were identified by overlying the land-use/cover onto flood inundation maps. LULC dataset was generated from the digital image classification of Landsat, satellite images of 1995, 2008, and 2017 with a spatial resolution of 30 × 30 m, downloaded from Global Land Cover Facility (https://glovis.usgs.gov/) (accessed 10 July 2018 (Table 1). Images from the same period (March–May), i.e., the first rainy season, were selected in order to minimize the seasonal effect on the classification results. In this study, supervised classification of the maximum likelihood algorithm was applied to classify Landsat images into discrete LULC categories. The area was classified into the following land-use/cover classes: built-up areas, bushlands, grassland, commercial farmland, subsistence farmland, tropical high forest, woodland, and wetland. Information collected during the field survey as ground-truthing point was used to assess the accuracy of classification. The elements at risk identified for the study areas included commercial farmland, subsistence farmland, and rural settlements (i.e., homesteads) because other land-use/cover classes were not important from a flood risk point of view. Finally, inundation layers were overlaid on the land-use/cover layer to obtain the overlaid zones. From the ArcGIS overlay analysis, different sorts of inundation statistics were generated. The land-use/cover areas under the influence of each flooding event were reclassified for the calculation of the total vulnerable areas.

Table 1. Summary of Satellite Imagery used for Land cover change analysis.

Satellite	Sensor	Path/Row	Date of Acquisition
Landsat 4–5	TM	171/059	2 April 1995
Landsat 7	ETM	171/059	12 March 2008
Landsat 8 OLI/TIRS	LANDSAT 8	171/059	14 April 2017

2.6. Flood Risk Analysis

The flood risk analysis included the combination of the results of both the vulnerability analysis and the hazard analysis by intersecting the flood depth polygons prepared during the hazard analysis with the land-use/cover vulnerability polygons. The resulting attribute tables were reclassified to develop the land-use/cover-flood depth relationship. Potential flood areas in terms of both the land cover vulnerability classes and water depth hazard classes were then presented. Flood risk maps were then prepared by overlaying the flood depth grids with the land-use/cover map. The following equation was used to generate the flood risk map of the Manafwa catchment in the raster calculator of ArcGIS. Finally, and based on Equation (4), flood risk was reclassified into five classes, as shown in Table 2

$$\text{Risk Map} = \text{Hazard Map} \times \text{Vulnerability Map} \quad (4)$$

Table 2. Classes of flood risk in Manafwa, which results from the product of hazard and vulnerability.

Flood Risk Value	Risk Class (RC)	Risk Level (RL)
<0.5	1	Very Low
0.5–1.0	2	Low
1.0–1.5	3	Moderate
1.5–2.0	4	Significant
>2.0	5	Extreme

3. Results

3.1. Land Cover Classification in Manafwa Catchment

There are eight land-use/cover types identified in the Manafwa catchment, which are built-up area, bushland, commercial farmland, grassland, subsistence farmland, tropical high forest, wetland, and woodland (Figure 6). Subsistence farmland, Tropical High forest, wetland, and woodland were the dominant LULC types at the beginning of the study period (Figure 6). However, bushland, wetland, and tropical high forests significantly declined whilst subsistence farmland, commercial farmland, and woodland increased during the 1995–2008 period. The period of 2008–2017 is characterized by an increase in bushland, commercial farmland, and subsistence farmland with a marked decrease in the tropical high forest, wetland, and woodland.

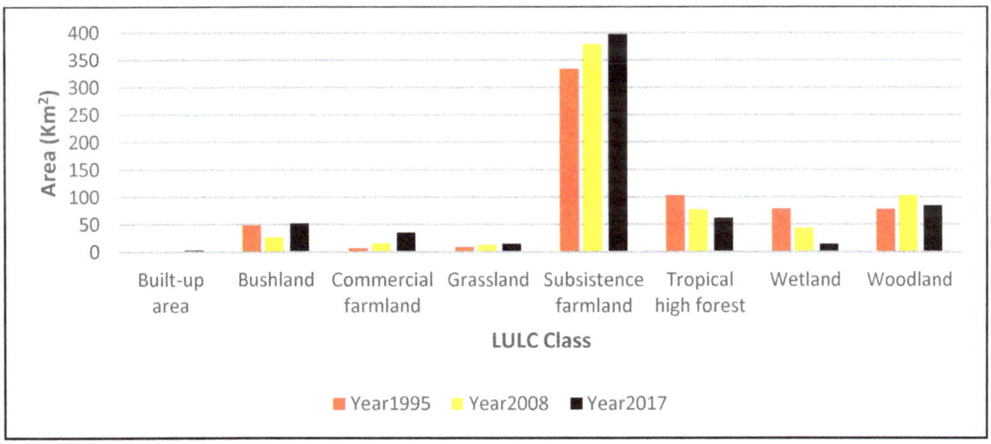

Figure 6. Temporal change of land cover types in Manafwa catchment between 1995 and 2017.

3.2. Model Calibration, Sensitivity, and Uncertainty Analysis

Through the sensitivity analysis of the SWAT model, 14 parameters with higher sensitivity were selected to calibrate and verify the model (Table 3). Sensitivity was evaluated based on t-stat values (a higher absolute value is more sensitive). Significance was also determined based on the *p*-value. Sequential Uncertainty Fitting program (SUFI-2) flow calibration was performed for the simulated results based on the sensitive parameters. This was conducted by simulating the flow for 26-year period, including two-year warm period from 1981–2013. The values of NSE and R^2 (Table 4) after calibration are greater than 0.65, which is the best predictor of the model. After calibrating (2000–2010) and obtaining acceptable values of NSE and R^2, validation of simulated stream flow for 3-year period, including one-year warm-up period from 2011 to 2013, was performed using monthly observed flows. The results after validation were also checked using NSE and R^2 and had magnitudes greater than 0.65 and 0.77, respectively, for the 2008 and 2017 (Table 5), except for 1995, which has an NSE value less than 1. The PBIAS also shows a good estimation since the values are less than ±25%, except for 1995. The calibrated and validated stream flow results showed a good agreement with the observed data (Table 4 and Figure 7) and therefore indicate that the SWAT model is a good predictor of stream flow of the Manafwa watershed.

Table 3. Flow sensitive parameters and their fitted value in SUFI2.

No	Parameter Name	Definition	Fitted Value	Min Value	Max Value	t-Stat
1	R__CN2.mgt	Initial SCS runoff curve number for moisture condition II	−0.24	−0.4	0.2	0.53
2	V__ALPHA_BF.gw	Baseflow alpha factor (days)	1.52	1.04	1.55	0.09
3	V__GW_DELAY.gw	Groundwater delay time (days)	35.26	−50.05	98.83	17.11
4	V__GWQMN.gw	Threshold depth of water in the shallow aquifer	−0.63	−0.77	0.12	−0.54
5	R__LAT_SED.hru	Sediment concentration in later and groundwater flow (mg/L)	59.56	43.68	74.52	−0.96
6	R__SOL_AWC(..).sol	Available water capacity of the soil layer (mm/mm)	−0.30	−0.49	−0.06	−0.44
7	R__CH_K2.rte	Effective hydraulic conductivity in main channel alluvium (mm/h)	46.80	1.07	62.45	−1.53
8	R__CH_N2.rte	Manning's "n" value for the main channel	0.05	0.05	0.12	0.55
9	R__ESCO.hru	Soil evaporation compensation factor	−0.04	−0.05	0.50	0.15
10	R__OV_N.hru	Manning's "n" value for overland flow	16.04	6.49	16.88	0.56
11	R__SURLAG.bsn	Surface runoff lag coefficient	15.19	12.34	17.29	0.91
12	R__RCHRG_DP.gw	Deep aquifer percolation factor	1.29	0.92	1.42	0.10
13	R__GW_REVAP.gw	Groundwater "revap" coefficient	0.47	0.28	0.76	1.72
14	R__SOL_K(..).sol	Saturated hydraulic conductivity (mm/h)	3.37	3.06	5.81	−2.94

Table 4. Summary of calibrated and validated performance criteria.

Performance Criteria	Calibration (2000–2010)			Validation (2011–2013)			Accepted Range [24].
	1995 LC	2008 LC	2017 LC	1995 LC	2008 LC	2017 LC	
R^2	0.94	0.79	0.94	0.81	0.78	0.79	$R^2 > 0.50$
NSE	0.65	0.79	0.74	−1.72	0.69	0.69	NSE > 0.50
PBIAS	−30.2	−12	−23.4	−68.8	−14.6	−11.7	PBIAS ≤ ±25%

Table 5. Flooded areas (km2) in different Land cover types (1995–2017).

Land Cover Type	Area (km^2)	Area (km^2)	Area (km^2)	% Change	% Change	% Change
	1995	2008	2017	1995–2008	2008–2017	1995–2017
Built-up area	0.0	0.1	0.3		200.0	
Bushland	2.0	1.8	3.2	−10.0	77.8	60.0
Commercial farmland	6.8	15.7	34.2	130.9	117.8	402.9
Grassland	0.0	0.1	0.01		−90.0	
Subsistence farmland	57.3	65.5	67.9	14.3	3.7	18.5
Tropical high forest	1.5	1.2	1.1	−20.0	−8.3	−26.7
Wetland	53.4	35.8	12.3	−33.0	−65.6	−77.0
Woodland	7.7	8.5	9.7	10.4	14.1	26.0

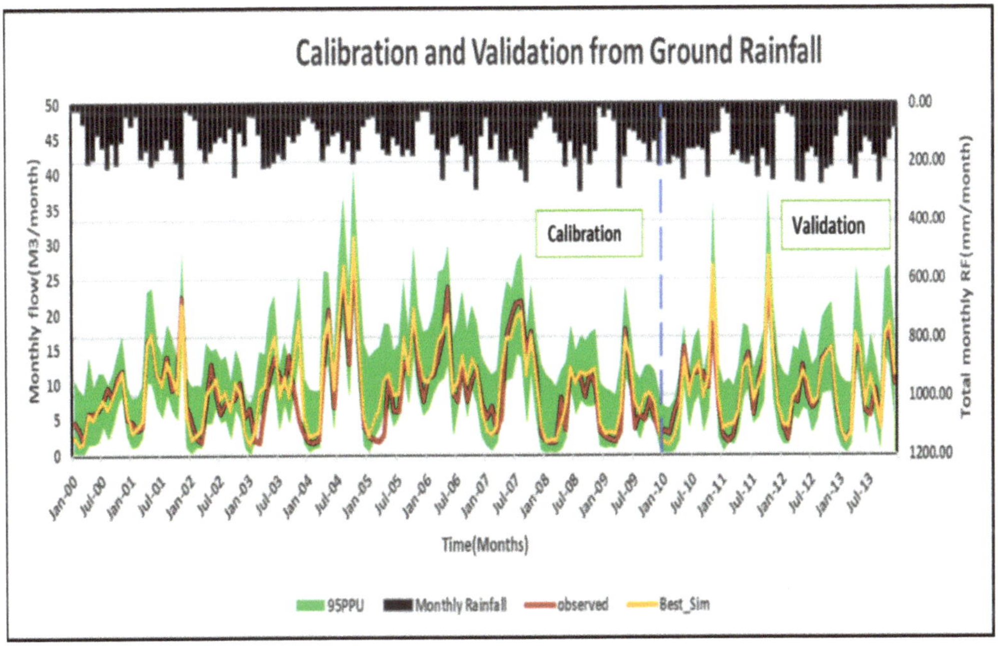

Figure 7. Observed and simulated monthly streamflow hydrographs for the calibration period of 2000–2010 and the validation period of 2011–2013 (separated by the vertical dashed line) for 2008 land Cover. Notes: Calibration; R^2 = 0.79, NSE= 0.79 & PBIAS= −12; Validation: R^2 = 0.78, NSE = 0.69, and PBIAS = −14.0.

3.3. Inundation Areas Mapped

The analysis of flood inundation area indicated that a considerable increase in flood inundation with increasing discharge of flood was shown from 5 years to 100 years return period (Figure 8). The classification of flood depth areas indicated that 13–19% of the total flooded areas had water depths greater than 2 m.

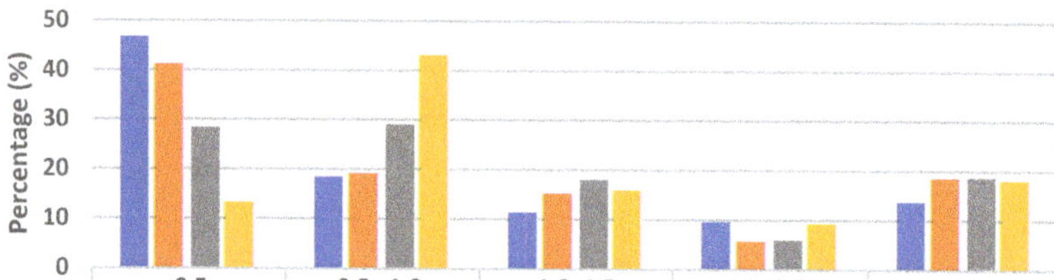

	<0.5	0.5– 1.0	1.0–1.5	1.5–2.0	>2.0
5 year Flood	46.7	18.5	11.4	9.6	13.8
10 year Flood	41.2	19.1	15.3	5.8	18.6
50 Year Flood	28.3	29	18	6	18.7
100 Year Flood	13.3	43.1	16	9.4	18.2

Water Depth (m)

Figure 8. Return Period–Flood Depth Relationship.

3.4. Floodplain Vulnerability

The land cover area under the influence of modeled flood showed that 42.7, 33.7, 10.8, and 5.3 km^2 of subsistence farming area, forest commercial farming area, wetland, and woodland area are respectively inundated by 5-year flood, and the total vulnerable area is 94 km^2 (Figure 9). The very high flood vulnerability areas covered 12.9 km^2, and high vulnerability areas occupied 9.0 km^2. Moderate, low, and very low vulnerability zones were 10.7, 17.4, and 43.9 km^2, respectively. Similarly, 67.9, 34.2, 12.3, and 9.7 km^2 of subsistence farming area, forest commercial farming area, wetland, and woodland area were respectively inundated by a 100-year flood, which showed flooded areas increased with an increase in flooding intensity; mostly subsistence farming area was inundated by different year floods, which was followed by commercial farming and wetland area. The flood vulnerability results for the 100 Yr. return period showed that the total vulnerable area is 128.7 km^2. The very high flood vulnerability areas covered 23.4 km^2, high vulnerability areas occupied 12.1 km^{2,} while moderate, low, and very low vulnerability zones were 20.6, 55.6, and 17.1 km^2, respectively.

According to [35], integrated flood management and land cover change, along with HECRAS hydraulic model simulations, are required for flood risk mitigation. Therefore, land cover change in the Manafwa basin was analyzed in two time periods, and the comparisons for the different two time periods shows that flooded area in Commercial farming, subsistence farming, Bushland, and Woodland has increased, but flooded areas in Wetland and Tropical High Forest decreased (Table 5).

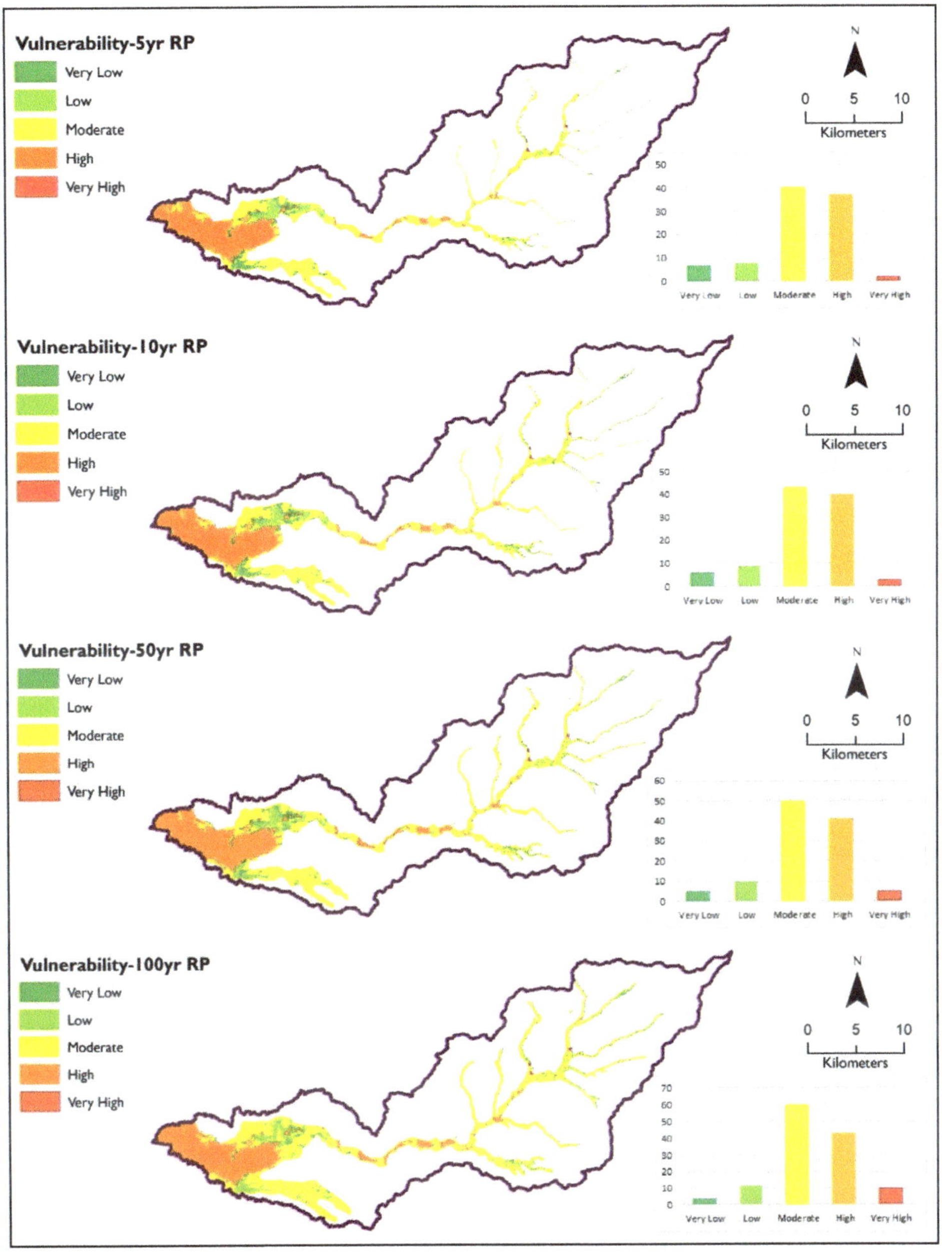

Figure 9. Flood Vulnerability Maps for different return periods.

3.5. Flood Risk Analysis

The classification of risk for the 5 YR and 100 YR return periods showed that commercial farmland, subsistence farmland, and Built-up/Settlement areas were under extreme risk of floods (Figure 10). The extreme flood risk areas covered 11.3 km^2; significant risk areas occupied 5.7 km^2, while moderate, low, and very low-risk zones were 12.9, 15.9, and 48.2 km^2, respectively. Similarly, for the 100 YR return period, the extreme flood risk areas covered 20.0 km^2, and significant flood risk areas occupied 15.0 km^2. Moderate, low, and very low-risk areas were 24.0, 46.0, and 23.7 km^2, respectively.

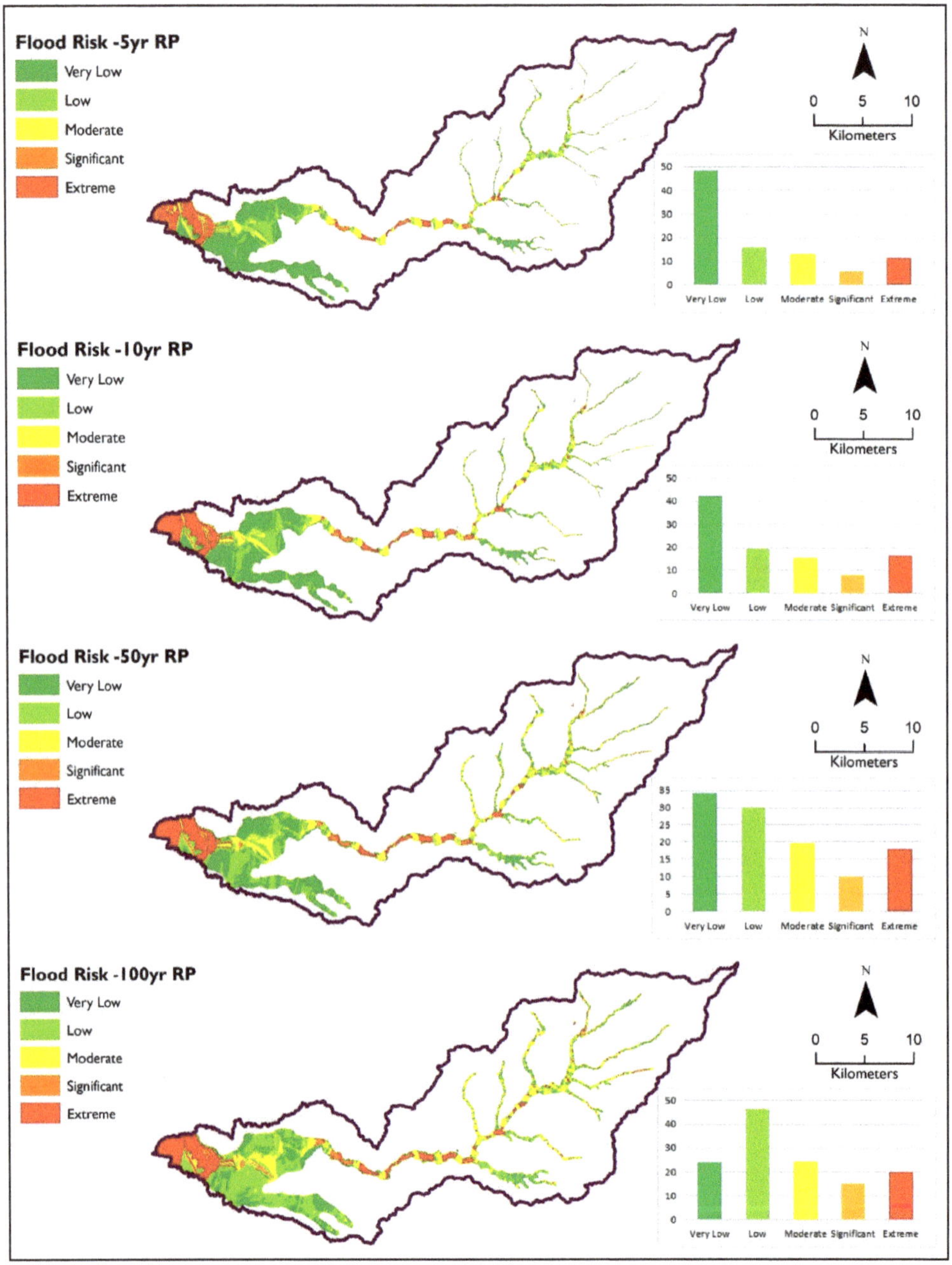

Figure 10. Flood Risk maps for different return periods.

The analysis of the relationship between the flood hazard level and settlement area (Figure 11) indicated a gradual increase in the significant and extreme hazard classes in all return periods. There is no change in the very low, low, and moderate hazard classes in all return periods.

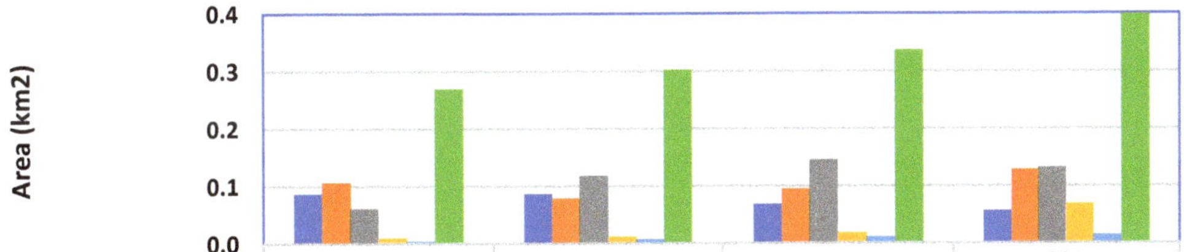

	5 Yr	10 Yr	50 Yr	100 Yr
<0.5 (Very Low)	0.1	0.1	0.1	0.1
0.5–1.0 (Low)	0.1	0.1	0.1	0.1
1.0–1.5(Moderate)	0.1	0.1	0.1	0.1
1.5–2.0 (Significant)	0.01	0.01	0.02	0.1
>2.0 (Extreme)	0.004	0.01	0.01	0.02
Total	0.3	0.3	0.3	0.4

Figure 11. Risk Classification of Settlement Land Cover Type.

Similarly, Subsistence farming area under a very low hazard class (<0.5 m) is 22.2, 22.6, 18.9, and 11.9 km^2 for return periods of 5–100 years (Figure 12). It also shows that there is a gradual increase in the low, moderate, and extreme hazard classes.

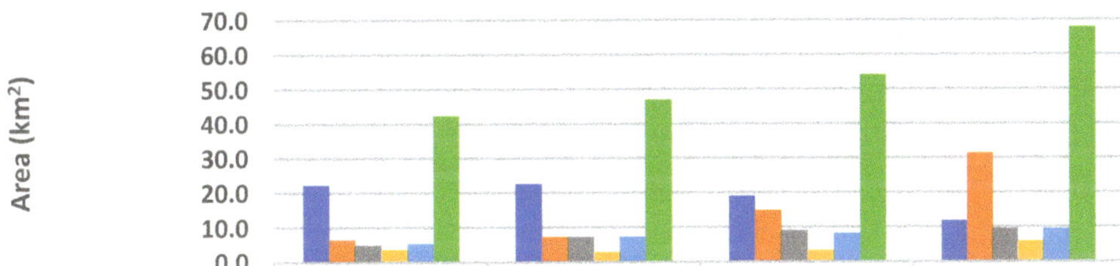

	5 Yr	10 Yr	50 Yr	100 Yr
<0.5 (Very Low)	22.2	22.6	18.9	11.9
0.5–1.0 (Low)	6.4	7.2	14.8	31.3
1.0–1.5(Moderate)	4.9	7.2	8.9	9.5
1.5–2.0 (Significant)	3.5	2.8	3.2	5.8
>2.0 (Extreme)	5.3	7.2	8.2	9.4
Total	42.4	46.9	54.0	67.9

Figure 12. Risk classification of Subsistence Farming Land Cover Type.

In the commercial farming land cover type, there is a gradual increase in every return period in the low, moderate, and extreme hazards and a gradual decrease in every return period in the very low hazard (Figure 13).

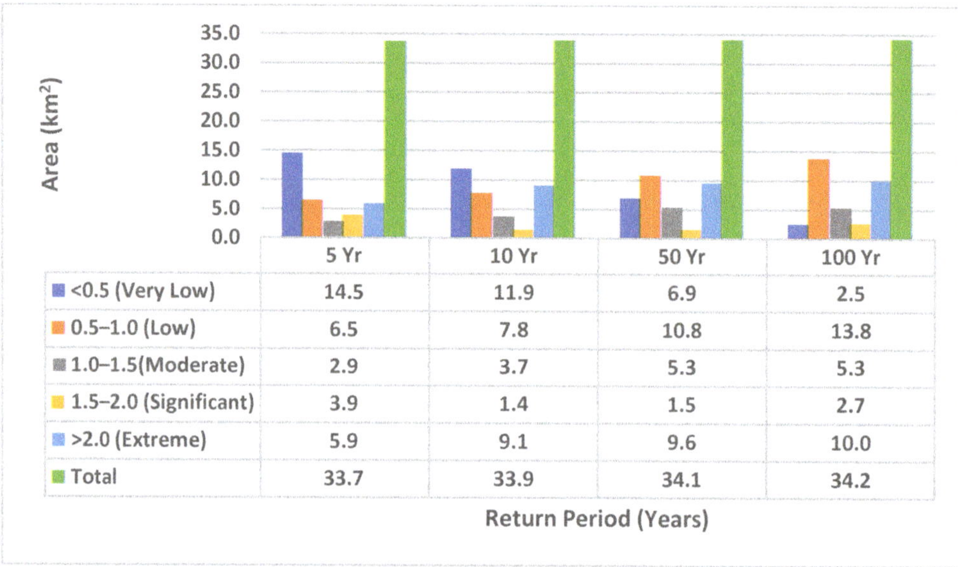

Figure 13. Risk classification of Commercial Farming Land Cover Type.

4. Discussion

SWAT is one of the most widely used models when simulating water balance within a basin [36]. However, the software has some limitations related mostly to a large number of input parameters. Sometimes, several parameters must be obtained or estimated from global databases, equations, or other computer software [37]. The information gave a satisfactory representation of the total flow behavior in the basin once the model was calibrated for the different land coverage scenarios. The error metrics for calibration and validation periods in the Manafwa catchment were "good", according to [29]. Ref [29] recommended that the general performance of objective functions on monthly time step calibration are satisfactory if NSE > 0.50 and RSR $\leq$ 0.70, and if PBIAS $\leq \pm$25% for streamflow.

The combination of Arc GIS and HECRAS 2-D flood simulation model indicates the capability of simulating flood events and spatially depicting the degree of exposure or vulnerability of the region towards a hazard event in terms of inundation extent and depth of water levels. The model can be said to have generated reliable quantified output. This hybrid approach provides quantified information on the water level depths and facilitates access to the data at any point of interest. As there are no quantified data on the inundation depths for flood hazards in the study region, the visualization and the quantification of the flood risks, as facilitated by this approach, can generate invaluable information and assist the decision-making authorities to making informed choices towards mitigating the catastrophic effects of flooding disasters.

Whereas in literature, there is considerable debate about whether a 1D or a 2D model provides a better representation of a flood event [21], it should be noted that even for the most sophisticated models, the performance of models is influenced by the quality of the source of information that is available for their parameterization, calibration, and validation. This is especially critical in undeveloped countries where financial and data sources are scarce. With regard to the calibration or validation of the model result, it could have been improved if it had been possible to compare it to an actual flood event, e.g., upstream and downstream flow hydrographs, mapped and recorded inundation extents, depths, or flow velocities. Such data were not available for the model area.

The lack of appropriate infrastructures and data makes the development of FHA difficult in African countries beyond the inclusion of this continent in global risk studies, such as those cited in [38]. In most African countries, the 30 m spatial resolution digital elevation model (DEM) from the SRTM project or the ASTER project, or the SRTM-derived 'Bare-Earth' DEM and Multi-Error-Removed Improved-Terrain (MERIT) DEM are the best options. The lack of detailed DEMs can be considered as one but not the only key factor of the limited FHA in Africa. Apart from the lack of DEMs for flood hazard analysis, the poor quality or limited availability of flow data must be kept in mind [39].

The vulnerability assessment approach used in this study for identifying and providing a vulnerability rank based on land use/cover in the flood area is a simple yet powerful approach. It not only identified the most to least vulnerable critical land-use/cover types but also provided enough information for flood preparedness processes that could significantly reduce the impact. The approach could easily be extended for the vulnerability evaluation of other infrastructures in order to estimate economic losses, the navigation route of people, including high-density areas, and other region-specific important factors. Moreover, futuristic higher magnitude flood events can be simulated to assess magnified vulnerability and associated risks. Land use planning decisions could be made based on flood inundation maps. Following such approaches will help save lives and resources at the same time and provide a proven and more accurate way to contest the uncertainties of the natural events causing floods.

It should be kept in mind that uncertainties exist in every stage of flood hazard mapping, from the beginning of the process (data collection, model selection, parameter selection, input data, model calibration, operation, and handling of the models) until the outcome is obtained [40]. The main limitations of this study were data quality and availability (e.g., missing rainfall and hydrologic data; unevenly distributed discharge and water level gauges with varying time series length and missing data; little field survey cross-section data and lack of hydraulic structure data along the river, such as bridges, weirs, etc.) contributing to uncertainties and inaccuracy in the results. The accuracy of the flood maps could be improved through the identification of possible sources of uncertainty and uncertainty analysis; the sources of uncertainty include the DEM resolution. The current 30 m is not sufficient and could lead to errors. We, therefore, recommended this for further research, as well as the integration of better-quality data into the models if they become available.

5. Conclusions

This study presented a systematic approach of coupling the hydrodynamic model HEC-RAS with the hydrologic model SWAT in delineating flood inundation zones and subsequently assessing the vulnerability of different land cover types in the Manafwa River Watershed Eastern Uganda. The HEC-RAS flood simulation model was found to be capable of simulating flood events and spatially depicting the vulnerability of the region towards a hazard event in terms of inundation extent, whereas SWAT was proven to be an appropriate tool in generating simulated flood hydrographs at desired locations. The calibration and validation results of the streamflow generally show good agreement with the observations in terms of R^2, PBIAS, and Nash-Sutcliffe efficiency coefficient. This study demonstrates a useful case study for applying the coupled hydrological and hydraulic models for flood hazard mapping. The integrated model used in this study could also be used for the analysis and design of possible structural measures and alternatives or be improved to establish a flood forecasting and warning system. The real-time inundation maps generated from forecasting systems are also an effective tool to inform relevant stakeholders and can significantly assist in communication with residents in areas susceptible to flooding. In the future, additional river survey data and high-resolution satellite images should be used for calibrating the model and improving the accuracy of flood hazard mapping. More hydro-meteorological observation stations are advocated to be installed in Manafwa and its surrounding area to provide first-hand hydrological information.

The results of the paper can be applied, especially in the areas of prevention, flood risk management, and crisis management. By incorporating flood maps into the local development plan of the catchment area, irresponsible expansion and densification of construction near the watercourse or in areas with moderate and high degrees of flood hazard could be prevented. In general, it is concluded that the use of integrated models to develop probabilistic flood hazard maps is an important step in the future flood protection of the Manafwa River Basin and similar river systems in Uganda and in the region. From the methodological point of view, the importance of the paper can be seen in the universality of the proposed steps to assess flood hazard and flood risk, which could be transferred to other similar flood-prone areas. However, further case studies in other regions should be undertaken to verify their general applicability. There are several potential research directions that can be mentioned as the next step, such as the comparison of HECRAS results with other 2D models such as Iber and BASEMENT. All three models are free, and such a comparison will highlight the advantages and disadvantages of each model's structure as well as the assumptions for the applied location.

Author Contributions: Conceptualization, G.E., I.K., A.G. and Y.B.; data curation, G.E.; formal analysis, G.E. and I.K.; funding acquisition, Y.B.; Investigation, G.E.; methodology, G.E. and I.K.; software, G.E.; supervision, A.G., Y.B., A.E. and I.K.; writing—original draft, G.E.; writing—review and editing, G.E., I.K., A.G., Y.B. and A.E. All authors contributed to the writing of the manuscript. All authors have read and agreed to the published version of the manuscript.

Funding: This research was funded by Swedish International Development Agency (SIDA) under the Building Resilient Ecosystems and Livelihoods to Climate Change and Disaster Risks (BREAD) project (Project, 331).

Institutional Review Board Statement: Not applicable.

Data Availability Statement: Not applicable.

Acknowledgments: The authors are grateful to the Ministry of Water and Environment for providing the hydrological data that were used for the simulation of the hydrological model. Further, the authors also extend gratitude to Nsiimire Peter for his contribution to field data collection.

Conflicts of Interest: The authors declare no conflict of interest. The funding sponsors had no role in the design of the study; in the collection, analyses, or interpretation of data; in the writing of the manuscript, and in the decision to publish the results.

References

1. Arnell, N.W.; Gosling, S.N. The Impacts of Climate Change on River Flood Risk at the Global Scale. *Clim. Chang.* **2016**, *134*, 387–401. [CrossRef]
2. Cred Crunch. *Natural Disasters in 2017: Lower Mortality, Higher Cost*; Technical Report; Centre for Research on the Epidemiology of Disasters: Brussels, Belgium, 2018; pp. 1–2.
3. Mubialiwo, A.; Onyutha, C.; Abebe, A. Historical rainfall and evapotranspiration changes over Mpologoma catchment in Uganda. *Adv. Meteorol.* **2020**, *2020*, 8870935. [CrossRef]
4. Ministry of Water and Environment. *Mpologoma Catchment Management Plan*; Ministry of Water and Environment: Kampala, Uganda, 2018.
5. Floodlist. Uganda—Deadly Floods and Landslides in Eastern Region (Updated). 2020. Available online: https://floodlist.com/africa/uganda-floods-bududa-sironko-december-2019 (accessed on 5 February 2021).
6. Pinos, J.; Timbe, L. Performance assessment of two-dimensional hydraulic models for generation of flood inundation maps in mountain river basins. *Water Sci. Eng.* **2019**, *12*, 11–18. [CrossRef]
7. Vojtek, M.; Vojteková, J. Flood hazard and flood risk assessment at the local spatial scale: A case study. *Geomat. Nat. Hazards Risk* **2016**, *7*, 1973–1992. [CrossRef]
8. Alahacoon, N.; Matheswaran, K.; Pani, P.; Amarnath, G.A. Decadal Historical Satellite Data and Rainfall Trend Analysis (2001–2016) for Flood Hazard Mapping in Sri Lanka. *Remote Sens.* **2018**, *10*, 448. [CrossRef]
9. Cao, C.; Xu, P.H.; Wang, Y.H.; Chen, J.P.; Zheng, L.J.; Niu, C.C. Flash Flood Hazard Susceptibility Mapping Using Frequency Ratio and Statistical Index Methods in Coalmine Subsidence Areas. *Sustainability* **2016**, *8*, 948. [CrossRef]
10. Nkiaka, E.; Nawaz, N.; Jon, C. Evaluating global reanalysis datasets as input for hydrological modelling in the Sudano-Sahel Region. *J. Hydrol.* **2017**, *4*, 13. [CrossRef]

11. Gorgoglione, A.; Gioia, A.; Iacobellis, V.; Piccinni, A.; Ranieri, E. A rationale for pollutograph evaluation in ungauged areas, using daily rainfall patterns: Case studies of the Apulian Region in Southern Italy. *Appl. Environ. Soil Sci.* **2016**, *2016*, 9327614. [CrossRef]
12. Liu, G.; He, Z.; Luan, Z.; Qi, S. Intercomparison of a lumped model and a distributed model for streamflow simulation in the Naoli River Watershed, Northeast China. *Water* **2018**, *10*, 1004. [CrossRef]
13. Daniel, R.; Todd, M.; Charlotte, M.; Arthur, T.; Zachary, M.E. Using the Climate Forecast System Reanalysis as Weather Input Data for Watershed Models. *Hydrol. Process.* **2014**, *28*, 5613–5623. [CrossRef]
14. Yu, Y.H.; Zhang, H.B.; Singh, V.P. Forward Prediction of Runoff Data in Data-Scarce Basins with an Improved Ensemble Empirical Mode Decomposition (EEMD) Model. *Water* **2018**, *10*, 388. [CrossRef]
15. Chomba, I.C.; Banda, K.E.; Winsemius, H.C.; Chomba, M.J.; Mataa, M.; Ngwenya, V.; Sichingabula, H.M.; Nyambe, I.A.; Ellender, B. A Review of Coupled Hydrologic-Hydraulic Models for Floodplain Assessments in Africa: Opportunities and Challenges for Floodplain Wetland Management. *Hydrology* **2021**, *8*, 44. [CrossRef]
16. Horrit, M.S.; Bates, P.D. Evaluation of 1D and 2D numerical models for predicting river flood inundation. *J. Hydrol.* **2002**, *268*, 87–99. [CrossRef]
17. Neal, J.C.; Fewtrell, T.J.; Bates, P.D.; Wright, N.G. A comparison of three parallelisation methods for 2D flood inundation models. *Environ. Model. Soft.* **2009**, *25*, 398–411. [CrossRef]
18. Cook, A.; Merwade, V. Effect of topographic data, geometric configuration and modeling approach on flood inundation mapping. *J. Hydrol.* **2009**, *377*, 131–142. [CrossRef]
19. Neal, J.C.; Villanueva, I.; Wright, N.; Willis, T.; Fewtrell, T.; Bates, P.D. How much physical complexity is needed to model flood inundation? *Hydrol. Process.* **2012**, *26*, 2264–2282. [CrossRef]
20. Papaioannou, G.; Loukas, A.; Vasiliades, L.; Aronica, G.T. Flood inundation mapping sensitivity to riverine spatial resolution and modelling approach. *Nat. Hazards.* **2016**, *83*, 117–132. [CrossRef]
21. Jha, M.K.; Afreen, S. Flooding Urban Landscapes: Analysis Using Combined Hydrodynamic and Hydrologic Modeling Approaches. *Water* **2020**, *12*, 1986. [CrossRef]
22. Arnold, J.G.; Moriasi, D.N.; Gassman, P.W.; Abbaspour, K.C.; White, M.J.; Srinivasan, R.; Santhi, C.; Harmel, R.D.; van Griensven, A.; van Liew, M.W.; et al. SWAT: Model use, calibration, and validation. *Trans. ASABE* **2012**, *55*, 1491–1508. [CrossRef]
23. Abbaspour, K.C.; Rouholahnejad, E.; Vaghefi, S.; Srinivasan, R.; Yang, H.; Kløve, B. A continental-scale hydrology and water quality model for Europe: Calibration and uncertainty of a high-resolution large-scale SWAT model. *J. Hydrol.* **2015**, *524*, 733–752. [CrossRef]
24. Yan, T.; Bai, J.; Lee Zhi Yi, A.; Shen, Z. SWAT-simulated streamflow responses to climate variability and human activities in the Miyun Reservoir basin by considering streamflow components. *Sustainability* **2018**, *10*, 941. [CrossRef]
25. Jabbar, F.K.; Grote, K. Evaluation of the predictive reliability of a new watershed health assessment method using the SWAT model. *Environ. Monit. Assess.* **2020**, *192*, 224. [CrossRef] [PubMed]
26. Thapa, S.; Shrestha, A.; Lamichane, S.; Adhikari, R.; Gautam, D. Catchment-scale flood hazard mapping and flood vulnerability analysis of residential buildings: The case of Khando River in eastern Nepal. *J. Hydrol. Reg. Stud.* **2020**, *30*, 100704. [CrossRef]
27. Getahun, Y.S.; Gebre, S.L. Flood Hazard Assessment and Mapping of Flood Inundation Area of the Awash River Basin in Ethiopia using GIS and HEC-GeoRAS/HEC-RAS Model. *J. Civil Environ. Eng.* **2015**, *5*, 179. [CrossRef]
28. Okirya, M.; Albert, R.; Janka, O. Application of HEC HMS/RAS AND GIS tools in flood modeling: A case study for River Sironko UGANDA. *Global J. Eng. Design Technol.* **2012**, *1*, 19–31.
29. Moriasi, D.N.; Arnold, J.G.; Van Liew, M.W.; Bingner, R.L.; Harmel, R.D.; Veith, T.L. Model evaluation guidelines for systematic quantification of accuracy in watershed simulations. *Trans. ASABE* **2007**, *50*, 885–900. [CrossRef]
30. HEC-RAS. River Analysis System: Hydraulic Reference Manual. In *USACE Version: 5.0*; US Army Corps of Engineers: Washington, DC, USA, 2016; p. CPD-68.
31. Dimitriadis, P.; Tegos, A.; Oikonomou, A.; Pagana, V.; Koukouvinos, A.; Mamassis, N.; Koutsoyiannis, D.; Efstratiadis, A. Comparative evaluation of 1D and quasi-2D hydraulic models based on benchmark and real-world applications for uncertainty assessment in flood mapping. *J. Hydrol.* **2016**, *534*, 478–492. [CrossRef]
32. Farooq, M.; Shafique, M.; Khattak, M.S. Flood Hazard Assessment and Mapping of River Swat Using HEC-RAS 2D Model and High-Resolution 12-m TanDEM-X DEM (WorldDEM). *Nat. Hazards* **2019**, *97*, 477–492. [CrossRef]
33. Yalcin, E. Assessing the impact of topography and land cover data resolutions on two-dimensional HEC-RAS hydrodynamic model simulations for urban flood hazard analysis. *Nat. Hazards* **2020**, *101*, 995–1017. [CrossRef]
34. Quiroga, V.M.; Kurea, S.; Udoa, K.; Manoa, A. Application of 2D numerical simulation for the analysis of the February 2014 Bolivian Amazonia flood: Application of the new HEC-RAS version 5. *Ribagua* **2016**, *3*, 25–33. [CrossRef]
35. Suriya, S.; Mudgal, B.V. Impact of urbanization on flooding: The Thirusoolam Sub-watershed: A case study. *J. Hydrol.* **2012**, *412*, 210–219. [CrossRef]
36. Thai, T.H.; Thao, N.P.; Dieu, B.T. Assessment and Simulation of Impacts of Climate Change on Erosion and Water Flow by Using the Soil and Water Assessment Tool and GIS: Case Study in Upper Cau River basin in Vietnam. *Vietnam. J. Earth Sci.* **2017**, *39*, 376–392. [CrossRef]
37. Nyeko, M. Hydrologic Modelling of Data Scarce Basin with SWAT Model: Capabilities and Limitations. *Water Resour. Manag.* **2014**, *29*, 81–94. [CrossRef]

38. Ward, P.J.; Blauhut, V.; Bloemendaal, N.; Daniell, J.E.; de Ruiter, M.C.; Duncan, M.J.; Emberson, R.; Jenkins, S.F.; Kirschbaum, D.; Kunz, M.; et al. Review article: Natural hazard risk assessments at the global scale. *Nat. Hazards Earth Syst. Sci.* **2020**, *20*, 1069–1096. [CrossRef]
39. Díez-Herrero, A.; Garrote, J. Flood Risk Assessments: Applications and Uncertainties. *Water* **2020**, *12*, 2096. [CrossRef]
40. Jha, A.K.; Bloch, R.; Lamond, J. *Cities and Flooding: A Guide to Integrated Urban Flood Risk Management for the 21st Century*; The World Bank: Washington, DC, USA, 2012.

hydrology MDPI

Article

Trivariate Joint Distribution Modelling of Compound Events Using the Nonparametric D-Vine Copula Developed Based on a Bernstein and Beta Kernel Copula Density Framework

Shahid Latif and Slobodan P. Simonovic *

Department of Civil and Environmental Engineering, Western University, London, ON N6A 5B8, Canada
* Correspondence: simonovic@uwo.ca

Abstract: Low-lying coastal communities are often threatened by compound flooding (CF), which can be determined through the joint occurrence of storm surges, rainfall and river discharge, either successively or in close succession. The trivariate distribution can demonstrate the risk of the compound phenomenon more realistically, rather than considering each contributing factor independently or in pairwise dependency relations. Recently, the vine copula has been recognized as a highly flexible approach to constructing a higher-dimensional joint density framework. In these, the parametric class copula with parametric univariate marginals is often involved. Its incorporation can lead to a lack of flexibility due to parametric functions that have prior distribution assumptions about their univariate marginal and/or copula joint density. This study introduces the vine copula approach in a nonparametric setting by introducing Bernstein and Beta kernel copula density in establishing trivariate flood dependence. The proposed model was applied to 46 years of flood characteristics collected on the west coast of Canada. The univariate flood marginal distribution was modelled using nonparametric kernel density estimation (KDE). The 2D Bernstein estimator and beta kernel copula estimator were tested independently in capturing pairwise dependencies to establish D-vine structure in a stage-wise nesting approach in three alternative ways, each by permutating the location of the conditioning variable. The best-fitted vine structure was selected using goodness-of-fit (GOF) test statistics. The performance of the nonparametric vine approach was also compared with those of vines constructed with a parametric and semiparametric fitting procedure. Investigation revealed that the D-vine copula constructed using a Bernstein copula with normal KDE marginals performed well nonparametrically in capturing the dependence of the compound events. Finally, the derived nonparametric model was used in the estimation of trivariate joint return periods, and further employed in estimating failure probability statistics.

Keywords: compound flooding; D-vine copula; trivariate joint analysis; Bernstein estimator; beta kernel estimator; parametric copulas; kernel density estimation; return periods

Citation: Latif, S.; Simonovic, S.P. Trivariate Joint Distribution Modelling of Compound Events Using the Nonparametric D-Vine Copula Developed Based on a Bernstein and Beta Kernel Copula Density Framework. *Hydrology* **2022**, *9*, 221. https://doi.org/10.3390/hydrology9120221

Academic Editors: Aristoteles Tegos, Alexandros Ziogas and Vasilis Bellos

Received: 25 October 2022
Accepted: 29 November 2022
Published: 7 December 2022

Publisher's Note: MDPI stays neutral with regard to jurisdictional claims in published maps and institutional affiliations.

1. Introduction

Compound events (CE) is a multidimensional phenomenon that can be defined by the joint probability occurrence of two or more extreme or non-extreme events, which may not be dangerous or devastating if considered individually [1–4]. However, CE can have severe consequences if their underlying variables co-occur or are in close succession. On the global scale, the flooding events in low-lying coastal cities or the risk of extreme compound phenomena have already been recorded and outlined in the previous literature [5–7]. Climate change has already triggered a rising coastal water level called sea level rise (SLR), increasing the frequency and severity of flooding, which threatens coastal communities worldwide [8–10]. Coastal flooding can be significantly defined and estimated by combining the driving forces, such as storm surge (oceanographic), rainfall (or pluvial flooding) and river discharge (or fluvial flooding). These events can be interlinked through a common forcing mechanism, such as tropical or extra-tropical cyclones (or a low-atmospheric-pressure scenario). Among the different coastal flood drivers, a storm surge event is often considered a significant flood-driving agent [11]. When combined with rainfall (e.g., [2]) or with high river discharge (e.g., [10]), it can result in a devastating situation.

Different mathematical or statistical frameworks are often pointed out in the demonstration of the compound phenomenon but still can lack a consistent or robust approach. The traditional statistical evaluation of the CEs is usually a multivariate framework that observes the number of extreme joint episodes by targeting the most justifiable flood drivers. For instance, take the studies performed by Coles [12], Coles et al. [13], Svensson and Jones [14], Cooley et al. [15], Zheng et al. [16] and Zheng et al. [17]. In reality, the validity of the univariate probability or frequency analysis (and return periods) is questionable. Due to the multidimensional behaviour, it must demand an efficient framework that can reduce the hydrologic risk much more efficiently.

In recent studies, copula functions gained more popularity than traditional multivariate models and are recognized as highly flexible tools in the bivariate or multivariate joint distribution analysis of hydro-meteorological observations [18–22]. In the modelling of CE or flooding, the adequacy of different parametric class 2D copulas is tested by targeting different contributing variables, for instance, storm surge (or storm tide) and rainfall, or storm surge (or storm tide) and river discharge [23–28]. Such incorporations are limited to bivariate joint cases employing 2D parametric class copulas to observe pairwise joint dependencies. However, the more realistic and practical flood risk can be obtained by compounding the joint distribution behaviour, including more relevant flood-driving agents (e.g., storm surge, rainfall, and river discharge) simultaneously instead of their pairwise dependencies. For instance, tropical cyclones in the coastal region can trigger storm surges, rainfall and possible high-river-discharge events simultaneously; thus, the complex interplay between them can exacerbate flooding in the coastal zones. Therefore, the risk of coastal flooding can be analysed much more efficiently by considering the above triplet variable simultaneously instead of just considering bivariate joint dependency.

The application of the 3D (or any higher dimension) copula in hydro-meteorological modelling is minimal. Few previous works highlighted, for instance, the asymmetric, fully nested Archimedean copula [29,30].; the meta-elliptical Student's t copula [31]; the Plackett copula [32]; and the entropy copula [33]. All such frameworks have some statistical constraints and limitations when projected into higher dimensions. For example, the 3D symmetric Archimedean copula models the dependencies between multiple random variables by employing single-dependence parameters or generator functions and thus cannot preserve all pairwise dependencies [32,34]. Besides this, an asymmetric or fully nested Archimedean (FNA) copula can be much more reliable than a symmetric structure. FNA can individually approximate each random attribute pair through multiple parametric joint asymmetric functions [35–37]. The faithful preservation of all the lower-level dependencies among the targeted variables is still challenging based on the FNA structure. This framework is only effective and practical when two correlation structures are identical or near and lesser than the third correlation structure and are limited to a positive range [31]. Additionally, when considering more variables, the asymmetric FNA structure permits a narrow range of mutual dependencies [18]. Therefore, to alleviate all such statistical issues, the vine or pair-copula construction (PCC) approach is highly flexible and is a much more practical way of constructing any higher-dimensional joint dependence by mixing multiple 2D (bivariate) copulas in a stage-wise hierarchical nesting procedure or conditional mixing procedure [38–41].

In CE modelling, Bevacqua et al. [42] introduced a 3D vine copula for evaluating flooding events in Ravenna, Italy. In a recent study, Jane et al. [43] introduced the vine framework in the trivariate joint analysis of rainfall, ocean-side water-level and groundwater-level observations in South Florida, USA. Besides the above two, other studies—for instance, Graler et al. [41], Saghafian and Mehdikhani [44], Tosunoglu et al., [45], Latif and Mustafa [46]—often incorporated a vine copula under parametric distribution settings, thereby fitting the parametric-class 2D copulas with parametric marginal pdfs in the parametric fitting procedure. In some previous literature, such as Silverman [47], Moon and Lall [48], Sharma et al. [49], Kim et al. [50] and Karmakar and Simonovic [51], the performance of nonparametric kernel density estimation (KDE) has been revealed to be much better than those of parametric family functions. Due to the absence of any prior distribution assumption about their marginals probability density function (PDF) type, KDE performed much more reliably, especially suited for multimodal random samples. However, the copula function eliminates the restriction to model any marginal distribution from the same family functions. The subjective assumption of the joint PDF type of the fitted parametric copulas in the traditional vine distribution framework is not much more effective at approximating joint structure, which would be questionable. In other words, fixing the joint PDF of the dependence structure to any specific or predefined copula class may fail to fully acknowledge the flexibility of the copula fitted in the vine tree structure. Parametric copulas are frequently used because of their simplicity. However, the parameter estimation procedures of the fitted parametric models are time consuming using standard statistical techniques [52]. Rauf and Zeephongsekul [53] claimed that it could lead to spurious inferences and be challenging if the underlying assumptions of the fitted parametric distribution are violated. Fitting an appropriate parametric copula demands much more attention and extra caution, which might bear the risk of uncertainty in their estimated joint exceedance values if an inappropriate dependence structure is selected.

To deal with all the above-raised issues, introducing the nonparametric copula density in the vine copula construction could be a better alternative where the 2D copulas could adapt to any dependence structure without having any specific or fixed joint PDF form. To do this, the Bernstein copula estimator and beta kernel copula density could be a good choice for modelling multivariate copula density in nonparametric settings [54–58] and reference therein. In reality, the Bernstein copula can provide higher consistency and lack boundary bias problems [59], resulting in a better estimation of the underlying dependence structure than an empirical copula estimate. Besides this, there is the performance of beta kernel density is already proved by Rauf and Zeephongsekul [53] and Latif and Mustafa [22]. The nonparametric copula density gained more attention in economics but is rarely accepted in hydro-meteorological studies. Additionally, all the above nonparametric frameworks are often limited to bivariate cases.

The main contribution of the present work is the first to incorporate the Bernstein estimator and Beta kernel copula estimator in the nonparametric estimation of the 3D vine copula density in the trivariate modelling of compound flooding (CF) events on the west coast of Canada. The objective of the present study is (i) to incorporate and test the efficacy of above-mentioned nonparametric copula densities in establishing the D-vine structure and in determining trivariate joint cumulative distribution functions (JCDF),(ii) comparing the performance with the semiparametric approach in the vine copula density, introducing parametric copulas with nonparametric marginal pdfs and the parametric approach in the vine copula. Finally, the selected best-fitted vine copula density is employed to estimate trivariate joint return periods and in assessing hydrologic risk. Our recent study is the first that incorporates the Bernstein estimator in flood modelling and confirms that this function performed well compared to Beta copula density in the bivariate dependence modelling of storm surge and rainfall events [60]. Our present study extends the previous bivariate approach by dealing with three variables, integrating the impact of river discharge events with storm surge and rainfall events in the risk of compound flooding (CF) events.

Pirani and Najafi's [61] study already shows that the higher risk of compound extreme on Canada's west coasts is due to the joint impact of precipitation, extreme water level (also, storm surge events) and streamflow discharge. Additionally, west or Pacific Canada's coast experienced higher coastal instability because of the higher risk of coastal water levels [62]. This paper is organized into four sections. After the introduction, the required theoretical background of the nonparametric copula density and in development of the 3D vine copula framework are discussed in Section 2. Section 3 of this manuscript presents the application of the developed trivariate distribution framework to a case study in compounding the joint impact of rainfall, storm surge and river discharge events. This section comprises, for instance, modelling univariate marginal distribution via nonparametric KDE, constructing D-vine structure in the nonparametric fitting procedure via Bernstein and Beta copula estimator, D-vine structure under the parametric fitting procedure and in the semiparametric settings. This section also compares the model adequacy and performance of all three developed D-vine structures in the trivariate CF dependence. Additionally, the best-fitted trivariate structure is employed in estimating primary joint return periods for both AND and OR-joint cases and also employed in the estimation of FP statistics. Finally, Section 4 provided the research summary and conclusions.

2. Methodology

2.1. Nonparametric Copula Density Estimator

Figure 1 illustrates the methodological workflow used in this study. Firstly, compound flood variables' marginal distributions are modelled using the nonparametric kernel density estimation (KDE). The best-fitted parametric family distributions are adapted from our previous study [63], and their performance is compared with the selected KDE in the present study. The D-vine copula framework is developed under parametric, semiparametric and nonparametric settings, and their performance is compared in describing the most parsimonious flood dependence. The nonparametric vine density comprises multiple 2-D copulas via the Bernstein and Beta kernel density with kernel density margins without having any prior assumption about their marginal pdf and joint density function. The parametric and semiparametric vine copula density defines through parametric class 2-D copulas (i.e., Archimedean and Extreme value) with parametric and nonparametric via KDE margins. The best-fitted trivariate structure is employed in the estimation of trivariate joint return periods for both OR-and AND-joint cases and is further employed in estimating FP statistics. In this study, the D-vine copula are developed for three different cases, each defined by permutating the location of the conditioning variable. For instance, in case 1, the river discharge event is a conditioning variable; in case 2, the storm surge event is a conditioning variable; in case 3, the rainfall event is a conditioning variable.

Mirror image modification, transformed kernels, boundary kernels, etc., are a few examples of nonparametric approaches in joint density estimation [64–66]. This study introduces the beta kernel copula and Bernstein copula estimator for developing the D-vine structure for the trivariate joint analysis of storm surge, rainfall and river discharge events in relation to flood risk in the coastal regions. The beta kernel copula density was discussed earlier by Brown and Chen [67], Harrell and Davis [54] and Chen [68]. It is naturally free of boundary bias problems which are often encountered in the standard kernel estimator. The consistency remains in the beta kernel density if the actual density is unbounded at the boundary [69].

The 1D beta kernel density function for the given univariate variables, $A_1, A_2, \ldots, A_t$, is estimated by:

$$s_h(a) = \frac{1}{t}\sum\nolimits_{i=1}^{t} K(A_i, \frac{a}{h}+1, \frac{1-a}{h}+1) \tag{1}$$

where "h" is the kernel's bandwidth.

In Equation (1), the density of the beta kernel function with parameters q and v is estimated by

$$K(a, q, v) = \frac{a^q(1-a)^v\Gamma(q)\Gamma(v)}{\Gamma(q+v)},\ a \in [0,1] \tag{2}$$

According to Charpentier et al. [52], multiplying the beta kernel densities can result in beta copula joint density, known as the beta kernel copula, at point (a, b), as given below.

$$c_h(a,b) = \frac{1}{ph^2}\sum\nolimits_{i=1}^{p} K(A_i, \frac{a}{h}+1, \frac{1-a}{h}+1) \times K\left(B_i, \frac{b}{h}+1, \frac{1-b}{h}+1\right) \tag{3}$$

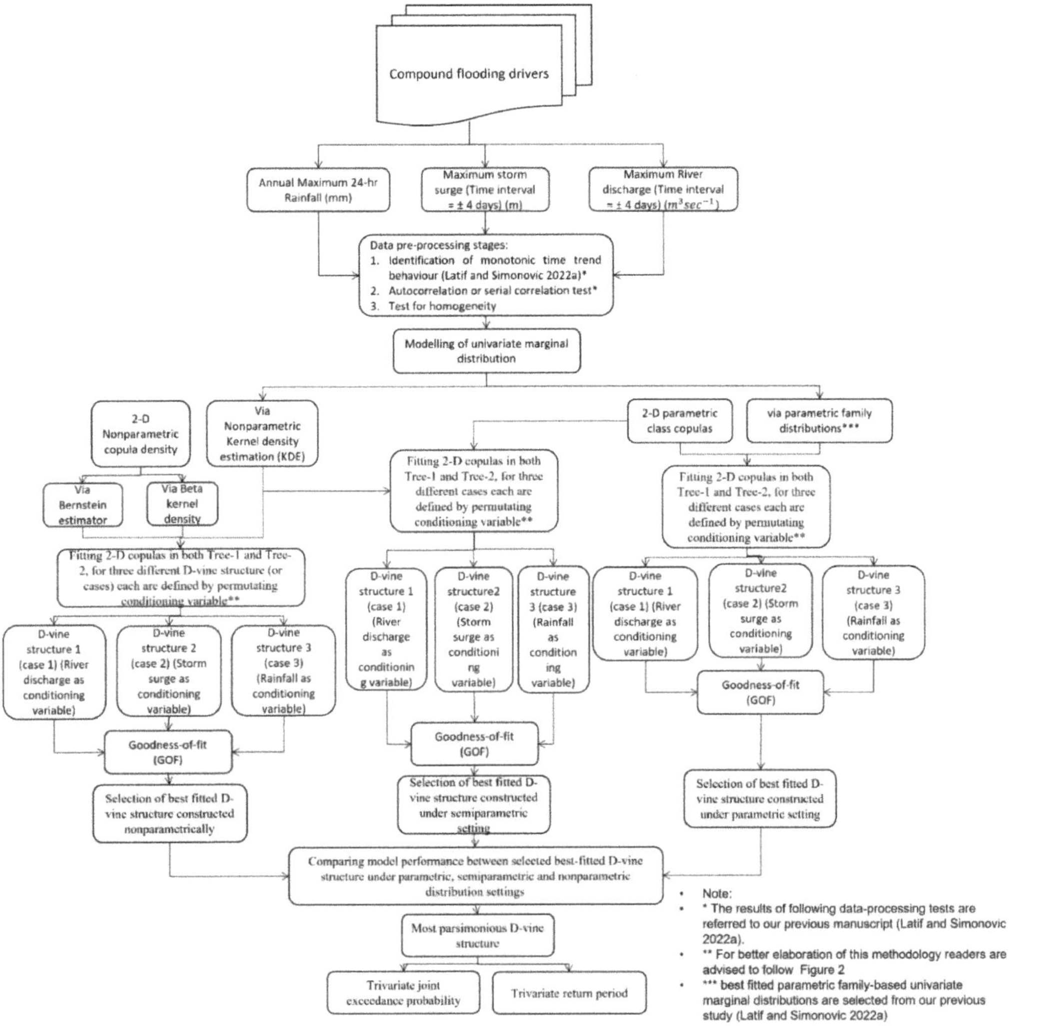

Figure 1. Workflow chart of the present study.

The bandwidth of Equation (3) is estimated by the rule of thumb (ROT) estimation procedure, which is based on minimizing the asymptotic mean-integrated-squared error (AMISE) statistics. For this, Nagler [70] pointed out the applicability of the Frank copula as the reference copula. The ROT bandwidth estimation for the fitted 2D beta kernel estimator of Equation (3) is estimated by

$$h = \left(\frac{1}{8\pi} \frac{\varsigma(c)}{\xi(c)} \right)^{\frac{1}{3}} n^{\frac{-1}{3}} \tag{4}$$

where "c" is assumed to be the Frank copula in Equation (4).

The efficacy of the Bernstein copula estimator is also tested and compared in constructing the D-vine structure together with beta kernel density. Lorentz [71] highlighted that the Bernstein polynomial could be used to approximate any continuous functions within a range of [0,1]. Tenbush [72] constructed bivariate joint density using the Bernstein estimator. Approximation of nonparametric joint density using the Bernstein copula can provide higher consistency and remove boundary bias problems [57,73]. Additionally, it can better estimate the underlying mutual correlation and good approximation with an asymmetric and extreme dependency compared to an empirical copula approach [69].

Mathematically, the n-degree Bernstein polynomial is estimated by [57,69]:

$$B(n, w, z) = \binom{n}{w} z^{w} (1 - z)^{n-w} \tag{5}$$

In Equation (5), $w = 0, 1, 2, \ldots, n \in \mathbb{N}; 0 \le z \le 1$.

Now, if $X = (X_1, X_2)$ illustrates bivariate observations having a uniform marginal distribution over $Y_i = \{0,\ 1,\ 2, \ldots,\ n_i\}$ with grid size $n_i \in \mathbb{N}$ and where $i = 1,\ 2$, then

$$y(w_1, w_2) = Y(\cap_{I=1}^{2}\{X_i = w_i\}), (w_1, w_2) \in [0,1]^2 \tag{6}$$

Hence, for the 2D joint distribution case, the Bernstein copula density is estimated by;

$$c(x_1, x_2) = \sum_{w_1=0}^{n_1-1} \sum_{w_2=0}^{n_2-1} y(w_1, w_2) \prod_{i=1}^{2} n_i B(n_i - 1, w_i, x_i), \tag{7}$$

where $(x_1, x_2) \in [0,1]^2$.

2.2. *Univariate Kernel Density Estimation of Flood Margins*

Parametric class functions are often restricted to prior distributional assumptions about their univariate marginal pdfs. However, on the other side, the parametric functions performed well if the given observation exhibited unimodality or symmetrical behaviour. A nonparametric kernel density estimation (KDE) is identified as much more robust and better performing in modelling the probability densities of different hydro-meteorological characteristics, especially when the given observation departed from the symmetrical behaviour or, say, bi- or multimodality [48,50,74,75]. Our present study tested the efficacy of different KDE functions and compared their performance with the selected best-fitted parametric distributions from our previous study [63].

Mathematically, the 1D kernel function can approximate a probability density structure having the following statistical property.

$$\int_{-\infty}^{+\infty} K(x)dx = 1 \tag{8}$$

Furthermore, the kernel function can be represented by a general function:

$$K_o(x) = \frac{1}{o} K\left(\frac{x}{o}\right) \tag{9}$$

where "o" is the bandwidth of the fitted univariate kernel function.

By taking the average of Equation (9), the univariate kernel density estimator $\hat{f}_o(x)$ of the probability density function is estimated by

$$\hat{f}_o(x) = \frac{1}{po} \sum_{i=1}^{p} K_o\left(\frac{x - X_i}{o}\right) \tag{10}$$

where "p" is the observation counts. In fitting the kernel density to the given observational samples, selecting an appropriate way of estimating kernel bandwidth is often essential; otherwise, it may be attributed to either over-smoothing or under- or insufficient smoothing (also called rough smoothing). For extended details about different statistical procedures in kernel bandwidth estimation, readers are advised to read Sharma et al. [49] and Jones et al. [76]. In our present analysis, the direct plug-in (DPI) method is used to estimate the bandwidth of the fitted kernel density [77–79]. Table 1 lists some kernel density functions which are used in this study.

Table 1. 1-D Kernel density estimation (KDE) tested in the modelling of flood marginals.

SI No.	Kernel Density Estimation	K(x)
1	Normal	$=(2\pi)^{\frac{-1}{2}}e^{-(x^2)/2}$
2	Epanechnikov (or parabolic)	$=0.75(1-x^2),\ \lvert X\rvert\leq 1$ $=0$, otherwise
3	Biweight (or Quartic)	$=0.9375\ (1-x^2)^2,\ \lvert x\rvert\leq 1$ $=0$, otherwise
4	Triweight	$1.09375\ (1-x^2)^3,\ \lvert x\rvert\leq 1$ $=0$, otherwise

2.3. D-Vine Copula Structure in the Trivariate Modelling

The vine or pair-copula construction (pcc) is based on the decomposition of full multivariate density into a cascade of local blocks of the best-fitted 2D copulas modelled between each random pair and their conditional and unconditional functions [38,39]. Two famous decomposition steps in the vine framework are the canonical or C-vine and drawable or D-vine [80,81]. The D-vine's structure is highly flexible, and this is accepted widely [40,41,82]. The traditional approach in the vine framework considers multiple 2D parametric class copulae in the stagewise hierarchy. A few statistical constraints with parametric copula joint density are highlighted in Section 1. Therefore, it could be problematic if the vine copula is constructed using the parametric class copulas. Due to this, we individually tested the efficacy of the nonparametric method via beta kernel copula density and Bernstein copula estimator in the 3D vine simulation for the given CF variables. This study also compares the performances of the parametric and semiparametric approaches in the vine simulation, where both frameworks are defined through multiple 2D parametric copulas. The univariate marginal distribution is modelled using the kernel density estimations (KDE) in both nonparametric and semiparametric and parametric class distributions in the parametric vine approach.

Due to the involvement of three flood characteristics in characterizing C.F. events in our study, the present 3D vine framework must demand three 2D copulae and two tree levels, Tree 1 and Tree 2 (refer to Figure 2). For trivariate variables (A, B, C), the D-vine structure can be mathematically expressed as

$$\mathrm{f(a,b,c)} = \mathrm{f(a)\cdot f(b|a)\cdot f(c|a,b)} \tag{11}$$

$$\mathrm{f(b|a)} = \frac{\mathrm{f(a,b)}}{\mathrm{f(a)}} = \mathrm{c_{ab}(F(a),\ F(b))\cdot f(b)} \tag{12}$$

$$\mathrm{f(c|a,b)} = \frac{\mathrm{f(b,\ c|a)}}{\mathrm{f(b|a)}} = \mathrm{c_{bc|a}(F(b|a),\ F(c|a))\cdot c_{ac}(F(a),\ F(c))\cdot f(c)} \tag{13}$$

In Equation (11), the conditional cumulative distribution functions $\mathrm{f(b|a)}$ and $\mathrm{f(c|a,b)}$ are estimated using the pair-copula densities. Additionally, F(a), F(b) and F(c) are the fitted univariate margins. In Equation (12), C_{ab} is the best-fitted 2D copula (parametric class or nonparametric) for variables A and B. Our proposed framework selects the D-vine with five nodes, three edges and two tree levels (Refer to Figure 2). We constructed a D-vine copula framework for three cases. Each case was defined based on permutating the conditioning variables (or variable placed at the centre of the selected D-vine structure; refer to Figure 2). For instance, the D-vine structure 1 (case-1) was defined by selecting the river discharge (RD) as a conditioning variable placed between storm surge (SS) and rainfall (R) events. Similarly, D-vine structure 2 (case 2) and D-vine structure 3 (case 3) are defined by placing storm surge and rainfall events as conditioning variables (refer to Figure 2). This permutation approach to considering each variable of interest as a conditioning variable and selecting the best-fitted vine structure using the fitness test statistics can provide a much more practical and flexible way to the vine copula approach.

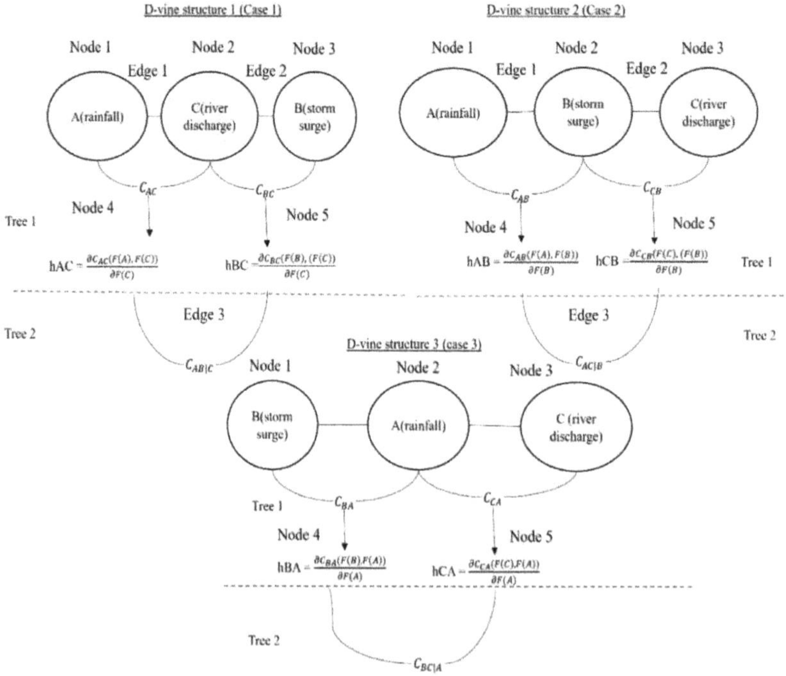

- F(A) = best fitted univariate marginal distribution (using parametric or univariate functions) of annual maximum 24-hr rainfall (mm)
- F(B) = best fitted univariate marginal distribution (using parametric or nonparametric functions) of maximum storm surge (Time interval = ± 4 days)
- F(C) = best fitted univariate marginal distribution (using parametric or nonparametric functions) of maximum river discharge (Time interval = ± 4 days)
- C_{AC} (best fitted 2-D copula between flood pair A and C; C_{BC} (between B and C); C_{AB} (between A and B).
- hAC = conditional cumulative distribution functions (CCDF) estimated using the partial derivatives of C_{AC} w.r.t F(C); hAB = CCDF using partial derivative of C_{AC} w.r.t to F(B); hBC = CCDF using partial derivative of C_{BC} w.r.t to F(C); hBC = CCDF using partial derivative of C_{BC} w.r.t to F(B); hBA = CCDF using partial derivative of C_{BA} w.r.t to F(A); hCA = CCDF using partial derivative of C_{CA} w.r.t to F(A);

Figure 2. Schematic diagram in the 3-D D-vine copula simulation for three different cases [Note: each case of the D-vine structure is defined by permutating the location of the conditional variable, for instance, D-vine structure-1 (River discharge as conditioning variable), D-vine structure-2 (Storm surge as conditioning variable), D-vine structure-3 (Rainfall events as conditioning variable)].

Refer to Figure 2 (consider either case) for illustration; the best-fitted univariate flood marginal distribution is selected after selecting the conditioning or centred variable (say B or A or C). In constructing the D-vine framework nonparametrically, kernel density estimation (KDE), obtained from Section 2.2, is selected to define flood marginal probability distribution. After that, nonparametric copula density (refer to Section 2.1) is introduced and tested via the Bernstein estimator and the beta kernel density estimator. Thus best-fitted models are selected using the fitness test statistics for different tree levels (Tree 1 and Tree 2, refer to Figure 2).

At first, using the most parsimonious 2D copulas, either parametric class (refer to Latif and Simonovic [63]) or nonparametric (refer to Section 2.1), are selected for each flood pair, say C_{AB} and $C_{C.B.}$, the conditional cumulative distribution function (CCDFs), also called the h-function, is estimated [41,82].

$$F_{A|B}(a, b) = h_{A.B.} = \frac{\partial C_{A\ B}(F(A),\ F(B))}{\partial F(B)} \text{ and } F_{C|B}(C, B) = h_{C.B.} = \frac{\partial C_{C\ B}(F(C),\ F(B))}{\partial F(B)} \quad (14)$$

In the second Tree 2 level (refer to Figure 2), the CCDFs statistics estimated from Tree 1 level, using copula C_{AB} and $C_{C.B.}$, is now input to describe another 2D copula in the modelling of joint dependence of conditional pair (AC | B), such as $C_{AC|B}$.

In the nonparametric vine copula approach, the 2D Bernstein copula estimator (refer to Equation (7)) and beta kernel copula density (refer to Equation (3)) are tested individually in both tree levels in the D-vine structures (for all three cases, refer to Figure 2). Our recent study tested the adequacy of different parametric copulas, for instance, mono-parametric Archimedean copulas, mixed or bi-parametric Archimedean copulas and rotated versions (by 180 degrees) of mixed Archimedean copulas, etc., for the same flood pairs [63]. The selected best-fitted 2D copulas from our previous study are now employed in fitting bivariate flood pairs and estimating CCDFs in the first tree level (Tree 1), of the parametric and semiparametric-based vine framework.

Finally, after selecting the most justifiable copula for each tree level for each D-vine structure (case 1, case 2 and case 3) finally, the full trivariate joint density is calculated by

$$C_{A\,B\,C}(a,b,c) = C_{A\,C|B}\left(F_{A\backslash B}(a,b),\ F_{C|B}(c,b)\right)\ \cdot C_{A\,B} \cdot C_{C\,B} \tag{15}$$

2.4. Trivariate Joint Return Periods

Frequency analysis provides a mathematical relationship between extreme events quantiles and their non-exceedance probabilities (or return period) by fitting the most justifiable univariate or multivariate probability distribution function [83,84]. The return period measures the mean or average inter-arrival time between the two design events [85]. The univariate return period's validity is questionable in multidimensional extremes like compound flooding due to the joint action of multiple drivers. In our current study, the developed 3D joint framework is applied to estimate primary return periods, which are further defined in two cases: OR-joint return period and AND-joint return period [86–89]. Different notations of return periods have their own importance that could depend upon the nature of the undertaken problem. For example, just considering an OR-joint return period or either AND-joint return period would be problematic [31]. A practical risk assessment approach must consider different approaches in the return period estimations; readers are advised to see Graler et al. [41] and Requena et al. [90].

Consider the trivariate events ($A \geq a$ OR $B \geq b$ OR $C \geq c$), where either of the events exceeds a specific threshold value; the OR-joint return periods are estimated using the trivariate joint exceedance probability given below.

$$T^{OR}_{A,\,B,C}(a,b,c) = \frac{1}{P\,(A \geq a\ \vee\ B \geq b\ \vee C \geq c)} = \frac{1}{(1 - C(F(a),\ F(b),\ F(c))} \tag{16}$$

where $C(F(a),\ F(b),\ F(c))$ is the trivariate joint cumulative distribution function (JCDF) estimated using the best-fitted 3D vine copula structure.

Similarly, consider another trivariate joint dependency case ($A \geq a$ AND $B \geq b$ AND $C \geq c$) where all the events exceed a specific threshold value simultaneously; the AND-joint return periods are estimated by considering both trivariate joint cumulative distribution function (JCDF) and bivariate JCDFs which are defined for each random flood pair given below.

$$T^{AND}_{A,\,B,\,C}(a,b,c) = \frac{1}{P\,(A \geq a\ \wedge\ B \geq b \wedge\ C \geq c)} = \frac{1}{(1 - F(a) - F(b) - F(c) + C(F(a),\ F(b)) + C(F(b),F(c)) + C(F(a),F(c)) - C(F(a),F(b),F(c)))} \tag{17}$$

In Equation (17), $C(F(a),\ F(b))$, $C(F(b),\ F(c))$ *and* $C(F(a),\ F(c))$ are the bivariates (JCDFs) obtained by fitting most parsimonious 2D copulas to targeted random pairs, and $C(F(a), F(b), F(c))$ is the JCDF using the fitted 3D copula density.

2.5. Failure Probability in the Hydrologic Risk Evaluation

In the hydrologic risk assessments of CE, consideration of only traditional joint primary return periods would be ineffective in describing the risk of potential flood events during the entire project design lifetime [10,91]. In recent studies, a hydrologic risk tool called failure probability (FP) statistics [92,93] is highlighted and used efficiently. FP usually defines the chance of potential flood hazards occurring at least once in a given project design lifetime. FP statistics can define the risk of CF events more appropriately than just visualizing their joint return periods. Few studies incorporated FP statistics in the bivariate hydrologic risk assessments [10,94]. This study incorporated FP statistics in the trivariate compound flood risk assessment, which can be mathematically expressed as

$$\mathrm{FP_T} = 1 - (1 - \mathrm{P})^{\mathrm{T}} \quad (18)$$

where T is the arbitrary project lifetime.

Similarly, for the trivariate flood hazard scenario, the risk of failure for the OR-joint case can be estimated by;

$$\mathrm{FP_T} = 1 - (1 - \mathrm{P}\ (\text{Rainfall} \geq \text{r OR Storm surge} \geq \text{s OR River discharge } \geq \text{rd})^{\mathrm{T}} \quad (19)$$

3. Application

3.1. Study Area and Defining the Compound Hazard Scenario

The complex interplay between oceanographic, fluvial and pluvial factors increases the risk of extreme devastation in low-lying coastal communities worldwide. This study introduces a nonparametric approach to constructing a 3D vine copula framework in compounding the collective impact of rainfall, storm surge and river discharge in flooding events. Our work introduces 46 years of selected flood characteristics collected at west Canada's coast in the trivariate joint probability analysis. The low-lying regions near the Pacific coast and Fraser River are highly susceptible to flooding and often encounter mature and extra-large tropical storms. When these storms are encountered in the coastal mountains, they can result in devastating disasters, forming the potential for prolonged impact. Fraser River is the longest river in the south of Metro Vancouver, BC, with an annual discharge at its river mouth of 3550 m^3s^{-1}. This river flows for 1375 km and finally drains out into the Strait of Georgia. Pirani and Najafi's [61] study already identified that the joint combination of tidal water extreme level, precipitation and river discharge can increase the risk of coastal flooding at the Pacific west coast of Canada. The risk of extreme water levels increases the risk of storm surge events. The same scenario can result in devastating hydrologic or compound flooding when combined with high river discharge and extreme rainfall events. The Environment Ministry of BC report [95] also reported the expectation of a rise in sea level by about half a meter by the end of this 2050 and one meter by the end of 2100. Besides this, according to Lemmen et al. [96], the impact of climate change across Canada significantly increases the risk of extreme events.

This study searches the dependency for the annual maximum 24 h rainfall events and their associated river discharge and storm surge events observed within a time lag of ±4 days from the date of annual maximum 24 h rainfall events. Our previous study [63] has already confirmed that more significant dependencies can be observed when considering the maximum storm surge and river discharge events within a time lag of ±4 days from the calendar date of the annual maximum 24 h Rainfall events. At first, the coastal water level (CWL) data were obtained of 1970 to 2018 from the New Westminster tidal gauge station (station id = 7654) with their geographical coordinates (49.2° N Lat and 122.9° W Lon), which were delivered by Fisheries and Ocean Canada. Secondly, the storm-surge data were estimated by differencing observed CWL data and predicted water level or astronomical tide data, which requires proper time matching between them. Canadian Hydrographic Services (CHS) delivered the predicted tide data. Similarly, the rainfall data were collected at by Haney UBC RF Admin gauge station (49°15′52.1″ N Lat and 122°34′400″ W Lon).

Both storm-surge data and rainfall data were collected for the same calendar year. Third, Environment and Climate Change Canada provided the streamflow discharge data collected at the Fraser River at Hope (49°23′09″ N Lat and 121°27′15″ W Lon). It should be noted that the nearest rainfall gauge station and streamflow discharge station were selected within a radial distance of 50 km centring the selected tidal gauge station.

The annual maximum 24 h rainfall data were defined for each year using the daily-basis rainfall events. The river discharge and storm surge data were selected by observing their maximum values within a time lag of ±4 days from the date of annual maximum 24 h rainfall events. Due to the missing data in the period between the years 1970 to 2018, we considered 46 years of data in establishing a trivariate compound relationship between the variable of interest. Supplementary Table S1 lists the descriptive statistics of targeted compound flood (CF) driving agents. Supplementary Figure S1, S2a–c and S3a–c illustrate the box plots, histogram plots and normal quantile-quantile (Q-Q) plots. From Figure S3c it was found that river discharge observations exhibited more deviation from normality (or straight line) compared to storm surge (Figure S3b) and rainfall events (Figure S3a).

3.2. Nonparametric Estimation in the Univariate Flood Marginals

Modelling the univariate flood marginal is a statistical procedure to infer the population based on a finite random sample and is often a mandatory pre-requisite. Our previous study, using the same dataset, confirmed that annual maximum 24 h rainfall and maximum river discharge events exhibit no serial correlation and zero monotonic trends within their time series [63]. Conversely, maximum storm surge (Time interval = ±4 days) events have zero serial correlation but exhibit monotonic time trend behaviour, which is estimated using the nonparametric Mann–Kendall (M-K) test [97,98] at 5% significance (or 95% confidence interval). Besides this, homogeneity tests for the given time series were examined to identify if changes occur within time series of flood characteristics, using Pettitt's test [99], the SNHT (standard normal homogeneity test) [100] and Buishand's test [101]; refer to Supplementary Table S2. It was found that both rainfall and river discharge events exhibited homogenous behaviour, but storm surge events showed non-homogenous characteristics. In the second row of Table S2, the estimated p-value for storm surge events is less than 0.05 for both the Pettit and SNHT tests. In conclusion, an independent and identical distribution (i.i.d.) is required before introducing it into the probability distribution framework. Thus, a differencing procedure was adopted to remove non-stationarity or de-trend storm surge observations [63].

Table 1 introduces some frequently used kernel density estimations (KDE), whose efficacy was tested in this study to model univariate flood margins. The bandwidth of the fitted KDE was estimated using the direct plug-in (DPI) algorithm; refer to Section 2.2. Table 2a–c list the fitted KDE and their estimated bandwidth. The adequacy of the fitted nonparametric KDE models was tested by comparing empirical and theoretical probabilities. The empirical probabilities were estimated using the Gringorten-based position-plotting approach for each flood characteristic [102]. The cumulative distribution function (CDF) of the fitted KDE was estimated via numerical integration technique or empirical approach because of the lack of a closed form of probability density and cumulative distribution [50]. The goodness-of-fit (GOF) tests, such as mean-square error (MSE), root mean-square error (RMSE), Akaike information criterion (AIC), Bayesian information criterion (B.I.C.), Hannan–Quinn information criterion (H.Q.C.) and mean absolute error (MAE), were estimated for each fitted model [103–107] refer to Table 2a–c. It was found that normal KDE performed best (minimum value of MSE., RMSE, AIC, BIC, HQC and MAE statistics) and was selected for defining the marginal probability density function (PDF) of the maximum 24 h rainfall, maximum storm surge (Time interval = ±4 days) and maximum storm surge (Time interval = ±4 days) events. The qualitative or graphical investigation, using the comparative C.D.F. plots and probability–probability (P-P) plots (refer to Supplementary Figure S4a–c and S5a–c), confirmed the suitability of the selected normal KDE.

Table 2. Fitting univariate kernel density estimation (KDE) and their goodness-of-fit (GOF) test for (a) Annual maximum 24 h Rainfall (mm) (b) Maximum Storm surge (Time interval = ±4 days) (m) (c) Maximum River discharge (Time interval = ±4 days) (m^3s^{-1}).

(a)

Nonparametric KDE	Estimated Bandwidth (via Direct-Plug-in Method)	MSE (Mean Square Error)	RMSE (Root Mean Square Error)	AIC (Akaike Information Criterion)	BIC (Bayesian Information Criterion)	HQC (Hannan-Quinn Information Criterion)	MAE (Mean Absolute Error)
Normal *	11.25	0.0003	0.0199	−358.22	−356.39	−357.53	0.015
Epanechnikov (or parabolic)	24.90	0.0007	0.0281	−326.36	−324.53	−325.67	0.023
Biweight (or Quartic)	29.50	0.0010	0.0316	−315.56	−313.73	−314.88	0.026
Triweight	33.50	0.0011	0.0342	−308.42	−306.59	−307.73	0.027
Parametric GEV (Latif and Simonovic 2022a [63]) **	Estimated parameters via Maximum likelihood estimation (MLE) location(mu = μ) = 1494.64; scale (sigma = σ) = 616.37; shape (xi = ξ) = 0.31	0.0009	0.0312	−312.97	−307.48	−310.91	0.024

(b)

Nonparametric KDE	Estimated Bandwidth (via Direct-Plug-in Method)	MSE (Mean Square Error)	RMSE (Root Mean Square Error)	AIC (Akaike Information Criterion)	BIC (Bayesian Information Criterion)	HQC (Hannan-Quinn Information Criterion)	MAE (Mean Absolute Error)
Normal *	**0.07**	**0.0003**	**0.0175**	**−369.83**	**−368.00**	**−369.15**	**0.014**
Epanechnikov (or parabolic)	0.16	0.0007	0.0265	−331.95	−330.12	−331.26	0.020
Biweight (or Quartic)	0.20	0.0008	0.0288	−324.17	−322.34	−323.48	0.022
Triweight	0.22	0.0009	0.0304	−319.30	−317.47	−318.61	0.023
Parametric Normal (Shahid and Simonovic 2022a [63]) **	Estimated parameters via Maximum likelihood estimation (MLE) mean (mu = μ) = 2.340757e−18; sd (sigma = σ) = 1.676386e−01	0.0011	0.034	−306.56	−302.90	−305.19	0.026

Table 2. *Cont.*

(c)							
Nonparametric KDE	**Estimated Bandwidth (via Direct-Plug-in Method)**	**MSE (Mean Square Error)**	**RMSE (Root Mean Square Error)**	**AIC (Akaike Information Criterion)**	**BIC (Bayesian Information Criterion)**	**HQC (Hannan-Quinn Information Criterion)**	**MAE (Mean Absolute Error)**
Normal *	**340.21**	**0.0005**	**0.0223**	**−347.61**	**−345.78**	**−346.93**	**0.017**
Epanechnikov (or parabolic)	753.16	0.0017	0.0415	−290.65	−288.82	−289.96	0.030
Biweight (or Quartic)	892.25	0.0019	0.0444	−284.42	−282.60	−283.74	0.033
Triweight	1013.19	0.0022	0.0477	−277.81	−275.99	−277.13	0.037
Parametric GEV (Shahid and Simonovic 2022a [63]) **	Estimated parameters via Maximum likelihood estimation (MLE) location (mu = μ) = 1494.64; scale (sigma = σ) 616.37; shape (xi = ξ) = 0.31	0.0008	0.0291	−319.15	−313.67	−317.10	0.022

(a) Note: Normal KDE (bold letter with single asterisk *) outperformed (minimum value of MSE, RMSE, AIC, BIC, HQC and MAE), thus selected in defining the univariate marginal distribution of Annual maximum 24 h Rainfall (mm) events. Additionally, the GEV distribution (double asterisk **) was selected as best-fitted when comparing the performance of different 1-D parametric family distributions in modelling Annual maximum 24 h Rainfall (mm) events (Latif and Simonovic 2022a [63]). (b) Note: Normal KDE (bold letter with single asterisk *) outperformed (minimum value of MSE, RMSE, AIC, BIC, HQC and MAE), thus selected in defining the univariate marginal distribution of Maximum Storm surge (Time interval = ±4 days). Additionally, Normal distribution (double asterisk **) selected as best-fitted when comparing the performance of different 1-D parametric family distributions in modelling storm surge events (Latif and Simonovic 2022a [63]). (c) Note: Normal KDE (bold letter with single asterisk *) outperformed (minimum value of MSE, RMSE, AIC, BIC, HQC and MAE), thus selected in defining the univariate marginal distribution of Maximum River discharge (Time interval = ±4 days). Additionally, GEV distribution (double asterisk **) was best fitted when comparing the performance of different 1-D parametric family distributions in modelling river discharge events (Latif and Simonovic 2022a [63]).

Our previous study selected the generalized extreme value (GEV), normal and GEV distribution fit that were best for the same dataset tested in the present study [63]). The nonparametric KDE outperformed the others (refer to Table 2).

3.3. Incorporation of Nonparametric Vine Structure in the Trivariate Flood Dependence

Our previous study [63] already confirmed the existence of positive dependence between flood attribute pairs, which was measured both parametrically via Pearson correlation coefficient and nonparametric via Kendall's tau (τ), and Spearman's rho (ρ) at a 5% significance level (95% confidence interval). At first, the nonparametric via 2D Bernstein estimator and beta kernel estimator (refer to Equations (7) and (3)) were employed in the bivariate dependence modelling of the rainfall–storm-surge, storm surge–river-discharge and rainfall–river-discharge pairs (refer to Table 3. The beta kernel density and Bernstein copula estimator can alleviate the risk of boundary bias problems. The Bernstein copula can facilitate higher consistency and better approximate joint structure than the empirical copula. The fitted beta kernel density bandwidth was examined using the rule of thumb (ROT) approach by minimizing the AMISE statistics (refer to Equation (4) of Section 2.1). Similarly, in fitting the 2D Bernstein copula estimator, their coefficient was adjusted by the approach discussed by Weiss and Scheffer [58].

Table 3. Fitting the nonparametric 2-D copula density and their goodness of fit (GOF) test to the given flood attribute pair.

Flood Attribute Pairs	Nonparametric 2-D Copula Density	Estimated Bandwidth (Only for Beta Kernel Copula Density)	MSE (Mean Square Error)	RMSE (Root Mean Square Error)	MAE (Mean Absolute Error)	K-S (Kolmogorov–Smirnov)	NSE (Nash–Sutcliffe model Efficiency Coefficient)
Annual Maximum 24 h Rainfall (mm)-Maximum Storm surge (Time interval = ±4 days) (m)	**Bernstein estimator ***	0.11	**0.0013**	**0.0360**	**0.0290**	**D = 0.0869, *p*-value = 0.99**	**0.980**
	Beta kernel density		0.0014	0.0371	0.0295	D = 0.1521, *p*-value = 0.66	0.979
Maximum Storm surge (Time interval = ±4 days) (m)-Maximum River discharge (Time interval = ±4 days) (m^3/s)	**Bernstein estimator ***	0.12	**0.0012**	**0.0350**	**0.0270**	**D = 0.1521, *p*-value = 0.66**	**0.981**
	Beta kernel density		0.0014	0.0375	0.0295	D = 0.1521, *p*-value = 0.66	0.978
Annual Maximum 24 h Rainfall (mm)-Maximum River discharge (Time interval = ±4 days) (m^3/s)	Bernstein estimator	0.17	0.0011	0.0338	0.0258	D = 0.1956, *p*-value = 0.34	0.981
	Beta kernel density *		**0.0008**	**0.0298**	**0.0221**	**D = 0.1521, *p*-value = 0.66**	**0.985**

Note: Bernstein copula estimator (bold letter with an asterisk) fitted best for flood pairs rainfall and storm surge, and storm surge and river discharge. Beta kernel copula density is most parsimonious for flood pair rainfall and river discharge.

The nonparametric models' performances were evaluated using various G.O.F. measures, for instance, MSE, RMSE, MAE, KS (Kolmogorov–Smirnov) [108] and NSE (Nash–Sutcliffe model efficiency coefficient) [109]; refer to Table 3. The Bernstein copula estimator's performance was better for flood pairs (rainfall and storm surge) and (storm surge and river discharge) (minimum values of MSE, RMSE, MAE and KS test and high NSE test statistic). However, according to Table 3, the beta kernel density outperformed Bernstein estimator for the rainfall and river discharge pair.

We constructed the D-vine copula for three cases. Each case defines a D-vine structure by permutating the locations of conditioning variables. All the computation involved in the establishment of 3D vine copula (also in fitting 2D nonparametric copula density) was carried out using R software [110] with the libraries "kdecopula" [111] and "kdevine" [70].

1. D-Vine structure 1 (case 1) considers river discharge observation as a conditioning variable by placing it at the centre of the vine structure (refer to Figure 2 and Tables 3 and 4). In this structure, at first, the 2D beta kernel copula density and 2D Bernstein copula estimator, which were selected as best-fitted from Table 3 for flood pair rainfall–river-discharge and storm surge–river-discharge in the first tree level (Tree 1), were now employed in the estimation of conditional cumulative distribution functions (CCDFs); $h_{\text{RAIN RIVER DISCHARGE}}$ and $h_{\text{STORM SURGE RIVER DISCHARGE}}$ (refer to Equation (14)). The copula in the second tree level (Tree 2) was then identified using the above estimated CCDDFs values as input. It was found that the Bernstein copula estimator outperformed the beta kernel density to model joint dependence for the flood pair (RAIN, STORM SURGE | RIVER DISCHARGE) $C_{\text{RAIN STORM SURGE}|\text{RIVER DISCHARGE}}$ (which exhibited the minimum value of MSE, RMSE, MAE and KS and the higher value of NSE test statistics). Finally, the full 3D trivariate joint density was obtained using Equation (15).
2. Similarly, D-vine structure 2 (case 2) comprises storm surge events as a conditioning variable (refer to Tables 3 and 4 and Figure 2). In this vine framework, at first, in the first tree level (Tree 1), the Bernstein estimator is identified as the most

justifiable and thus is employed in the estimation of CCDFs $h_{\text{RAIN, STORM SURGE}}$ and $h_{\text{RIVER DISCHARGE, STORM SURGE}}$, followed by Equation (14). Secondly, in the second tree level (Tree 2), the Bernstein copula estimator is selected as the most parsimonious in establishing the dependence between of flood pair (RAIN RIVER DSICHARGE | STORM SURGE) $C_{\text{RAIN RIVER DSICHARGE|STORM SURGE}}$. Finally, using Equation (15), the full trivariate joint density of the fitted vine structure is estimated.

3. D-Vine structure 3 (case 3) is defined by considering rainfall events as a conditioning variable placed in the centre of the selected D-vine structure (refer to Figure 2, and Tables 3 and 4). The Bernstein estimator and beta kernel density were identified as most justifiable and thus employed in the estimation the CCDFs $h_{\text{STORM SURGE, RAINFALL}}$ and $h_{\text{RIVER DISCHARGE, RAIN}}$ (followed by Equation (14)) in Tree 1. In the second level (Tree 2), again the Bernstein copula estimator was identified as the most parsimonious in modelling joint dependence of flood pair (STORM SURGE RIVER DSICHARGE | RAINFALL) (refer to Table 4). Finally, followed by Equation (15), the full trivariate joint density of the fitted vine structure was defined.

After approximating three different D-vine structures (case 1, case 2 and case 3), their performances were compared using the fitness test statistics (MSE, RMSE, MAE, NSE and K-S). The theoretical probability (CDF) was estimated using a developed 3D vine structure for each case and compared with empirical observations for estimating the GOF test statistics. Table 4 shows that the D-vine structure for case-2, considering storm surge as a conditioning variable, performed better than other D-vine structures. The selected structure exhibited the minimum MSE, RMSE and MAE values and a high NSE test value. The above approach in the vine copula provided much better flexibility in selecting the best vine model, not just by fixing the conditioning variable but by switching or permutating the conditioning variable. For example, in the above case, when considering storm surges as conditioning variables, the performance of the fitted D-vine copula got better than considering either rainfall or river discharge events. Supplementary Figure S8 illustrates the vine tree plot of the developed D-vine structure in the nonparametric fitting procedure.

Table 4. Summary statistics in the nonparametric D-vine structure for three different cases by incorporating the Bernstein estimator and Beta kernel density (each D-vine structure is defined by permutating the location of the conditioning variable).

	Vine Structure (Conditioning Variable)	Tree Level	Flood Attribute Pairs	Fitted Non-parametric Copula Estimator	Estimated Bandwidth (for Beta Kernel Copula Density)	MSE (Mean Square Error)	RMSE (Root Mean Square Error)	MAE (Mean Absolute Error)	K-S (Kolmogorov–Smirnov)	NSE (Nash–Sutcliffe Model Efficiency Coefficient)
Case 1 (1-3-2)	3 or Maximum River Discharge (Time interval = ±4 days) placed in the centre	Tree 1	1-3 (Rain-River discharge)	Beta kernel	0.11	0.0004	0.0223	0.017	D = 0.086, *p*-value = 0.99	0.992
			2-3 (storm surge-river discharge)	Bernstein						
		Tree 2	12 ∣ 3	Bernstein		0.0006	0.0249	0.019	D = 0.087, *p*-value = 0.99	0.990
				Beta kernel						
Case 2 (1-2-3) *	2 or Maximum Storm Surge (Time interval = ±4 days) placed in the centre	Tree 1	1-2 (Rain–Storm Surge)	Bernstein	0.17	**0.0002**	**0.0153**	**0.013**	**D = 0.152, *p*-value = 0.66**	**0.995**
			3-2 (Storm surge–river discharge)	Bernstein						
		Tree 2	13 ∣ 2	**Bernstein**		0.0003	0.0154	0.013	D = 0.152, *p*-value = 0.66	0.995
				Beta kernel						
Case 3 (2-1-3)	1 or Annual Maximum 24 h Rainfall placed in the centre	Tree 1	2-1 (Storm surge-Rain)	Bernstein	0.12	0.0004	0.0203	0.015	D = 0.152, *p*-value = 0.66	0.993
			3-1 (River discharge-Rainfall)	Beta kernel						
		Tree 2	23 ∣ 1	Bernstein		0.0005	0.0214	0.016	D = 0.086, *p*-value = 0.99	0.992
				Beta kernel						

Note: D-vine structure-2 (case-2) (bold letter with an asterisk) selected the most parsimonious vine density (exhibits the minimum value of MSE, RMSE MAE, K-S and high value of NSE test statistics).

3.4. *Comparing the Adequacy of Fitted Nonparametric D-Vine with Parametric and Semiparametric Approaches in the D-Vine Copula Framework*

3.4.1. Constructing D-Vine Structure in the Parametric Fitting Procedure

In the parametric vine approach, at first, the best fitted univariate marginal pdfs, for instance, GEV (for rainfall), normal (for storm surge) and GEV (for river discharge) distribution, were selected (refer to Table 2). This was followed by the same steps we discussed in the last section. Three different D-vine structures (case 1, case 2 and case 3) were considered by permutating the conditioning variables (refer to Figure 2). Our previous study confirmed that Survival BB7 fit best for flood pair rain and river discharge, Survival BB1 for storm surge–river-discharge and BB1 copula for the rain and storm surge pair [63].

For vine structure 1 (case 1, river discharge as conditioning variable; refer to Figure 2 and Table 5), both the selected 2D copulas (Survival BB7 and Survival BB1) were employed in the estimation CCDFs, which became the input to define another 2D copula in the second tree level (Tree 1). The present study tested different parametric copulas to fit the D-vine structure's second tree level (Tree 2) (refer to Supplementary Table S3a–c). The parameters of the fitted copulas were estimated using maximum pseudo-likelihood estimation (MPL) [112,113], and the performances of the fitted models were compared using the Cramer–von Mises functional test statistics S_n, with the parametric bootstrap procedure (N is the number of bootstrap samples = 1000) [114,115]. From Table S3a–c, the Frank copula was identified as best for Tree 2 (D-vine structure 1, case 1), rotated BB6 270-degree copula for Tree 2 (D-vine structure 2, case 2) and Frank copula overall (D-vine structure 3, case 3). The full trivariate D-vine structure (parametric settings) for each case was obtained using Equation (15).

After developing vine structures for the given flood characteristics, their performances were compared to select the most efficient D-vine structure developed under parametric settings for three cases. From Table 5, it was found that D-vine structure-3 (case-3), with rainfall as a conditioning variable, outperformed other vine structures (minimum values of AIC, BIC, MSE, RMSE, MAE and K-S statistics and with high values of log-likelihood (L-L) and NSE statistics).

Table 5. Summary statistics of the fitted D-vine structure under parametric distribution settings (via parametric copulas with parametric marginal distributions).

	Vine Structure (Conditioning Variable)	Tree Level	Flood Attribute Pairs	Most Parsimonious Fitted Copula	Copula Dependence Parameters (θ) via MPL	MSE	RMSE	MAE	K-S	NSE	*Log-Likelihood (LL)*	*AIC*	*BIC*
Case 1 (1-3-2)	3 or Maximum River Discharge (Time interval = ±4 days) placed in the centre	Tree 1	1-3 (Rain-River discharge)	Survival BB7	θ = par = 1.142; δ = par2 = 0.19	0.00086	0.0294	0.021	D = 0.152 (0.66)	0.982	10.51	−11.02	−1.87
			3-2 (storm surge-river discharge)	Survival BB1	θ = par = 0.23; δ = par2 = 1.29								
		Tree 2	12 | 3	Frank	θ = par = **2.92**								
Case 2 (1-2-3)	2 or Maximum Storm Surge (Time interval = ±4 days) placed in the centre	Tree 1	1-2 (Rain–Storm Surge)	BB1 (Clayton-Gumbel)	θ = par = 0.19; δ = par2 = 1.36	0.00088	0.0297	0.022	D = 0.173 (0.48)	0.982	10.50	−11.01	−1.87
			2-3 (Storm surge–river discharge)	Survival BB1	θ = par = 0.23; δ = par2 = 1.29								
		Tree 2	13 | 2	Rotated BB6 270 degrees	θ = par = −1; δ = par2 = −1								
Case 3 (2-1-3) *	1 or Annual Maximum 24 h Rainfall placed in the centre	Tree 1	2-1 (Storm surge-Rain)	BB1 (Clayton-Gumbel)	θ = par = 0.19; δ = par2 = 1.36	**0.00084**	**0.0290**	**0.021**	**D = 0.152 (0.67)**	**0.982**	**10.85**	**−11.71**	**−2.57**
			3-1 (River discharge-Rainfall)	Survival BB7	θ = par = 1.142; δ = par2 = 0.19								
		Tree 2	23 | 1(Storm-River/Rainfall)	Frank	θ = par = 2.97								

Note: D-vine structure-3 (case 3, indicated by bold letter with an asterisk), considering rainfall events as a conditioning variable, perform better (minimum value of AIC, BIC, HQC, MSE, RMSE, K-S and high value of model's L-L and NSE statistics).

3.4.2. Constructing a D-Vine Structure with the Semiparametric Settings

2D parametric class copulas were incorporated with the nonparametric marginal pdf in the semiparametric D-vine structure. Firstly, the best-fitted 2D parametric copulas for Tree-1 in all three cases of the D-vine structure (refer to Figure 2) were selected from our previous study [63]. Refer to Section 3.4.1. Supplementary Table S4a–c shows different parametric class 2D copulas fitted with an MPL-based parameter estimation procedure for estimating the most justifiable bivariate density fitted to the second tree level (Tree 2). The investigation found that the Frank copula was best for the D-vine structure-1 (case-1), the rotated BB6 270 degree one for vine structure-2 (case-2) and the Frank copula for D-vine structure-3 (case-3). Using Equation (15), the full trivariate vine copula joint density was estimated for each fitted D-vine structure. The most justifiable semiparametric-based vine structure was selected by comparing the performances of three different cases of D-vine structure. Table 6 provides the summary details of the fitted D-vine structures. It was found that D-vine structure 3 (case-3), considering rainfall as a conditioning variable, outperformed all other possible D-vine structures (case-1 and case-2); it exhibited minimum values of MSE, RMSE, MAE, K-S, AIC and BIC statistics and high values of model likelihood (L-L) and NSE statistics.

Table 6. Details of the fitted 3-D vine structure in the semiparametric distribution settings for three different cases.

	Vine Structure (Conditioning Variable)	Tree Level	Flood Attribute Pairs	Most Parsimonious or Best-Fitted Copula	Copula Dependence Parameters (θ)	MSE	RMSE	MAE	K-S	NSE	Log-Likelihood (LL)	AIC (Akaike Information Criterion)	BIC (Bayesian Information Criterion)
Case 1 (1-3-2)	3 or Maximum River Discharge (Time interval = ±4 Days) placed in the centre	Tree 1	1-3 (Rain-River discharge)	Survival BB7	θ = par = 1.142; δ = par2 = 0.19	0.0006	0.0236	0.0185	D = 0.152 (0.66)	0.9886	10.68	−11.37	−2.23
			3-2 (storm surge-river discharge)	Survival BB1	θ = par = **0.23**; δ = par2 = **1.29**								
		Tree 2	12∣3	Frank	θ = par = **2.89**								
Case 2 (1-2-3)	2 or Maximum Storm Surge (Time interval = ±4 days) placed in the center	Tree 1	1-2 (Rain–Storm Surge)	BB1 (Clayton-Gumbel)	θ = par =0.19; δ = par2 = 1.36	0.0006	0.0239	0.0192	D = 0.130 (0.82)	0.9883	10.81	−9.62	1.34
			2-3 (Storm surge–river discharge)	Survival BB1	θ = par = **0.23**; δ = par2 = **1.29**								
		Tree 2	13∣2	Rotated BB6 270 degrees	θ = par = −1; δ = par2 = −1								
Case 3 (2-1-3) *	1 or Annual Maximum 24 h Rainfall placed in the center	Tree 1	2-1 (Storm surge-Rain)	BB1 (Clayton-Gumbel)	θ = par =0.19; δ = par2 = 1.360	**0.0005**	**0.0232**	**0.0185**	**D = 0.1303 (0.82)**	**0.9890**	**11.38**	**−12.77**	**−3.62**
			3-1 (River discharge-Rainfall)	Survival BB7	θ = par = 1.14; δ = par2 = 0.19								
		Tree 2	23∣1	Frank	θ = par =2.75								

Note: D-vine structure 3 (case 3) fitted best (bold letter with an asterisk) (minimum value of AIC, BIC, MSE, RMSE, MAE, K-S and high value of NSE and L-L statistics.

3.4.3. Comparison of the Models' Performances (Nonparametric vs. Semiparametric vs. Parametric Vine)

Section 3.3, Section 3.4.1, and Section 3.4.2 recognized the most justifiable vine structures, D-vine structure-2 (obtained via nonparametric vine approach), D-vine structure-3 (via parametric vine approach) and D-vine structure-3 (via semiparametric framework). We performed an analytical and graphical comparison to check the adequacy of the selected nonparametric D-vine density with parametric and semiparametric vine approaches fitted to the given triplet flood variables. Refer to Table 7. The D-vine structure (case-2) defined in the nonparametric setting outperformed the others (minimum values for MSE, RMSE, K-S, MAE and high NSE test statistics). It was also observed that the performance of the selected semiparametric D-vine structure (case-2) was better than that of the parametric D-vine structure (case-2) in the trivariate flood dependence. These results can further reveal how well the performance of the incorporated vine getting improves when switching its marginal distribution from parametric to non-parametric and the copula joint density from parametric to nonparametric joint pdf. The reliability and suitability of the selected D-vine structures were examined further by comparing Kendall's τ correlation coefficient estimated from the simulated flood events (sample size N = 1000) using the best-fitted nonparametric vine copula (D-vine structure-2), parametric vine (D-vine structure-2) and semiparametric vine copula (D-vine structure-3) and compared with the empirical Kendall's τ coefficient estimated from the historical flood events (refer to Table 8). It was found that the obtained nonparametric D-vine structure (case-2) exhibited a minimum gap or difference between the empirical and theoretical Kendall's tau statistics. These results confirm that the selected nonparametric vine structure regenerates the historical flood dependence structure (or correlation) much more efficiently. The same table also revealed that the semiparametric vine approach better captures and regenerates flood dependence than parametric vine copula density.

Table 7. Comparing the performance of the selected nonparametric D-vine structure with parametric and semiparametric vine copula density.

Best-Fitted D-Vine Structure	MSE	RMSE	MAE	K-S	NSE
Nonparametric settings (D-vine structure-2 (case-2) *	**0.0002**	**0.0153**	**0.0130**	**D = 0.152, *p*-value = 0.66**	**0.995**
Semiparametric settings (D-vine structure-3 (case-3)	0.0005	0.0232	0.0185	D = 0.130 (0.82)	0.989
Parametric settings (D-vine structure-3 (case-3)	0.00084	0.0290	0.0218	D = 0.152 (0.67)	0.982

Note: D-vine structure-2 derived in the nonparametric settings (bold letter with an asterisk) outperformed both parametric and semiparametric approaches in the D-vine structure for trivariate CF events.

A graphical visual inspection was carried out to crosscheck the adequacy of the selected D-vine structure-2 obtained nonparametrically. The overlapped scatterplots between the observed samples (via historical flood) and simulated samples (using D-vine structure-2, case-2) of sample size (N = 1000) were obtained; refer to Supplementary Figure S6a–c. It was found that D-vine structure-2 (under nonparametric settings) performs adequately since the simulated random sample (indicated by light grey colour) overlapped with the natural mutual concurrency of the historical flood samples (red colour); refer to Figure S6a–c. Supplementary Figure S7 illustrates a 3D scatterplot matrix of the generated flood events (sample size N = 1000) using the selected nonparametric vine model. Supplementary Figure S8 illustrates the vine tree structure of the most justifiable D-vine structure in the nonparametric setting.

Table 8. Examining the reliability of the developed D-vine structure (nonparametric settings) vs. parametric D-vine vs. semiparametric D-vine copula framework by comparing Kendall's τ correlation coefficient estimated using the generated random samples (size N = 1000) obtained from the above-selected model with Empirical Kendall's τ values estimated from historical observations.

CF Pairs	Kendall's Tau Estimated from Historical Flood Events (Empirical Estimates)	Kendall's Tau Estimated from Best-Fitted D-Vine Copula (D-Vine Structure 3) in a Parametric Setting (Theoretical Estimates)	Kendall's Tau Calculated from Best-Fitted D-Vine Copula (D-Vine Structure-3) in a Semiparametric Setting (Theoretical Estimates)	Kendall's Tau Estimated from Best-Fitted D-Vine Copula (D-Vine Structure-2) in the Nonparametric Setting (Theoretical Estimates)
Annual maximum 24 h rainfall (mm) − Maximum Storm Surge (m) (Time interval = ±4 days)	0.30	0.29	0.31	29
Maximum Storm Surge (m) (Time interval = ±4 days) − Maximum River discharge (m^3/s) (Time interval = ±4 days)	0.29	0.34	0.31	29
Annual maximum 24 h rainfall (mm) − Maximum Storm Surge (m) (Time interval = ±4 days) − Maximum River discharge (m^3/s) (Time interval = ±4 days)	0.14	0.16	0.15	0.14

Note: It is clearly found that the best-fitted D-vine structure-2 constructed in the nonparametric distribution setting are very much closer to its empirical estimates.

In conclusion, the above investigations confirm that the nonparametric vine density is much better in trivariate dependence analysis of CF events. This framework has no prior distributional assumption about their copula joint density and its marginal behaviour. Finally, the selected nonparametric vine density is employed to estimate trivariate joint cumulative distribution functions (JCDF) and joint return periods.

3.5. Compound Flooding Events Risk Assessments

Flood frequency analysis (FFA) establishes an interrelationship between the flood design quantiles and their non-exceedance probabilities by fitting the best-fitted univariate or multivariate probability distribution function. Primary return periods comprise two different joint cases, OR- and AND-joint. The fitted nonparametric D-vine structure was employed in estimating trivariate joint cumulative distribution function (JCDF) and trivariate return periods for OR- and AND-joint cases for different possible combinations of flood events; refer to Table 9 and Equations (16) and (17). Table 9 also shows estimations of the bivariate joint return periods using the best-fitted 2D nonparametric copula density (refer to Table 3). It was found that the trivariate return periods for the AND-joint case were higher than for the OR-joint case for the same flood combination. Similarly, the bivariate AND-joint case was higher than the OR-joint case for the same flood pair combinations. These results further reveal that the occurrence of trivariate flood events simultaneously is less frequent in the "AND" case and more frequent in the "OR" joint case. The same observations are also valid for the bivariate case. For instance, refer to Table 9: a 1-in-100-year flood event with the following characteristics—rainfall = 147.541 mm, storm surge = 0.337486 m and river discharge = 5951.523 m^3s^{-1}—the trivariate OR- and AND-joint return periods are 33.66 years and 3486.75 years. For the same flood characteristics mentioned above, the bivariate return periods for OR- and AND-joint cases are 50.92 years and 2744.23 years for flood pair rainfall and storm surge events; 50.26 years and 9633.91 years for storm surge and river discharge; and 50.29 years and 8517.88 years for rainfall and river discharge pair events.

Table 9. Comparing primary return periods (univariate vs. bivariate vs. trivariate) for a different possible combination of triplet flood events.

Estimated Flood Quantiles Using the Inverse Cumulative Distribution Functions (CDFs) of Best-Fitted Marginal Distribution via KDE				**Bivariate Joint Return Periods (JRPs)**						**Trivariate Joint Return Periods (JRPs) Estimated Using the Best-Fitted D-Vine Structure (Case-2)**	
Return Period (Years), T	**Annual Maximum 24 h Rainfall (R) (mm)**	**Maximum Storm Surge (m) (SS) (Time Interval = ±4 Days)**	**Maximum River Discharge (RD) (m^3s^{-1}) (Time Interval = ±4 Days))**	**OR-JRP, T^{OR}_{RS}**	**AND-JRP T^{AND}_{RS}**	**OR-JRP, T^{OR}_{SRD}**	**AND-JRP T^{AND}_{SRD}**	**OR-JRP, T^{OR}_{RRD}**	**AND-JRP, T^{AND}_{RRD}**	**OR-JRP, T^{OR}_{RSRD}**	**AND-JRP, T^{AND}_{RSRD}**
5	97.04	0.151	2573.15	3.08	13.17	2.84	20.52	2.88	18.91	2.06	15.95
10	110.73	0.228	3367.54	5.69	40.97	5.32	81.46	5.34	77.23	3.70	50.42
20	128.04	0.277	5408.36	10.77	139.73	10.31	328.39	10.32	318.89	7.02	173.06
50	143.42	0.316	5854.64	25.85	760.16	25.29	2162.62	25.30	2063.98	17.00	950.29
100	147.54	0.337	5951.52	50.92	2744.23	50.26	9633.91	50.29	8517.88	33.66	3486.75

It is observed from the above-estimated return periods (refer to Table 9) that it would be preferable in practice to use trivariate return periods instead of the bivariate (or univariate). The above results also reveal that the accountability of both primary joint return periods is essential; just considering either AND-joint or OR-joint case would be problematic in the hydrologic risk evaluation. They also depend on the nature of water-related problems, which usually decide the importance of the different types of return periods.

In our current study, the developed nonparametric D-vine model was employed further in estimating the failure probability (FP) statistics. This risk approach examined variation in the trivariate, bivariate and univariate flood hazard events measured by service design lifetime for different return periods, such as 100 years, 50 years, 20 years, 10 years, and 5 years; refer to Figure 3a–e. A trivariate flood hazard scenario was found to result in higher-value FP than bivariate and univariate events. Both the trivariate and bivariate (also univariate) hydrologic risk or FP statistics are reduced when the return increases. FP statistics increase when there is an increase in the service design lifetime of the hydraulic infrastructure under consideration. For instance, at the return period of 100 years, the estimated value of FP statistics is 0.778 (for trivariate hazard scenario) and 0.629 (for bivariate hazard scenario) at a 50 year design lifetime. When considering a higher design lifetime, say 100 years, for the same return periods (100 years), the estimated value is 0.951 (for the trivariate hazard scenario) and 0.862 (for the bivariate scenario). Similarly, at 100-year return periods, the trivariate and bivariate hazard scenario is 0.933 and 0.832 (at 90 years design lifetime). When reducing the return periods to 50 years, the value is 0.995 and 0.971 at the same design lifetime (90 years).

From the above results, it is inferred that observing the joint behaviour of the storm surge event, rainfall event, and river discharge is essential in reducing the risk of coastal flood hazard. Considering the univariate probability analysis or even bivariate joint behaviour may underestimate the level of risk In conclusion, ignoring the trivariate probability analysis would be a problem which could result in the underestimation of FP. Their joint probability occurrence facilitates a better understanding and realization of extreme compound scenarios. All the above-discussed analytical and graphical investigations are crucial for sustainable design and planning in coastal flood management strategies.

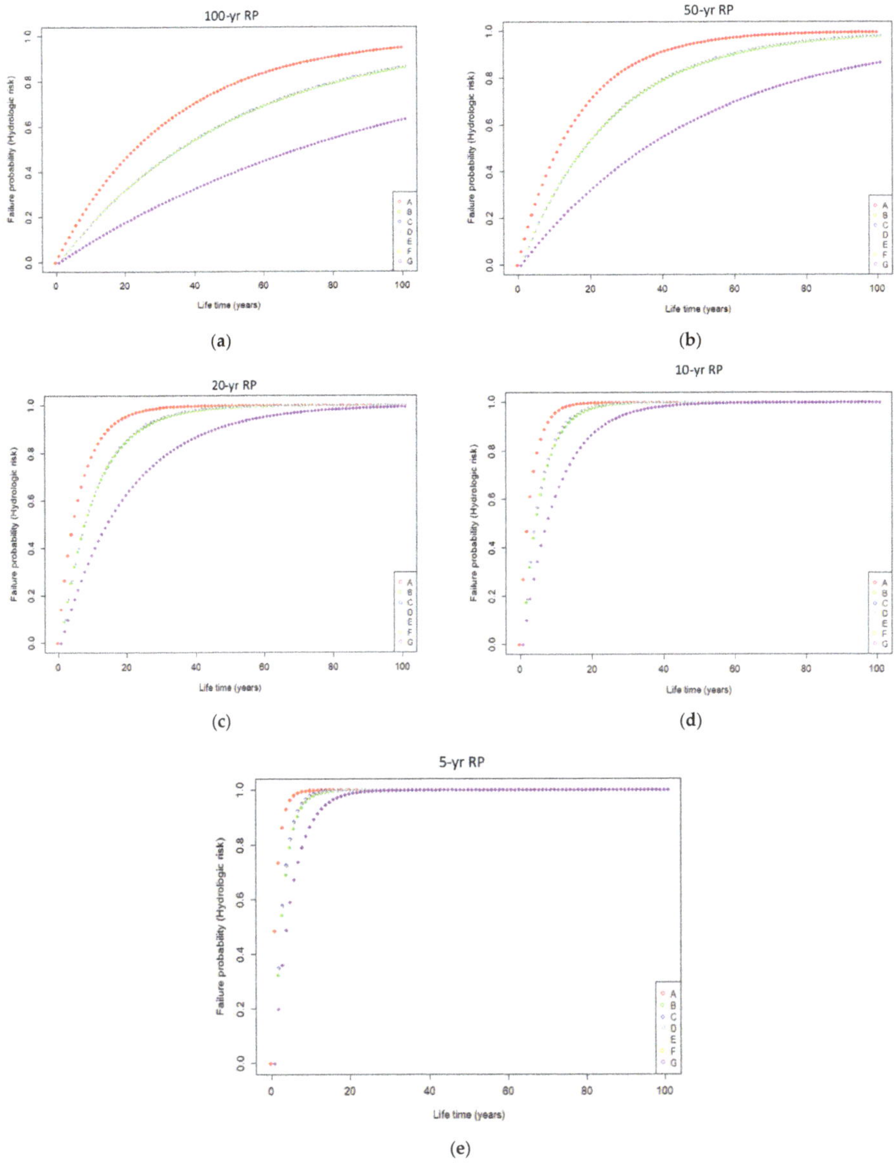

Figure 3. Assessments in the hydrologic risk of CF events for return periods (RPs) (**a**) 100 years, (**b**) 50 years, (**c**) 20 years (**d**) 10 years, (**e**) 5 years [Note: A [red colour]—describing trivariate CF hazard scenario for OR-joint case; B [green colour]—describing bivariate CF hazard scenario between flood pair rainfall and storm surge for OR-joint case; C [blue colour]—describing bivariate CF hazard scenario for flood pair storm surge and river discharge for OR-joint case; D [grey colour]—describing bivariate CF hazard scenario for flood pair rainfall and river discharge for OR-joint case; E [pink colour]—describing univariate hazard scenario through rainfall events; F [yellow colour]—describing univariate hazard scenario through storm surge events; G [purple colour]—describing univariate hazard scenario through river discharge events].

4. Research Summary and Conclusions

This study incorporated the D-vine copula in the nonparametric fitting procedure to model trivariate joint probability analysis of the storm surge, river discharge and rainfall in the compound flood risk assessments. The common forcing mechanisms that can derive multiple extreme events either successively or in close succession in the coastal regions can exacerbate the impact of flooding events. A comprehensive compound flood risk understanding can demand the accountability of multiple flood-driving agents simultaneously because the complex interplay between them can be devastating. The performance of the parametric and semiparametric approach in the vine framework was also compared with the proposed nonparametric vine density in the CF dependence. The traditional vine framework was defined by incorporating multiple parametric class 2D copula densities with parametric 1D univariate margins. This parametric density (and their marginal distribution) approximation had some statistical constraints already discussed in Section 1. The nonparametric via the Bernstein estimator and beta kernel copula estimator is a much more comprehensive way of vine construction, where the fitted 2D copula densities between each flood pair can adapt to any dependence structure without the requirement of any specific or fixed probability density structure. Conversely, the semiparametric vine framework integrated multiple 2D parametric class copulas with nonparametric marginal pdfs via the Kernel density estimation (KDE). The main findings of this study are summarized below:

1. This study compounded the joint relationship between annual maximum 24 h rainfall and their associated storm surge and river discharge observed within a time lag of ±4 days from the date of highest rainfall events. Our previous study [63] already examined the degree of mutual concurrencies and confirmed a significant positive correlation between selected flood contributing-variables.
2. Our previous study confirmed that rainfall and storm surge events did not exhibit any significant trend or serial correlation (or autocorrelation). From our present study [63], it was also confirmed that both variables are homogenous. However, the storm surge events exhibited nonstationary behaviour (time trend with non-homogeneity), but no serial correlation was identified.
3. The nonparametric Normal KDE is selected as the most parsimonious for all three targeted flood variables (refer to Table 2a–c). Additionally, the results were the same when comparing the performance with the best-fitted parametric function (GEV, NORMAL, GEV [63]; refer to Table 2a–c. This further reveals that a lack of prior distributional assumption can result in a better explanation of the targeted flood marginal distribution behaviour.
4. The 3-D vine copula was constructed by permutating the conditioning variable's location, which resulted in three different D-vine structures. In constructing the D-vine copula nonparametrically, it was found that the Bernstein copula fit best for flood pairs rainfall–storm surge and storm surge and river discharge, and the beta kernel estimator fit best for pair rainfall–river-discharge. All the selected nonparametric 2D copulas were employed in the 3D vine construction for three different D-vine structures. The fitness test statistics confirmed that nonparametric D-vine structure-2 (case-2) performs better when considering storm surge as a conditioning variable (with the Bernstein copula estimator fitted in both the first and second tree levels). It is important to note that it is much more practical to consider each targeted variable separately as a conditioning variable instead of just fixing it in the vine construction. This approach can generate multiple possible structures for selecting the most justifiable one.
5. Similarly, the parametric and semiparametric vine fitting approaches selected D-vine structure-3 (rainfall as a conditioning variable; refer to Tables 5 and 6) as the most justifiable density based on different GOF test statistics.
6. Best-fitted models have been compared analytically, such as nonparametric, semiparametric, and parametric fitting-based D-vine structures. Results confirmed the adequacy of the proposed nonparametric vine density. The model's reliability was

investigated analytically by comparing the theoretical and empirical Kendall's tau. Results revealed that the selected D-vine structure-2, in the nonparametric fitting procedure, outperformed the others. In other words, the selected vine structure regenerates historical flood events efficiently. The adequacy of D-vine structure-2 (Nonparametric framework) was further investigated graphically through overlapped scatterplots between historical observation and generated samples. It is clearly noted that the fitted model effectively captured the natural mutual dependencies of historical flood events. In conclusion, our proposed vine copula density in the nonparametric fitting is a much better alternative to the traditional parametric vine approach.

7. The best-fitted nonparametric vine density was employed to estimate trivariate primary joint return periods for OR- and AND-joint cases. The OR-joint case resulted in lower return periods than the AND-joint case for the same flood combinations. It was noted how important it is to take accountability for trivariate return periods rather than just considering a bivariate (or univariate) approach, which would be problematic and less efficient for solving different water-related decision-making.
8. The trivariate and bivariate joint CDFs were employed in estimating failure probability (FP) statistics which highlight the hydrologic risk due to the compound effect of rainfall, storm surge and river discharge events in the trivariate flood events. Investigation revealed that FP statistics could be underestimated if neglecting the trivariate joint probability analysis between targeted flood characteristics compared to when considering the same flood variables in pairwise joint modelling. The FP statistics were higher when considering trivariate joint distribution for the OR-joint event than when considering bivariate joint dependency between flood pairs. The hydrologic risk (trivariate, bivariate and univariate events) decreases with an increase in the return periods. At the same time, hydrologic risk increases, followed by the service design lifetime of hydraulic infrastructure under consideration. The same investigation also found that the FP of univariate flood events is much lower than trivariate (and bivariate) events. These further reveals that compound events may not be devastating if each flood source variable is considered separately.

This study has a few limitations. Firstly, this study only considered 46 years of the observational dataset. It might cause a source of uncertainty in the estimated outcomes. It could be preferred to take long-term data to reduce or minimize the risk of inheritance uncertainty. Secondly, our proposed model considers nonparametric distribution, both univariate marginals and multivariate nonparametric copula density in the vine construction. The model compatibility and performance have already been compared thoroughly with the existing parametric (or semiparametric) vine framework. It could be denied that there is much scope for applying this nonparametric framework to model the joint behaviour of different possible extreme events across the world. Even though it is possible to extend this proposed model to higher dimensional modelling for more than three variables. However, on the other side, it might not be easy to extrapolate to high return levels. It needs to be addressed in extreme event modelling. Our present study is not tackling this issue, which will be considered in our further study.

Supplementary Materials: The following supporting information can be downloaded at: https://www.mdpi.com/article/10.3390/hydrology9120221/s1.

Author Contributions: Project focus and supervision, S.P.S.; methodology, software, formal analysis, S.L.; writing—original draft preparation, S.L.; writing—review and editing, S.L. and S.P.S.; project administration, S.P.S.; funding acquisition, S.P.S. All authors have read and agreed to the published version of the manuscript.

Funding: This research was funded by the Natural Sciences and Engineering Research Council of Canada (NSERC) collaborative grant with the Institute for Catastrophic Loss Reduction (ICLR) to the second author [Collaborative Research and Development (CRD) Grant-CRDPJ 472152-14].

Data Availability Statement: Data used in the presented research are available at https://tides.gc.ca/eng/data (CWL data) (accessed on 9 June 2021).; https://wateroffice.ec.gc.ca/search/historical_e.html (Streamflow discharge records) (accessed on 15 June 2021); https://climate.weather.gc.ca/ (rainfall data) (accessed on 22 June 2021).

Acknowledgments: We are thankful for Canada's Fisheries and Ocean assistance for the coastal water level (CWL) data and Environment and Climate Change Canada for daily river discharge data. Special thanks to the Canadian Hydrographic Service (CHS) for providing the tide data. We are special thanks for the funding provided by NSERC (Natural Sciences and Engineering Research Council) and ICLR (Institute for Catastrophic Loss Reduction) in Canada.

Conflicts of Interest: The authors declare no conflict of interest.

References

1. Seneviratne, S.; Nicholls, N.; Easterling, D.; Goodess, C.; Kanae, S.; Kossin, J.; Luo, Y.; Marengo, J.; McInnes, K.; Rahimi, M.; et al. Changes in climate extremes and their impacts on the natural physical environment, Manag. Risk Extrem. *Events Disasters Adv. Clim. Change Adapt.* **2012**, 109–230. Available online: https://www.ipcc.ch/pdf/special-reports/srex/SREX-Chap3_FINAL.pdf (accessed on 6 July 2021).
2. Wahl, T.; Jain, S.; Bender, J.; Meyers, S.D.; Luther, M.E. Increasing risk of compound flooding from storm surge and rainfall for major US cities. *Nat. Clim. Chang.* **2015**, *5*, 1093–1097. [CrossRef]
3. Hendry, A.; Haigh, I.D.; Nicholls, R.J.; Winter, H.; Neal, R.; Wahl, T.; Joly-Laugel, A.; Darby, S.E. Assessing the characteristics and drivers of compound flooding events around the U.K. coast. *Hydrol. Earth Syst. Sci.* **2019**, *23*, 3117–3139. [CrossRef]
4. Lucey, J.T.; Gallien, T.W. Characterizing multivariate coastal flooding events in a semi-arid region: The implications of copula choice, sampling, and infrastructure. *Nat. Hazards Earth Syst. Sci.* **2022**, *22*, 2145–2167. [CrossRef]
5. Fritz, H.M.; Blount, C.D.; Thwin, S.; Thu, M.K.; Chan, N. Cyclone Nargis storm surge in Myanmar. *Nat. Geosci.* **2009**, *2*, 448–449. [CrossRef]
6. Jonkman, S.N.; Maaskant, B.; Boyd, E.; Levitan, M.L. Loss of life caused by the flooding of New Orleans after hurricane Katrina: Analysis of the relationship between flood characteristics and mortality. *Risk Anal.* **2009**, *29*, 676–698. [CrossRef]
7. Emanuel, K. Assessing the present and future probability of Hurricane Harvey's rainfall. *Proc. Natl. Acad. Sci. USA* **2017**, *114*, 12681–12684. [CrossRef]
8. Sweet, W.V.; Park, J. From the extreme to the mean: Acceleration and tipping points of coastal inundation from sea level rise. *Earths Future* **2014**, *2*, 579–600. [CrossRef]
9. Kemp, A.C.; Horton, B.P. Contribution of relative sea-level rise to historical hurricane flooding in New York city: Historical hurricane flooding in New York City. *J. Quat. Sci.* **2013**, *28*, 537–541. [CrossRef]
10. Moftakhari, H.R.; Salvadori, G.; Aghakouchak, A.; Sanders, B.F.; Matthew, R.A. Compounding effects of sea level rise and fluvial flooding. *Proc. Natl. Acad. Sci. USA* **2017**, *114*, 9785–9790. [CrossRef]
11. Resio, D.T.; Westerink, J.J. Modeling the Physics of Storm Surges. *Phys. Today* **2008**, *61*, 33. [CrossRef]
12. Coles, S.G. *An Introduction to Statistical Modelling of Extreme Values*; Springer: London, UK, 2001.
13. Coles, S.; Heffernan, J.; Tawn, J. Dependence measures for extreme value analyses. *Extremes* **1999**, *2*, 339–365. [CrossRef]
14. Svensson, C.; Jones, D.A. Dependence between extreme sea surge, river flow and precipitation in eastern Britain. *Int. J. Climatol.* **2002**, *22*, 1149–1168. [CrossRef]
15. Cooley, D.; Davis, R.A.; Naveau, P. The pairwise beta distribution: A flexible parametric multivariate model for extremes. *J. Multivar. Anal.* **2010**, *101*, 2103–2117. [CrossRef]
16. Zheng, F.; Westra, S.; Sisson, S.A. Quantifying the between extreme rainfall and storm surge in the coastal zone. *J. Hydrol.* **2013**, *505*, 172–187. [CrossRef]
17. Zheng, F.; Seth, W.; Michael, L.; Sisson, S.A. Modeling dependence between extreme rainfall and storm surge to estimate coastal flooding risk. *Water Resour. Res.* **2014**, *50*, 2050–2071. [CrossRef]
18. Joe, H. *Multivariate Models and Dependence Concept*; CRC Press: Boca Raton, FL, USA, 1997.
19. Nelsen, R.B. *An Introduction to Copulas*; Springer: New York, NY, USA, 2006.
20. Karmakar, S.; Simonovic, S.P. Bivariate flood frequency analysis. Part-2: A copula-based approach with mixed marginal distributions. *J. Flood Risk Manag.* **2009**, *2*, 32–44. [CrossRef]
21. Latif, S.; Mustafa, F. Copula-based multivariate flood probability construction: A review. *Arab J. Geosci.* **2020**, *13*, 132. [CrossRef]
22. Latif, S.; Mustafa, F. A nonparametric copula distribution framework for bivariate joint distribution analysis of flood characteristics for the Kelantan River basin in Malaysia. *AIMS Geosci.* **2020**, *6*, 171–198. [CrossRef]
23. Lian, J.J.; Xu, K.; Ma, C. Joint impact of rainfall and tidal level on flood risk in a coastal city with a complex river network: A case study of Fuzhou City. *China Hydrol. Earth Syst. Sci.* **2013**, *17*, 679–689. [CrossRef]
24. Xu, K.; Ma, C.; Lian, J.; Bin, L. Joint probability analysis of extreme precipitation and storm tide in a coastal city under changing environment. *PLoS ONE* **2014**, *9*, e109341. [CrossRef]
25. Masina, M.; Lamberti, A.; Archetti, R. Coastal flooding: A copula-based approach for estimating the joint probability of water levels and waves. *Coast. Eng.* **2015**, *97*, 37–52. [CrossRef]

26. Paprotny, D.; Vousdoukas, M.I.; Morales-Nápoles, O.; Jonkman, S.N.; Feyen, L. Compound flood potential in Europe. *Hydrol. Earth Syst. Sci. Discuss.* **2018**. [CrossRef]
27. Zellou, B.; Rahali, H. Assessment of the joint impact of extreme rainfall and storm surge on the risk of flooding in a coastal area. *J. Hydrol.* **2019**, *569*, 647–665. [CrossRef]
28. Ghanbari, M.; Arabi, M.; Kao, S.; Obeysekera, J.; Sweet, W. Climate Change and Changes in Compound Coastal-Riverine Flooding Hazard Along the U.S. Coasts. *Earth's Future* **2021**, *9*, e2021EF002055. [CrossRef]
29. Serinaldi, F.; Grimaldi, S. Fully nested 3-copula procedure and application on hydrological data. *J. Hydrol. Eng.* **2007**, *12*, 420–430. [CrossRef]
30. Grimaldi, S.; Serinaldi, F. Asymmetric copula in multivariate flood frequency analysis. *Adv. Water Resour.* **2006**, *29*, 1155–1167. [CrossRef]
31. Reddy, M.J.; Ganguli, P. Probabilistic assessments of flood risks using trivariate copulas. *Theor. Appl. Climatol.* **2013**, *111*, 341–360.
32. Kao, S.; Govindaraju, R. Trivariate statistical analysis of extreme rainfall events via the Plackett family copulas. *Water Resour. Res.* **2008**, *44*. [CrossRef]
33. Fan, L.; Zheng, Q. Probabilistic modelling of flood events using the entropy copula. *Adv. Water Resour.* **2016**, *97*, 233–240.
34. Genest, C.; Favre, A.C.; Beliveau, J.; Jacques, C. Metaelliptical copulas and their use in frequency analysis of multivariate hydrological data. *Water Resour. Res.* **2007**, *43*, W09401. [CrossRef]
35. Whelan, N. Sampling from Archimedean copulas. *Quant. Financ.* **2004**, *4*, 339–352. [CrossRef]
36. Savu, C.; Trede, M. Hierarchies of Archimedean copulas. *Quant Financ.* **2010**, *10*, 95–304. [CrossRef]
37. Hofert, M.; Pham, D. Densities of nested Archimedean copulas. *J. Multivar. Anal.* **2013**, *118*, 37–52. [CrossRef]
38. Bedford, T.; Cook, R.M. Vines-a new graphical model for dependent random variables. *Ann. Stat.* **2002**, *30*, 1031–1068. [CrossRef]
39. Aas, K.; Berg, D. Models for construction of multivariate dependence—A comparison study. *Eur. J. Financ.* **2009**, *15*, 639–659. [CrossRef]
40. Aas, K.; Czado, K.C.; Frigessi, A.; Bakken, H. Pair-copula constructions of multiple dependence. *Insur. Math. Econ.* **2009**, *44*, 182–198. [CrossRef]
41. Graler, B.; Berg, M.J.V.; Vandenberg, S.; Petroselli, A.; Grimaldi, S.; Baets, B.D.; Verhost, N.E.C. Multivariate return periods in hydrology: A critical and practical review focusing on synthetic design hydrograph estimation. *Hydrol. Earth Sys. Sci.* **2013**, *17*, 1281–1296. [CrossRef]
42. Bevacqua, E.; Maraun, D.; Hobæk Haff, I.; Widmann, M.; Vrac, M. Multivariate statistical modelling of compound events via pair-copula constructions: Analysis of floods in Ravenna (Italy). *Hydrol. Earth Syst. Sci.* **2017**, *21*, 2701–2723. [CrossRef]
43. Jane, R.; Cadavid, L.; Obeysekera, J.; Wahl, T. Multivariate statistical modelling of the drivers of compound flood events in South Florida. *Nat. Hazards Earth Syst. Sci. Discuss.* **2020**, *20*, 2681–2699. [CrossRef]
44. Saghafian, B.; Mehdikhani, H. Drought characteristics using new copula-based trivariate approach. *Nat. Hazards* **2014**, *72*, 1391–1407. [CrossRef]
45. Tosunoglu, F.; Gürbüz, F.; İspirli, M.N. Multivariate modeling of flood characteristics using Vine copulas. *Environ. Earth Sci.* **2020**, *79*, 459. [CrossRef]
46. Latif, S.; Mustafa, F. Parametric vine copula construction for flood analysis for Kelantan River basin in Malaysia. *Civ. Eng. J.* **2020**, *6*, 1470–1491. [CrossRef]
47. Silverman, B.W. *Density Estimation for Statistics and Data Analysis*, 1st ed.; Chapman and Hall: London, UK, 1986.
48. Moon, Y.-I.; Lall, U. Kernel function estimator for flood frequency analysis. *Water Resour. Res.* **1994**, *30*, 3095–3103. [CrossRef]
49. Sharma, A.; Lall, U.; Tarboton, D.G. Kernel bandwidth selection for a first order nonparametric streamflow simulation model. *Stoch. Hydrol. Hydraul.* **1998**, *12*, 33–52. [CrossRef]
50. Kim, T.W.; Valdes, J.B.; Yoo, C. Nonparametric approach for bivariate drought characterization using Palmer drought index. *J. Hydrol. Eng.* **2006**, *11*, 134–143. [CrossRef]
51. Karmakar, S.; Simonovic, S.P. Bivariate food frequency analysis. Part-1: Determination of marginal by parametric and nonparametric techniques. *J. Flood Risk Manag.* **2008**, *1*, 190–200. [CrossRef]
52. Charpentier, A.; Fermanian, J.; Scaillet, O. *Copulas: From Theory to Application in Finance*, 1st ed.; Chapter The Estimation of Copulas: Theory and Practice; Risk Books: Torquay, UK, 2006.
53. Rauf, U.F.A.; Zeephongsekul, P. Analysis of Rainfall Severity and Duration in Victoria, Australia using Nonparametric Copulas and Marginal Distributions. *Water Resour. Manag.* **2014**, *28*, 4835–4856. [CrossRef]
54. Harrell, F.E.; Davis, C.E. A new distribution-free quantile estimator. *Biometrika* **1982**, *69*, 635–640. [CrossRef]
55. Chen, S.X. Beta kernel estimators for density functions. *Comput. Stat. Data Anal.* **1999**, *31*, 131–145. [CrossRef]
56. Pfeifer, D.; Strassburger, D.; Philipps, J. *Modelling and Simulation of Dependence Structures in Nonlife Insurance with Bernstein Copulas*; Working Paper; Carl von Ossietzky University: Oldenburg, Germany, 2009.
57. Sancetta, A.; Satchell, S. The Bernstein copula and its applications tomodeling and approximations of multivariate distributions. *Econom. Theory* **2004**, *20*, 535–562. [CrossRef]
58. Weiss, G.N.F.; Scheffer, M. Smooth Nonparametric Bernstein Vine Copulas. *SSRN Electron. J.* **2012**. [CrossRef]
59. Kulpa, T. On approximation of copulas. *Int. J. Math. Math. Sci.* **1999**, *22*, 259–269. [CrossRef]
60. Latif, S.; Simonovic, S.P. Nonparametric approach to copula estimation in compounding the joint impact of storm surge and rainfall events in coastal flood analysis. *Res. Sq.* **2022**. [CrossRef]

61. Pirani, F.J.; Najafi, M.R. Recent trends in individual and multivariate flood drivers in Canada's Coasts. *Water Resour. Res.* **2020**, *56*. [CrossRef]
62. Atkinson, D.E.; Forbes, D.L.; James, T.S. Dynamic coasts in a changing climate. In *Canada's Marine Coasts in a Changing Climate*; Lemmen, D.S., Warren, F.J., James, T.S., Clarke, C.S.L.M., Eds.; Government of Canada: Ottawa, ON, Canada, 2016; pp. 27–68.
63. Latif, S.; Simonovic, S.P. Compounding joint impact of rainfall, storm surge and river discharge on coastal flood risk: An approach based on 3D Fully Nested Archimedean Copulas. *Res. Sq.* **2022**. [CrossRef]
64. Wand, M.P.; Marron, J.S.; Ruppert, D. Transformations in Density Estimation: Rejoinder (in Theory and Methods). *J. Am. Stat. Assoc.* **1991**, *86*, 360–361.
65. Müller, H.G. Smooth Optimum Kernel Estimators near Endpoints. *Biometrika* **1991**, *78*, 521–530. [CrossRef]
66. Schuster, E. Incorporating Support Constraints into Nonparametric Estimators of Densities. *Commun. Stat. Theory Methods* **1985**, *14*, 1123–1136. [CrossRef]
67. Brown, B.M.; Chen, S.X. Beta-bernstein smoothing for regression curves with compact support. *Scand. J. Stat.* **1999**, *26*, 47–59. [CrossRef]
68. Chen, S.X. Beta Kernel for Regression Curve. *Stat. Sin.* **2000**, *10*, 73–92.
69. Bouezmarni, T.; Ghouch, E.; Taamouti, A. Bernstein estimator for unbounded copula densities. *Stat. Risk Model.* **2013**, *30*, 343–360. [CrossRef]
70. Nagler, T.; Czado, C. Evading the curse of dimensionality in nonparametric density estimation with simplified vine copulas. *J. Multivar. Anal.* **2016**, *151*, 69–89. [CrossRef]
71. Lorentz, G. *Bernstein Polynomials*; University of Toronto Press: Toronto, ON, Canada, 1953.
72. Tenbusch, A. Two-dimensional Bernstein polynomial density estimation. *Metrika* **1994**, *41*, 233–253. [CrossRef]
73. Diers, D.; Eling, M.; Marek, S. Dependence modeling in non-life insurance using the Bernstein copula. *Insur. Math. Econ.* **2012**, *50*, 430–436. [CrossRef]
74. Santhosh, D.; Srinivas, V. Bivariate frequency analysis of flood using a diffusion kernel density estimator. *Water Resour. Res.* **2013**, *49*, 8328–8343. [CrossRef]
75. Latif, S.; Mustafa, F. A nonparametric statistical framework using a kernel density estimator to approximate flood marginal distributions—A case study for the Kelantan River Basin in Malaysia. *Water Supply* **2020**, *20*, 1509–1533. [CrossRef]
76. Jones, M.C.; Marron, J.S.; Sheather, S.J. A brief survey of bandwidth selection for density estimation. *J. Am. Stat. Assoc.* **1996**, *91*, 401–407. [CrossRef]
77. Sheather, S.J.; Jones, M.C. A reliable data-based bandwidth selection method for kernel density estimation. *J. R. Stat. Soc. Ser. B* **1991**, *53*, 683–690. [CrossRef]
78. Wand, M.P.; Jones, M.C. *Kernel Smoothing*; Chapman and Hall: London, UK, 1995.
79. Chen, S. Optimal Bandwidth Selection for Kernel Density Functionals Estimation. *J. Probab. Stat.* **2015**, *2015*, 242683. [CrossRef]
80. Czado, C.; Jeske, S.; Hofmann, M. Selection strategies for regular vine copulae. *J. French Soc. Stat.* **2013**, *154*, 174–190.
81. Kurowicka, D.; Cooke, R. *Uncertainty Analysis with High Dimensional Dependence Modelling*; John Wiley: Hoboken, NJ, USA, 2006.
82. Vernieuwe, H.; Vandenberghe, S.; Baets, B.D.; Verhost, N.E.C. A continuous rainfall model based on vine copulas. *Hydrol. Earth Syst. Sci.* **2015**, *19*, 2685–2699. [CrossRef]
83. Yue, S.; Rasmussen, P. Bivariate frequency analysis: Discussion of some useful concepts in hydrological applications. *Hydrol. Process* **2002**, *16*, 2881–2898. [CrossRef]
84. Shiau, J.T. Return period of bivariate distributed hydrological events. *Stoch. Environ. Res. Risk Assess.* **2003**, *17*, 42–57. [CrossRef]
85. Salvadori, G. Bivariate return periods via-2 copulas. *J. R. Stat. Soc. Ser. B* **2004**, *1*, 129–144. [CrossRef]
86. Salvadori, G.; De Michele, C. Frequency analysis via copulas: Theoretical aspects and applications to hydrological events. *Water Resour. Res.* **2004**, *40*, W12511. [CrossRef]
87. Zhang, L.; Singh, V.P. Trivariate flood frequency analysis using the Gumbel-Hougaard copula. *J. Hydrol. Eng.* **2007**, *12*, 431–439. [CrossRef]
88. Zhang, L.; Singh, V.P. Bivariate flood frequency analysis using copula method. *J. Hydrol. Eng.* **2006**, *11*, 150. [CrossRef]
89. Reddy, M.J.; Ganguli, P. Bivariate flood frequency analysis of Upper Godavari River flows using Archimedean copulas. *Water Resour. Manag.* **2012**, *26*, 3995–4018. [CrossRef]
90. Requena, A.; Flores, I.; Mediero, L.; Garrote, L. Extension of observed flood series by combining a distributed hydro-meteorological model and a copula based model. *Stoch. Environ. Res. Risk Assess.* **2016**, *30*, 1363–1378. [CrossRef]
91. Read, L.K.; Vogel, R.M. Reliability, return periods, and risk under nonstationarity. *Water Resour. Res.* **2015**, *51*, 6381–6398. [CrossRef]
92. Salvadori, G.; Durante, F.; de Michele, C.; Bernardi, M.; Petrella, L. A Multivariate Copula-Based Framework for Dealing with Hazard Scenarios and Failure Probabilities. *Water Resour. Res.* **2016**, *52*, 3701–3721. [CrossRef]
93. Serinaldi, F. Dismissing return periods! Stoch. *Hydrol. Hydraul.* **2014**, *29*, 1179–1189.
94. Xu, Y.; Huang, G.; Fan, Y. Multivariate flood risk analysis for Wei River. *Stoch. Hydrol. Hydraul.* **2015**, *31*, 225–242. [CrossRef]
95. British Columbia Ministry of Environment. Sea Level Rise Adaptation Primer: A Toolkit to Build Adaptive Capacity on Canada's South Coasts. 2013. Available online: https://www2.gov.bc.ca/assets/gov/environment/climate-change/adaptation/resources/slr-primer.pdf (accessed on 15 December 2019).

96. Lemmen, D.S.; Warren, F.J.; James, T.S.; Mercer Clarke, C.S.L. (Eds.) *Canada's Marine Coasts in a Changing Climate*; Government of Canada: Ottawa, ON, Canada, 2016; 274p.
97. Mann, H.B. Nonparametric test against trend. *Econometrics* **1945**, *13*, 245–259. [CrossRef]
98. Kendall, M.G. *Rank Correlation Methods*, 4th ed.; Charles Griffn: London, UK, 1975; p. 1975.
99. Pettitt, A.N. A Non-Parametric Approach to the Change-Point Problem. *Appl. Stat.* **1979**, *28*, 126–135. [CrossRef]
100. Alexandersson, H. A homogeneity test applied to precipitation data. *J. Clim.* **1986**, *6*, 661–675. [CrossRef]
101. Buishand, T.A. Some methods for testing the homogeneity of rainfall records. *J. Hydrol.* **1982**, *58*, 11–27. [CrossRef]
102. Gringorten, I.I. A plotting rule of extreme probability paper. *J. Geophys. Res.* **1963**, *68*, 813–814. [CrossRef]
103. Moriasi, D.N.; Arnold, J.G.; Van Liew, M.W.; Bingner, R.L.; Harmel, R.D.; Veith, T.L. Model evaluation guidelines for systematic quantification of accuracy in watershed simulations. *Trans. ASABE* **2007**, *50*, 885–900. [CrossRef]
104. Akaike, H. A new look at the statistical model identification. *IEEE Trans. Autom. Control* **1974**, *19*, 716–723. [CrossRef]
105. Schwarz, G.E. Estimating the dimension of a model. *Ann. Stat.* **1978**, *6*, e461e–e464. [CrossRef]
106. Hannan, E.J.; Quinn, B.G. The Determination of the order of an autoregression. *J. R. Stat. Soc. Ser. B* **1979**, *41*, 190–195. [CrossRef]
107. Willmott, C.; Matsuura, K. Advantage of the Mean Absolute Error (M.A.E.) OVER THE Root Mean Square Error (RMSE) in assessing average model performance. *Clim. Res.* **2005**, *30*, 79–82. [CrossRef]
108. Farrel, P.J.; Stewart, K.R. Comprehensive study of tests for normality and symmetry: Extending the Spiegelhalter test. *J. Stat. Comput. Simul.* **2006**, *76*, 803–816. [CrossRef]
109. Nash, J.; Sutcliffe, J. River flow forecasting through conceptual models part i e a discussion of principles. *J. Hydrol.* **1970**, *10*, e282–e290. [CrossRef]
110. R Core Team. *R: A Language and Environment for Statistical Computing*; R Foundation for Statistical Computing: Vienna, Austria, 2021; Available online: https://www.R-project.org/ (accessed on 10 October 2022).
111. Nagler, T. kdecopula: An R Package for the Kernel Estimation of Bivariate Copula Densities. *J. Stat. Softw.* **2018**, *84*, 1–22. [CrossRef]
112. De Michele, C.; Salvadori, G.; Canossi, M.; Petaccia, A.; Rosso, R. Bivariate statistical approach to check the adequacy of dam spillway. *J. Hydrol. Eng.* **2005**, *10*, 50–57. [CrossRef]
113. Klein, B.; Pahlow, M.; Hundecha, Y.; Schumann, A. Probability analysis of hydrological loads for the design of food control system using copulas. *J. Hydrol. Eng. ASCE* **2010**, *15*, 360–369. [CrossRef]
114. Genest, C.; Rémillard, B. Validity of the parametric bootstrap for goodness-of-ft testing in semiparametric models. *Ann l'Inst. Henri. Poincare Prob. Stat.* **2008**, *44*, 1096–1127.
115. Genest, C.; Rémillard, B.; Beaudoin, D. Goodness-of-ft tests for copulas: A review and a power study. *Insur. Math Econ.* **2009**, *44*, 199–214. [CrossRef]

hydrology

Article

Numerical and Physical Modeling of Ponte Liscione (Guardialfiera, Molise) Dam Spillways and Stilling Basin

Monica Moroni *, Myrta Castellino and Paolo De Girolamo

Dipartimento di Ingegneria Civile Edile e Ambientale (DICEA), Sapienza University of Rome, Via Eudossiana 18, 00184 Rome, Italy
* Correspondence: monica.moroni@uniroma1.it

Abstract: Issues such as the design or reauditing of dams due to the occurrence of extreme events caused by climatic change are mandatory to address to ensure the safety of territories. These topics may be tackled numerically with Computational Fluid Dynamics and experimentally with physical models. This paper describes the 1:60 Froude-scaled numerical model of the Liscione (Guardialfiera, Molise, Italy) dam spillway and the downstream stilling basin. The k-ω SST turbulence model was chosen to close the Reynolds-averaged Navier–Stokes equations (RANS) implemented in the commercial software Ansys Fluent ®. The computation domain was discretized using a grid with hexagonal meshes. Experimental data for model validation were gathered from the 1:60 scale physical model of the Liscione dam spillways and the downstream riverbed of the Biferno river built at the Laboratory of Hydraulic and Maritime Constructions of the Sapienza University of Rome. The model was scaled according to the Froude number and fully developed turbulent flow conditions were reproduced at the model scale (Re > 10,000). From the analysis of the results of both the physical and the numerical models, it is clear that the stilling basin is undersized and therefore insufficient to manage the energy content of the fluid output to the river, with a significant impact on the erodible downstream river bottom in terms of scour depths. Furthermore, the numerical model showed that a less vigorous jet-like flow is obtained by removing one of the sills the dam is supplied with.

Keywords: dams; numerical simulations; physical modeling; water management

Citation: Moroni, M.; Castellino, M.; De Girolamo, P. Numerical and Physical Modeling of Ponte Liscione (Guardialfiera, Molise) Dam Spillways and Stilling Basin. *Hydrology* **2022**, *9*, 214. https://doi.org/10.3390/hydrology9120214

Academic Editors: Aristoteles Tegos, Alexandros Ziogas and Vasilis Bellos

Received: 23 October 2022
Accepted: 23 November 2022
Published: 28 November 2022

1. Introduction

Water resource management in hydrology involves the processes of planning, developing and managing water resources. Climate change is making these processes more difficult to deal with [1]. Water storage has always represented an essential task for human activity, with significant implications for flood control or the generation of electricity. From this point of view, dams represent a suitable system to divert water, control flooding and produce hydroelectricity.

All these processes are sensitive to the complex three-dimensional flow effects involved in dam hydrodynamics. To accurately study the hydrodynamics and the fluid–structure interaction issues, Computational Fluid Dynamics (CFD) numerical tests together with experimental models are considered within the present research study as mandatory tools (as shown by [2,3]). CFD solves the governing equations of fluid-flow problems, i.e., the continuity, the Navier–Stokes and the energy equations. Because of the nonlinear terms in these equations, analytical methods yield very few solutions. Then, numerical methods, i.e., CFD, are used to obtain the required solutions. Numerical models prescribe the discretization of the domain. The continuous spatial and temporal domain of the problem must be replaced by a discrete one made up of grid points or cells and time levels. The governing equations of the problem must be replaced by a set of algebraic equations with the grid points/cells and the time levels as their domain. Finally, the solutions at each grid point/cell are obtained when advancing from one time level to the next. Conversely, a physical model consists of the "physical" reproduction of a scaled artifact and the phenomena that occur in it. Experimental tests performed on physical models provide useful information on the entity and behavior of the variables involved in the phenomena under investigation in a controlled environment. In general, those quantities may be measured in a limited number of points within the domain.

The construction of a physical model can be time- and cost-ineffective. Moreover, a physical model may be affected by scale effects, since not all physical conditions present in nature are reproducible at a laboratory scale. This is especially true in the case of turbulent phenomena. Then, in some cases, a numerical model is the only tool to answer questions related, for instance, to the suitability of existing dams to manage discharge increases with respect to the design values or modifications of the dam geometry. In addition, numerical models make it possible to easily evaluate and compare different scenarios. Nevertheless, the physical model, when available, represents an important tool for verifying and calibrating the results provided by numerical models [4].

The remarkable technological advances of recent decades have made it possible to develop increasingly refined numerical models, allowing the study of the temporal evolution of the fluid features with a spatial resolution which can be very high. The authors in [5–8] presented some of the first examples of numerical simulation applied to the reconstruction of flow over a spillway with a 3D Reynolds-averaged Navier–Stokes (RANS) model. The reliability of numerical models in capturing the water surface profile along dam spillways located in different parts of the world is demonstrated in a few contributions [9–15]. Ref. [16] investigated the hydraulic characteristics of the dam discharge flow and its downstream impact by employing Reynolds-averaged Navier–Stokes equations with the RNG k-ε eddy viscosity model for its turbulence closure, as well as the volume of fluid method. Complex turbulent flow patterns, including collision, reflection and vortices, were captured by three-dimensional simulation.

The published results encourage the use of numerical models for assessing the hydraulic performance of structures. Furthermore, [17] shows how 3D flood numerical simulations can qualitatively and quantitatively assess flood hazards and serve as a visual reference for the development of flood control schemes, providing an important foundation for flood forecasting, dam design and flood control system application.

This paper describes the 1:60 Froude-scaled numerical model of the Liscione dam spillway and the downstream stilling basin. The k-ω SST turbulence model was chosen to close the Reynolds-averaged Navier–Stokes equations (RANS), due to its remarkable robustness and reliability in simulations involving similar geometries. The Autocad ® software was used to construct the geometry of the computational domain, whereas the simulations were performed with the commercial software ANSYS Fluent ®. The discretization of the domain was performed via the software provided by ANSYS (Fluent Meshing), which guarantees the generation of a simply connected domain (Watertight Geometry). Data for validating the numerical model were gathered by means of a 1:60 scale physical model built at the Laboratory of Hydraulic and Maritime Constructions of the Sapienza University of Rome. The model was scaled according to the Froude number and fully developed turbulent flow conditions were reproduced at the model scale (Re > 10,000) [18]. In [19], the physical model, the experimental campaign conducted to investigate the key hydrodynamical parameters such as hydraulic levels and hydraulic jump location are described in detail. Furthermore, an innovative technical solution suitable to protect the riverbed located just downstream of the stilling basin by means of artificial Antifer blocks is also illustrated.

The Liscione dam was affected between 24 and 25 January 2003 by a serious rainfall event that caused extensive damage. The rain intensity of the event was measured by two weather stations and the related inflow and outflow rates were quantified. The outflow rates turned out to be 830.0 m^3/s, which caused the maximum allowed elevation into the reservoir, i.e., 125.5 m a.s.l., to be overcome. The return period was 30 years. The event caused extensive damage both upstream and downstream of the stilling basin: the failure and breakage of some concrete elements at the end of the dam chute on the right side of the river; damaged and removed gabions on both the left and right banks, and displaced the bottom protection in the central area of the riverbed.

The event demonstrated that the stilling basin of the Liscione dam was ineffective in dissipating the flow energy content with potential severe effects on the stability of the downstream unprotected riverbed due to massive erosion phenomena. To tackle the scour issue, a few measures are available: the redesign of the existing stilling basin, the redesign of the dam as a whole, replacing the dam elements that contribute to the formation of jet-like flows downstream from the chute; and the implementation of protection strategies employing boulders properly arranged in the riverbed downstream of the stilling basin. The aim of this paper was to demonstrate that a validated numerical model, efficiently implemented using a commercial software and a reasonably powerful PC, can be usefully employed for the reconstruction of the hydrodynamics of existing dams which may need either maintenance or upgrading works, such as in the case of flood discharge increments, but also for the design of novel dams [20].

2. Materials and Methods

2.1. The Liscione Dam

The Liscione dam is located in the municipality of Guardialfiera in Molise (central Italy). Its construction, which took place between 1967 and 1973, had as a main objective the creation of an artificial reservoir, namely Lake Guardialfiera, by collecting water from the Biferno river (Figure 1a).

The reservoir (Figure 1b), obtained with a barrier in loose materials sealed with a bituminous conglomerate coating, was aimed at flood retention and water storage for irrigation purposes. Figure 1c presents a detailed view of the main dam elements, i.e., the surface spillway, chute, stilling basin and bottom outlet.

Important features of both the reservoir (Lake Guardialfiera) and the Liscione dam are listed in Table 1. Characteristic discharge values and return periods of the catchment area of the dam are shown in Table 2.

The dam surface spillway consists of an ungated ogee weir, 92 m long with a crest elevation at 125.5 m a.s.l. (Figure 2a; detailed view in Figure 2b) and a gated weir with three 13 m wide openings, equipped with automatic flap gates with counterweights pivoted at the sill (Figure 2a; detailed view in Figure 2c). The gate configurations are either open (minimum elevation of 122.0 m a.s.l.) or closed, sharing the same elevation as the ogee weir (i.e., 125.5 m a.s.l.). The gate drop takes place automatically and progressively as soon as the reservoir water level reaches an elevation of 125.5 m a.s.l.

Water collected by the surface spillway is conveyed into the stilling basin via a chute. The channel has a uniform rectangular section of 25 m in length and a horizontal development that is 180 m long. If the water stored in Lake Guardialfiera reaches an elevation of 129 m a.s.l., the ungated ogee spillway and the gated spillway release discharge values of 1080 m^3/s and 1174 m^3/s, respectively.

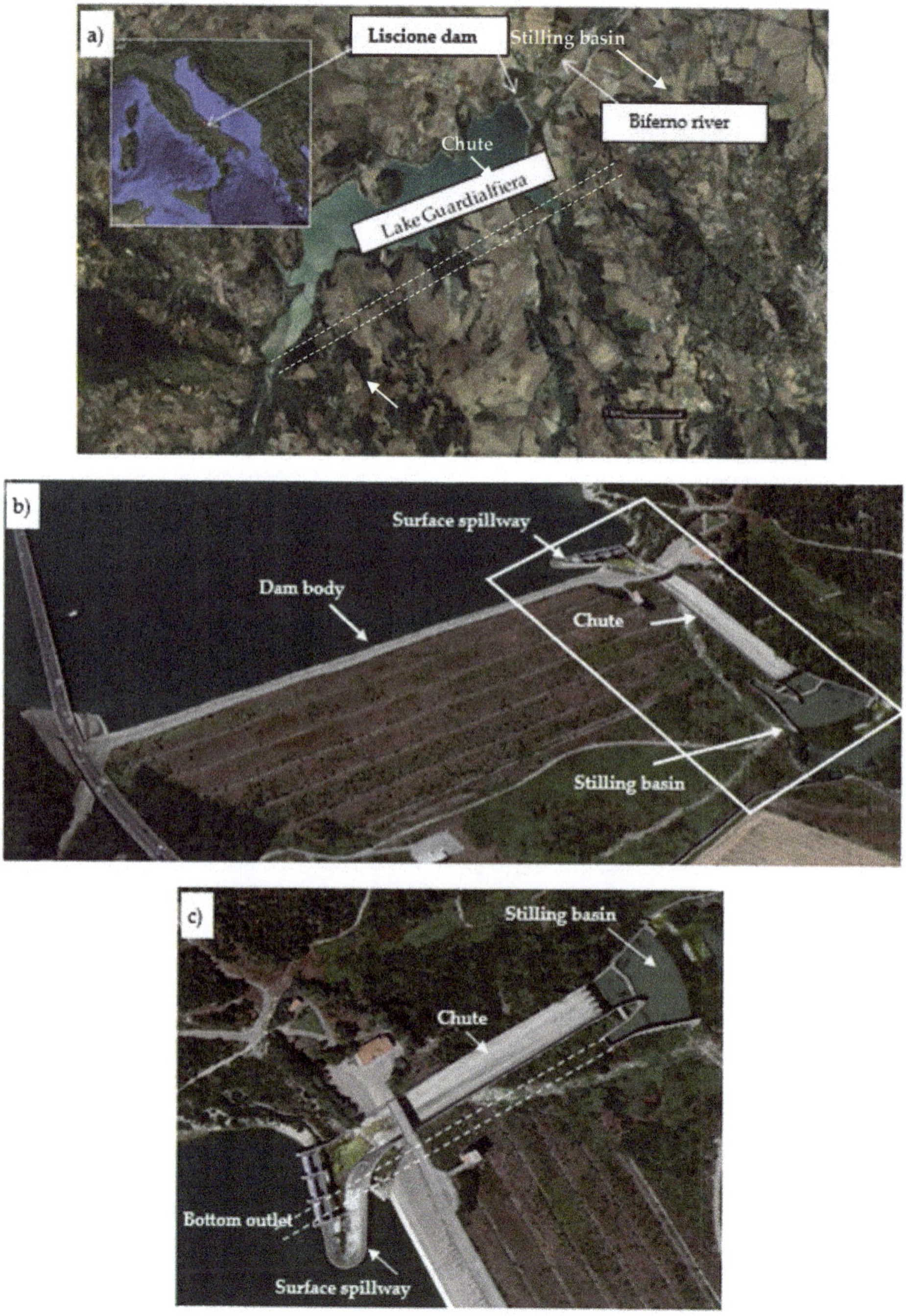

Figure 1. (**a**) The investigated area: geographical framework [Map data: Google]; (**b**) overview of the dam body and main elements; (**c**) detailed view of the main elements of the dam, i.e., surface spillway, chute, stilling basin and bottom outlet.

Figure 2. Surface spillway: (**a**) top view of the spillway; (**b**) ungated ogee weir (free Creager-type sill); and (**c**) gated weir with three 13 m wide openings, equipped with automatic flap gates.

Table 1. Important features of the reservoir (Lake Guardialfiera) and the Liscione dam.

Reservoir	Total volume (millions of m^3)	Useful storage (millions of m^3)	Dead storage capacity (millions of m^3)	Reservoir maximum surface (km^2)	Surface of the catchment area (km^2)
	173.0	137.0	11.0	7.45	1043
Dam	Management upper storage elevation (m a.s.l.)	Maximum allowed water elevation (m a.s.l.)	Dam crest (m a.s.l.)	Management minimum operating level (m a.s.l.)	Minimum foundation height (m a.s.l.)
	125.5	129.0	131.5	92.0	71.5

Table 2. Characteristic discharge values and return periods of the Liscione dam catchment area.

Discharge inlet to the reservoir (m^3/s)	1050	1800	2300	2650
Discharge at the spillways (m^3/s)	830	1450	1850	2250
Return period (y)	30	200	500	1000

To prevent dam overtopping, a bottom outlet is realized to convey water into the stilling basin (Figure 1c). It consists of a tunnel with an internal diameter of 7.2 m, a length of 309.5 m and a slope of 1%. Its intake is placed at an elevation of 76.4 m a.s.l. and the outlet is at 73.5 m a.s.l. If the water stored in Lake Guardialfiera reaches an elevation of 129 m a.s.l., the bottom outlet is activated to drain a flow rate of approximately 500 m^3/s.

To reduce the kinetic energy of water drained into the stilling basin, four nappe splitters are placed at the end of the chute (Figure 3). To the same aim, the stilling basin is equipped with four sills: the first one is placed downstream of the bottom outlet (sill#1), two other sills are placed at the downstream boundary of the chute (sill#2, a sky-jump-like sill, and sill#3) and the last one is at the end of the stilling basin (sill#4). Sill#4 is higher at the hydraulic left of the stilling basin to better dissipate the energy, which in that area, due to the slight curvature of the riverbed, may lead to massive erosion.

Figure 3. Dam elements aimed at reducing the kinetic energy of water drained into the stilling basin.

Downstream of the stilling basin, the central area of the first 500.0 m of the riverbed was covered with 0.3 m thick Reno type bottom protection, and the right and left banks of the riverbed protected by 1.0 m high gabions. After that distance, the riverbed presents the natural waterway.

2.2. Experimental Investigation

The physical model was realized in the DICEA-Sapienza University of Rome Hydraulic and Maritime Construction Laboratory. Referring to Figure 4, the physical model was designed in such a way that the following requirements were met:

- The model tank (mimicking the prototype reservoir) dimensions made it possible to include the surface spillway and to ensure a constant water level in the tank up to the maximum tested flow rate, as occurs in reality, due to the large size of the artificial basin. A preliminary investigation demonstrated that a tank with dimensions shown in Figure 4, i.e., approximately 150 × 210 m, was sufficient to guarantee the above requirement;

- The physical model downstream section was set considering the area wherein the river protection interventions were planned to take place. In addition, the model includes the river portion downstream of the stilling basin characterized by an irregular planimetric geometry. Based on these requirements, the downstream closure section of the model was set after the first river bend, as highlighted in Figure 4.

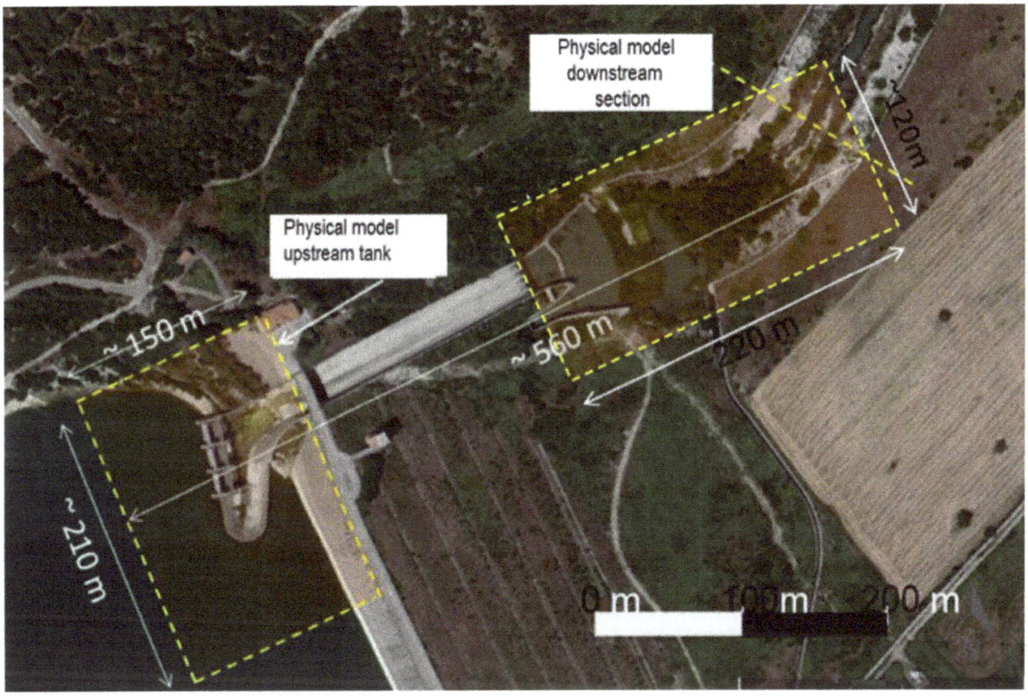

Figure 4. Area reproduced with the physical model.

Due the above requirements, the prototype dimensions of the area reproduced with the physical model were 560.0 m as the longitudinal extension and roughly 210.0 m as the maximum width (see Figure 4). A geometric reduction scale of 1:60 results from the adoption of the dimensions reported above.

The components of the spillway, i.e., the three gates (in their lowered configuration), the ogee weir, the chute, the bottom outlet terminus, the sills, and the stilling basin, mimicking the prototype counterparts, are shown in Figure 5. A detailed description of the physical model can be found in [19].

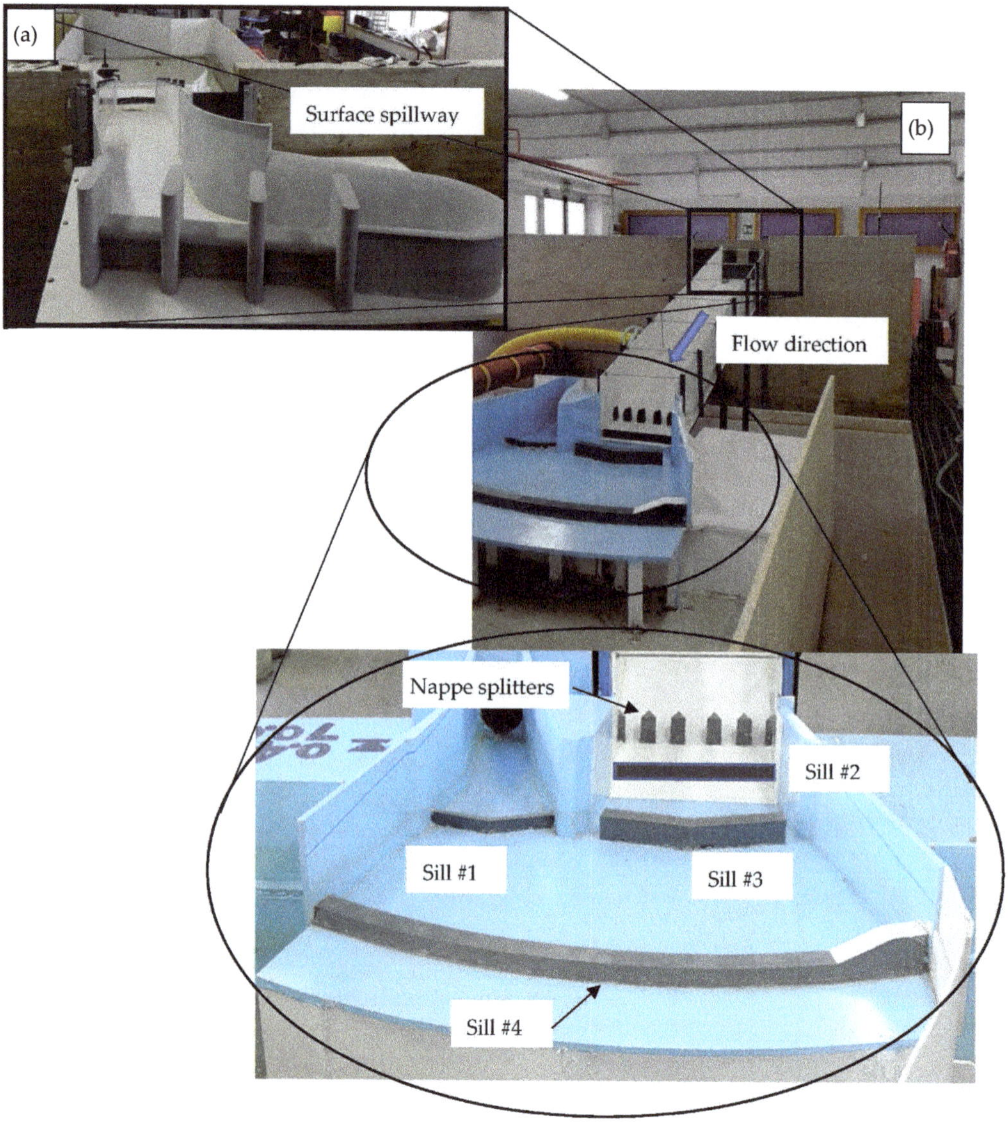

Figure 5. Physical mode: (**a**) surface spillway; (**b**) chute, bottom outlet terminus, sills and stilling basin.

2.3. Numerical Simulations

A geometry similar to that tested experimentally was investigated with the numerical model. Two sets of numerical simulations were carried out and will be described in the following sections, namely the upstream tank that reproduces the Lake Guardialfiera and the surface spillway (Model #1) and the complete model enclosing the upstream tank, the surface spillway, the chute, the stilling basin, and a small portion of the riverbed downstream of the stilling basin (Model #2). Runoff conditions on the surface spillway were numerically reproduced by imposing a constant water level in the tank. In some preliminary numerical tests, an upstream tank of dimensions larger than those employed for the physical model were tested. No remarkable differences were noticed in terms of the fluid-free surface features and fluid height above the surface spillway. Model #1 outcomes were employed to determine the stage–discharge rate curve, which was compared to the experimental one. Model #2 outcomes made it possible to investigate the impact area of the jet impinging into and downstream of the stilling basin. Simulations were performed both considering the presence of the ski-jump sill (sill#2) at the foot of the chute and without it to evaluate the distance of impingement of the jet outflowing from the spillway chute in both cases.

To perform the numerical simulations of Model #1 and Model #2, the 3D drawings of the spillway and of the whole dam in the model scale were realized with AutoCAD, starting from the planimetric and cross-sections provided by the dam concessionaire and appropriately compared with the technical drawings realized by the designer. These drawings reported in Figure 6 were imported in Fluent in the Geometry component section of the software.

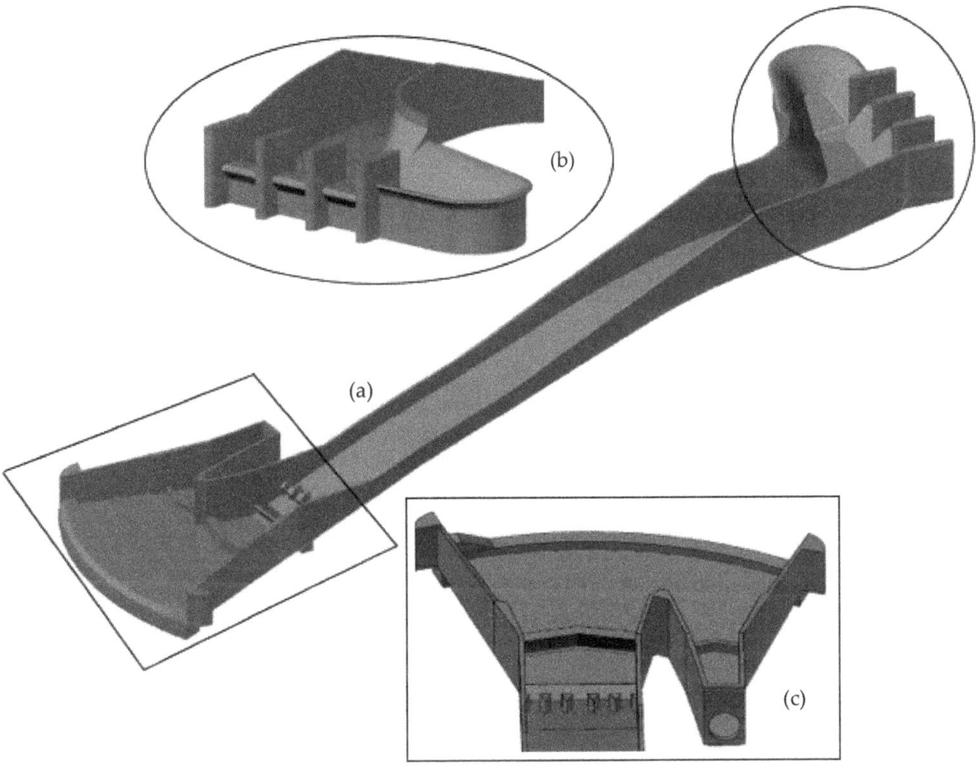

Figure 6. (**a**) Three-dimensional drawing of the infrastructure in the model scale carried out with AutoCAD; (**b**) AutoCAD model of the surface spillway; and (**c**) AutoCAD model of the stilling basin.

Initially, the bathymetry was imported as an STL file built from information gathered from the area Digital Terrain Model. Preliminary tests demonstrated that its influence on the reconstructed water levels was negligible with respect to the analogous simulations performed without implementing the lake bottom profile. For this reason, the bathymetry was not included in the final configuration of both Models #1 and #2.

For all models, the computational domain was discretized using a grid with hexagonal meshes. To verify the independence of the results from the mesh size, several simulations were carried out, doubling the number of elements or, when an excessive computational burden was expected, reducing the mesh size by at least 20% in each direction. Table 3 shows the main features of the discretization adopted for the models listed above, specifically the minimum and maximum size of the grids and the total number of elements.

Table 3. Main features of the grids employed for both models.

Model #	Model Description	Resolution	Minimum Size (mm)	Maximum Size (mm)	Number of Elements
1	Surface spillway	Low	10	20	259,148
		High	5	12	887,386
2	The whole dam	Low	10	20	485,258
		Medium	8	16	879,883
		High	5	12	1,024,524

The computational domains for the geometries listed above are shown in Figure 7 for Model #1 and Figure 8 for Model #2.

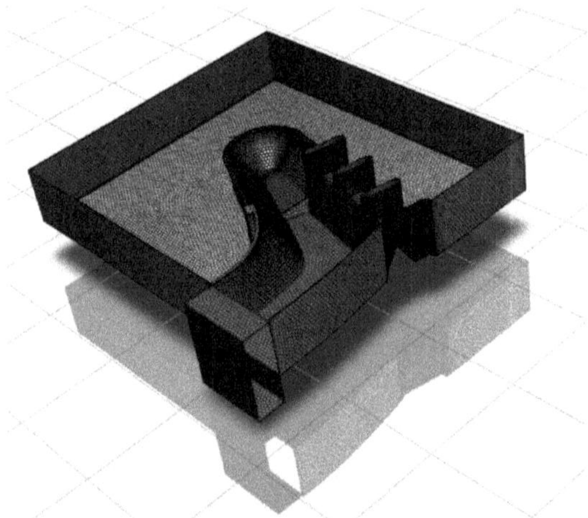

Figure 7. Computational domain for Model #1.

For both models, the inlet boundary condition was of the "pressure inlet" type. It was provided by assigning the height of the free surface inside the upstream tank. The "pressure outlet" boundary condition was set at the surfaces in contact with the atmosphere. It was also applied on the walls of the step at the toe of the stilling basin. The step was introduced in the numerical model following [21] since it makes the results more consistent with the experimental evidence. The "wall" boundary condition was assigned to the walls of the dam body including the sills, the chute, and the bottom surface of the stilling basin. The "no slip" condition was set, which prescribes the fluid to adhere to the interface with the wall and moves with the same velocity, and zero velocity in our models since the walls are fixed.

The Fluent software requires an initial value of the water volume fraction (WVF). Inside the upstream tank, the WVF of a certain number of cells was assigned the value 1. Those cells were selected, ensuring a water level slightly above the free surface height provided by the "Inlet" boundary condition. This made it possible to provide initial conditions that were not too far from those of the final solution, considerably reducing the computational cost of the simulations.

Figure 8. Computational domain for Model #2.

Velocity values measured with a Pitot tube were used to validate the numerical models (Figure 9).

Figure 9. Pitot tube placed at (roughly) half the chute length.

These measurements were performed at a discharge rate of 1450 m^3/s, i.e., the design flow rate for the hydraulic structure. The velocity profiles in two different points were measured:

(1) At 0.14 m upstream from the chute, within the surface spillway volume;
(2) At 1.639 m downstream from the chute (see Figure 9).

3. Results

Model #1: Surface Spillway

The numerical model including the surface spillway was implemented using two grids of different resolution. For both grids, the simulation was interrupted up to the achievement of the steady-state condition for the flow field.

Table 4 presents the complete set of numerical simulations performed for Model #1, namely the height of the free surface inside the upstream tank, the expected flow rate, the turbulence model adopted and the grid resolution. The expected flow rate for a given height of the free surface was determined from the experimentally achieved stage–discharge rate curve.

Table 4. Height of the free surface inside the upstream tank, expected flow rate and turbulence model adopted for Model #1.

Height of the Free Surface Inside the Upstream Tank (m)	Expected Discharge (m^3/s)	Turbulence Model	Grid Resolution
0.21327	304	k-omega	Low
0.22357	424	k-omega	Low
0.22937	530	k-omega	Low
0.24077	830	k-omega	Low
0.24077	830	k-omega	High
0.25697	1450	k-omega	Low
0.25697	1450	k-omega	High
0.26237	1650	k-omega	Low
0.26237	1650	k-eps	Low
0.26747	1850	k-omega	Low
0.27897	2250	k-omega	Low

Low discharges were simulated to characterize the initial portion of the stage-discharge rate curve. Q = 830 m^3/s was the maximum flow rate discharged from the spillway during the event which occurred in January 2003, characterizing a rainfall event with a return period of 30 years. Q = 1450 m^3/s, Q = 1650 m^3/s, Q = 1850 m^3/s and Q = 2250 m^3/s correspond to a return period of 100 years, 200 years, 500 years and 1000 years, respectively.

Figure 10 qualitatively compares the hydrodynamics that occurred during the flood event of January 2003 (Figure 10a), the outcome of the laboratory experiment (Figure 10b) and the numerical model (Figure 10c).

Though the air entrainment at the prototype scale appears more evident, the physical and numerical models present similar flow features.

To determine the best parameters to employ within the simulations, the numerical model for a prototype discharge value equal to 1450.0 m^3/s was run for both a low- and a high-grid resolution. The k-omega SST model was used as a turbulent model. The comparison between the hydrodynamics resulting from the numerical model (for the high grid resolution case) and the physical model is presented in Figure 11. The numerical model output of the low resolution case looks very similar to Figure 11a, and for this reason, it was not presented. Moreover, the numerical model outputs are very similar to the water-free surface provided by the physical model. This suggests that the low-resolution grid was sufficiently refined to describe the phenomenon under investigation. The grid independence of the solution is further demonstrated by analyzing the velocity profile located 0.14 m upstream from the chute, within the surface spillway volume where the Pitot tube measurements were available for the discharge value under investigation. Figure 11c presents the comparison between the velocity values at different heights calculated with the high- and low-resolution models and corresponding values measured with the Pitot tube. No remarkable difference can be noticed between the numerical profiles which appear to be slightly overestimated with respect to the measured velocity values, as was to be expected due to the intrusive nature of the measurement with the Pitot tube, which may affect the magnitude of the velocity value detected. Considering this result, the low resolution grid was employed for all simulations.

Figure 10. Comparison among the 2003 flow event (**a**) with a picture of the real event, (**b**) the physical model output and (**c**) the result of the numerical simulation.

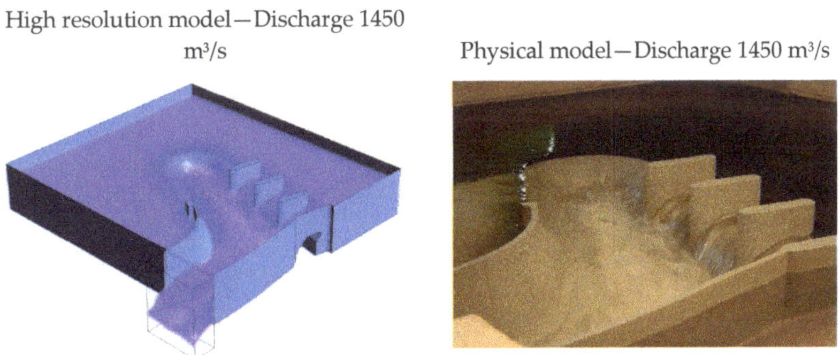

Figure 11. *Cont.*

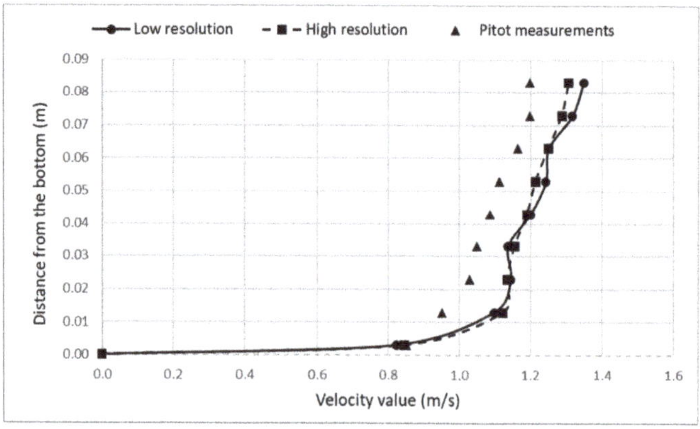

Figure 11. Comparison between the hydrodynamics resulting from the numerical model (for both the tested grid resolutions) and the physical model.

A further test was conducted to choose the turbulent model providing the best results in terms of similarity between the numerical output and physical model. The k-omega SST and k-eps models were employed and, as shown in Figure 12, no remarkable differences can be noticed between the output of the numerical model employing the k-omega SST turbulence model and the output of the physical model. Once again, the free surfaces output by the numerical models employing the k-omega SST and the k-eps turbulence models look very similar. For this reason, the k-omega SST model was used for the simulations.

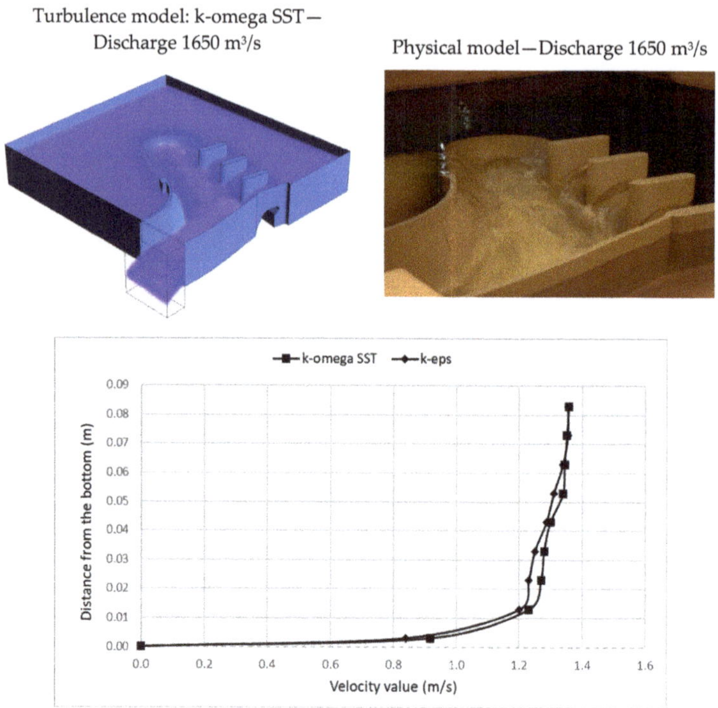

Figure 12. Comparison between the hydrodynamics resulting from the numerical model (for both the tested turbulence models) and the physical model.

In order to quantitatively check the reliability of the numerical model, the discharge computed for each model (and the corresponding water stage employed as the inlet boundary condition) was compared to that employed within the experimental investigation. The experimental procedure consisted in varying the discharge flowing through the model and contextually measuring the water level within the upstream tank that reproduces the modeled Lake Guardialfiera. Eighteen experiments were run with different values of the discharge. The experimentally obtained stage–discharge rate curve was then compared to that one used at the dam design stage and validated. The comparison was satisfactory as the maximum error was roughly equal to 3% (refer to [19] for the complete procedure employed to construct the experimental rating curve).

For the numerical model, the discharge was computed assuming a threshold value for the simulated water volume fraction and computing the flow rate value as the product of the average velocity and the area of the water section. Different threshold values were considered, ranging from 0.3 to 0.7 with step 0.1, and the value 0.6 appeared to be the one providing the best agreement between the experimentally detected rating curve and the numerical one. The two curves are displayed in Figure 13.

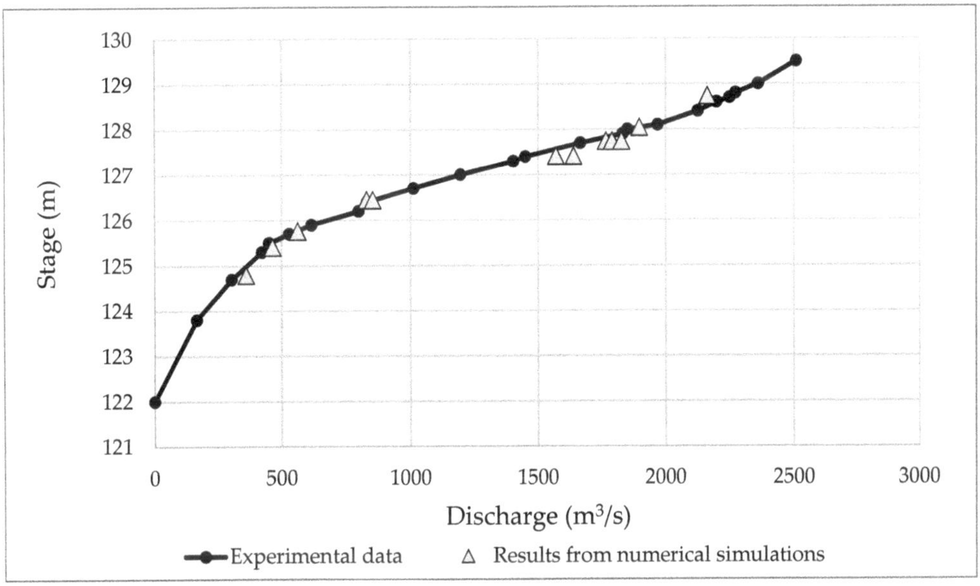

Figure 13. Rating curve from numerical simulations (Model #1) and experiments.

Simulations for Model #2 were conducted using three different sizes of the calculation grid (10–20 mm, 8–16 mm, 5–12 mm). For all cases, the simulation continued until the steady state was reached inside the dissipation tank.

As mentioned earlier, velocity values measured with a Pitot tube were used to determine the grid resolution to employ for further analysis. The investigated discharge value was 1450 m^3/s and the grid resolutions were those previously defined as low, medium, and high. According to the results obtained for Model #1, the k-omega SST model was used for the simulations.

Figure 14 presents the comparison between the velocity profiles reconstructed with Model #2 at three different resolutions and the Pitot measurements at location 0.14 m upstream the chute, within the surface spillway volume. No remarkable differences can be noted for the different resolutions adopted for the simulations. Conversely, Figure 15 which presents the same comparison at a location of 1.639 m downstream from the chute suggests that higher resolution is required for a better match. For this reason, further simulations were conducted, employing the higher resolution.

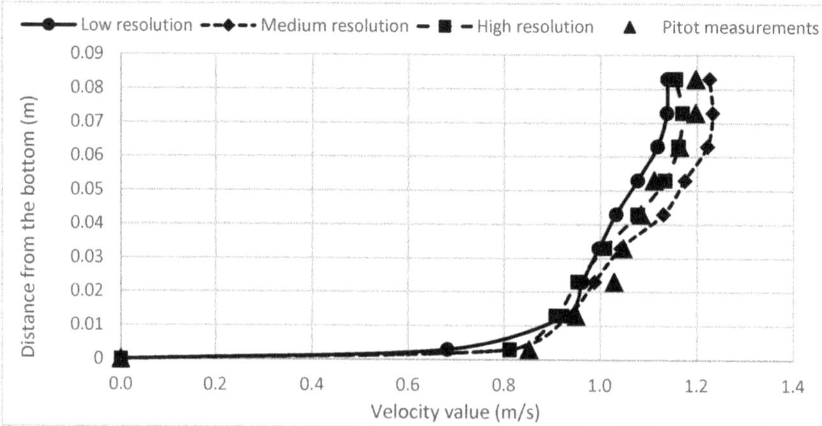

Figure 14. Comparison between the velocity profiles reconstructed with Model #2 at three different resolutions and Pitot measurements at location 0.14 m upstream from the chute and within the surface spillway volume.

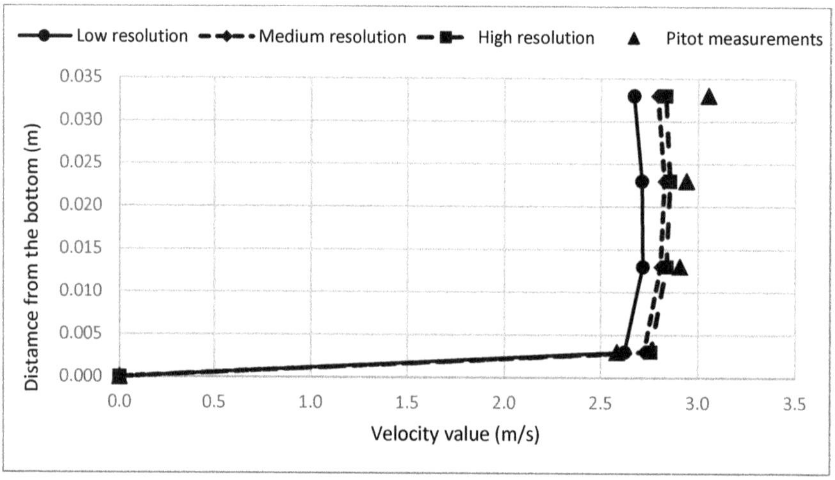

Figure 15. Comparison between the velocity profiles reconstructed with Model #2 at three different resolutions and Pitot measurements at location 1.639 m downstream from the chute.

Table 5 presents the details of the simulations conducted recalling that Q = 830 m^3/s was the maximum flow rate discharged from the spillway during the event which occurred in January 2003, characterizing a rainfall event with a return period of 30 years; whereas Q = 1450 m^3/s and Q = 1650 m^3/s correspond to a return period of 100 and 200 years.

Table 5. Height of the free surface inside the upstream tank, expected flow rate and the presence of sill #2 for Model #2.

Height of the Free Surface Inside the Upstream Tank (m)	Expected Discharge (m^3/s)	Sill #2
0.24077	830	yes
0.25697	1450	yes
0.25697	1450	no
0.26237	1650	yes
0.26237	1650	no

Figure 16 presents the water volume fraction reconstructed with Model #2 for a discharge value of 1650 m^3/s when sill #2 is included in the numerical model.

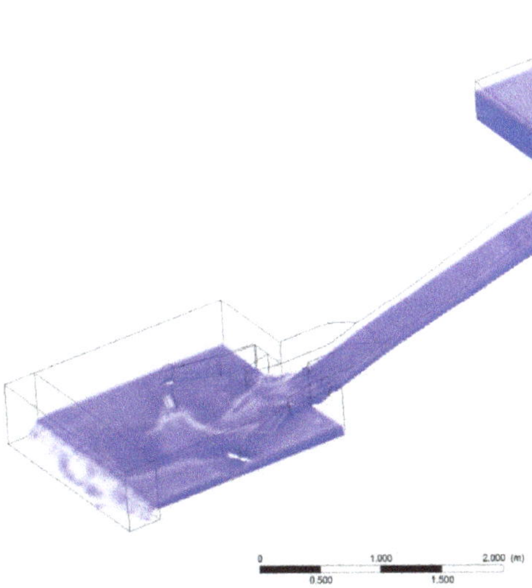

Figure 16. Water volume fraction reconstructed with Model #2 for a discharge value of 1650 m^3/s and the presence of sill #2.

The effects of sill #2 on the hydrodynamics in the stilling basin and downstream areas are quite evident and may be quantitatively appreciated in Figure 17 where a zoom in the dissipation tank area is presented and compared to the experimental outcomes.

Figure 17 shows the images of the hydrodynamics in the stilling basin and the riverbed downstream from the lateral view. The corresponding images below present the water volume fraction computed by the numerical model. In each experimentally gathered image, a 5 × 5 cm^2 mesh (model scale) corresponding to a 3 × 3 m^2 mesh at a prototype scale was overlapped over the investigated area. The blue line corresponds to the end of sill #4 while the red line defines the area of impact of the jet outflowing from the spillway chute. The comparison between the images, both the experimental and numerical ones, clearly shows the effect of the ski-jump sill removal, which produces a decrease in the distance between the jet impact area and the stilling basin of roughly 12 m.

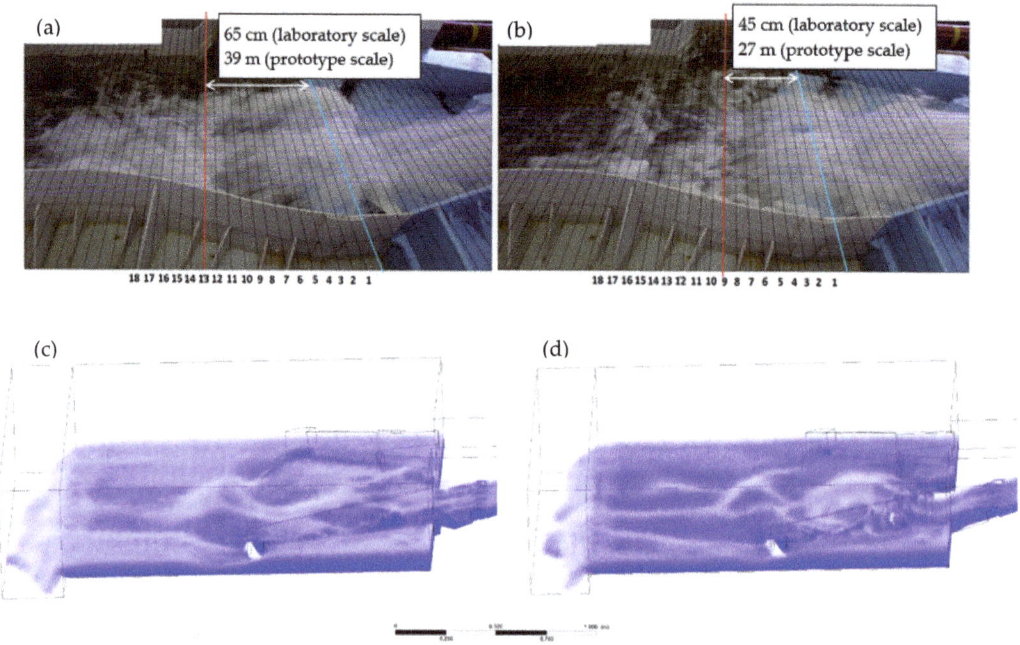

Figure 17. Hydrodynamics in the stilling basin and the riverbed downstream from the lateral view for a discharge value of 1650 m^3/s and (**a**) ski-jump-like sill in place (**c**) is the output of the numerical model); and (**b**) ski-jump-like sill removed (**d**) is the output of the numerical model).

4. Conclusions

In this paper, a comparison between a numerical and experimental model of the Liscione dam was presented. The velocity and free surface elevation were the physical quantities compared. It is not always possible to reproduce a large infrastructure (i.e., a dam) in a laboratory. For this purpose, numerical models can be implemented as a useful alternative. Numerically reproducing a dam also enables it to be made independent of the scale (as it can be reproduced in a prototype scale). Nevertheless, it is mandatory to verify the numerical results with experimental ones to validate the numerical simulations.

In the research presented herein, the results of the numerical simulations confirm the outcomes of the experimental investigation, i.e., the dissipation tank is undersized and therefore insufficient to contain the jet-like flow outflowing from the spillway chute. Due to the high energy content of the current, a further jet-like flow is generated and introduced into the riverbed downstream from the dam with important effects in terms of river bottom erosion. The numerical model also made it possible to compare the hydrodynamics when the ski-jump-like sill is kept or removed from the bottom of the chute. The model clearly shows the beneficial effect achievable with the removal of this sill. It is worth underlining the effectiveness of the commercial software employed in this investigation for the Computational Fluid Dynamics simulations. The numerical model, properly validated with the experimental outcomes, describes the hydrodynamics of the current for the various discharge values under investigation fairly well. Such a validated model can then be employed in the design stage to provide a qualitative visualization of the current for different discharge values and quantitative information on the hydrodynamic features of the flow.

Author Contributions: Conceptualization, M.M., M.C. and P.D.G.; methodology, M.M.; validation, M.C., M.M. and P.D.G.; investigation, M.C. and M.M.; data curation, M.C. and M.M.; writing—original draft preparation, M.M.; funding acquisition, P.D.G. All authors have read and agreed to the published version of the manuscript.

Funding: This research was partially funded by MoliseAcque.

Institutional Review Board Statement: Not applicable.

Informed Consent Statement: Not applicable.

Data Availability Statement: Data are available upon request.

Acknowledgments: The authors wish to thank Molise Acque and Carlo Tatti for funding the present research. The authors also wish to thank Fabio Sammartino for his immeasurable contribution in building the laboratory model; Cosmo Cimorelli and Valerio Ricceri for their invaluable help during the model setup, experimental campaign, and numerical simulations.

Conflicts of Interest: The authors declare no conflict of interest. The funders had no role in the design of the study; in the collection, analyses, or interpretation of data; in the writing of the manuscript, or in the decision to publish the results.

References

1. Muir, M.J.; Luce, C.H.; Gurrieri, J.T.; Matyjasik, M.; Bruggink, J.L.; Weems, S.L.; Hurja, J.C.; Marr, D.B.; Leahy, S.D. Effects of climate change on hydrology, water resources, and soil [Chapter 4]. In *Climate Change Vulnerability and Adaptation in the Intermountain Region [Part 1]*; Halofsky, J., Peterson, D., Ho, J.J., Little Natalie, J., Joyce Linda, A., Eds.; Department of Agriculture, Forest Service, Rocky Mountain Research Station: Fort Collins, CO, USA, 2018; pp. 60–88.
2. Savage, B.M.; Johnson, M.C. Flow over ogee spillway: Physical and numerical model case study. *J. Hydraul. Eng.* **2001**, *127*, 640–649. [CrossRef]
3. Ghafourian, A.; Jahromi, H.M.; Bajestan, M.S. Hydraulic of siphon spillway by physical and computational fluid dynamics. *World Appl. Sci. J.* **2011**, *14*, 1240–1245.
4. Begam, S.; Sen, D.; Dey, S. Moraine dam breach and glacial lake outburst flood generation by physical and numerical models. *J. Hydrol.* **2018**, *563*, 694–710. [CrossRef]
5. Castellino, M.; Sammarco, P.; Romano, A.; Martinelli, L.; Ruol, P.; Franco, L.; De Girolamo, P. Large impulsive forces on recurved parapets under non-breaking waves. A numerical study. *Coast. Eng.* **2018**, *136*, 1–15. [CrossRef]
6. Olsen, N.R.B.; Kjellesvig, H.M. Three-dimensional numerical flow modelling for estimation of spillway capacity. *J. Hydraul. Res.* **1998**, *36*, 775–784. [CrossRef]
7. Damarnegara, S.; Wardoyo, W.; Perkins, R.; Vincens, E. Computational fluid dynamics (CFD) simulation on the hydraulics of a spillway. In *IOP Conference Series: Earth and Environmental Science*; IOP Publishing: Bristol, UK, 2020.
8. Chatila, J.; Tabbara, M. Computational modeling of flow over an ogee spillway. *Comput. Struct.* **2004**, *82*, 1805. [CrossRef]
9. Cook, C.B.; Richmond, M.C.; Serkowski, J.A. *The Dalles Dam, Columbia River: Spillway Improvement CFD Study*; Technical Report PNNL-14768, 969745; Pacific Northwest National Laboratory: Richland, WA, USA, 2006.
10. Ho, D.K.H.; Riddette, K.M. Application of computational fluid dynamics to evaluate hydraulic performance of spillways in Australia. *Austral. J. Civ. Eng.* **2010**, *6*, 81–104. [CrossRef]
11. Maghrebi, M.; Alizadeh, S.; Lotfi, R. Numerical simulation of flow over rectangular broad crested weir (real case study). In Proceedings of the First International Conference on Dams & Hydropowers, Tehran, Iran, 8–13 February 2012.
12. Gadhe, V.; Patil, R.; Bhosekar, V.V. Performance assessment of upgraded spillway—Case study. *J. Hydraul. Eng.* **2021**, *27*, 327–335. [CrossRef]
13. Demeke, G.K.; Asfaw, D.H.; Shiferaw, Y.S. 3D Hydrodynamic Modelling Enhances the Design of Tendaho Dam Spillway, Ethiopia. *Water* **2019**, *11*, 82. [CrossRef]
14. Yang, J.; Andreasson, P.; Teng, P.; Xie, Q. The Past and Present of Discharge Capacity Modeling for Spillways—A Swedish Perspective. *Fluids* **2019**, *4*, 10. [CrossRef]
15. Badas, M.G.; Rossi, R.; Garau, M. May a Standard VOF Numerical Simulation Adequately Complete Spillway Laboratory Measurements in an Operational Context? The Case of Sa Stria Dam. *Water* **2020**, *12*, 1606. [CrossRef]
16. Rong, T.Z.; Peng, L.; Feng, P. Three-Dimensional Numerical Simulation of Dam Discharge and Flood Routing in Wudu Reservoir Youtong. *Water* **2019**, *11*, 2157. [CrossRef]
17. Han, M.; Chen, M. Three-dimension visualization research of flood routing simulation. *Comput. Appl.* **2005**, *25*, 1906–1907.
18. Heller, V. Scale effects in physical hydraulic engineering models. *J. Hydraul. Res.* **2011**, *49*, 293–306. [CrossRef]
19. Castellino, M.; Moroni, M.; Cimorelli, C.; Di Risio, M.; De Girolamo, P. Riverbed Protection Downstream of an Undersized Stilling Basin by Means of Antifer Artificial Blocks. *Water* **2021**, *13*, 619. [CrossRef]
20. Li, S.; Cain, S.; Wosnik, M.; Miller, C.; Kocahan, H.; Wyckoff, R. Numerical modeling of probable maximum flood flowing through a system of spillways. *J. Hydraul. Eng.* **2011**, *137*, 66–74. [CrossRef]

21. Moroni, M.; Lorino, S.; Cicci, A.; Bravi, M. Design and Bench-Scale Hydrodynamic Testing of Thin-Layer Wavy Photobioreactors. *Water* **2018**, *11*, 1521. [CrossRef]

MDPI
St. Alban-Anlage 66
4052 Basel
Switzerland
Tel. +41 61 683 77 34
Fax +41 61 302 89 18
www.mdpi.com

Hydrology Editorial Office
E-mail: hydrology@mdpi.com
www.mdpi.com/journal/hydrology

www.ingramcontent.com/pod-product-compliance
Lightning Source LLC
Chambersburg PA
CBHW061801210926
51850CB00006B/609
9783036578095